International Conference on Computer Vision, High Performance Computing, Smart Devices and Networks (CHSN 2020)

IOP Conference Series: Materials Science and Engineering
Volume 1074

Kakinada, India
28 - 29 December 2020

Editors:

Ch. Satyanarayana
Naresh Babu Muppalaneni
M. Radhika Mani
B.N. Jagadesh

ISBN: 978-1-7138-2694-1
ISSN: 1757-8981

Printed from e-media with permission by:

Curran Associates, Inc.
57 Morehouse Lane
Red Hook, NY 12571

Some format issues inherent in the e-media version may also appear in this print version.

Institute of Physics
Dirac House, Temple Back
Bristol BS1 6BE UK

Phone: 44 1 17 929 7481
Fax: 44 1 17 920 0979

techtracking@iop.org

Additional copies of this publication are available from:

Curran Associates, Inc.
57 Morehouse Lane
Red Hook, NY 12571 USA
Phone: 845-758-0400
Fax: 845-758-2633
Email: curran@proceedings.com
Web: www.proceedings.com

TABLE OF CONTENTS

HYBRID SHUFFLED FROG LEAPING ALGORITHM WITH PROBABILITY DISPERSAL METHOD FOR TUMOR DETECTION IN 3D MRI BRAINTUMOR IMAGES 1
P. G. K. Sirisha, D. Haritha

A REVIEW ON DETECTION OF LAND USE AND LAND COVER FROM AN OPTICAL REMOTE SENSING IMAGE. 14
A. V. Kavitha, A. Srikrishna, Ch. Satyanarayana

CBCA: CONSIGNMENT BASED COMMUNAL AUTHENTICATION AND ENCRYPTION SCHEME FOR INTERNET OF THINGS USING DIGITAL SIGNATURE ALGORITHM 36
Arun. M, Praveenkumar. S, Rajakumar. P. S, Thamizhikkavi. P

CANCER CLASSIFICATION USING ENSEMBLE FEATURE SELECTION AND RANDOM FOREST CLASSIFIER 52
Nimrita Koul, Sunilkumar S Manvi

ADVANCED CONVOLUTIONAL NEURAL NETWORK CLASSIFICATION FOR AUTOMATIC SEIZURE EPILEPSY DETECTION IN EEG SIGNAL 60
Venkata Ramana Mancha, Srinivasa Reddy E, Satyanarayana Ch

AN EFFECTIVE ANALYTICS USING MACHINE LEARNING INTEGRATED APPROACHES FOR DIAGNOSIS, SEVERITY ESTIMATION ANDPREDICTION OF HEART DISEASE 71
N Satyanandam, Ch Satyanarayana

SENTIMENT ANALYSIS USING NEURAL NETWORK AND LSTM 82
Akana Chandra Mouli Venkata Srinivas, Ch. Satyanarayana, Ch. Divakar, Katikireddy Phani Sirisha

CLASSIFICATION OF BREAST CANCER USING HISTOLOGY IMAGES: HANDCRAFTED AND PRE-TRAINED FEATURES BASED APPROACH 89
Jyoti Kundale, Sudhir Dhage

ENSEMBLE DEEP LEARNING FOR BRAZIL CURRENCY COIN PREDICTION 107
Naresh Babu Muppalaneni, Ch. Prathima, Akula Chandra Sekhar

VISUALIZING UML'S SEQUENCE AND CLASS DIAGRAMS USING GRAPH-BASED CLUSTERS 116
Nakul Sharma, Prasanth Yalla

DETECTION OF INTRUDER USING THE IMPROVED DRAGONFLY OPTIMIZATION ALGORITHM 133
Nagaraju Devarakonda, S. Anandarao, Raviteja Kamarajugadda

AUTOMATIC VEHICLE NUMBER PLATE RECOGNITION SYSTEM USING MACHINE LEARNING 139
J M S V Ravi Kumar, B Sujatha, N Leelavathi

CONTEXT-AWARE MULTI-DATA FUSION TO ACHIEVE FINEST RIDE-SHARE SCHEDULE USING BIG DATA 148
Avinash Potluri, N. V. Narendra Kumar

DEEP RECURRENT Q REINFORCEMENT LEARNING MODEL TO PREDICT THE ALZHEIMER DISEASE USING SMART HOME SENSOR DATA 155
C. Dhanusha, A. V. Senthil Kumar

AN EFFICIENT MACHINE LEARNING TECHNIQUES BASED ON IOT FOR EFFECTIVE LOAD FORECASTING 164
K. Rameshwaraiah, Sankarpu Sateesh Kumar, K. Srinivasa Babu, T. Madhu

FAST HYBRID ADABOOST BINARY CLASSIFIER FOR BRAIN TUMOR CLASSIFICATION 169
S. Jayaprada, G. Jayalakshmi, L. Kanyakumari

SAMS: SMART AGRICULTURE MANAGEMENT SYSTEM USING EMERGING TECHNOLOGIES IOT, AI -A STUDY 178
B. Srinivas Kumar, S. G. Santhi, K. Kranthi Kumar

ANALYZING AND PREDICTING COVID-19 EPIDEMIC USING MACHINE LEARNING TECHNIQUES 184
A. Lakshmanarao, Ch. Seshadri Rao, Sridevi Gadde

TIMELY AND ACCURATELY PREDICT RAINFALL BY USING ENSEMBLE PREDICTIVE MODELS 192
Veera Ankalu. Vuyyuru, Giduturi. Apparao, S. Anuradha

UNSUPERVISED LEARNING APPROACH FOR CLUSTERING LEAF IMAGES 216
G Chamundeswari

AN APPROACH FOR SUGGESTION MINING BASED ON DEEP LEARNING TECHNIQUES 224
T Raghunadha Reddy, P Vijayapal Reddy, T Murali Mohan, Raju Dara

PREDICTION AND ANALYSIS OF CORONA VIRUS DISEASE (COVID-19) USING CUBIST AND ONER 234
R. V. S. Lalitha, J. Divya Lalitha, K. Kavitha, T Ramareddy., Rayudu Srinivas, Challapalli Sujana

TRANSFORM DOMAIN ANALYSIS OF MULTIMODAL MEDICAL IMAGE FUSION 244
Ch. Hima Bindu, G. Sridevi

EXPEDITING INTERNATIONAL STUDENT ADMISSION PROCESS USING DATA ANALYTICS 253
Mounicasri Valavala, Wasim Alhamdani, Hemalatha Indukuri

EPOCH INTERROGATION FOR SKIN CANCER DETECTION USING CONVOLUTIONAL NEURAL NETWORK MODELS 260
A Poornima, M Shyamala Devi, M Sumithra, Mullaguri Venkata Bharath, Swathi, S Sathishkumar, K Yogesh, Shubham Santosh Upadhyay, Nitesh Kumar Sah

CANONICAL VARIATE FEATURE SELECTION BASED ADAPTIVE ENHANCED WINNOW MAP REDUCE CLASSIFICATION FOR PREDICTIVE ANALYTICS 270
R. Sivakkolundu, V. Kavitha

KEYPHRASE EXTRACTION AND SOURCE CODE SIMILARITY DETECTION-A SURVEY 289
Nakul Sharma, Prasanth Yalla

ICT-BASED LEARNER-CENTRIC EVOLUTIONARY LEARNING MODEL: AN EFFECTIVE SOLUTION TO TEACHING-LEARNING PROCESS 303
Smita Vinit Bhoir, Sunita Patil

DESIGN OF IMPLANTABLE CARDIOVERTER DEFIBRILLATOR USING LOW POWER SUBTHRESHOLD DIGITAL CIRCUIT 322
T Vasudeva Reddy, B Nareshkumar, R Anirudh Reddy, P Kavitha Reddy

ACTIVATION LAYERS IMPLICATION OF CNN SEQUENTIAL MODELS FOR FACIAL EXPRESSION RECOGNITION 336
M. Shyamala Devi, Ankita Sagar, Karan Thapa, Mreenmoy Hazarika, P. Swathi

PREDICTING USER'S WEB NAVIGATION BEHAVIOUR USING AMD AND HMM APPROACHES 347
Rita Roy, G Appa Rao

REDUCING COMPLEXITY IN PTS SCHEME USING OPTIMIZATION TECHNIQUES TO REDUCE PAPR IN OFDM SYSTEMS 359
R. Sarankumar, K. Sumathi, G. Sai Sravanthi, B. Venkateswarlu, P. Rajasekar

MULTI FACIAL EXPRESSION RECOGNITION (MFER) FOR IDENTIFYING CUSTOMER SATISFACTION ON PRODUCTS USING DEEP CNN AND HAAR CASCADE CLASSIFIER 369
Dnvsls Indira, L Sumalatha, Babu Rao Markapudi

EXTRACTING FACIAL FEATURES IN FILTER MODIFIED IMAGES USING NEURAL NETWORKS 379
M Sumithra, M Shyamala Devi, A Poornima, Shatadru Sarkar, Harish Patel, Preethi Mahankali, Boggula Navya

A NEW FEATURE BASED ONLINE RATING SYSTEM FOR E-COMMERCE WEBSITES 387
D. R. Kumar Raja, M. Ramu, Akula Chandra Sekhar

DESIGN AND DEVELOPMENT OF ADAPTIVE FILTER FOR REAL TIME SIGNAL PROCESSING 394
Dr Krishna Samalla, S. P. V Subbarao, G Mallikarjuna Rao, B. N. Jagadeesh

Author Index

IOP Conf. Series: Materials Science and Engineering 1074 (2021) 011001 doi:10.1088/1757-899X/1074/1/011001

International Conference on Computer Vision, High Performance Computing, Smart Devices and Networks (CHSN-2020) 28-29 December, 2020, Kakinada, INDIA

Selected, peer reviewed papers from the **International Conference on Computer Vision, High Performance Computing, Smart Devices and Networks (CHSN-2020)**

Edited by:

Dr. Ch.Satyanarayana
JNTUK, Kakinada
chsatyanarayana@yahoo.com

Dr. Naresh Babu Muppalaneni
NIT Silchar
nareshmuppalaneni@gmail.com

Dr. M. Radhika Mani
Pragati Engineering College, Surampalem
drradhikamani@gmail.com

Dr. B.N. Jagadesh
Srinivasa Institute of Engineering and Technology, Amalapuram.
nagajagadesh@gmail.com

IOP Conf. Series: Materials Science and Engineering 1074 (2021) 011001 doi:10.1088/1757-899X/1074/1/011001

Preface

The Department of CSE, Jawaharlal Nehru Technological University Kakinada (JNTUK), Kakinada, Andhra Pradesh has organized an International Conference on Computer Vision, High Performance Computing, Smart Devices and Networks (CHSN-2020) during 28th-29th December 2020. Principal focus of the conference is to encourage the enthusiasm among researchers, academicians and the scientists of premier research laboratories across the globe to submit and present the basic and applied work in various specialized fields like Communication Engineering, Information Theory, Signal, Image and Speech Processing, Wireless and Mobile Communication, Internet-of-Things and Cyber Security etc.

The main objective of the conference is to bring researchers together on to a common platform to acquire the knowledge, discuss and share the emerging trends which help in achieving the best results. Hence, we solemnly invite all to join and get acquainted with recent advances and trends in the fields of Computer Vision, High Performance Computing, Smart Devices and Networks.

CHSN 2020 invited Technical Papers, Short Papers and Research Articles in the domains of:

- 3D Computer Vision
- Action Recognition
- Big data and Large Scale Methods
- Biometrics, face and gesture
- Biomedical image analysis
- Deep Learning
- Emotion recognition
- Face Recognition
- Fingerprint recognition
- Graphical user interfaces
- Handwriting recognition
- Image acquisition
- Image coding and compression
- Image processing
- Image segmentation
- Iris recognition
- Low-level vision and Image Processing
- Man-machine systems
- Machine Learning
- Medical image processing
- Motion Detection
- Motion and Tracking
- Multi-channel Filtering
- Multidimensional Signal Processing
- Multiple Filtering and Filter Banks
- Neural Networks
- Optimization methods
- Radar Image Processing
- Recognition: detection, categorization, indexing and matching
- Robot Vision
- Segmentation, grouping and shape representation
- Sonar Image Processing
- Statistical learning
- Video: events, activities and surveillance

IOP Conf. Series: Materials Science and Engineering 1074 (2021) 011001 doi:10.1088/1757-899X/1074/1/011001

In this conference, topics of contemporary interested would be discussed to provide a holistic vision on latest technologies for computer science and engineering. The scope includes data science, machine learning, computer vision, deep learning, artificial intelligence, artificial neural networks, mobile applications development and Internet of Things etc; Conference participants are expected to gain relevant knowledge and better understanding of the applications of computer science in various fields.

CHSN-2020 would be both stimulating and informative with the active participation of galaxy of keynote speakers. We would like to thank all the authors who submitted the papers, because of which the conference became a story of success. We also would like to express our gratitude to the reviewers, for their contributions to enhance the quality of the papers. We are very grateful to the Keynote Speakers, Session Chairs and Committee members who selflessly contributed to the success of CHSN-2020.

Dr. D. HARITHA

CONVENER

CHSN-2020

IOP Conf. Series: Materials Science and Engineering 1074 (2021) 011001 doi:10.1088/1757-899X/1074/1/011001

International Conference on Computer Vision, High Performance Computing, Smart Devices and Networks (CHSN-2020)

Chief Patron

Prof.M.Ramalinga Raju
Hon'ble Vice-Chancellor, JNTUK, Kakinada

Patrons

Prof.G.V.R.Prasada Raju
Rector, JNTUK, Kakinada

Prof.B.Bala Krishna
Principal, UCEK, JNTUK, Kakinada

Honorary Chair

Prof.Allam Appa Rao
Ph.D(Engg).,D.Sc(h.c).
Former Vice-Chancellor, JNTUK, Kakinada
Chairman, IBCB, Visakhapatnam

General Chair

Prof.Ch.Satyanarayana
Registrar, JNTUK, Kakinada

Prof.S.V.Rao
Head-CSE, IIT Guwahati

Prof.Xiao-Zhi Gao
School of Computing, University of Eastern Finland, Finland

Program Chair

Dr.Ch.Sudhakar
National Institute of Technology Warangal, Telangana

Dr.Naresh Babu Muppalaneni
National Institute of Technology Silchar, Assam

Convener

Dr. D. Haritha
Professor & Head-CSE
JNTUK, Kakinada

Organizing Secretary

Dr. B.N. Jagadesh
Professor & Head-CSE
Srinivasa Institute of Engineering and Technology, Amalapuram

IOP Conf. Series: Materials Science and Engineering 1074 (2021) 011002 doi:10.1088/1757-899X/1074/1/011002

Peer review declaration

All papers published in this volume of *IOP Conference Series: Materials Science and Engineering* have been peer reviewed through processes administered by the Editors. Reviews were conducted by expert referees to the professional and scientific standards expected of a proceedings journal published by IOP Publishing.

- **Type of peer review:** Double blind
- **Conference submission management system:** Easy Chair
- **Number of submissions received: 144**
- **Number of submissions sent for review: 144**
- **Number of submissions accepted: 55**
- **Acceptance Rate (Number of Submissions Accepted / Number of Submissions Received X 100): 38.19%**
- **Average number of reviews per paper: 2**
- **Total number of reviewers involved: 48**
- **Any additional info on review process:** Reviewers from premier organizations such as IITs, NITs, Central Universities and State Universities.
- **Contact person for queries: Dr.D.Haritha, Convenor, CHSN-2020 (icchsn2020@gmail.com)**

IOP Conf. Series: Materials Science and Engineering 1074 (2021) 011002 doi:10.1088/1757-899X/1074/1/011002

PEER REVIEW POLICY Adopted by CHSN-2020

Reviewers were asked to consider the following key points related to scientific content, quality and presentation.

Technical Criteria

- Formal Structure of the review
- Language
- Scientific merit: Originality and Topicality, Materials and Methods
- Clarity of expression; communication of ideas; readability and discussion
- References

Quality Criteria

- Originality: Novelty of the proposed approaches/methods
- Results: Clarity of the illustration
- Repetition: Have significant parts of the manuscript already been published?
- Length: Is the content of the work of sufficient scientific interest to justify its length?

Presentation Criteria

- Title: Is it adequate and appropriate for the content of the article?
- Abstract: Does it contain the essential information of the article? Is it complete? Is it suitable for inclusion by itself in an abstracting service?
- Diagrams, figures, tables and captions: Are they essential and clear?
- Text and mathematics: Are they brief but still clear? If you recommend shortening, please suggest what should be omitted.
- Conclusion: Does the paper contain a carefully written conclusion, summarizing what has been learned and why it is interesting and useful?

The conference coordinators reviewed the submitted papers relevance to the conference topics and checked whether the papers meet the IOP format requirements. The independent reviewers evaluated the submitted papers according to the following criteria as the relevance to meet the study, novelty of work, quality and scientific knowledge, paper structure for IOP format and adequate references etc. Based on the reviewers report, the authors are suggested to revise the submitted manuscript for further reviews.

HYBRID SHUFFLED FROG LEAPING ALGORITHM WITH PROBABILITY DISPERSAL METHOD FOR TUMOR DETECTION IN 3D MRI BRAINTUMOR IMAGES

P.G.K.Sirisha, Research scholar, CSE Department, UCE, JNTUK

D. Haritha, Professor & HOD, CSE Department, UCE, JNTUK

ABSTRACT

In the medical image study, the brain tumor classification using MRIs is difficult due to the brain's complicated structure and the high variance in tumor tissues' position. So, the requirement for useful and specific tumor identification methods is developing for medical recognition and regular medical applications. The conventional brain tumor identification performs anatomical knowledge of irregular tissues in the brain, helping the doctor design approach. The research proposes several techniques for brain tumor identification. This work aims to present brain tumor identification methods based on evolutional intelligence and segmentation. Unusual areas in the brain are identified by using the Expectation-Maximization (EM) algorithm. For segmenting the 3D brain MRI data, this work presents a novel hybrid optimization meta-heuristic called the Shuffled Frog Leaping Algorithm (SFLA) with probability dispersal (i.e., SFLA - Stochastic Diffusion Search (SDS)). The efficacy of the suggested 3D SFLA probability dispersal EM in enhancing the performance of the 3D SFLA tabu EM has been proven by empirical outcomes.

***Keywords**: Medical Imaging, Magnetic Resonance Imaging (MRI), Image Segmentation, Expectation Maximization (EM), Shuffled Frog Leaping Algorithm (SFLA) and Stochastic Diffusion Search(SDS).*

1 INTRODUCTION

A brain tumor is a mass of tissue such as a tumor or cancer depends on the slow addition of irregular and atypical cells and unusual tissue proliferation, causing damage to the vital system and increased pressure inside the skull. A brain tumor can be both benign (not cancerous) and malignant (cancerous). The most common brain tumors are glioma and low-stage meningioma, which is a type of malignancy. A homogeneous form of a benign brain tumor that no longer contains most of the cancer cells or has been removed entirely by surgery can be monitored radio logically. If a malignant mental tumor is present, it carries a cancerous cell phone treated with chemotherapy and radiotherapy. MRI is primarily a clinical imaging method [1] that provides useful records of the analysis. It is mainly used by surgical radiologists who plan and visualize the frame's function and the frame's internal structure. Magnetic resonance imaging plays a vital role in detecting most cancers

IOP Conf. Series: Materials Science and Engineering 1074 (2021) 012001 doi:10.1088/1757-899X/1074/1/012001

and tissue abnormalities with affected tissues' dimensions and area. Further evaluation is offered using MRI rather than CT images. It makes it especially useful in diseases of the cardiovascular system and the nervous system (mind).

The development of image processing into unique areas of interest has been connected with the latest clinical imaging techniques, as more acquisition models and larger images are being produced. It makes it insufficient to see the photos have received without image processing software. Thus, image processing packages have a important influence on the method of analysis by serving the physician investigate all volumetric information to make a treatment verdict and wait for disease consequences.

In just a few decades, there has been a variety of research involving the topic of segmentation of mental tumors because it is essential to understand the points obtained with MRI to predict and monitor treatment effectiveness [2].

The subdivision of mental tumors from magnetic resonance imaging presents an intrinsic and challenging task in assessing and planning solutions. Image segmentation is a vivid neighborhood in the clinical picture and consists of extracting one or more regions of the image that make up a hobby site on the Internet. Numerous algorithms have been developed in the literature to find intellectual neoplasms and methods that in most cases depend on thresholds, techniques that rely specifically on regions, deformable techniques, separation strategies and deep understanding. Distorted models are the most popular of the various strategies used to remove intellectual tumors through the use of MRI. It is represented with the help of curves (2D) or surfaces (3-D) specified in Fig [3] which rotate through the influence of two forces, internal or close specific within the curve to keep calm at some moments. From the deformation model, at the same time, external forces are calculated from photographic facts to pass the angle towards the boundary of the desired object. He distinguishes between two primary classes, parametric distorted or snake fashions and distorted geometric patterns in deformed techniques. The benefit of these patterns is their ability to topological seizures at some point in the curve diffusion.

The holistic rule-based method uses a set of optimization rules [4] to optimize the segmentation outcomes. Initially, brain tissue is improved by some methods and an optimization algorithm is applied. The most commonly applied algorithms are nature-inspired or meta-heuristic techniques due to their simplicity and flexibility. Several sets of descriptive properties have been proposed for entirely rule-based segmentation techniques. In the hybrid strategies, additional tactics for the tumor area class of brain MRI are incorporated. This work presents a 3-D EM approach to SFLA potential dispersion (SFLA-SDS hybrid) that optimizes the 3D MRI category. The final part of the investigation is prepared in the following sections. The second section analyzes related works in the literature. Section 3 explains the different methods used in creation. Section IV analyzes the experimental results, and Section 5 concludes with panels.

2 RELATED WORKS

The 3D brain tumor segmentation is a medical need for mental tumor analysis and radiotherapy planning. It is a difficult challenge because of the difference in genre. Several techniques, along with a set of Particle Clustering Optimization (PSO) rules, the size, location, and shape of the tumors have formed a topological association of slides that convert 2D images into 3D MRI images that no longer afford good results and depend on the number of segments introduced and the positions and shape of the MRI images. Gtifa et al. [5] proposed a practical approach to 3D mental tumor segmentation known as modified PSO. Moreover, the segmentation results are compared to the Darwinian PSO (DPSO) process and Fractional-Order DPSO (FODPSO).

The segmentation of 3D brain tumors is a clinical necessity for brain tumor analysis and radiotherapy planning. It is a difficult task due to the gender discrimination. Various strategies, along with stringent and rapid guidelines for particle agglomeration (PSO) improvement, have resulted in the establishment of dimensions, area, and shape of tumors a topological affiliation of slices that convert 2D images into 3D MRI snapshots that are not now. Correct outcomes are for adequate money and are dependent on the variety of pooled sectors and the positions and shape of the MRI image. Gtifa et al. [5] He proposed a sensitive 3D staging technique known as modified PSO. Moreover, the segmentation results were compared with the Darwinian form of PSO (DPSO) and the Fractional-Order DPSO (FODPSO).

Experimental effects showed that the technique achieved 97.6% 3D segmentation.

When compared to DPSO and FODPSO techniques with 78.1% and 70.21% for the case of the T1 - C method, there is an additional accuracy rate.

Badura [6] presented new flexible swarm intelligence optimization method has been provided to different segment systems into 3D or 2D images. Merchants in a self-regulating colony discover their host, use stigma to communicate, and mark areas of interest that lead to element extraction. In this work, detailed specifications of a bacterial colony segmentation technique (BCS) in terms of individual and social behavior are defined. The approach has been demonstrated and evaluated through various experiments, including synthetic information, tomography, and ultrasound research. The results and observations obtained in terms of parameter settings and the approach's potential utility are discussed in the various segmentation tasks..

Mehmood et al., [7] developed intelligent computer-assisted diagnostic device specializing in magnetic resonance imaging of the human mind. These panels provided the ability to detect brain tumors, their fragmentation and the 3D display device, and offered good medical deals regardless of the geographical area and the degree of knowledge of scientific experts in this research. Malignant and benign based primarily on the BoW version of the Resistant Supporting Vector Machine

(SVM). The BoW function extraction technology is further amplified by the powerful accelerated features (SURF) that incorporate a selection factor of interest factor. Finally, 3D visualization of the brain and tumor was implemented using a set of volume rules that are used to represent clinical events. The efficacy of the proposed system was constructed on a dataset collected from 30 patients and implemented with an accuracy of 99%.

Sumathi et al., [8] a unique set of rules has been provided that has evolved and relied entirely on entropy-based morphological reconstructions and Kapoor's cuckoo search optimization filters. The former is used to locate and stage tumor boundaries, while the latter is used to eliminateunnecessary pixels within the cropped footage. The proposed method produces up to 97% accuracy in determining the tumor site's exact topographic region. If we need less computing time (about 3ms, combined) for this method. Therefore, the proposed method can assist radiologists in quickly determining the exact topographic area of tumor areas, even assuming versions of excessive depth and serious limitations.

To arrive at and develop a completely unique dependency framework for prominently perceptions of tumor regions and division of tissue systems, particularly within the human brain, Narayanan et al. He proposed a completely unique set of synthesis rules [9]. Combined with two optimization technologies, PSO and Bacteria Foraging Optimization (BFO), the algorithm enables the unique function of bacteria worldwide for the BFO. It supports the modified Fuzzy C-mean (MFCM) set by moving the improved block heads finally; the MFCM divides tissue regions and identifies the tumor component, minimizing the reaction and problem encountered with the help of a radiologist while reading the affected man or woman.

Krishna et al., [10] A version of the local radial linear function neural network (LLRBFNN) that relies primarily on PSO to classify and find mental tumors into malignant (precancerous) and benign (non-cancerous) tumors is provided. In these paints, the wavelets are reshaped to improve Magnetic resonance image segmentation and post-extraction performance. Machine learning techniques, the SVM device, and the least-squares mean (LMS) classifier were also examined to validate the proposed PSO-based LLRBFNN model. Whiteboards follow steps that involve extracting features that are viable functions for research work. In a second step, the characteristics are entered into the rating company's PSO-based LLRBFNN model. In a third step, the instrument was implemented to determine the LMS-based SVM and LLRBFNN approach to the category responsibilities and compare the results.

Sahu et al. [11] presented a hybrid version of LLRBFNN that mainly relies on ant-colony optimization - mimetic annealing (ACO-SA) to detect mental tumor tissue and its type of blurred images. Blurring the blurry images affected by the software becomes a complex and challenging task. The local ambiguous C (FLICM) segmentation algorithm was taken into account from the start. The algorithm's characteristic value was modified for a better final segmentation result to eliminate the MRI noise. Images are split using a modified FLICM algorithm. Capacitances are extracted with gray matrix matching feature extraction (GLCM) technology, and fed as input to the fully ACO-based LLRBFNN model due to the elegance of the malignant and benign neoplasms of the MRI. The proposed model is altered compared to PSO-LLRBFNN and the Adaptive Particle Swarm optimization(APSO) -LLRBFNN improved, and evaluation effects are provided. The proposed algorithm fully based on ACO-SA LLRBFNN to indicate superior effects over conventional techniques was identified..

3 METHODOLOGY

Automated 3D segmentation of a brain tumor can store clinicians' time and provide an accurate, repeatable solution for further tumor evaluation and follow-up. The Expectation-Maximization (EM) algorithm [12] is used to define the predictable parameters with the highest or maximum likelihood later in aarithmetical copy when facts are "incomplete" or absent. The parameter is frequently estimated in two steps. It contains steps, the first one being step E, which calculates the expected price of the logarithmic probability property, and in step M the parameters to maximize the value of M based on the observed rate in step E. Then to discover the distribution of the underlying variables in the next step E. In the 3-d SFLA tabu EM, This hybrid algorithm maintains the framework of the SFLA rule set. However, it uses the Community Structure approach of Tabu Search (TS) and avoids the neighborhood that was already explored within the solution area to move it. Towards a significant response within the evolution of local memory. The SFLA, SDS, and 3-D SFLA by chance EM (SFLA-SDS) methods are discussed in these panels.

Shuffled Frog Leaping Algorithm(SFLA)

The SFLA implementation mechanism aims to end the problem-solving process by simulating the cooperation and data-sharing performance of frog flocks by frog in the wild. Each frog is described as a problem option. The entire frog organization is divided into unique subgroups by compiling a meme to simulate frog crowd behavior, and each subset is known as a meme institution. Each frog within a meme establishment strives to get close to the target and determine the area from food assets—meanwhile, every frog was encouraged by other frogs, which is known by tradition here. Each meme organization has its lifestyle, which affects different people and develops with the memes foundation [13].

While making each evolutionary step of the meme set, discover frogs with good neighborhood and worst area. The frog with the worst zone organizes its region by implementing the neighborhood zone replacement and adopting the gear shift release trigger similar to the PSO. After the positive times of meme development for a meme enterprise, the various meme companies' frog reorganizes the entire organization and understands the exchange and exchange of information between the meme's organizations so that the pre-defined cases of population development are done using the algorithm. The rule set conveys facts according to ethnic grouping and instead implements neighborhood evolution and reintegration. It efficiently combines the exchange of international events and the search for neighborhood development. It's incredibly high overall computational performance and global search ability to solve optimization problems. The mathematical version of SFLA (Equation (1 to 3)) can be briefly presented here.

$$Min\, f\,(x) = \;[f_1\,(x), f_2(x),, f_k(x)] \qquad \text{-----(1)}$$

$$s.t. g_i \leq 0\;\;(i \quad 1,2,,m) \qquad \text{-----(2)}$$

$$h_j\,(x) = 0\,(\,j \quad 1,2,,p) \qquad \text{----(3)}$$

Basic flow of SFLA can be described as follows:

Step 1: Configure the algorithm parameters, along with group P length, number of meme M deals, meme G development times, a maximum permissible distance of frog Dmax, and full population generation range MaxIter.

Step 2: Randomly initialize population.

Step 3: Arrange the sequence of the frog population, which is divided into several memegroups according to grouping operators.

Step 4: Execute local location on updating operators in every meme group.

Step 5: Frogs leap between different groups and remix to form new populations.

Step 6: Check whether the ending condition is satisfied. If yes, output the solution ; otherwise, jump to Step 3 and repeat.

Stochastic Diffusion Search (SDS)Algorithm

SDS is a global multi-agent research and improvement foundations that rely entirely on simple retail interaction. A hyperbolic description of the SDS level is provided in the form of a social metaphor describing the movements through which the SDS identifies sources. SDS provided a new generation perspective to solve the problems of sample recognition and exact shape matching. SDS, as an international multi-agent research and development group, is mainly based on population in custom account mode which appropriately uses agent interaction [14]. The SDS algorithm starts searching or optimizing by configuring the population (for example, miners, within the mining game metaphor). In any SDS research, each agent maintains a hypothesis, h, that identifies a potential solution to the problem. In a mining analogy, you determine the dealers' guess a hill. After initialization, two stages are followed:

- Test Phase (e.g. testing goldavailability)
- Diffusion Phase (e.g. congregation and exchanging of information

The SDS algorithm:

Initializing agents ()
While (stopping condition is not met)
Testing hypotheses ()
Diffusion hypotheses ()
End

In the test segment, SDS examines whether or not predictive analysis is successful using a partial hypothesis evaluation that returns the logical cost. Subsequently, under iteration, depending on the appropriate recruitment approach used, the hypotheses of compliance are disseminated to the majority of the population. Therefore, statistics regarding solutions suitable for capacity are sent at some point over the service provider's network. In the validation stage, each agent plays a partial characteristic evaluation, pFE, which are some of the agent's guessing functions; pFE = f (h). In mining recreation, the assessment of micro-characteristics involves extracting a randomly defined area on the hill, described by the agent's speculation (rather than excavating all of the regions on this hill). In the broadcast segment, each agent recruits another factor to interact and connect to the hypothesis ability. In the metaphor of mining recreation, diffusion ends with the discourse of the hill hypothesis.

Proposed 3D SFLA Probability Dispersal EM (i.e. Hybrid SFLA-SDSAlgorithm)

SFLA is an evolutionary set of rules similar to the PSO algorithm and the memetic algorithm used to search for a universal answer, that is, in real number problems. Each SFLA frog represents a solution [15], and each frog is sent to a different memeplex. Each memeplex represents a small portion of the response area. A separate live search is performed on each memeplex. After a rigorous and rapid number of memeplex evolutionary steps, a shuffle is performed where frogs can exchange messages between memeplexes and aggregate towards a global response.

IOP Conf. Series: Materials Science and Engineering 1074 (2021) 012001 doi:10.1088/1757-899X/1074/1/012001

Traditionally, SFLA allows for quick searches; however, similar to other optimization algorithms, you can get stuck with nearby optimizations, which can also protect you from finding a more reliable solution globally. To avoid such pitfalls, he mixed it with SDS, used as an alternative to the SFLA update equation, i.e., the SDS does careful research and avoids falling to a close front while still adapting to the evolutionary memeplex of SFLA. For this algorithm, each memeplex will independently evolve to perform local searches in different areas of the response area. Flow chart of the proposed method is shown in Figure 1.

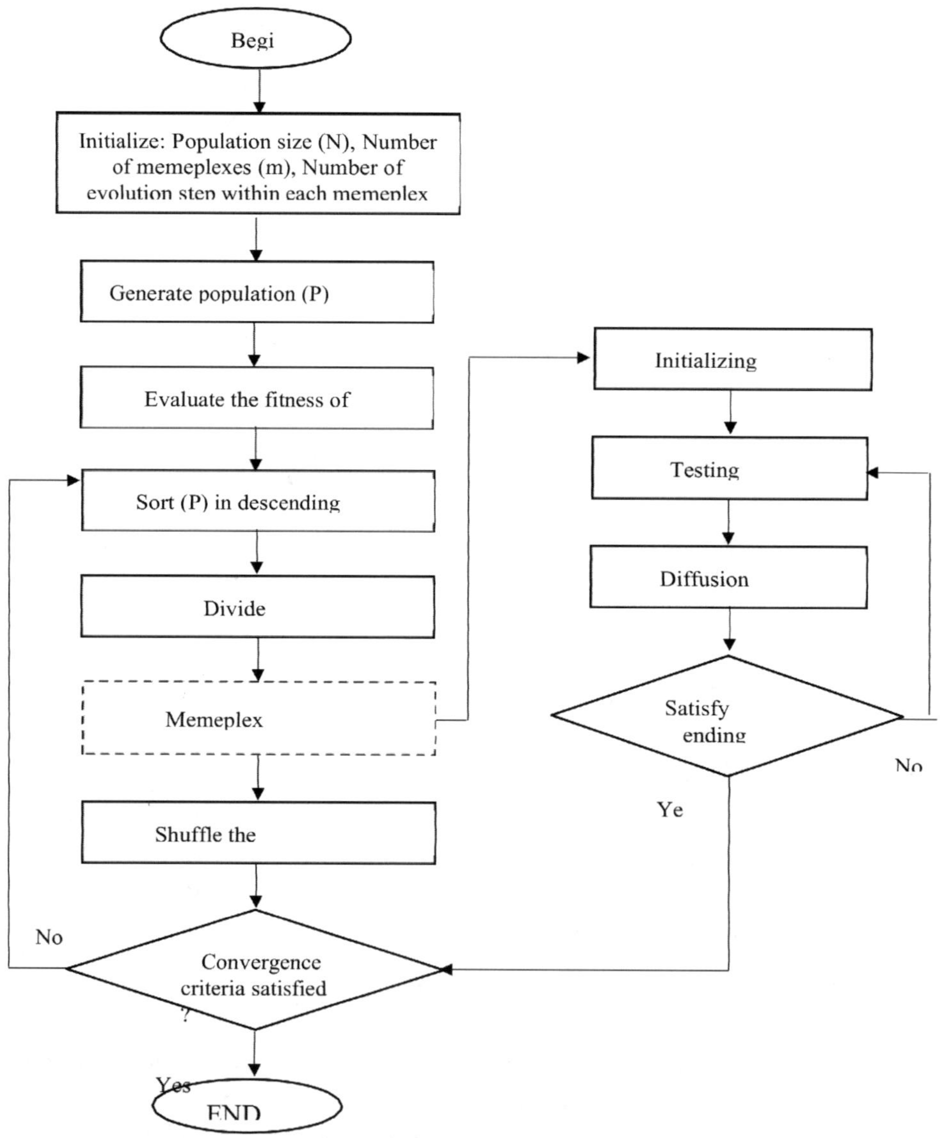

Figure 1 Flowchart for Hybrid SFLA-SDS Algorithm

4 RESULTS ANDDISCUSSION

In this stage, the professional strategies of 3D SFLA dispensability EM and 3D SFLA tabu EM are used. Experiments were performed using 250 images. Total correct fraction means common sensitivity, common specificity, common cube coefficient, and common volume errors, as shown in Figures 2 to 6. Table 1 shows a summary of the results.

	3DSFLA TABU EM	3DSFLA Probability Dispersal EM
Average Total Correct Fraction	0.957358	0.969295
Average Sensitivity	0.971577	0.981855
Average Specificity	0.964429	0.975928
Average Dice Coefficient	95.16225	96.42342
Average Volume Error	11.69856	11.51542

Table 1 Summary of Results

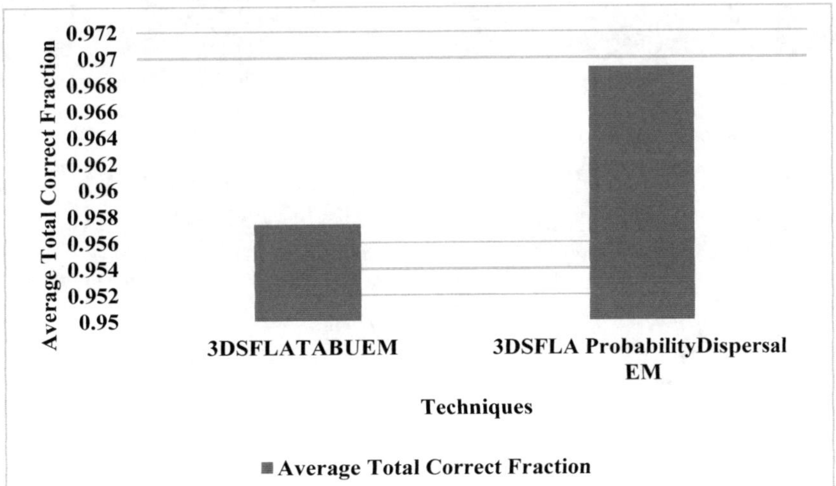

Figure 2 Average Total Correct Fraction for 3D SFLA Probability Dispersal EM

From figure 2, it can be seen that the dispersion potential of 3D SFLA probability dispersal EM has a higher global average fraction resolution with the help of 1.24% compared to the SFLA 3D tabu EM.

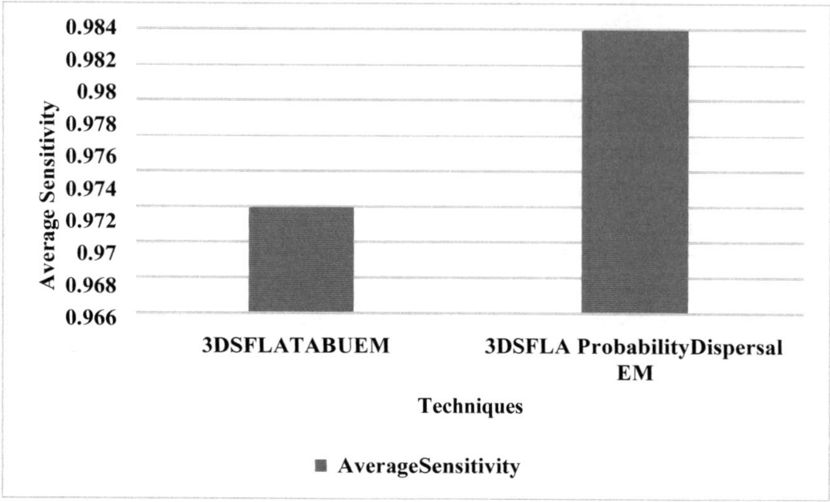

Figure 3 Average Sensitivity for 3D SFLA Probability Dispersal EM

From figure 3, it can be found that SFLA 3-D randomized dispersion MS has a higher mean sensitivity of up to 1.05% compared to SFLA 3-D tabu EM.

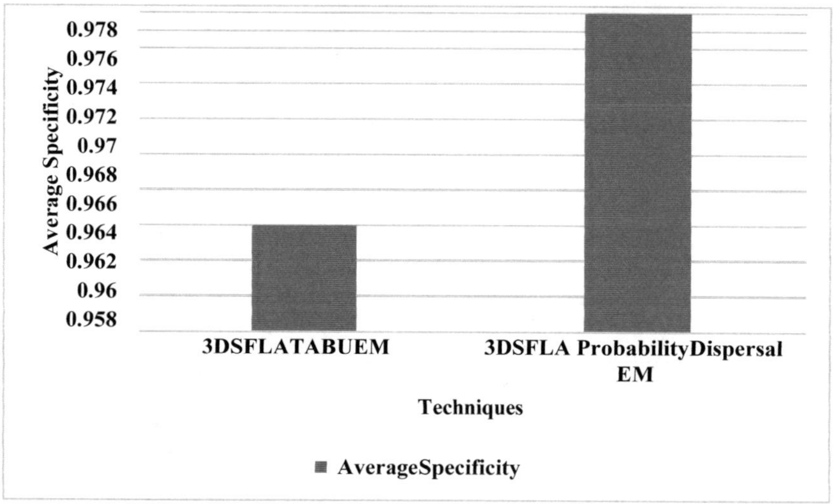

Figure 4 Average Specificity for 3D SFLA Probability Dispersal EM

Figure 4 shows that 3D SFLA probability dispersal EM has a better average specificity of 1.18% compared to the 3D SFLA tabu EM.

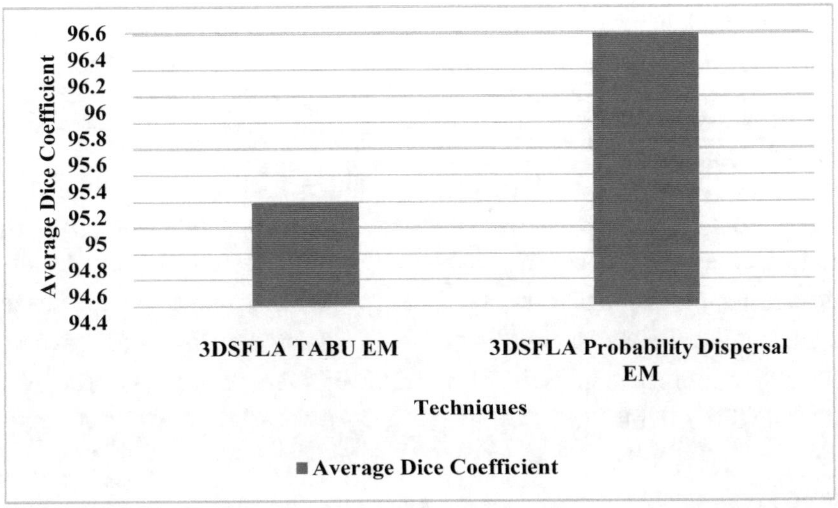

Figure 5 Average Dice Coefficient for 3D SFLA Probability Dispersal EM

Figure 5 shows that the SFLA 3D probability EM has the best average dice coefficient across 1.32% compared to the SFLA 3D tabu EM.

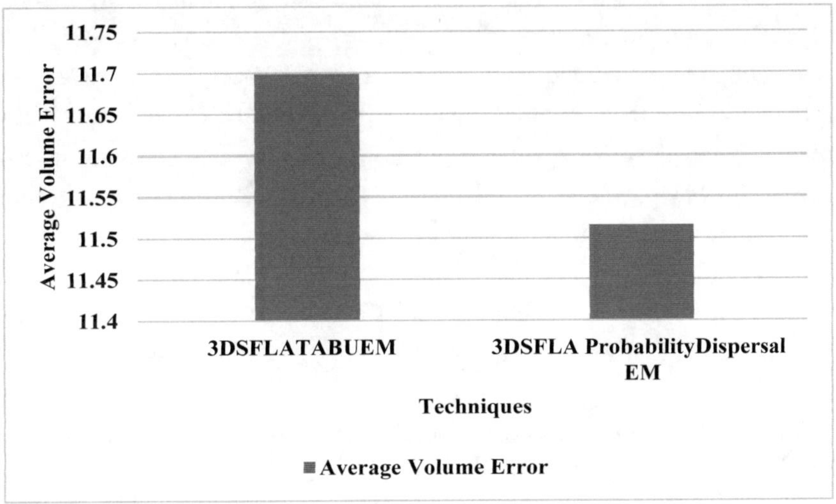

Figure 6 Average Volume Error for 3D SFLA Probability Dispersal EM

From figure 6, it is possible to determine that 3D SFLA probability EM has a lower mean extension error across 1.98% compared to 3D SFLA tabu EM.

5 CONCLUSION

The Automated segmentation of brain tumors from 3D MRI images is essential to identify, informant, and handle the disease. For this purpose, parametric representations are typically considered online, i.e., together with an affiliated segmentation maximizing an a posteriori likelihood population frequency using EM. In this work, the hybrid SFLA- SDS is recommended for the brain tumor 3D image segmentation method. The SFLA connects with local search and global knowledge interchange. The most qualified particular leaps out of the local. Then, with the shuffle approach, it can convoy an international exchange of local information. Through SDS, every agent can obtain the whole of the search space and carry with it information about the total of its target design. SFLA-SDS connects the benefits of the SFLA, specifically the interchange of data between particular parts of the group (frogs), and executes a local search using an SDS to assess the process results. The suggested work results show that the 3D SFLA probability dispersal EM has a more extensive average total exact section by 1.24% associated with 3D SFLA tabu EM. The 3D SFLA probability dispersal EM has a lower average volume error by 1.58% compared to 3D SFLA tabu EM.

REFERENCES

1. Sonavane and P Sonar, 2016,"Classification and segmentation of brain tumor using Adaboost classifier", pp. 396-403, IEEE.
2. A. M Hasan, 2018, "A hybrid approach of using particle swarm optimization and volumetric active contour without edge for segmenting brain tumors in MRI scan", page no.292-300.
3. M, Raihani, Bouattane A, 2019, "Towards Reinforced Brain Tumor Segmentation on MRI Images Based on Temperature Changes on Pathologic Area". IJBM.
4. Mahesh, K. M, Renjit, J. A, 2018, "Evolutionary intelligence for brain tumor recognition from MRI images: a critical study and review", page no.19-30.
5. Gtifa, W andA Sakly, 2019, "3D brain tumor segmentation in MRI images based on a modified PSO technique",(IJIST).
6. P Badura, 2018, "Virtual bacterium colony in 3D image segmentation", pp.152-166.
7. M, Shoaib ,Mehmood, I, 2018,"An efficient computerized decision support system for the analysis and 3D visualization of brain tumor",pp.1- 26.
8. P, Venkatesulu and S Arjunan, ,2018, "Extracting tumor in MR brain and breast image with Kapur's entropy based Cuckoo Search Optimization and morphological reconstruction filters", pp.918-930.
9. A, Narayanan and Thiyagarajan, 2019, "Multi-channeled MR brain image segmentation: A novel double optimization approach combined with clustering technique for tumor identification and tissue segmentation", pp.350-381.
10. S, Krishna and Mishra, 2018, "Detection and classification of brain tumor from MRI medical image using wavelet transform and PSO based LLRBFNN algorithm", IJCSE, 6(1).
11. Kalla. H and S.Ayane, 2019, "Detection and Classification of Brain tumor tissues from Noisy MR Images using hybrid ACO-SA based LLRBFNN

model and modified FLIFCM algorithm",ICACCP, pp. 1-6,IEEE.

12. G ,Aswathy, GlanDevadhas, 2015, "Quick detection of brain tumor using a combination of EM and level set method", pp.74-82.
13. Peng, X, Li, K, 2016, "An efficient hybrid algorithm based on hs and sfla", IJPRAI 30(05),pp.1659012.
14. Blackwell, T M, al-Rifaie, 2011, "An investigation into the merger of stochastic diffusion search and particle swarm optimization", pp.37-44.
15. Yang, C. H, Yang, C , 2008, "A combination ofshuffled frog-leaping algorithm and genetic algorithm for gene selection"

IOP Conf. Series: Materials Science and Engineering 1074 (2021) 012002 doi:10.1088/1757-899X/1074/1/012002

A Review on Detection of Land Use and Land Cover from an Optical Remote Sensing Image.

A.V.Kavitha[1,3] A.Srikrishna[2] and Ch.Satyanarayana[3]

[1]Sri. A.B.R. Government degree college, Repalle, Guntur(dt), Andhra pradesh, India. [2]Department of Information and Technology, RVR JC College of engineering, Chowdavaram, Guntur, Andhra pradesh, India. [2]Department of Computer science, JNTUK, Kakinda, Andhra pradesh, India.

anubrolukavitha@yahoo.com

Abstract. Detection of land use and land cover from an optical remote sensing image is an essential research area from the inception of a remote sensing image. Land use land cover maps have numerous applications in agriculture, environment monitoring, urban planning, etc, along with managing various catastrophic events like floods, tsunamis, forest fires, etc. This paper reviewed major techniques for detection of land use and land cover from an optical remote sensing image. Many techniques based on only spectral information, spatio-contextual information and knowledge based methods have been discussed, finally arguing the importance of the techniques based on spatio-contextual information and Mathematical Morphology.

1. Introduction

Land use land cover detection from a remote sensing image has attracted the attention of several researchers since the first remote sensing image has been captured. It forms the basis for many socio, economic and environmental applications. Few of them include urban planning, agricultural monitoring, crop yield estimation, forest monitoring, oil spills detection, sea coast monitoring, natural resources conservation & management. Today, with the availability of hyper spectral, high spatial resolution images with high temporal frequency, the opportunities and challenges of detecting and extracting land use and land cover data is also ever increasing. Plenty of studies have been conducted over the past decades and numerous techniques have been investigated and proposed by several researchers.

Figure 1 presents the taxonomy of LULC detection techniques from an optical remote sensing image. All the techniques presented in this section are not necessarily independent and some times the same technique may exist in two or more categories. In this paper, different approaches to detect land use and land cover from high resolution optical remote sensing images are analyzed with available books, survey papers and other reviews. Basic taxonomy of LULC detection from an optical remote sensing image has been presented in section 2.2. Further, Spatio-Contextual techniques have been explained in section III, Statistical methods in section IV, Methods of Mathematical Morphology in section V and finally section VII concludes the paper.

2. Taxonomy of LULC detection from an optical remote sensing image

2.1 Spectral information

A broad array of feature extraction and classification techniques for extraction of LCLU have been mentioned in the literature [53], [58]. But, most of the methods have used only spectral variables [53].

2.1.1 Spectral intensities as feature space

In literature, many works can be found where the image space has been used as the features space for classification or segmentation of the image. In [108], features of every pixel are calculated based on HSV model and then are classified with the help of k-means algorithm. They have applied this algorithm on the ETM remote sensing image. In [33], Classification of hyperspectral remote sensing images has been implemented using only spectral values of the image and SVM classifier. All the spectral features have been used without any dimensionality reduction. High dimensionality of this data has been taken care by the SVM classifier. In these cases, only individual pixel spectral values of the image have been considered. The context of the pixel has been neglected.

2.1.2 *Colour histograms*

Global colour histogram is another traditional spectral technique for retrieving remote sensing images [94]. The algorithm is very easy to implement and is effective to some extent and has been widely applied to remote sensing imagery [32]. It is a widely used algorithm [52], [24] and is transnational and rotational invariant. But, colour histogram algorithm is sensitive to mild changes in illumination and noise. Also spatial context has not been taken into consideration.

2.1.3 *Vegetation indices*

Many vegetation indices like NDVI, NWDI, VIN, etc have been considered in the later years to make best use of various spectral bands available in the multispectral and hyperspectral images for the extraction of LCLU information [4]. The quantitative measurements which reflects the vigor of vegetation are called as Vegetation indices [11]. Many vegetation indices have been proposed in the literature like Ratio Vegetation Index (RVI), Vegetation Index Number (VIN), Normal Difference Vegetation Index (NDVI), Soil Brightness Index (SBI), Normal Difference Index (NDI), etc. Out of them most popular are NDI, NDVI, NWVI, etc. To present the formulae for some indices,

$$RG = (R-G)/(R+G) \quad (1)$$

$$NDI = (NIR - MIR)/(NIR + MIR) \quad (2)$$

$$NDVI = (NIR - R)/(NIR + R) \quad (3)$$

where,
R - Red band
G - Green band
NIR - Near infrared band
MIR - Mid infrared band

Vegetation indices are more sensitive to green vegetation detection, when compared to individual spectral values of the band [4]. Also, they are more suitable to extract greenery than built up areas. Thus, another type of indices called built up indices are proposed in the literature. Unlike vegetation indices, built up indices proposed in literature are very limited. Normalized Difference Built-up Index (NDBI) [112] and Index based Built-up Index (IBI) [110] are two among them. In [110], Xu, et.al have proposed a new index called "Index based Built-up Index", which has been derived from the existing indices to extract built-up areas from the remote sensing images. This index has been derived from three indices known as the "Modified Normalized Difference Water Index" (MNDWI), "Soil Adjusted Vegetation Index" (SAVI), and the "Normalized Difference Built-up Index" (NDBI). In general indices are built from original images, where as the proposed index IBI has been built-up on various indices. This algorithm has been applied on Landsat ETM+ image. Fuzhou city of China. The proposed method was able to extract built-up areas effectively and also by surpassing the background noise. It also reduced the data dimensionality. Hence, vegetation and built up indices have contributed well for the detection of LULC detection, forest density, crop determination, etc. But, they do not have a fixed universal value for all satellite images. They show different values for different sensors. Thus, sensor calibration, soil moisture, illumination, etc affect the vegetation indices. Also, classification of LULC using vegetation or built up indices is more difficult and complicated in heterogeneous images, where there is a mixture of many elements like crops, water, mountains, buildings, etc [4]. Vegetation indices are also application specific. Related vegetation indices are to be selected for an application.

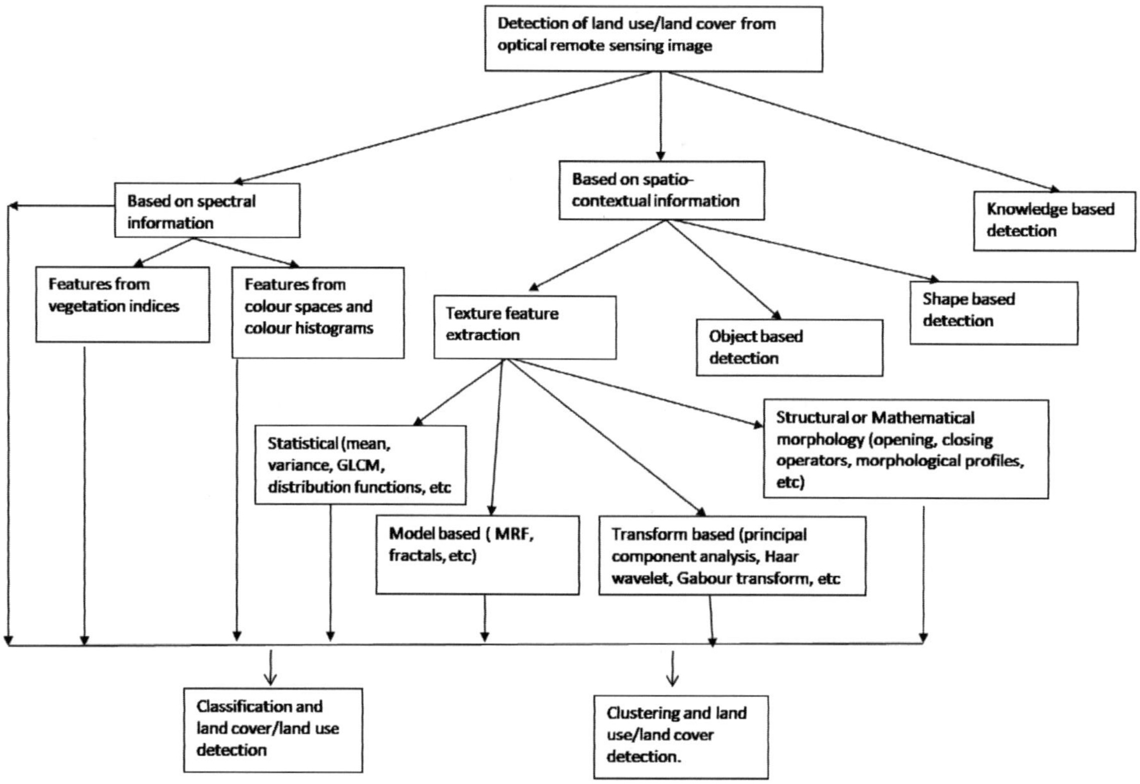

Figure 1: Taxonomy of LULC detection techniques from an optical remote sensing image

2.2 Spatio-contextual information

Spectral signatures, vegetation indices and built up indices are simple to understand and easy to implement. But, they have neglected the spatial dependencies of the pixels in the remote sensing images [53], [65]. Hence, spectral based features are not complete to detect the LULC information from a remote sensing image. This problem is more worse with the availability of high spatial resolution images [53] like Quick bird, World view, etc. High spatial resolution images have higher within class spectral variability and thus only spectral variables based classifiers could not produce satisfactory results [69]. This paved way to spatio-contextual variables. The spatial information extracted from a panchromatic image, or from a particular band of a multispectral image or from the first principal component analysis of the image has been given as the input for classification, This is called as spatio-contextual information [10]. More details about Spatio-contextual techniques have been presented in section 3.

2.3 Knowledge based detection of land use/land cover

Extraction of land use/land cover data from remote sensing image could be done more effectively with the help of knowledge based techniques, as different kinds of data from multiple sources could be integrated with the help of these techniques [58]. For example, digital elevation data along with spatial information from an optical remote sensing image could extract the buildings more easily. Urban parcel imperviouness has been detected in [42] using high resolution ortho colour photography and lidar data. The surface height has been extracted with the help of lidar data and has been integrated with spectral clustering from colour photography, to extract the land imperviousness. Similarly soil maps, road networks, information from toposheets, temperature, precipitation data, well inventory, population density, etc could be incorporated easily in the classification process of remote sensing images using knowledge based methods. In [115], object based image analysis technique along with domain knowledge has been used for remote sensing image land cover classification. Image segmentation, feature extraction and classification has been used in this paper for Object based land cover detection and has used semantic networks to incorporate domain knowledge into the classification. Forestier and his collaborators [29] have used knowledge based techniques to automatically detect urban objects. Here, the knowledge-base is used to assign segmented regions into semantic objects. Similarly, Frick and his collaborators have used these rule based techniques for vegetation monitoring in QuickBird imagery [30]. Thus, this is a growing field, where LULC can be detected and extracted more promisingly.

3. Spatio-contextual techniques

Spatio-contextual techniques could be categorized as

- Texture based techniques.
- Object based image analysis techniques.
- Shape based techniques.

3.1 Texture based techniques

Texture plays a major role in detecting objects or regions from an image. The concept of colour could be explained as shades of gray of a particular pixel, where as texture could be explained as the spatial distribution of gray shades [39]. In an other way, texture is the visual effect which is produced by patial distribution of tonal variables over relatively small areas [38]. Generally, texture feature or texture value of a pixel will be calculated by considering a window around the pixel and doing some mathematical calculations on that window [14]. The ability of the texture extraction method to differentiate between various land covers greatly depends on the image spatial resolution and the structural differences within a class and between classes of the image. Remarkable amounts of literature is available on texture feature extraction. Numerous amounts of researchers have proposed various methods to extract texture features from a remote sensing image [53], [58], [69]. In a broad sense, the texture feature extraction techniques could be classified as

- Statistical Methods.
- Structural Methods.
- Model based Methods. and
- Transform based Methods.

3.1.1 Statistical methods

Most popular and widely used texture features extraction methods are statistical methods. Human eye can detect the differences in intensity distributions of any texture [47]. Futher, Julez says that the perception of texture by a human being is the result of statistical analysis of the intensities. Thus, texture patterns can be featured by the statistics of their intensities. In one way, statistical texture extraction methods can be classified as first order statistics (One pixel), second order statistics (pair of pixels) and third order statistics (three pixels).Various methods like first order statistics, GLCM, LBP, etc have been presented in literature. These methods have been presented in more detail in section 4.

IOP Conf. Series: Materials Science and Engineering 1074 (2021) 012002 doi:10.1088/1757-899X/1074/1/012002

3.1.2 Structural methods

Haralick, et.al [36] have proposed that structural approaches are based on the idea that textures are made up of various primitive elements, which are found in regular spatial arrangements. Thus, the texture of an image could be defined with the help of primitives and their placement rules [53]. The choice of the primitive, whether it is a line segment, open polygon, close polygon, etc, and the probability of this primitive to be at a particular location could be a strong or weak function of the location [36].

In [12], Carlucci have suggested a texture model where line segments and polygons have been used as primitives and graph like methods have been used to implement the placement rules. In [60], Lu and his collaborators have used tree structures for implementing placement rules. Later with the advancements in Mathematical Morphology, the structural texture extraction methods have also been advanced. In Mathematical Morphology, non linear operators could be employed with the help of structuring elements [37]. Mathematical Morphology has been presented in detail in section 5.

3.1.3 Model based texture feature extraction

Model based texture feature extraction techniques are yet another family of texture extraction techniques. Best examples of this class of techniques are Markov Random Field (MRF) models, fractal models, etc [65]. Structure of the image forms the basis for model based texture extraction techniques. Linear combination of a set of few basic functions will be used to describe these methods. Model based texture feature extraction techniques incorporate the contextual information into the features by incorporating spatial correlation term into the discriminant function [28].

MRF models have been applied on remote sensing images for accomplishing various tasks like classification of land use land cover, change detection, segmentation, etc [28], [65], [116]. Tso and his collaborators [97] have applied MRF model based techniques on remote sensing image for classification and reported that these MRF based methods have improved the classification accuracy and visual interpretation, when compared to non spatio contextual algorithms. Later in [82], authors have implemented a MRF model based segmentation approach for classifying a multispectral image. It captures both spectral and textural features of the image. Over segmented image and the multispectral image are taken as inputs for this algorithm. It then models a MRF over Spatial Adjacency Graph build on the input images. This method was an unsupervised approach and was able to classify various land use and land cover elements with no prior knowledge about the classes. Zhang, et al in [113] have applied MRF based models on hyper spectral imagery and classified the features with SVM, which significantly improved the classification accuracies. Another frequently used model based texture feature extraction method is fractal model. Mandelbrot has introduced fractal model. Fractal models have been applied on several remote sensing images for texture feature extraction [69], [92]. Estimation of fractal dimension is the essential issue in fractal geometry. Fractal geometry presents many tools to extract features from complicated objects and patterns. Thus, it could be said that model based texture feature extraction techniques have been used for classification of remote sensing images, but they are considered difficult to implement by many researchers. They also involve many computational problems [65].

3.1.4 Transform based texture feature extraction

Transform based texture feature extraction techniques transforms the original image into a new form by utilizing various filters. The energy of the transformed image has then been calculated to represent various features of the image [16]. The higher frequency variations detect the edges in the image and lower frequency variations detects the flat regions of the image. Various techniques like Fourier transforms, Gabor transforms, wavelet transforms, etc are available in the literature to extract the texture features from a remote sensing image [53], [58]. When compared to Fourier transforms, Gabor transforms performed better as they provide better spatial localization. In [50], various techniques using Gabor filters have been compared. In [16], features from Gabor filters and gray level co-occurrence probabilities have been integrated to extract the texture features. Features from Gabor filters have been applied to remote sensing image analysis [16], [48], [41]. In [41], authors have used statistical features like mean, variance and features from co-occurance matrices and Gabor filters for texture feature extraction. The features so extracted are then encoded and normalized using Self Organizing Maps (SOM). A subset of them has been selected using Classification And Regression Trees based feature selection method. Later, the selected features have been classified using K-Nearest Neighbourhood classifier. This algorithm can be applied on various types of remote sensing images and can be used with various classifiers for classification. But, this is a very complicated algorithm which yields in high computational complexity. Gabor filters yield in better spatial localization, but their usage is limited as there is no single filter resolution to localize a structure in natural texture images. Thus, wavelets performed better than Gabor filters [53] as they are based on multiple spatial resolutions. In literature, few papers have been presented where wavelets have been used for remote sensing image analysis. In a new method for crop classification has been proposed. A data base of different crop images has been generated by assembling clips from satellite imagery. Discrete Cosine Transforms (DCT) and Discrete Wavelet Transform (DWT) features from every crop image have been extracted. Later they are classified with the help of various classifiers and their performance has been studied. Authors concluded that DWT has performed better than DCT. In [45], Gabor wavelet has been integrated with Gradient Field Distribution (GFD) to extract feature vectors based on their directional texture properties for weed image classification.

3.2 Object based detection from remote sensing images

Object detection from optical remote sensing images deals with detecting objects like car, buildings, aeroplanes, ships, etc which have crisp boundaries and also detecting land use/land cover parcels which have a vague and ambiguous boundaries. Object Based Image Analysis (OBIA) is one of the popular techniques, where many efforts have been made in the past decade for the detection of various objects from optical remote sensing images [32]. It consists of two steps. Image segmentation and object classification. By selecting the shape, compactness, smoothness , scale and various other features, initially the image is segmented into homogeneous regions called objects [68]. These objects are then classified in the second step. Instead of considering the per pixel algorithms for classification, here objects are to be classified. With the availability of submeter resolution of images, each object falls in more number of pixels. Earlier an object used to be a part of pixel. Hence, the importance of OBIA techniques has been growing from the past decade, and number of researchers have contributed their works in OBIA [10], [32]. Earlier only spectral properties have been considered to group the pixels into an object. Later many parameters like shape, smoothness, compactness, scale, etc are being considered for segmentation of the image into objects. Also, it was proved that considering the shape and spatial properties have increased the efficiency of the algorithm [68]. Land use /land cover elements like roads, buildings, trees, crops etc have been successfully detected with this technique [3], [63], [109], [52], [74], [101]. Definiens Developer 7.0 or e-cognition software is used mostly for OBIA. Multi Resolution Segmentation (MRS) algorithm is being used more frequently for segmentation in this process. Membership function classifier and Nearest neighbour classifier are the two mostly used approaches for object classification [68], [32].

3.3 Shape based LULC detection

Template matching is one of the frequently used shape based image analysis techniques, where roads, trees, buildings, ships, aeroplanes, etc [32] have been detected successfully from an optical remote sensing image in the past decade. Techniques like rigid template matching and deformable template matching have been used for the detection of roads [13], [66], buildings [1], [89], trees crowns [44],

IOP Conf. Series: Materials Science and Engineering 1074 (2021) 012002 doi:10.1088/1757-899X/1074/1/012002

etc. Rigid templates have been used in the past for detecting comparatively simple objects. Hit or Miss transform is a popular technique used for rigid template matching [32]. These algorithms are very much sensitive to shape and are rotation and scale dependent. Deformable template matching is the extension to rigid template matching. Snake based template methods have proved to detect various land use and land cover objects like buildings, trees, roads, coast lines, ships, aeroplanes, etc [89], [32], [44]. Deformable template matching algorithms are more powerful when compared to rigid template matching algorithms. But, they are computationally expensive and needs more prior parameters of geometrical shape to be detected.

4 Statistical Methods

4.1 First order statistics methods

The first order statistics estimate the properties using individual pixels. For example: mean, histogram, variance, etc.

The second order statistics estimate the properties of two pixel values occurring at specific locations relative to each other [88]. Most popular method using second order statistics is Gray Level Co-occurrence Matrix (GLCM).

4.2 Gray Level Co-occurrence Matrix (GLCM)

Haralick et al [39] proposed Gray level co-occurrence matrix which uses second order statistical measures and could be used to estimate the texture properties of an image. GLCM has been used in various applications [64], [78], [91]. It is expressed in the form of a matrix, and represents how often different combinations of various gray values occurs in the image. If 'G' is the number of gray level values available in the image, than a gray level co-occurrence matrix contains 'G' rows and 'G' columns. Every matrix element of GLCM p(I,j / d, ɵ) is the estimation value of the frequency with which the two pixels with distance 'd' and angle ɵ having intensity values 'i' and 'j' appear in the image. Based on these values of GLCM, Harlick et al have proposed 14 texture estimation measures – contrast, angular second moment, correlation, sum of squares, inverse difference moments, sum average, sum variance, sum entropy, entropy, difference variance, difference entropy, info. Measure of correlation1, info. Measure of correlation2, max correlation coefficient [39]. Some of the measures proposed by Harlick like contrast and variance are related to first order statistics concepts. Later, R.Cossu investigated the fourteen parameters proposed by Harlick and found that six parameters among these fourteen parameters are more effective [20]. These six measure are energy, contrast, variance, correlation, entropy and inverse difference.These six parameters of GLCM have been applied successfully on a high resolution remote sensing image [5] for clustering and found that the two parameters energy and contrast are most efficient for discriminating texture features. GLCM has been widely applied on various types of remote sensing images and obtained reasonably satisfactory results [19], [71], [61], [78]. Coburn and his collaborators have used first and second order statistical features to perform supervised classification of forest region from an airborne remotely sensed image [19]. Pacifici and his collaborators have used nural networks and GLCM to classify urban scenes from very high resolution satellite images like Quickbird and WorldView [71]. Maillard and his collaborators have compared various techniques like GLCM, forier transforms, variograms for classification of texture features [62]. Thus GLCM is a well established traditional texture feature extraction method from remote sensing images. Later, various variants of GLCM like Gray Level Difference Method (GLDM)[59], Gray level Co-occurrence Linked Lists (GLCLL) [17], etc have been proposed to make it invariant to the changes in the overall luminance, to decrease the time complexity, etc.

4.3 Local Binary Patterns

Local Binary Patterns (LBP) is a recent and proven technique to identify the texture features of an image[34]. LBP technique summarizes the texture pattern details of every pixel into a histogram. LBP values are calculated for every pixel.It works by calculating the frequencies of local patterns in the image. LBP algorithm is very easy to implement, at the same time, it is an efficient approach. LBP has been derived from an other texture analysis method called as "Texture Spectrum" [40], [105]. The major disadvantage of Texture spectrum method is, it consists of large number of texture units. LBP has been introduced in 1996 [70] and has been applied on various remote sensing images [52], [43], [9].

5. Mathematical morphology

Mathematical Morphology makes best use of set theory concepts and finds applications in various fields of image processing like image filtering, image segmentation, image measurement, extracting geometrical structures [84], [85] etc. Mathematical Morphology is first introduced by Matheron and Serra [84], [85] to study the geometry of binary porous media such as sand stone [84] and has been efficiently applied on remote sensing images for object extraction and segmentation [86]. Morphology provides a wide variety of tools and transformations for featuring textured objects [83], [102] and suits very well for applying to remote sensing images [8], [23], [57], [80], [111], [35], [87], [67]. Figure 2 present the taxonomy of various techniques of Mathematical Morphology to detect LULC from an optical remote sensing image.

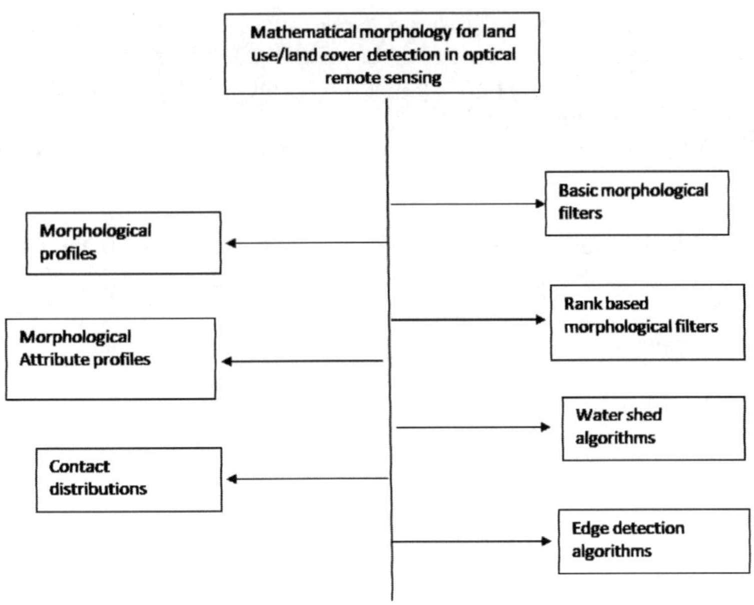

Figure 2: Taxonomy of LULC detection techniques from an optical remote sensing image using Mathematical Morphology

Morphology is a broad set of image processing operations that process images based on shapes[81]. Any morphological operation uses two sets of images. One set is the original image and the other set is called as structuring element. Structuring element is a matrix which consists of some shape, used to probe or interact with a given image, with the purpose of drawing conclusions on how this shape fits or misses the shapes in the processing image [81]. For example the following are some of the structuring elements.

IOP Conf. Series: Materials Science and Engineering 1074 (2021) 012002 doi:10.1088/1757-899X/1074/1/012002

$$S1 = \begin{bmatrix} 1 & 1 & 1 \\ 1 & 1* & 1 \\ 1 & 1 & 1 \end{bmatrix} \quad S2 = \begin{bmatrix} 0 & 1 & 0 \\ 1 & 1* & 1 \\ 0 & 1 & 0 \end{bmatrix} \quad S3 = \begin{bmatrix} 0* & 1 & 0 \\ 0 & 1 & 0 \\ 0 & 1 & 0 \end{bmatrix}$$

The element with '*' symbol acts as the origin. Any Mathematical Morphology operation works by moving a structuring element over the binary image to be modified in such a way that it is centered (placing the origin of the structured element) over every image pixel at some point [46]. When the structuring element is placed over a region of the image, a logical operation is performed on the pixels covered by the structuring element, yielding a binary output. The morphological process is like a convolution process [46]. A morphological operation could be constructed that is sensitive to specific shapes in the input image.

5.1 Morphological Filters

All basic morphological operators could be used as filters overs the given input image [86]. Few of the basic morphological operators have been presented in the coming section.

5.1.1 Dilation and Erosion

The most basic morphological operations are dilation and erosion [84]. Dilation adds pixels to the boundaries of objects in an image , while erosion removes pixels on object boundaries. Size and shape of the structuring element determines the number of pixels added or removed from the objects in an image. When an Erosion operation has been applied on an input image, it produces an output image, which shows where this structuring element fits the objects in the image. Similarly, when a Dilation operation has been applied on an input image, it produces an out put image, which shows where this structuring element hits the objects in the image.

In morphological image analysis, a 2-D digital image could be defined as a subset of Z x Z [86]. If 'I' is an image such that $I \subseteq Z \times Z$, and 'u' is a point where $u \in Z \times Z$,

Translation of I by u is given as

$$\text{Translation of } I \text{ by u} \quad = (I)_u = \{x + u | x \in I\} \qquad (4)$$

Let 'F' is a structuring element. Erosion of I by 'F' is given as

$$\text{Erosion of } I \text{ by F} \quad = \varepsilon_F(I)(x) = \bigcap_{f \in F} (I)_{-f} \qquad (5)$$

Dilation of I by 'F' is given as

$$\text{Dilation of } I \text{ by F} \quad = \delta_F(I)(x) = \bigcup_{f \in F} (I)_f \qquad (6)$$

5.1.2 Gradient operations

External gradient of the image is given as

$$\text{External gradient of the image} = \delta_F(I) - I \tag{7}$$

Internal gradient of the image is given as

$$\text{Internal gradient of the image} = I - \varepsilon_F(I) \tag{8}$$

Symmetric gradient of the image is given as

$$\text{Symmetric gradient of the image} = \delta_F(I) - \varepsilon_F(I) \tag{9}$$

5.1.3 Opening and Closing

Opening and Closing are the two most commonly used Morphological operators. Opening dilates an eroded image, whereas Closing erodes a dilated image. Opening smoothens the edges in an image, whereas Closing fills the gaps in the contours of the image, thus eliminating the holes and combining the narrow breaks. In general Open operations are used to extract bright structures of the image, whereas close operations are used to isolate dark structures of the image. The Opening of I By 'F' is given as

$$\text{Opening of } I \text{ by F} = I \circ F = \delta_F(\varepsilon_F(I)(x) \tag{10}$$

Similarly, the Closing of I By 'F' is given as

$$\text{Closing of } I \text{ by F} = I \bullet F = \varepsilon_F(\delta_F(I)(x) \tag{11}$$

5.1.4 Tophat transforms

In digital image processing, if thinner elements than a structuring element are to be extracted, residues of the opening and closing operations could be considered. These actions are accomplished with top-hat and bot-hat operations. Top-hat transform can be defined as the difference between the opening of the image and the image itself. It highlights the narrow brighter areas between the regions. Inverse top-hat is also called as bot-hat transform. The difference between closing operation on the image and the image itself is called as bot-hat transform [46].

The top-hat of I is given as

$$\text{Top-hat of } I = TH(I) = I - I \circ F \tag{12}$$

The bot-hat of I is given as

$$\text{Bot-hat of } I = BH(I) = I \bullet F - I \tag{13}$$

5.1.5 Applications of Morphological filters

All basic morphological operators dilation, erosion, opening and closing could be used as filters to enlarge, shrink, remove or fill in image boundaries or regions of the image. Erosion and dilation operators could be used in different combinations to perform various operations like thickening, thinning, skeletonizing, etc [86]. similarly, top-hat transformations could be used for image processing tasks like image enhancement, feature extraction, etc. The morphological operators like (AB)_B have been used extensively to eliminate salt and pepper noise, where 'A' is the original source image and 'B' is the structuring element. Gradient operators have been used for extracting edges. Dilations,

IOP Conf. Series: Materials Science and Engineering 1074 (2021) 012002 doi:10.1088/1757-899X/1074/1/012002

complementations and intersections could be used iteratively to accomplish the tasks of regions fillings.

In remote sensing, Morphological operators have been used for classifying the remotely sensed images of high resolutions and have been mentioned as promising tools for remote sensing data analysis [86], [73]. In [90] and [103], various basic Morphological operators like opening and closing have been applied on panchromatic images. In [98], authors investigated the use of Mathematical Morphological operators in the land use classification of a remotely sensed optical image. Algorithm was applied on two Quick Bird images. The study proved the potential use of basic mathematical operators like opening, closing, opening by reconstruction, closing by reconstruction, opening by top hat and closing by top hat in the land cover/land use classification. Texture features have been extracted from the remote sensing image by applying these operators with structuring elements of various sizes and shapes and concluded that the simple operators like opening and closing have performed more efficiently. As a good number of features have been extracted from each pixel, a feature dimensionality reduction algorithm has been employed. SVM has been used for classification of the features extracted by the said operators. It is a highly computational algorithm, as the feature selection algorithm was a greedy algorithm. In [22], authors have proposed an algorithm to extract various land cover elements like forest, water, urban land, hills, etc. They have utilized various techniques namely principal component analysis, morphological operators like erosion, dilation, opening and closing and vegetation indices like NDVI and NDWI to classify a LANDSAT-8 multispectral image. They have applied principal component analysis for dimensionality reduction. Proposed method obtained better results when compared to that of direct multispectral image analysis. In [76] and [25], authors have used morphological operators like tophat, geodesic morphological operators, opening and closing operations for segmentation of remotely sensed image. In [25], experiments have been conducted on ortho photos and multispectral images. Reconstruction filters have also been used in the literature, for remote sensing image analysis [25].

Extraction of structures from a remote sensing image highly depends on the size and shape of the structuring elements used. Structures in an image may respond to a given structuring element size, and some more similar structures may not respond to this particular size of the structural element. They may react to a structuring element of different size. Sometimes we know the size of the structures to be extracted from a remote sensing image, but more often structures of different sizes are to be extracted. In this case, single size structuring elements could not help. A range of structuring elements with various sizes could be think of in such complex scenes giving rise to multiscale approach [8]. In multiscale approach, various sizes of structuring elements could be explored, and the best suited one for the pixel could be selected as the feature for that particular pixel in the image.

5.2 Edge detection techniques through Morphological filters

Edges separate two different regions of an image. Edges could be detected by the sharp change in the intensity of the image [46]. Morphological gradients are the first term edge detection techniques of Mathematical Morphology. Equation 7 refers to gradient by dilation, equation 8 refers to gradient by erosion and equation 9 refers to symmetric gradient or morphological gradient [84]. Edges could be detected using any of the three gradients. Several other filters like sobel edge detector, Canny edge detector, Prewitz edge detector, etc are available in the literature for detecting the edges. In [72], gradient operators have been used to classify built area from multisensor images. In [55], multiscale approach has been applied to gradient operator for remote sensing image segmentation. Mohamed Ali and his collaborators [18] have used Canny edge detector to detect edges in the remote sensing image with noise. When compared to the other edge detection techniques, authors proved that Canny edge detector performed well under noise conditions. They argued that neither the Roberts Cross, the Sobel operator, nor Prewitt operator are able to detect the edges of the object while removing the noise from the image. Later various morphological operators have been implemented on remote sensing images for edge detection.

5.3 Watershed algorithms

Watershed algorithm is a classical region based segmentation approach. Classification of the image if needed, could be obtained from the segmented image. Watershed algorithm do not assume any primitive shapes beforehand. So, image classification is not biased [106]. Watershed transformations do not consists of any primitive morphological operations. In this algorithm, any digital image is treated as a topographic surface, where the gray values of the image represent the elevation of the land parcel. Pixels with higher gray values are considered as hills and mountains and the pixels with low values are considered to be valleys. Watershed algorithm divides the image into regions called as catchment basins. Watershed is the line separating two catchment basins. Watershed algorithm thus allocates each pixel of the image either to one catchment basin or to a watershed line, resulting in segmentation of an image. But, most of the times watershed algorithms result in over segmentation [104]. To overcome this many techniques like merging, marker based water shed algorithms, etc have been proposed in the literature [93], [51], [104].

5.3.1 Applications of Watershed algorithms to remote sensing images

In [106], watershed algorithm has been applied on a remote sensing image. Authors argued that as no shape is assumed before hand, classification will be unbiased. In [107], watershed algorithm has been applied on a satellite image having complex mountainous regions. In [104], [54], the algorithm has been extended to remote sensing multispectral images. The efficiency of image segmentation through watershed algorithms mainly depend on the gradient images used. If the quality of gradient images increases, the output of the watershed algorithm also increases. To utilize structuring elements of various sizes, in [104], multi scale approach has been used to derive the gradient images from the original images. This gradient image has been given as input to the watershed algorithm. The drawback of watershed algorithm is, it resulted in over segmentation. In [54], watershed algorithm has been applied on Quick-Bird and IKONOS images. Here, the gradient magnitude has been calculated from multispectral image using vector based morphological approach. Gradient magnitude has been computed at multiple scales and irrelevant local minima has been eliminated before giving it as an input to watershed algorithm. Gradient magnitude image performs better than Principal component analysis image, as the gradient magnitude image uses all the information from all the spectral bands and hence no data loss. In [56], watershed algorithm has been applied on multispectral SPOT-5 image for land cover classification. Multiscale gradiant has been used in this here, where component wise approach has been used to calculate the multiscale gradient image. To reduce over segmentation, local minima has been eliminated using morphological gray scale reconstruction. Watershed algorithm has been used for segmentation and pixel wise classification has been applied for land cover classification. In [93], [51], marker based watershed algorithms have been implemented on multispectral high resolution remote sensing images. As they are semi automatic, procedure to select optimal parameters for segmentation automatically has to be studied. In [95], watershed algorithm has been extended to hyperspectral images to segment and classify various crop covers.

5.4 Morphological Profiles

Composition of morphological openings and closings with increasing size of structuring element is called as a Morphological Profile (MP) [6]. Each set of opening and closing operation results in a transformation of the original image into a new image. The objects smaller than the structuring element are deleted in the new image. For every pixel x, the vector containing the pixel values in openings and closings of different scales can also be called as MP of the pixel x. Thus, the MP of a pixel contains the information about the size and shape of the objects in the image [86].Later Differential Morphological Profiles (DMP) have been proposed to extract the objects of interest. Differences in the respective values of the MP can be called as DMP. The large value of DMP at scale s, says that the object of size s is present in the image [6]. Extended Morphological Profiles (EMP), MP with linear structuring elements, etc have been proposed in the literature [53].

5.4.1 Applications of Morphological Profiles to remote sensing images

In the recent years, MP and DMP have been implemented extensively on remotely sensed images for extraction of various objects [6], [15], [2]. In [15], the standard MP computed by openings and closings have been applied on panchromatic image to study the effect of Bam Earthquake. Pesaresi and his collaborators [73] have used DMP for panchromatic image segmentation. In [2], authors have proposed an automatic hierarchical segmentation technique using DMP. They have used both spatial and spectral properties for segmentation. In [8], authors have used DMP's for classification of an IKONOS panchromatic image. Reconstruction DMP's along with neural network classifier have been used and feature extraction techniques have been used on DMP's to reduce the dimentionality. In [6], authors have proposed a method to classify a very high resolution urban land cover remote sensing imagery. As spectral characteristics of urban land cover are so similar, they can not be classified using only spectral information. Thus, authors have used morphological profiles to incorporate geometric information into the classification. Instead of using disk structuring elements, in this work, authors have used linear structuring elements also along with disk structuring elements to aid in extracting roads, rivers, etc. Also instead of using traditional morphological reconstruction, authors have used partial morphological reconstruction for retaining the shape of the objects. But, it resulted in over reconstruction. In [100], authors have used advanced directional morphological profiles for extracting roads from very high resolution remote sensing data. The method works on creating morphological profiles for every pixel. They have used path openings and path closings as morphological filters. This method efficiently extracted road networks, but needs two types of thresholds and hence is not automatic. Sascha Klemenjak and his collaborators [49] have proposed automatic detection of rivers from Terra SAR-X data by extending the work of [100]. In [43], morphological texture features are extracted from multispectral IKONOS imagery to evaluate mangrove forest mapping using MP based on Opening by reconstruction and closing by reconstruction. Discrimination of the species of mangrove forests has also been done and acquired promising accuracies. In [7] and [27], MP have been used to classify hyperspectral images. In [7], Extended Morphological Profiles (EMP) have been utilized instead of MP. MP calculated on the principal component transform of the image are called as EMP.

As MP and DMP are to be calculated for every pixel with various structuring elements of different sizes and shapes, feature space of the image could be very huge. This is called as Hughes phenomenon [98]. To overcome this phenomenon various algorithms have been implemented. Few of them deal with reducing the feature set reduction [8], [98]. In [67], Morphological Attribute Profiles (MAP) have been used to analyse very high resolution images. Later, this is extended to extract buildings from SAR images in [111]. The sparse representation of Extended Multi Attribute Profile (EMAP) has been implemented in [87] to classify Avris data. EMAP provide multilevel characterization of the image, with the help of which different kinds of structural information present in the remote sensing image could be modeled. EMAP have been extended to hyperspectral images in [31]. An automatic framework has been established to classify the ROSIS-03 university scene and the AVIRIS Indian Pines image based on supervised feature extraction and morphological profiles. EMAPs have been used instead of DMPs (Differential Morphological Profiles) for extracting spatial features. Main difficulties associated with morphological profiles are how to initialize each attribute profile and to discriminate which attribute profiles lead to better classification of various classes. EMAPs have been generated automatically and "Discriminant Analysis Feature Extraction" and "Discriminant Boundary Feature Extraction" have been used for feature extraction. Final classification map has been produced using Random Forest generator. Later in [75], an optical remotely sensed image has been classified with the help of Local Feature based Attribute Profiles (LFAP), which are the extension of AP. LFAPs have been constructed by extracting few first-order statistical features of the local patch around each APs' sample.

5.5 Contact distributions

Any binary image or gray scale image could be represented in the form of random closed set. Uni variate random close sets are used to represent binary images and multivariate random closed sets are used to represent gray scale images. Any subset of a binary image can also be represented as a random closed set and many descriptors are given in the literature to describe these subsets. If 'W' is a window of pixels surrounding a pixel 'p' and if 'Φ' is the set of pixels with value '1', in window 'W', then many functions such as volume fraction, Co-variance, K-function, contact distributions, etc are used to describe the subset 'W' [26]. Contact distributions are of two types, namely linear contact distributions and spherical contact distributions.

Volume fraction of Φ is given by

$$\mathcal{VF} = \frac{A(\phi)}{A(W)} \quad (15)$$

Co-variance function is given by

$$\mathcal{C}(t) = \frac{A(\varepsilon_{(0,t)}(\phi))}{A(\varepsilon_{(0,t)}(W))} \quad (16)$$

Linear contact distribution with line segment 't' and orientation 'θ' is given by

$$\mathcal{H}^S(t) = 1 - \frac{A(\varepsilon_{B(t,0)}(\phi)^c).A(W)}{A(\varepsilon_{B(t,0)}(W)).A((\phi)^c)} \quad (17)$$

Where $\varepsilon_{S(t,\theta)}(W)$ represents erosion of 'W' with structuring element 'S(t, θ)'. Similarly, the estimator for spherical contact distribution is given by

$$\mathcal{H}^S(t) = 1 - \frac{A(\varepsilon_{B(t,0)}(\phi)^c).A(W)}{A(\varepsilon_{B(t,0)}(W)).A((\phi)^c)} \quad (18)$$

5.5.1 Applications of Contact distributions to remote sensing images

Irene Epifanio and Guillermo Ayala have used these descriptors for identifying textures [26]. They have implemented them on synthetic textures database and proved that contact distributions are more effective than other texture descriptors like GLCM and Markov random fields. They have also derived that spherical contact distributions and linear contact distributions performed well for some databases and linear contact distributions performed well when compared to spherical contact distributions for some other data bases.

As remote sensing images have real textures, it is more difficult to implement the texture descriptors on them. Later, Irene Epifanio and Peirre Soille have used LCD for extracting texture features for segmentation of a remote sensing image [25]. They have proposed both supervised and unsupervised approaches. In the supervised approach, the necessary thresholds to acquire binary images are to be estimated manually and finally need some training data to segment the image. Thus the method utilized to extract features can not be used for unsupervised classification. Unsupervised approach proposed in [25], was very complicated, where the procedure to estimate the seed values of every cluster is very time consuming, complicated and computationally intensive.

6 Hybrid algorithms

Many hybrid algorithms which used more than one technique have been proposed in the literature. In [77], features using statistics, vegetation indices NDVI and NDWI have been applied on PCI transformed image. In [41], authors have used statistical features like mean, variance and features calculated from co-occurrence matrices and Gabour filters for feature extraction. The features so extracted are then encoded and normalized using Self Organizing Maps (SOM). A subset of them has been selected using CART (Classification And Regression Trees) based feature selection method. Later, the selected features have been classified using KNN classifier. But, this is a very complicated algorithm which yields in high computational complexity. In [22], authors have used various techniques like vegetation indices, morphological operators to LANDSAT multispectral image classification. In [96], authors have considered both spectral and context based feature space for clustering the hyperspectral remote sensing image. It integrates the results of pixel wise SVM classification and the segmented image resulted from partitioning clustering using majority voting method. For this integration they have used ISODATA algorithm and Guassian mixture resolving techniques. Experiments have been carried out on ROSIS-03 optical sensor image. This method performed well when compared to traditional pixel based classification algorithms. It was able to classify various land cover data and in particular is very good where large spatial structures are involved. But, for the images where small spatial structures are involved, this algorithm is not very good at classifying various land cover elements. In [99], a new method has been proposed to integrate spectral and textural information to classify an agricultural image. It takes a multispectral image and a panchromatic image as inputs. It extracts texture features from the panchromatic image and parallel extracts edge information from the multispectral image. Finally fusion of classified image based on texture features and edge information image gives classified land use agriculture image. Texture features are extracted based on Linear Fourier Histograms (LFH) as it suits for extracting various tree crops. The Nearest Neighbour (NN) classifier has been used to classify the texture features. Similarly, the multispectral image has been pre processed with Guassian low pass filter and then using the three spectral values of each pixel, clustered the image with the help of k-means clustering. Later these two images have been fused. In [114], authors have taken the advantage of both spectral and shape features of a High Resolution Remotely Sensed Image for extracting the land use and land cover elements. They have presented a novel index, known as Pixel Shape Index (PSI), which features the shape information of the image. PSI measures the gray similarity distance in all the directions. Along with PSI, the image has been transformed with the help of Individual Component Analysis (ICA). These shape and spectral features are then given as input to SVM classifier for classification. Proposed algorithm has been implemented on QuickBird image of Beijing city. Later, the results have been compared with the results of the algorithms like water shed algorithm, GLCM, and the length-width and was proved that the integration of shape and spectral features resulted in better classification accuracies. Similarly many hybrid algorithms have been presented in literature for LULC classification / segmentation [21], [58], [79], [32], [56] and acquired efficient accuracies. But, one must be careful that because of different capabilities of various feature vectors, utilization of too many feature variables may result in poor classification accuracies [58].

7. Conclusion

This review presents that a good amount of work has been carried out in the filed of detection of LULC from optical remote sensing images. Many techniques discussed in this paper have applications in detection of LULC. However, incorporation of texture in the features has encouraged the classification / segmentation / detection / extraction of LULC to a large extent. Thus spatio-contextual approaches with emphasis on structural and statistical methods can be applied to optical remote sensing image mapping.

As the property of connectivity between the pixels, plays an essential role in texture feature extraction, it is proposed to utilize the concepts of "Contact distributions" applied with the help of "Mathematical Morphology", for the detection of LULC from optical remote sensing images. This method can tackle images having groups of pixels of any shape and even though they do not have well defined center. This review reveals that algorithms based on contact distributions have not been extensively used for the detection of LULC from optical remote sensing images. But, as it appears promising, algorithms based on contact distributions applied through Mathematical Morphology could be studied further to investigate feature extraction techniques.

References

[1] S. Ahmadi, M.V. Zoej, H. Ebadi H.A. Moghaddam, and A. Mohammadzadeh. Automatic urban building boundary extraction from high resolution aerial images using an innovative model of active contours. International journal of applied Earth Observation and Geoinformation, 12:150–157, 2010.

[2] H. Akcay and S. Aksoy. Automatic detection of geospatial objects using multiple hierarchical segmentations. IEEE transactions on Geoscience and remote sensing, 46(7):2097–2111, July 2008.

[3] B.A. Baker, T.A Warner, J.F. Conley, and B.E. McNeil. Does spatial resolution matter? a multi-scale comparison of object-based and pixel-based methods for detecting change associated with gas well drilling operations. Internationa journal of remote sensing, 34:1633– 1651, 2013.

[4] A. Bannari, D. Morin, F. Bonn, and A.R. Huete. A review of vegetation indices. Remote Sensing Reviews, 13:95–120, October 2009.

[5] A. Baraldi and F. Parmiggiani. An investigation of the textural characteristics associated with graly level co-occurrence matrix statistical parameters. IEEE trans on Geo science and remote sensing, 33(2):293–304, March 1995.

[6] Rik Bellens, Sidharta Gautama, Leyden Martinez-Fonte, Wilfried Philips, Jonathan Cheung-Wai Chan, and Frank Canters. Improved classification of vhr images of urban areas using directional morphological profiles. IEEE transactions on Geoscience and remote sensing, 46(10):2803–2813, October 2008.

[7] J.A. Benediktsson, J.A. Palmason, and J. R. Sveinsson. Classification of hyperspectral data from urban areas based on extended morphological profiles. IEEE transactions on Geoscience and remote sensing, 43(3):480–491, March 2005.

[8] J.A. Benediktsson, Pesaresi, and Kolbeinn Arnason. Classification and feature extraction for remote sensing images from urban areas based on morphological transformations. IEEE trans. on Geo science and remote sensing, 41(9):1942–1949, 2003.

[9] S. Bhagavathy and B. S. Manjunath. Modeling and detection of geospatial objects using texture motifs. IEEE Transactions on Geoscience and remote sensing, 44(12):3706–3715, 2006.

[10] T. Blaschke. Object based image analysis for remote sensing. ISPRS Journal of Photogrammetry and Remote Sensing, 65:2–16, 2010.

[11] J. B. Campbell. Introduction to Remote Sensing. The Guilford Press, New York, USA, 551, 1987.

[12] L. Carlucci. A formal system for texture languages. Pattern recognition, 4:53–72, 1972.

[13] D. Chaudhuri, N. Kushwaha, and A. Samal. Semi-automated road detection from high resolution satellite images by directional morphological enhancement and segmentation techniques. IEEE journal of selected topics in applied Earth observations and remote sensing, 5:1538–1544, 2012.

[14] D. Chen, D.A. Stow, and P. Gong. Examining the effect of spatial resolution and texture window size on classification accuracy: An urban environment case. International Journal of Remote Sensing, 25(11):2177–2192, 2004.
[15] M. Chini, N. Pierdicca, and W. Emery. Exploiting sar and vhr optical images to quantify damage caused by the 2003 bam earthquake. IEEE transactions on Geoscience and remote sensing, 47(1):145–152, January 2009.
[16] D.A. Clausi and Huang Deng. Design-based texture feature fusion using gabor filters and co-occurrence probabilities. IEEE trans on image processing, 14(7):925–936, June 2005.
[17] D.A. Clausi and M. Jernigan. A fast method to determine cooccurrence texture features using a linked list implementation. Remote Sensing of Environment, 36:506–509, 1996.
[18] Mohamed Ali David Clausi. Using the canny edge detector for feature extraction and enhancement of remote sensing images. Proceedings. IEEE International Geoscience and Remote Sensing Symposium, 1:2298–2300, July 2001.
[19] C.A. Coburn and A.C.B. Roberts. A multiscale texture analysis procedure for improved forest and stand classification. International journal of remote sensing, 25(20):4287–4308, october 2004.
[20] R. Cossu. Segmentation by means of textural analysis. Pixel, 1(2):21– 24, 1988.
[21] V. Dev, Y. Zang, and M.Zhong. A review on image segmentation techniques with remote sensing perspective. ISPRS TC VII Symposium, pages 31–42, Jul 2010.
[22] M. Dharani and Dr.G. Sreenivasulu. Land classification of land sat multispectral image using principal component analysis and morphological operations. Journal of advanced research in Dynamical and Control Systems, 10(6):164–174, 2018.
[23] P.C. Doraiswamy and et al. Crop classification in the u.s. corn belt using modis imagery. IEEE proceedings of Geoscience and Remote Sensing Symp, 45(4):1074 – 1083, 2007.
[24] J.A. dos Santos, O.A.B. Penatti, and R. Da Silva Torres. Evaluating the potential of texture and color descriptors for remote sensing image retrieval and classification. Proceedings of VISAPP, page 203, January 2010.
[25] I. Epifanio and P. Soille. Morphological texture features for unsupervised and supervised segmentations of natural landscapes. IEEE trans on Geo science and Remote sensing, 45(4):1074 – 1083, 2007.
[26] Irene Epifanio and Guillermo Ayala. A random set view of texture classification. IEEE trans of image processing, 11(8):859–867, August 2002.
[27] M. Fauvel, J. Benediktsson, J. Chanussot, and J. Sveinsson. Spectral and spatial classification of hyperspectral data using svms and morphological profiles. IEEE transactions on Geoscience and remote sensing, 46(11):3804–3814, November 2008.
[28] M. Fauvel, Y. Tarabalka, J.A. Benediktsson, J. Chanussot, and J.C. Tilton. Advances in spectral-spatial classification of hyperspectral images. Proceedings of the IEEE, 101:652–675, 2012.
[29] G. Forestier, A. Puissant, C. Wemmert, and P. Ganc¸arski. Knowledge based region labeling for remote sensing image interpretation. Computers, Environment and Urban Systems, 36(5):470–480, September 2012.
[30] A. Frick, G. Weyer, H. Kenneweg, and B. Kleinschmit. A knowledge based approach to vegetation monitoring with quickbird imagery. ISPRS workshop, pages 1–6, 2005.
[31] Pedram Ghamisi, Jon Atli Benediktsson, Gabriele Cavallaro, and Antonio Plaza. Automatic framework for spectral-spatial classification based on supervised feature extraction and morphological attribute profiles. IEEE Journal of Selected Topics in Applied Earth observations and Remote Sensing, 7(6):2147 – 2160, June 2014.
[32] Junwei Han Gong Cheng. A survey on object detection in optical remote sensing images. Journal of photogrammetry and remote sensing, 117:11–28, July 2016.
[33] Anthony Gualtie and R. F. Cromp. Support vector machines for hyperspectral remote sensing classification. https://ntrs.nasa.gov/archive/nasa/casi.ntrs.nasa.gov/19990021532.pdf, 1, 1998.
[34] Zhenhua Guo, Lei Zhang, and David Zhang. A completed modelling of local binary pattern operator for texture classification. IEEE Transactions on Image Processing, 19, 2010.

[35] O. Hagner and H. Reese. A method for calibrated maximum likelihood classification of forest types. Remote sensing of environment, 110:438–444, 2007.
[36] R.M. Haralick. Statistical and structural approaches to texture. Proceedings of the IEEE Transactions on Geoscience and Remote Sensing, 5:786–804, 1979.
[37] R.M. Haralick, S.R. Sternberg, and X. Zhuang. Image analysis using mathematical morphology. IEEE Transactions on Pattern Analysis and Machine Intelligence, 4:532–550, 1987.
[38] R.M. Harlick. Glossary and index of remotely sensed image pattern recognition concepts. Pattern recognition, 5:391–403, 1973.
[39] R.M. Harlick, K. Shanmugam, and I. Dinstein. Texture features for image classification. IEEE trans on systems, Man. and Cybernetics, SMC-3:610–621, November 1973.
[40] DC. He and L. Wang. Texture unit, texture spectrum, and texture analysis. IEEE transactions on Geoscience and Remote Sensing, 28:509–512, 1990.
[41] Jukka heikkonen and Aristide varfis. Land cover/land use classification of urban areas: A remote sensing approach. International Journal of Pattern Recognition and Artificial Intelligence, 12(4):475–489, 1998.
[42] M.E. Hodgson, Jensen, J.A. Tullis, K.D. Riordan, and C.M. Archer. Synergistic use lidar and color aerial photography for mapping urban parcel imperviousness. Photogrammetric Engineering and Remote Sensing, 69:973–980, September 2003.
[43] Xin Huang, Liangpei Zhang, and Le Wang. Evaluation of morphological texture features for mangrove forest mapping and species discrimination using multispectral ikonos imagery. IEEE transactions on Geoscience and remote sensing, 6(3):393–397, July 2009.
[44] C. Hung, M. Bryson, and S. Sukkarieh. Multi-class predictive template for tree crown detection. ISPRS Journal of photogrammetry and remote sensing, 68:170–183, 2012.
[45] Asnor Juraiza Ishak, Aini Hussain, and Mohd Marzuki Mustafa. Weed image classification using gabor wavelet and gradient field distribution. Computers and Electronics in Agriculture, 66(1):53–61, April 2009.
[46] S. Jayaraman, S. Esakkirajan, and T. Veera Kumar. Digital image processing. Tata McGraw Hill, 7 West patel nagar, New delhi, 1 edition, 2012.
[47] B. Julez. Texture and visual perception. Scientific American, 222:38– 48, 1965.
[48] U. Kandaswamy, D.A. Adjeroh, and M.C. Lee. Efficient texture analysis of sar imagery. IEEE Transactions on Geo science and Remote Sensing, 43(9):2075–2083, September 2005.
[49] Sascha Klemenjak, Bj¨orn Waske, Silvia Valero, and Jocelyn Chanussot. Automatic detection of rivers in high resolution sar data. IEEE Journal of selected topics in applied earth observations and remote sensing, 5(2):1–9, October 2012.
[50] P. Kruizinga, N. Petkov, and S.E. Grigorescu. Comparison of texture features based on gabor filters. Proc. International Conference on Image Analysis Process, pages 142–147, 1999.
[51] Deren Li, Guifeng Zhang, Zhaocong Wu, and Lina Yi. An edge embedded marker-based watershed algorithm for high spatial resolution remote sensing image segmentation. IEEE trans on image processing, 19(10):2781 – 2787, May 2010.
[52] H. Li, H. Gu, Y. Han, and J. Yang. Object-oriented classification of high-resolution remote sensing imagery based on an improved colour structure code and a support vector machine. International journal of remote sensing, 31(6):1453–1470, 2010.
[53] Miao Li, Shuying Zang, Bing Zhang, ShanshannLi, and Changshan Wu. A review of remote sensing image classification techniques: the role of spatio-contextual information. European journal of remote sensing, 47(1):389–411, May 2014.
[54] P. LI and X. XIAO. Multispectral image segmentation by a multichannel watershed-based approach. International Journal of Remote Sensing, 28(19):4429–4452, October 2007.
[55] Peijun Li and Xiaobai Xiao. Evaluation of multiscale morphological segmentation of multispectral imagery for land cover classification. IEEE International Geoscience and Remote Sensing Symposium, September 2004.

[56] Peijun Li and Xiaobai Xiao. Evaluation of multiscale morphological segmentation of multispectral imagery for land cover classification. IGARSS-IEEE International Geoscience and Remote Sensing Symposium, pages 2676–2679, 2004.
[57] Runsheng Li and Fanzhi Cao. Road network extraction from high resolution remote sensing image using homogenous property and shape feature. Journal of Indian society of remote sensing, 46(1):51– 58, January 2018.
[58] D. Lu and Q. Weng. Survey of image classification methods and techniques for improving classification performance. International Journal of Remote Sensing, 28:823–870, 2007.
[59] Dengsheng Lu, Scott Hetrick, and Emilio Moran. Land cover classification in a complex urban-rural landscape with quickbird imagery. Photogrammetric Engineering and Remote Sensing, 76(10):1159– 1168, October 2010.
[60] S.Y. Lu and K.S. Fu. A syntactic approach for texture analysis. Computer graphics. image processing, 7:303–330, 1978.
[61] Danielle J. Marceau, Philip J. Howrath, M. Jeam Marrie Dubois, and Denis J. Gratton. Evaluation of the grey-level co-occurrence matrix method for land-cover classification using spot imagery. IEEE transactions on Geoscience and remote sensing, 28(4):513–519, July 1990.
[62] P. Millard. Comparing texture analysis methods through classification. PE and RS, 69(4):357–367, 2003.
[63] N.B. Mishra and K.A. Crews. Mapping vegetation morphology types in a dry savanna ecosystem: integrating hierarchical object-based image analysis with random forest. International journal of remote Sensing, 35:1175–1198, 2014.
[64] P. Mohanaiyah, P. Satyanarayana, and L. GuruKumar. Image texture feature extraction using glcm approach. International journal of scientific and research publications, 3(5):1–5, 2013.
[65] G. Moser, S.B. Serpico, and J.A. Benediktsson. Land-cover mapping by markov modeling of spatio-contextual information in very-highresolution remote sensing images. Proceedings of the IEEE, 101:631– 651, 2013.
[66] S. Movaghati and A. Moghaddamjoo A. Tavakoli. Road extraction from satellite images using particle filtering and extended Kalman filtering. IEEE transactions on Geoscience and remote sensing, 48:2807–2817, 2010.
[67] Mauro Dalla Mura, Jon Atli Benediktsson, and et al. Morphological attribute profiles for the analysis of very high resolution images. IEEE trans on geosicence and remote sensing, 48(10):3747–3762, October 2010.
[68] Soe W. Myint, Patricia Gober, Anthony Brazel, Susanne Grossman- Clarke, and Qihao Weng. Per-pixel vs. object-based classification of urban land cover extraction using high spatial resolution imagery. Remote Sensing of Environment, 115:1145–1161, 2011.
[69] S.W. Myint, P. Gober, A. Brazel, S. Grossman-Clarke, and Q. Weng. Per-pixel vs object-based classification of urban land cover extraction using high spatial resolution imagery. Remote Sensing of Environment, 115:1145–1161, 2011.
[70] T. Ojala, M. Pietik¨ainen, and D. Harwood. A comparative study of texture measures with classification based on feature distributions. Pattern recognition, 29:51–59, 1996.
[71] F. Pacifici, M. Chini, and W.J. Emery. A neural network approach using multi-scale textural metrics from very high-resolution panchromatic imagery for urban land-use classification. Remote Sens. Environ, 113(6):1276–1292, June 2009.
[72] M. Pesaresi. A mixed morphological and spectral procedure to improve the accuracy of the built area classification from multisensory satellite data. Proceedings of ISMM, 1:43–44, September 1994.
[73] M. Pesaresi and J.A. Benediktsson. A new approach for the morphological segmentation of high-resolution satellite images. IEEE Transactions on Geo science and Remote Sensing, 39(2):309–320, February 2001.
[74] Jose M Pe˜na-Barragan, Moffatt K Ngugi, Richard E Plant, and Johan Six. Object-based crop identification using multiple vegetation indices, textural features and crop phenology. Remote Sensing of Environment, 115:1301–1316, 2011.

[75] Minh-Tan Pham, S´ebastien Lef`evre, and Erchan Aptoula. Local feature-based attribute profiles for optical remote sensing image classification. IEEE Transactions on Geoscience and Remote Sensing, 56(2):1199 – 1212, February 2018.
[76] P. Pina and T. Barata. Classification by mathematical morphology. Procedings of IGARSS, 1:3516–3518, 2003.
[77] Yuguo Qian, Weiqi Zhou, Jingli Yan, Weifeng Li, and Lijian Han. Comparing machine learning classifiers for object-based land cover classification using very high resolution imagery. Remote sensing, 7:153–168, 2015.
[78] Andrik Rampun, Harik strange, and et al. Texture segmentation using different orientations of glcm features. In 6th International Conference on Computer Vision / Computer Graphics Collaboration Techniques and Applications, MIRAGE, Germany, June 2013.
[79] RezaKhatami, Giorgos Mountrakis, and Stephen V. Stehman. A meta-analysis of remote sensing research on supervised pixel-based land-cover image classification processes: General guidelines for practitioners and future research. Remote Sensing of Environment, 177:89–100, May 2016.
[80] N. Robbe and T. Hengstermann. Remote sensing of marine oil spills from airborne platforms using multi-sensor systems. WIT trans on Ecology and the environment, 95:9, 2006.
[81] N. Santhi. An enhanced approach for morphological shape representation and image retrieval, 2015.
[82] Anjan Sarkar, Manoj Kumar Biswas, B. Kartikeyan, Vikash Kumar, K. L. Majumder, and D.K. Pal. A mrf model-based segmentation approach to classification for multispectral imagery. IEEE trans on geoscience and remote sensing, 40:1102–1113, May 2002.
[83] Michael Schr¨oder and Alex Dimai. Texture information in remote sensing images: A case study, 1998.
[84] J. Serra. Image Analysis and Mathematical Morphology. Academic Press, New York, 1982.
[85] J. Serra and P. Soille. Mathematical morphology and its applications to image processing. Springer science and Bussiness media., 1984.
[86] P. Soille and M. Pesaresi. Advances in mathematical morphology applied to geo science and remote sensing. IEEE trans. on Geo science and Remote sensing, 40(9):2042–2055., 2002.
[87] Benqin Song, Jun Li, and et al. Remotely sensed image classification using sparse representations of morphological attribute profiles. IEEE trans on geo sicence and remote sensing, 52(8):5123–5136, august 2014.
[88] G.N. Srinivasan and G. Shobha. Statistical texture analysis. Proceedings of world academy of science, Engineering and technology, 36:1264–1269, December 2008.
[89] K. Stankov and D.C. He. Building detection in very high spatial resolution multispectral images using the hit-or-miss transform. IEEE transactions on Geoscience and Remote sensing letters, 10:86–90, 2013.
[90] S.R. Sternberg. Grayscale morphology. Computer vision graphics and image process, 35(3):333–355, September 1986.
[91] Safaa Muthana Sulaiman and N. Anupaman. Texture and gabour wavelet based classification of hysteroscopy images of endometrium. International journal of engineering science and computing, 7(5):12138–12142, May 2017.
[92] W. Sun, G. Xu, P. Gong, and S. Liang. Fractal analysis of remotely sensed images: A review of methods and applications. International journal of remote sensing, 27:4963–4990, February 2006.
[93] Ying Sun and Guo jin He. Segmentation of high-resolution remote sensing image based on marker-based watershed algorithm. IEEE conf procs at shandong, chaina., pages 201–206, 2008.
[94] M.J. Swain and D.H. Ballard. Color indexing. International journal of computer vision, 7(1):11–32, 1991.
[95] Y. Tarabalka, J. Chanussot, and J.A. Benediktsson. Segmentation and classification of hyperspectral images using watershed transformation. Pattern Recognition, 43(7):2367–2379, July 2010.

[96] Yuliya Tarabalka, Jon Atli Benediktsson, and Jocelyn Chanussot. Spectral–spatial classification of hyperspectral imagery based on partitional clustering techniques. IEEE transactions on Geoscience and Remote sensing, 47(8):2973–2987, August 2009.
[97] B. Tso and R.C. Olsen. A contextual classification scheme based on mrf model with improved parameter estimation and multiscale fuzzy line process. Remote Sensing of Environment, 97:127–136, 2005.
[98] Devis Tuia, Fabio Pacifici, Mikhail Kanevski, and William J. Emery. Classification of very high spatial resolution imagery using mathematical morphology and support vector machines. IEEE transactions on Geo science and remote sensing, 47(11):3866–3879, November 2009.
[99] Ahsan Ahmad Ursani, Kidiyo Kpalma, Camille C. D. Lelong, and Joseph Ronsin. Fusion of textural and spectral information for tree crop and other agricultural cover mapping with very-high resolution satellite images. IEEE Journal of Selected Topics in Applied Earth Observations and Remote Sensing, 5(1):225–235, February 2012.
[100] S. Valero, J. chanussot, J.A. Benediktsson, H. Talbot, and B. Waske. Advanced directional mathematical morphology for the detection of the road network in very high resolution remote sensing images. Pattern recognition letters, 31(10):1120–1127, 2010.
[101] Matheus Alves Vieira, Antonio Roberto Formaggio, Camilo Daleles Renn´o, Clement Atzberger, and Marcio Pupin Mello Daniel Alves Aguiar. Object based image analysis and data mining applied to a remotely sensed landsat time-series to map sugarcane over large areas. Remote Sensing of Environment, 123:553–562, August 2012.
[102] L. Vincent. Morphological area openings and closings for gray scale images. NATO ASi series, 126:197–208., 1992.
[103] D. Wang, D.C. He, and D. Morin. Classification of remotely sensed images using mathematical morphology. Procedings of IGARSS, 1:1615–1617, 1994.
[104] DEMINWANG*. A multiscale gradient algorithm for image segmentation using watersheds. Pattern recognition letters, 30(12):2043– 2052, January 1997.
[105] L. Wang and DC. He. Texture classification using texture spectrum. Pattern recognition, 23(8):905–910, 1990.
[106] A. Watson. A new method of classification for landsat data using the watershed algorithm. Pattern Recognition Letters, 6:15–19, 1987.
[107] A. Watson, R. Vaugham, and M. Powell. Classification using the watershed method. International Journal of Remote Sensing, 13(10):1881–1890, 1992.
[108] ShuleiWu, Huandong Chen, Zhizhong Zhao, Haixia Long, and Chunhui Song. An improved remote sensing image classification based on k-means using hsv color feature. Proceedings of Tenth International Conference on Computational Intelligence and Security, 1:201–204, 2014.
[109] Z. Xie, C. Roberts, and B. Johnson. Object-based target search using remotely sensed data: A case study in detecting invasive exotic australian pine in south florida. ISPRS Journal of Photogrammetry and remote Sensing, 63:647–660, 2008.
[110] H. XU. A new index for delineating built-up land features from satellite imagery. International journal of remote sensing, 29(14):4269– 4276, July 2008.
[111] L. Xue, X. Yang, and et al. Building extraction of sar images using morphological attribute profiles. Springer Communications signal processing and systems, 202:13–21, 2012.
[112] Y. Zha, J. Goa, and S. Ni. Use of normalised difference built up index in automatically mapping urban areas from tm imagery. International journal of remote sensing, 24:583–584, 2003.
[113] B. Zhang, S. Li, X. Jia, L. Gao, and M. Peng. Adaptive markov random field approach for classification of hyperspectral imagery. IEEE Geoscience and Remote Sensing Letters, 8:973–977, 2011.
[114] Liangpei Zhang, Xin Huang, Bo Huang, and Pingxiang Li. A pixel shape index coupled with spectral information for classification of high spatial resolution remotely sensed imagery. IEEE transactions on Geoscience and Remote sensing, 44(10):2950–2961, October 2006.

[115] Limin Zhang, Tao Xu, Jianting Zhang, and Yingfu Yu. A knowledge based procedure for remote sensing image classification. IEEE International Conference on Fuzzy Systems and Knowledge Discovery, pages 72–76, 2014.
[116] X. Zhang, R. Sun, B. Zhang, and Q. Tong. Land cover classification of the north china plain using modis evi time series. Journal of Photogrammetry and Remote Sensing, 63:476–484, February 2008

CBCA: Consignment based communal authentication and encryption scheme for Internet of things using Digital Signature Algorithm

Arun.M , PraveenKumar.S , Rajakumar.P.S , Thamizhikkavi.P
Dr.M.G.R Educational and Research Institute, Chennai.

arun.cse@drmgrdu.ac.in, praveen.cse@drmgrdu.ac.in,
rajakumar.subramanian@drmgrdu.ac.in, thamizhikkavi.cse@drmgrdu.ac.in

Abstract. IoT (Internet of things) gives a dream that communicate the extent of web by incorporating physical articles to distinguish themselves among the participating people. This contemporary observation encourages physical objects to represent them in the advanced world. There are heaps of hypotheses and future conjecture about different physical substances associated with the web, programs, Most of the devices needs security features from the wide range of attacks that are present in the environment. The computationally complex and asset expending methods have a restricted support over these installed physical objects and miniature sensor hubs. In this paper we propose a CBCA (Consignment based communal authentication) conspire for real objects that are moderately associated with the IoT condition. Client server correspondence model is used by present reality substances to communicate with each other. In our proposed framework we use lightweight features of CoAP(Constrained Application Protocol) to screen servers, regardless of whether the assets from the server are empowered on client in a vitality proficient way. Digital signature algorithm (DSA) with a key length of 512 bits is used to set up a protected meeting for the resource assessment. Our plan is assessed for real world situation utilizing NetDuino Plus 2 sheets. Our proposed framework gives an overwhelming obstruction against various attacks and its computationally proficient, get less association overhead simultaneously.

Keywords: IoT, Constraint application protocol, Payload Encryption, Lightweight Authentication, Intrusion Detection, Resource Observation, digital signature algorithm(DSA).

1. Introduction

Innovative development in the field of remote advances, mobiles and sensor network laid a stage for the presentation of IoT. By utilizing the unique addressing format every devices interface with one another in the physical world dependent on a novel methodology [1]. Before the finish of 2020 its been evaluated that around 50 billion gadgets would be associated over web. These devices would be used to detect, procedure and control different occasions in physical world with various marvelous viewpoints. These internet based devices creates a huge amountof information that must be put away, investigated and transmitted in proficient way [2]. Development of these gadgets will lead us to IoE(Internet of Everything), where our day by day lives will be incorporated with object, information and procedure. By incorporating virtual world information with physical articles, devices will get inescapable by implanting the objects in internet. Various correspondence models are required for incorporating physical articles with internet. The requirement for various amounts of models will probably give a route to some innovative and imaginative models for the future internet [3]. Since each physical maneuver in IoT has its distinctive features consequently security provisioning is a significant challenge in IoT system. Each physical devices, items and hubs that are associated over internet must be substantiated. In the event that the confirmation isn't been performed intruders will access the gadgets that are associated over internet and will perform vindictive activities. Because of

IOP Conf. Series: Materials Science and Engineering 1074 (2021) 012003 doi:10.1088/1757-899X/1074/1/012003

the pernicious activities performed by the interloper over the network will prompt the different results that are assorted in nature with oscillating application

from crippling the security systems, transmitting bogus messages and so on., in spite of these security threats IoT expedients that is accessible currently are absence of security features. Therefore the IoT gadgets that we are going to utilize are defenseless to a catholic scope of security breaks. Indeed, even these devices are giving different favorable circumstances; this will lead our lives to new time of cybercrimes which will lead IoT tools into Inter of Vulnerabilities (IoV). The physical appliances that are associated over network are been included into digital battering was revealed by the main security firm by verification point inc.2. This is the significant security penetration in IoT. Roughly around 100,000 tools transmitted around 750,000 malignant email were transmitted over a time of under about fourteen days. Among the appliances associated over internet in excess of 25 percent of those appliances were certifiable gadgets including TV, fridges and different machines. There are various application associated through internet for those applications a small threat available was misconfiguration and default passwords which was satisfactory for such attacks. The IoT gadgets were presented to various security hazards which may incorporate botnets alongside the thingnets. As on the above conversation, we propose a light weight validation framework through OTP generation dependent on attribute based encryption. Four way contrivances are used for asset allotment. The significant commitment of work is as per the following.

1. The client and server communication with one another is checked utilizing our proposed scheme. The connection is established by mutual authentication scheme which establish a computationally simple and robust connection. For client sever authentication round trip message exchange is done using lightweight handshake mechanism. Encrypted payload of maximum 256 bits is generated for authentication process by client and server.
2. During the communication for the resource exchange between client and server CoAP is utilized along with the lightweight features in our proposed system. in transport layer the data security is provided with using three way communication which in turn also provided security relying on the features of DTLS[5] Datagram Transport Layer Security. By incorporating all the necessary features DTLS is replaced and we do not use separate protocol for reducing the computation and communication cost in our proposed system. the efficiency of proposed system is determined with the result obtained.
3. The possible fact of resource consumption and DoS attacks is eliminated by providing the server a certain time to establish the multiple connections, by which the malicious nodes are restricted from connection establishment. By our proposed methodology the resources are effectively utilized and by allowing the client to establish one connection for legitimate user with the server.
4. Report form the server is used to detect the relay attacks, our proposed system is resistant against DoS attacks and resource exhaustion.

The paper is organized as follows. In section II, literature survey is provided based on the related works. In section III, problem statement is discussed followed by the light weight consignment based communal authentication scheme and design, objective of the proposed system is also discussed. In section IV, analysis of our proposed system is done during data transfer and mutual authentication process. The Vth section the tentative result of the projected system id done. Conclusion is provided in section VI.

2. Related Work:

Securing the IOT devices and different substantiation procedures for the conserving correspondence between these devices are discussed in this session. Initial review of CoAP convention is provided trailed by comprehensive discussion on different authorization and cryptographic procedures are used to maintain the confidentiality and decurity challenges in IoT.

REST (Representation State Transfer) is used by HTTP based web innovations [6]. For resource providing a conventional need for comprehensive memory and computational capacities. The devices or the objects that are associated with an IoT domain is provided with resources compel so the device abilities are restricted to sustenance HTTP-based web administrations. An extraordinary work group was framed IETF (Internet Engineering Task Force) 3 known as CoRE (Constrained RESTful Environments). To give a web resource to IoT [7] the work cluster was responsible for creating lightweight convention. By obtaining the subnet of HTTP the prerequisites of the resource limitation in IOT is done by CoAP convention. For interchanging resources among client and server CoAP is used which deal with the information transmission utilizing simple request/response connection model. For resource perception every single client must be register itself with a specific server [8]. A 24-piece arrangement number is used to confirmable (CON) message for enrolling demand. With a coordinating response the server acknowledge (ACK) the CON demand. Four kinds of messages are supported by the CoAP: confirmable (CON), Non-Confirmable (NON), Acknowledgment (ACK) and REST (RST). For resource control manipulation techniques are bolstered by conventions, for example, GET, POST, DELETE and PUT. The request or response messages may convey a CON and NON with a specific or proper strategy applied to it Only few resource is used since number bytes used for header length is fixed by CoAP. To develop a perfect system for IOT deployment Low Lossy Netowrks(LLNs) is combined with low complexity protocol header, asynchronous transaction model and message parsing. As a result CoAP is perfect alternate for existing IoT protocols like MQTT[9] and XMPP[10]. Subsequently, CoAP is conveyed in different projects including conveyance coordination's [11], residential robotization framework [12], brilliant urban communities [13] and cargo management [14].

In IOT environment d different types of attacks are done in each layer. The authors of [15] featured different security challenges for this system. The use of TLS(transport Layer Security)and its limitations are discussed by the authors[16]. Moreover, the packets get in faulty, and can be absent or conceivably spoiled. Along these lines, the DTLS is a certain need for ensuring about the correspondence in a CoAP-based IoT. In each datagram overhead 25 bytes of data is achieved during the handshake and record layer of DTLS. Maximum transmission unit of 127 byte is specified by LR-WPANs. After the addition of DTLS, MAC and higher layer headers 75 bytes are left beyond for payload [17].

To make DTLS friendlier towards the aid restricted networks its activities have to be recorded [18]. To set up a unicast channel for communication a light weight authentication scheme and in order to reduce the power and computation price of IoT device a symmetric encryption algorithm was proposed by Bhattacharyya et al [19]. Granjal et al. [20] suggested providing security features in different layers starting from IPSec based on their studies using DTLS, in order to provide security for the data exchange in CoAP based on IoT surroundings. Kothmayr et al. [21] have proposed architecture to achieve high interoperability and communication overhead via their handshake but it consumes high power due to the use of RSA based algorithms. The overall performance of the DTLS handshaking system is evaluated for aid restricted system by using Elliptical Curve cryptography (ECC) which results in higher energy consumption in [22]. In [23] the authors have implemented DTLS for smart phones using CoAP known as INDIGO. This system use is restricted by most effective clever phones and uses a huge ingesting cryptographic suites. For aid remark in CoAP based IoT system the authors [24] have proposed an light weight authentication scheme. The individualities cooperating customers and server are set up sooner than building up a confirmation meeting. The proposed scheme needs test results for deciding the presentation and precision of the confirmation calculation.

In spite of the fact that, DTLS-based scheme provision a wide scope of figure suites for security provisioning, be that as it may, DTLS transformed into from the outset intended for systems having plentiful of assets. The guide ingesting complex cipher suits of DTLS do no longer remember the message time frame as an imperative structure standard for making sure about a system. Along these lines, the utilization of DTLS for an IoT usage is a costly decision and may not be a most reasonable answer for making sure about the system.

3. A Lightweight Consignment-based collective validation Scheme:

As discussed in section 2, for the authenticated transmission of asset between various entities, IoT depend on DTLS dependent on CoAP based usage. The computational and correspondence cost is expanded for security reason in DTLS empowered where CoAP stack presents an extra convention layer. The security messages traded among the customer and server isn't undermined if the extra convention layer is absent. The session key is transmitted inside the payload of transmitted message among the customer and server where the confirmation is given at the hour of request and response. In figure 1, examination of the proposed framework is finished with the current DTLS-empowered CoAP stack.

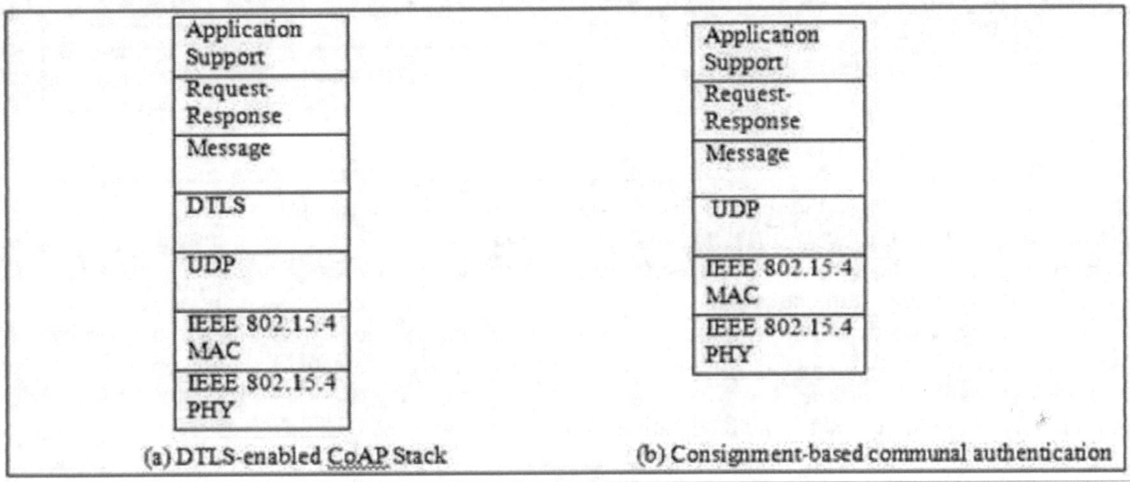

Figure 1:Stack vs. Our Proposed Scheme

In this session the light weight transfer based mutual correspondence validation scheme is accomplished for distinguish the characters of this real world objects. At first, the issue is distinguished which acts the establishment of our proposed verification plot.

4. Problem Statement:

The physical objects that are incorporated with IoT differ from one another in terms of operational behaviors and numerous resources. The seamless and interoperable communication is done when the sensor nodes that are available in each device are enabled. For each IoT devices a unique ID is provided for the sensors. As long as the devices can identify the network by itself for any IoT devices. IP address and an embedded sensor node is required to make the objects smart [25].

In figure 2 small scope IoT correspondence model comprise of different physical objects. Here the IoT objects are the clients which communicate to the server (NetDuino4) for monitoring the temperature of the resources and details are provided to the client. Each devices have certain conditions for temperature readings. In the absence of IP address the client and server will not communicate with one another. RESTful operation can be performed by IOT devices only if the devices have IP address..

The asset obliged arrangement of Figure 2 says how the devices are vulnerable when devices are connected with WIFI. The hackers may block the IOT device data, control it and rerun in various system. Even attackers can mix the data of their own with real data. Thusly, an enormous proportion of data is in danger which may achieve the falling flat of the whole system. Like a real device, a dangerous item furthermore requires an ID to check out system correspondence. Each contributing item ought to be affirmed to develop its actual character in the system. Without ID endorsement and check, an aggressor will reliably have the choice to direct a wide extent of vindictive exercises. A gatecrasher may set up various relationships with a server at a given time so unique system operations will be affected or neutralized by threatening article. Then the final output of the system will be affected [26]. Since devices use wireless medium for the transmission, the unauthorized user can square authentication to the system assets by continually exuding adhering signs to interface with bona fide transmissions [27]. The closeness of remote medium can without quite a bit of stretch open things to spying moreover.

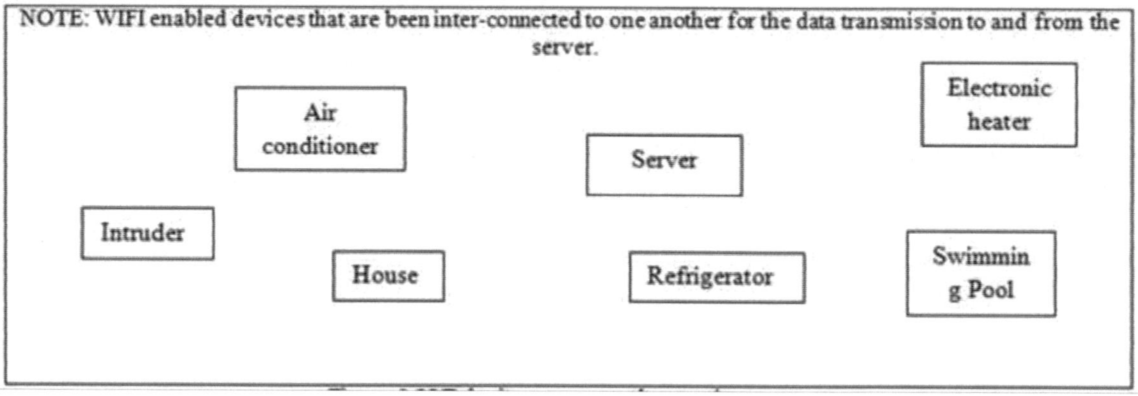

Figure 2. IOT devices connected to environment

So as to detect and relieve different kinds of attacks in a system, very lightweight yet secure conventions should be structured considering the constrained resources of installed sensor hubs. For securing the existing authentication and authorization standard protocol such as TLS [16] and Kerberos[28] a considerable effort has been spend. If the existing protocols are customized for resource constrained networks it would save lots of effort. Therefore the existing protocol will not provide optimal solution[29].

Practically the entirety of the authentication and encryption scheme in the resouce compelled IoT depend on DTLS convention. For the IoT devices to provide a secure solution DTLS uses an expensive computationally complex and resource consuming cipher suit. If the stateless cookies are absent the server will be exposed to replay attacks [30]. For the devices in IoT paradigm, as of the above discussion our goal is to develop an light weight CoAp based authentication scheme. Most importantly we have to decide how thus light weight protocol should be. If there is enough resources available for the task then the protocol is said to be as light weight protocol. Our aim is to develop an authentication scheme based on the above requirements. In our proposed authentication scheme we use basic principles of CoAP for RESTful interaction. If any client wants to interact with the server our proposed scheme will establish a secure communication channel. Upon shared verification, for resource allocation the client communicate with server. Just the validated customers are permitted to watch the resources.

5. Consignment-based communal authentication

Authentication, for the process of authentication process security features are add along with CoAp in order to make more efficient, robust and secure against various unauthorized activities. Authentication is provided by our proposed system by utilizing the payload message that is transmitted with client and server. The authentication process is performed by using four way handshake mechanisms were each message is of length 256 bits as shown in figure3. For encryption of data Attribute based encryption with a length of 256 bits and cipher block chaining mode [31] is utilized. For the devices that are connected to IoT paradigm 128 bit key is used due to limited resources available in server nodes. The four way handshake consist of the following steps a) connection establishment b)Server challenge c) client challenge d) server response e) client acceptance .

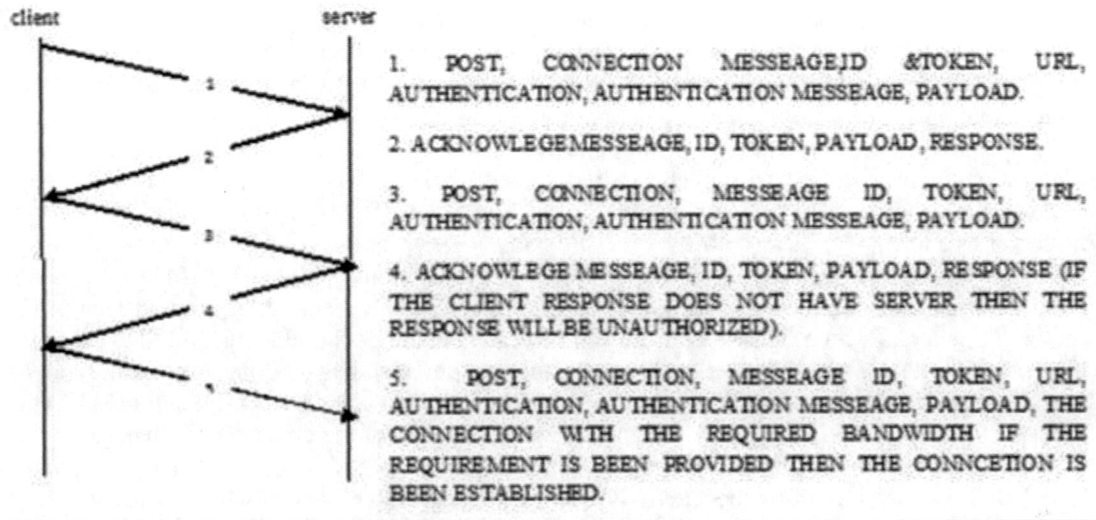

Figure 3. Authentication Handshake

First phase is authentication phase in this phase 128 bit private key λ_i is shared by client to server. The identity of the key is known only to the client and server, the identity of the key is stored in a table by the server by using an unique identifier that enables the server to perform lookup for the identity verification. Once after successful authentication both client and server will exchange the session key. The pre shared secret key is accepted and hardcore implanted the sensor nodes with the physical devices. If an intruder tries to access these devices an alarm is generated to notify the security breach.

In the session inception stage, every customer transmits a solicitation message to the server like a Invitation message. It is a comprehensible solicitation for the formation of a meeting as resource at the server. Server receives the request. Each request received by the server has a specific token correlating with the matching response. The object ID is transmitted along with the message with 2 options, Auth and Auth-Msg- Type where these values indicate the types of operation to be performed by the server. If the verification successful for the client ID, resource authorization is created in server. If the auth is true and auth-msg-type=0 determines the request to the server.

This is the second phase during this phase the object ID is retrieved by the server form the message. By using this ID server verifies the table for a match λi. If the match is found encrypted message is transmitted as response from server by using ABE. A separate session will be created for customer a reaction code 2.01 is remembered for the message. On the off chance that the server can't locate a coordinating _i, unauthorized reaction code is transferred. The server incorporate message ID and token that are identical and available based on the session demand. This can be done by duplicating the server challenge. Server creates 128-bits random pseudo code, μ_{server}, and a prospective session key, μ_{key}. The random key is a transitory number used one time by the article in cryptographic correspondence. Server creates a encrypted payload at this time. XOR operation is performed on λ_i and μ_{key} utilizing Equation 1. In complex ciphers XOR operation is cheap and amazing. This activity is not spill the data from the plaintext and to recovery the unique text the operation has to be performed twice. The plaintext is used as session key.

$$resultant = \lambda\text{i} \oplus \mu\text{key} \qquad (1)$$

The 128 bit resultant is added to server and encrypted with λi to produce the transfer of 512 bit utilizing equation 2. The server transfer is transmitted an as a trial to the client.

$$server\ consingment = \text{ABE}\{\ \lambda\text{i}\ ,(\text{resultant}|\text{server})\} \qquad (2)$$

To recover the session key μkey the client need to transfer the encrypted data for the client reaction and challenges. In the event that the client is fruitful to do as such, it will have the right server and μkey. The server and the μkey are known uniquely to the server and the λi has a place with the particular client. Just an authentic client can disentangle this test. An interloper can just listen in on the server and the μkey, however not the λi as per the Internet Threat model. The customer utilizes its λi to translate the payload. Upon effective deciphering, the customer has verified itself. As shared validation requires the two gatherings to be checked, the server additionally needs to verify itself. The customer creates another encoded payload like the server. Initial, a XOR is performed on server and λi utilizing Equation 3.

$$resultant = server \oplus \lambda i \quad (3)$$

The 256 resultant is added with client and encrypted with µkey to create an encoded payload as depicted in the condition 4. The 512 piece encrypted data is transferred; customer data is transmitted to the server as a test.

$$client\ consignment = ABE\{\text{µkey}, (\text{resultant}|\text{client})\} \quad (4)$$

At last the server reaction stage, the server decodes the scramble transfer, customer transfer of the customer challenge to watch the server in it. The server understands that the customer has effectively verified itself if its present. The server recovers the customer and makes its very own scrambled payload by annexing the customer to μ_{key} and encoding with λ_i as appeared in Equation 5. Next, the 512-piece encoded payload, server-payload, is transmitted in light of the customer challenge.

$$server\ consignment = ABE\{\lambda\text{i}, (\text{client}|\text{µkey})\} \quad (5)$$

The POST connection from the client is been provided to the server with the request of the bandwidth needed for the communication with the server and the other devices if the bandwidth is available for the transmission of the data to the server and the other devices the bandwidth is allotted for the data transmission with other devices and the payload from the server side is been provided with the encrypted format with load of 512-bit of cipher format. Once the request is provided for the connection and client request is accepted by the server with an acknowledgement for the connection for the client.

$$client\ consignment = ABE\{\text{µkey}, (\text{resultant}|\text{client}|\text{request})\} \quad (6)$$

The resource status is changed to authenticate in the server so client can effectively access and verify the resources within certain time. The client within the payload is observed by decrypting the transmitted message, yet the authenticity of client is verified by the server. The mutual authentication between the client and server is done by sing the session key µkey for transmitting data between client and server.

Authentication and authentication message type plays an important role during the authentication process. While transferring data to the endpoint the request message is not understandable since it has a critical option, so the sender is responded with a bad response. There is no default values assigned for these options. Each option is assigned a number based on IANA. 0 byte is allotted as header length since the if the authentication is true there is no need to occupy the space. Simple message format is essential for light weight authentication scheme. The format is shown in table 1

Table 1. Format of the Authentication Options

No	A cute	Un safe	Cache Key	Name	Form at	Leng th	Deaf ult
T BD	1	1	0	Auth	Unit	1	None
T BD	1	1	0	Auth-msg-type	Unit	1	None

POST method is used for the validation for the request message when the authentication alternative is demonstrated. Authentication message type is utilized to mix the authentication and confirmation demand by the client is demonstrated.

On the off chance that its worth is 0, it demonstrates a meeting commencement demand and if its worth is 1, it shows a customer reaction and challenge individually. The different stride for our transfer common verification alongside the information exchange is appeared in Algorithm 1.

5.1 *Algorithm: Lightweight consignment based communal authentication:*

Initialization:

a)unique ID is provided for every client C_i as objects such that ID_i and λ_i.

b) Array Ar[][] is utilized on the server side for the storage of all client ID_i and λ_i.

Step 1:
Input:(ID_i , λ_{i}.)
for i=1: N do
for j=1:2 do
input(Ar[i][j])
end for
end for
Step 2:
C_i sends a Connection message along with ID_i in the consignment to server S.
Step 3:
S retrieves the ID_i to find the harmonizing λ_i
If ID_i==Ar[i][j] then temporary session created
Step 4:
S responds with an ciphered consignment, ABE{ λ_i,(λ_i XOR μ_{key}|server)}
Else C_i unauthorized
End if
Step 5:
C_i decrypts the message and responds with ciphered content to server, ABE{ μ_{key} ,(server XOR λ_i|client)}
Step 6:
S checks server in the client challenge by comparing the server generated challenge.
If equals then access granted-C_i authorized.
Step7:
Server responds as ABE{ λ_i,(client|μ_{ke})} else
Access denied unauthorized endif
Step 8:
C_i compare with client of Step5 and Step 6
If both matches authenticated C_i request for the bandwidth to be allotted
Step 9:
S reads the request for the C_i if bandwidth available responds to the client
Else S server waits and once the bandwidth is available it respond to the client
S is authenticated
Step 10:
Mutual data exchange between S and C_i takes place.
Else
S is unauthenticated
End if

6. Resilience of the proposed system with various attacks:

At a particular amount of time each client is allowed to establish only one connection with the server this eliminated the resource problem of resource exhaustion. This problem occurs when the resource is entirely consumed. Therefore client establish multiple connections with the server which lead to scarcity of resource to the clients. This lead to partial sharing of resources which leads to DoS attacks.

The client wants to access the resource might be authenticated user or intruder. In order to get authentication form the server client sends a session request to authenticate itself. After the successful authentication the server verifies the database for object ID whether there is an ongoing session for exchange data, if there exist an ongoing session the request for establishing a new session will be rejected. For the request message that has been transmitted from the client in a CON message ACK is necessary. Once the CON message is transmitted the client initiates the timer and waits for the

IOP Conf. Series: Materials Science and Engineering 1074 (2021) 012003 doi:10.1088/1757-899X/1074/1/012003

response ack*t where ack is acknowledgement and t is time. The amount time taken for acknowledgement for the client is called as acknowledgement timeout.

If client is an unauthorized used then there will be a mismatch in object ID and server respond with RST message indicating invalid object ID but the server will not know whether the request is given by an authorized or unauthorized user. A particular object ID, message ID and token will be used for retransmission attempt by an intruder, after failing in all the attempts the client will be declared as an intruder and any further attacks will be considered as attacks. Server will terminate all the communication from that particular client. A new transmission will be done with different ID that will generate a response RST indicating the no further communication is established with the terminated session.

In wireless network there is a greater possibility of hacking the connection and steal the object ID that has been shared between the client and server for establishing the communication with the server. The server will generate a challenge to check the match in ID. The intruder will not be able to decipher the challenge given by the server since intruder will not have valid λi. As in case of authorized user a match will be found in the database table with object ID. Client can retransmit the communication CON message after waiting for certain amount of time. If the ongoing session exists with the client, the server evaluates the request when the IDs are matched. Server responds back if there exist a session with an ACK refereeing clients' needs to abstain for retransmission.

During the process of resource allocation and utilization object ID plays a vital role. Each object ID is linked with a key that is required for deciphering the challenge and these data's are stored in server table. Even when an invalid object ID is used to generate the response using RST will be unsuccessful attempt for the server challenge to deciphering the data. In algorithm 2 is a procedure to detect the resource exhaustion. The unauthorized user fails in both cases unable to use the resource.

Our proposed system is secure from all attacks. A pseudo random number Ri is generated for authentication purpose by client and server based on the timer Ti. Here Ri and Ti are used to assure client and server are secure and ensure that both are unpredictable. It difficult for the intruder to launch the relay attach due to its unpredictable nature of Ri and Ti. The intruder may intercept the transmission data between client and server to capture the μkey to perform relay attack and most importantly these types of attacks can be only for a short duration of specific time.

6.1 Algorithm: Detection of Resource Exhaustion Algorithm

Initialization: Ar[][], ID_i, Re_Try
Step 1: [Input:(ID_i, λ_i)]
for i=1 to:N do
    for j=1:2 do
        input(Ar[i][j])
    end for
end for
Step 2: initiation request send by client C_i.
Step 3: from the session initiation request ID server S retrieves the data.
for i:1:N do
    if ID_i==Ar[i][j] then
Step 4: check for ongoing session
If session=true then
Discard session request
Step 5: Send ACK response to C_i
else
generate a server challenge and send to C_i
if C_i deciphers the challenge then
C_i is a legitimate client
Else
C_i is an intruder

End if
Endif
If ID does not amtch Ar[i][j] then
Step 6:
Server respond with RST
Step 7:
C_i does not proceed with re-transmission attempts
Else
Step 8:
C_i retransmits after waiting for seconds a*r
Retry++
End if
If retry>4 then
C_i is an intruder and its blacklisted
Else
C_i retransmits again
End if
End for

Upon effective mutual confirmation, both the customer and the server are approved to utilize µkey to transfer data frames. For resource allocation every client transfers an enlistment request to server. The Registration Request Message (RRM) has a watch alternative, which has a 24-piece consecutively gradual succession number. RRM is provide with a time stamp when the sever realizes, it enrolls customer and modification in the resource are informed.

To transfer information from server to client the recipient determine certain conditions which in term know as warming updates. When the data is transmitted to the server each client keeps in track of transmitted RRM, which is stored utilizing a token and message ID for connecting notice from the server. Sequence number, token and message ID like RRM will be present in the warning message from the server. Private RRM is much secure for warning updates. When the conditions determined by a customer for notice refreshes are satisfied at the server, information packets, βdata, are transmitted to the given clients as appeared in Equation 6.

$$K_{client-server} = AES\{\mu_{key}, \beta_{data}\} \quad (6)$$

In Equation 6, βdata is encoded with µkey. This encoded information must be deciphered by the customer which has legitimate λi and µkey.

In Figure 4, A full duplex wireless connection between client and server is monitored by an intruder namely A. By using the session key and server message the intruder hacks the messages and replay them to clients connected to that network. Client checks the message ID and token due to this replay attacks. The customer utilize a simple rationale by contrasting the approaching notice and the recently gotten notice from a server as appeared in Equation 7.

$$\Omega_{new} = new\ when[(V_i < V_j)\wedge(V_j - V_i < 2^{23})]\vee[(V_i > V_j)\wedge(V_i - V_j > 2^{23})] \quad (7)$$

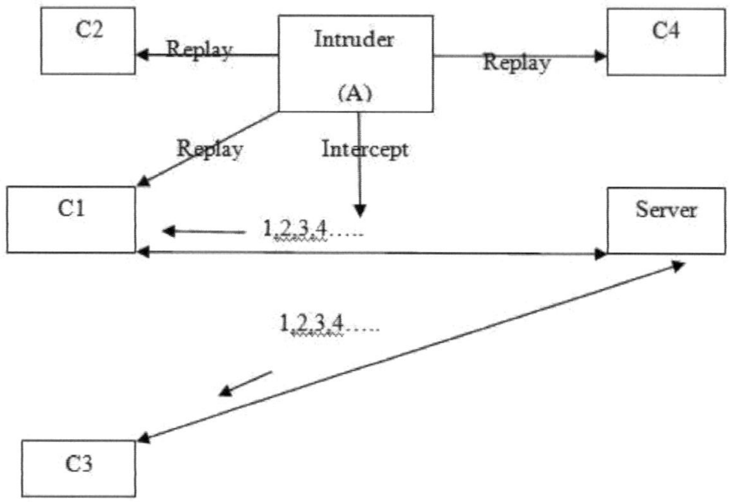

Figure 4. The Replay Attack.

In Equation 7, we have utilized the newness of the warning updates, Ω_{new}, to identify the reiteration attacks from a server. In this condition, V_j & V_i are 24-piece arrangement quantities of a watch choice in the approaching and the recently gotten warnings. A statement is new and most recent if V_j is more noteworthy than V_i and their distinction is under 2^{23}. An approaching notice is additionally new if Vj is littler than Vi and their distinction is more noteworthy than 2^{23}. The last is the situation when the estimation of V^j turns over [33].

In Figure 4, the impostor might possibly adjust the succession quantities of the approaching message from server. Independent adjustment, the customers C2 and C4 without much of a stretch distinguish a reiteration assault, because these approaching notices are for C1 as it were. For the request that are approaching other clients , there occur a mismatch between token and message ID in RRM. In the event that the grouping quantities of the warnings are adjusted, C1 can without much of a stretch distinguish it by contrasting the notices with the previous one. In any case, if the interloper replays the warnings with no alteration, it turns into a dubious circumstance as C1 is surely sitting tight for such notices from the server. Between RRM and approaching warning there occur a match between the token and message IDs.

In addition, the arrangement numbers are likewise in a similar request true to form by this customer. To understand this riddle, Client C1 uses time stamp as an attribute and updates the warning in RRM [34]. Inside the affirmation break client C1 will update the information that is received form the server. Client does not have any knowledge whether the warning message has updated or not. In this way, it authorize the arrangement number of notices compared to the recently gotten notices in a successive request. For instance, the approaching warning, Vn, is checked against Vn−1 to decide whether it was gotten inside the affirmation break after the fruitful gathering of Vn−1. The whole component of replay assault recognition is appeared in following advances.

Step 1: Start

Step 2: Transmit the registration request message to the server.

Step 3: Server registers the registration request message.

Step 4: Notification is generated and transmitted to the client and server side.

Step 5: Condition is been evaluated whether the terms and conditions are fulfilled by the server.

Step 6: If the condition is satisfied notification is transmitted to the client. Else replay attacks occur.

Step 7: The secondary validation is performed to check message token and token match.

Step 8: Once validated the newness is fulfilled the notification is validated.

Step 9: if it's not validated time checkout is performed and to validate the request is received within time frame.

Step 10: If notification received within time frame then notification is valid else replay attack occurs

Our proposed conspire; a hacker just dispatches a reiteration attack while trading of information between a genuine sender and receiver. Reiteration attack is beyond the realm of imagination at the hour of setting up a session, i.e., common verification, on the grounds each clients has its own λi called uniquely to the customer and the server. A hacker may catch the server payload challenge and recover server and µkey from it. Be that as it may, it despite everything require λi for encoding the payload. The absence of a legitimate λi will produce a dubious payload which will be distinguished effectively by the customers.

7. Experimental result:

Our proposed system is validated in this segment. For communication between client and server we have used NetDuino Plus sheet. DotNet micro framework is utilized for experimental results, by using dotnet an event with 256KB of data and with 64KB of RAM. A sensor with the service provider is used to monitor the temperature and acquire the reading. Clients are allowed to monitor the temperature after the successful verification. For resource exchange between the correspondences COAPSharp6 libraries are used. The libraries used are more effective and powerful. The comparison of our proposed system with the existing system is done.

7.1 Validation

Every user has validate for perception of assets. In Figure 6(a), determines the successful handshake between client and server. The verification and validation are the all-inclusive choices utilized uniquely for our verification reason. Yet there occur relegated numbers which are unrecognized. We utilize random numbers 0x101(257) and 0x102(258) for these choices in library utilizing choice vault system. By using session key the client and server have settled to initiate the session and the handshake span alongside the full circle reaction time is acquired. In Figure 6(b), the client can't unravel the server challenge as it doesn't have the necessary mystery, λi. The result is an erroneous session key which makes the forswearing of access the asset. In the two cases, the nearness or nonattendance of the pre-shared mystery decides the result of the verification.

(a) Access Granted

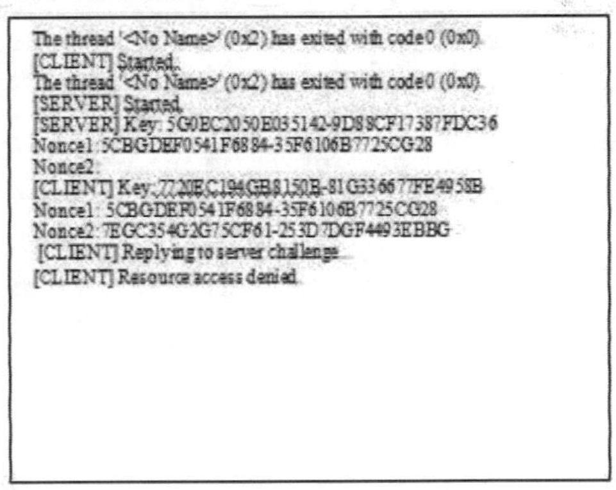

(b) Access Denied

Figure 5. The Authentication Process

7.2 Handshake Duration

The time taken is processed as total of time taken by two full circle messages. The client convention commencement demand is recognized by server trial while the receiver reaction and threat is recognized by server reaction. The handshake term, $d_{handshake}$, is figured at the client end utilizing Equation 8.

$$d_{hands\,hake} = T_{session} + T_{challenge} + \delta_{proc} \tag{8}$$

In this condition, $T_{session}$ is the full circle time taken by meeting commencement demand, $T_{challenge}$ is the full circle used by customer reaction and threat, δ_{proc} prepares the amount of time taken by the customer. Two full circle messages is used to handle time at the server side.

To calculate the shake time taken, we perform 50 irregular shakes among client and server. To decide the changeability and precision of evaluations, the standard deviation utilizing Equation 9

$$\sigma = \sqrt{\frac{1}{N}\sum_{i=1}^{N}(x_i - \mu)^2} \quad (9)$$

Here, σ is the standard deviation, N determines all out values of readings, μ describes mean worth and xi is the real hake length of every individual perusing.

In Figure 7, our plan and the CoAP-based DTLS usage for advanced cells. DTLS* speaks to the communication between a PDA and PC. For this situation, the PDA goes about as a client though PC goes as server. DTLS+ represents advanced mobile phone as sender and the PC receiver.

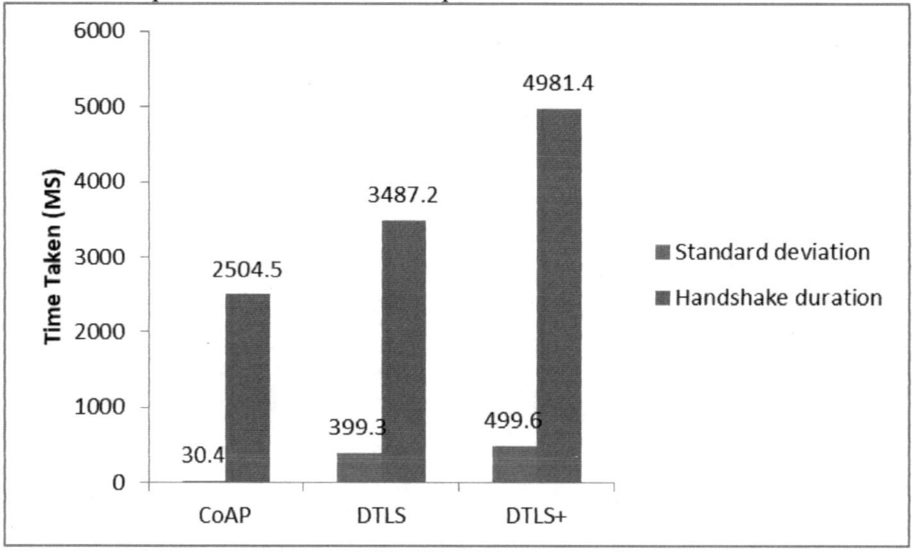

Figure 6. The Handshake Duration

7.3 Average Response Time

For exchange of information based on COAP wireless protocol is used in our proposed system. On monitoring the traffic in network every client associated with network keeps track of the communication CON message. Server handles numerous requests at a time and in figure 7(A) represents the response time taken for every message. If the size of transmitted messages increases the time taken for the response will also be incremented. When the number of request is higher the time taken for response will be lower. For example if there are 100 request to the server response time for first twenty to fifty numbers request the response time might be normal as estimated. In figure 8(b) the reaction time is normal for message of 1 byte is contrasted to DTLS and the CoAP convention with no safe. As clarified before, the imminence endorsements, crude open keys and costly flight-based confirmation settles on DTLS as a costly decision for the objects of an IoT.

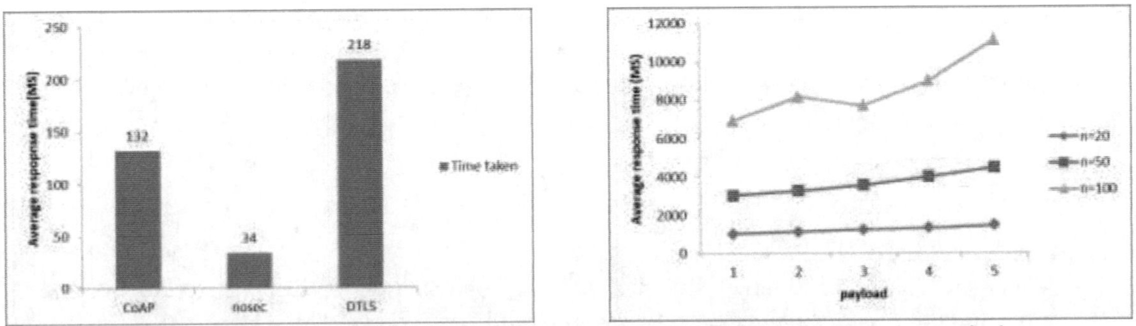

A) Response time at payload B)Simultaneous transmission

Figure 7. Response time

7.4 Memory consumption

By using a function debug.gc memory usage time is computed. The average memory usage is registered by altering the packet size. Whenever the request is generated the packet from each node will have a minor effect on the memory usage. The memory usage of our proposed system is tested and results are shown in figure 9(A & B). From the figure its shows that our proposed system has an efficient memory usage.

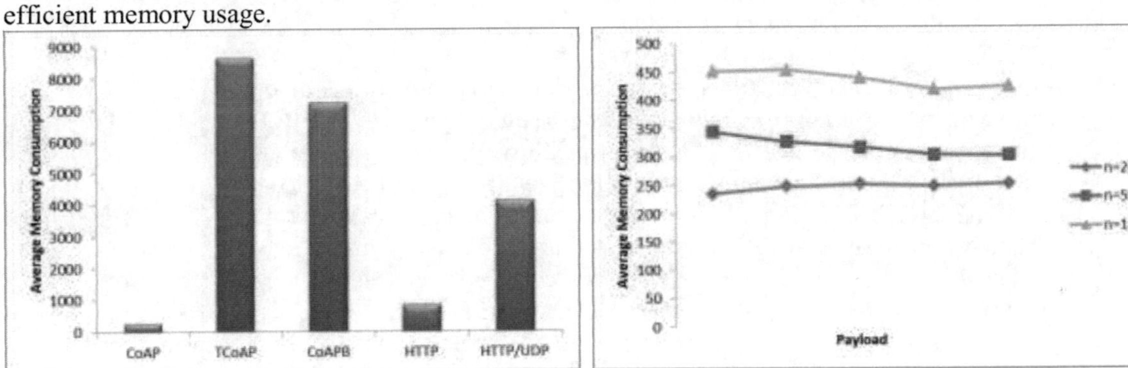

a)Consumption at payload b)Simultaneous transmission

Figure 8. Average memory consumption

7.5 Recognition of replay attacks

Our proposed system is exposed to DoS attacks during communication in application layer. During the process of verification unauthorized used tries to access the system regularly. The quality of system is improved based on exercises of attacks by intruders. The number of attacks occurring during the communication is registered over a time period of 60 seconds in Table 2. In a stream of communication only few packets are communicated in the relay attacks.

Table 2. Number of detected replay attacks

Time	Intruder			
	A'	B'	A''	B''
10 sec	4	6	10	2
The 20 sec	8	3	5	5
30 sec	5	6	9	11
40 sec	6	0	6	5
50 sec	2	4	5	1
60 sec	9	8	0	3

8. Conclusion

In this paper, we have proposed a Consignment communal certification plan to check the characters of imparting client, server in system. It utilizes a basic verification technique to trade key for asset perception. The proposed system utilizes insider facts for character confirmation of articles. These confidences are known uniquely to genuine client, server and they can't be acquired illegally considering the Internet Threat Model. For resource

IOP Conf. Series: Materials Science and Engineering 1074 (2021) 012003 doi:10.1088/1757-899X/1074/1/012003

allocation process our proposed system shows better improvement in our experimental result. Against DOS attacks, relay attacks and various attacks our proposed system is more effective. Eventually, our proposed system is more effective and powerful in contradiction to various kinds of attacks. At the provisioning stage our proposed system allocates pre-shared key for fixed data transmitting. For further enhancement of our proposed system adaptability and portability can be added. For light weight authentication process resource allocation nature for the sensor hubs is a difficult task for providing services.

References

[1] Al-Fuqaha, Ala, et al. "Internet of things: A survey on enabling technologies, protocols, and applications." *IEEE communications surveys & tutorials* 17.4 (2015): 2347-2376.
[2] Botta, Alessio, et al. "Integration of cloud computing and internet of things: a survey." *Future generation computer systems* 56 (2016): 684-700.
[3] Görmüş, Sedat, Hakan Aydın, and Güzin Ulutaş. "Security for the internet of things: a survey of existing mechanisms, protocols and open research issues." *Journal of the Faculty of Engineering and Architecture of Gazi University* 33.4 (2018): 1247-1272.
[4] Bormann, Carsten, Angelo P. Castellani, and Zach Shelby. "Coap: An application protocol for billions of tiny internet nodes." *IEEE Internet Computing* 2 (2012): 62-67.
[5] Rescorla, Eric, and Nagendra Modadugu. "Datagram transport layer security version 1.2." (2012).
[6] Fielding, Roy T., and Richard N. Taylor. "Principled design of the modern Web architecture." *ACM Transactions on Internet Technology (TOIT)* 2.2 (2002): 115-150.
[7] Kim, Jaewoo, et al. "M2M service platforms: Survey, issues, and enabling technologies." *IEEE Communications Surveys & Tutorials* 16.1 (2013): 61-76.
[8] Hartke, Klaus. "Observing resources in the constrained application protocol (CoAP)." (2015).
[9] Thangavel, Dinesh, et al. "Performance evaluation of MQTT and CoAP via a common middleware." *2014 IEEE ninth international conference on intelligent sensors, sensor networks and information processing (ISSNIP)*. IEEE, 2014.
[10] Saint-Andre, Peter. "Mapping the extensible messaging and presence protocol (xmpp) to common presence and instant messaging (cpim)." *Jabber Software Foundation* (2004).
[11] Becker, Markus, et al. "Deployment of coap in transport logistics." *Proceedings of 36th IEEE Conference on Local Computer Networks (LCN), Bonn, Germany*. Vol. 47. 2011.
[12] Son, Seung-Chul, et al. "A time synchronization technique for coap-based home automation systems." *IEEE Transactions on Consumer Electronics* 62.1 (2016): 10-16.
[13] Centenaro, Marco, et al. "Long-range communications in unlicensed bands: The rising stars in the IoT and smart city scenarios." *IEEE Wireless Communications* 23.5 (2016): 60-67.
[14] Becker, Markus, et al. "Wireless freight supervision using open standards." *5 th International Workshop Cold Chain Management. June*. 2013.
[15] Trappe, Wade, Richard Howard, and Robert S. Moore. "Low-energy security: Limits and opportunities in the internet of things." *IEEE Security & Privacy* 13.1 (2015): 14-21.
[16] Holmberg, Christer, and Jonathan Lennox. "Connection-Oriented Media Transport over the Transport Layer Security (TLS) Protocol in the Session Description Protocol (SDP)." (2017).
[17] Jan, Mian Ahmad, et al. "A payload-based mutual authentication scheme for Internet of Things." *Future Generation Computer Systems* 92 (2019): 1028-1039.
[18] Karthik, P. N. V., et al. "Energy-Efficient Elliptic Curve Cryptography-Based DTLS Key Establishment Protocol for IoT Communication." *Wireless Communication Networks and Internet of Things*. Springer, Singapore, 2019. 69-76.
[19] Cohen, Adam, et al. "Additive manufacturing of active devices using dielectric, conductive and magnetic materials." U.S. Patent Application No. 15/669,263.
[20] Stankovic, John A., et al. "Hardware/Software Security Patches for Internet of Trillions of Things." *arXiv preprint arXiv:1903.05266* (2019).

[21] Bhattacharjya, Aniruddha, et al. "CoAP—application layer connection-less lightweight protocol for the Internet of Things (IoT) and CoAP-IPSEC Security with DTLS Supporting CoAP." *Digital Twin Technologies and Smart Cities*. Springer, Cham, 2020. 151-175.

[22] Abosata, Nasr Rajb A., Andrew H. Kemp, and Mohsen Razavi. "Secure Smart-Home Application Based on IoTCoAP protocol." *2019 Sixth International Conference on Internet of Things: Systems, Management and Security (IOTSMS)*. IEEE, 2019.

[23] Jan, Mian Ahmad, et al. "A payload-based mutual authentication scheme for Internet of Things." *Future Generation Computer Systems* 92 (2019): 1028-1039.

[24] Jan, Syed Roohullah, et al. "Using CoAP protocol for resource observation in IoT." *International Journal of Emerging Technology in Computer Science & Electronics, ISSN: 0976* 1353 (2016).

[25] Nayyar, Anand, et al. "Smart-IoUT 1.0: A Smart Aquatic Monitoring Network Based on Internet of Underwater Things (IoUT)." *International Conference on Industrial Networks and Intelligent Systems*. Springer, Cham, 2018.

[26] Osanaiye, Opeyemi A., Attahiru S. Alfa, and Gerhard P. Hancke. "Denial of service defence for resource availability in wireless sensor networks." *IEEE Access* 6 (2018): 6975-7004.

[27] Abass, Ahmed A. Alabdel, et al. "Evolutionary game theoretic analysis of distributed denial of service attacks in a wireless network." *2016 Annual Conference on Information Science and Systems (CISS)*. IEEE, 2016.

[28] Ma, Yi, and Hongyun Ning. "The improvement of wireless LAN security authentication mechanism based on Kerberos." *2018 international conference on electronics technology (ICET)*. IEEE, 2018.

[29] Seitz, L., and G. Selander. "Design considerations for security protocols in constrained environments, draft-seitz-ace-design-considerations-00 (WiP)." (2014).

[30] Li, Chun-Ta, et al. "A Secure Dynamic Identity Based Authentication Protocol with Smart Cards for Multi-Server Architecture." *J. Inf. Sci. Eng.* 31.6 (2015): 1975-1992.

[31] Arunkumar, B., and G. Kousalya. "Analysis of AES-GCM Cipher Suites in TLS." *The International Symposium on Intelligent Systems Technologies and Applications*. Springer, Cham, 2017.

[32] Narten, Thomas, Barry Leiba, and Michelle Cotton. "Guidelines for writing an IANA Considerations Section in RFCs." (2017).

[33] Gustafsson, Oscar, and Lars Wanhammar. "Basic Arithmetic Circuits." *Arithmetic Circuits for DSP Applications* (2017).

[34] Rahman, Reem Abdul, and Babar Shah. "Security analysis of IoT protocols: A focus in CoAP." *2016 3rd MEC international conference on big data and smart city (ICBDSC)*. IEEE, 2016.

[35] Rahman, Reem Abdul, and Babar Shah. "Security analysis of IoT protocols: A focus in CoAP." *2016 3rd MEC international conference on big data and smart city (ICBDSC)*. IEEE, 2016.

[36] Ludovici, Alessandro, Pol Moreno, and Anna Calveras. "TinyCoAP: A novel constrained application protocol (CoAP) implementation for embedding RESTful web services in wireless sensor networks based on TinyOS." *Journal of Sensor and Actuator Networks* 2.2 (2013): 288-315.

IOP Conf. Series: Materials Science and Engineering 1074 (2021) 012004 doi:10.1088/1757-899X/1074/1/012004

Cancer Classification using Ensemble Feature Selection and Random Forest Classifier

Nimrita Koul, Sunilkumar S Manvi

REVA University, Bangalore, Karnataka, India - 560064

Emailnk1@gmail.com

Abstract. High volumes of genomic data made available by high through put gene expression sequencing technologies like next generation sequencing, microarray gene expression data have made it possible to develop models to computationally analyse this data and infer meaningful insights like presence of a disease, nature of disease, place of localization of the tumour in cancers etc. Since gene expression data is very high dimensional, each gene stands for one dimension, and has very small number of observations, it is imperative to apply feature selection on the data before using it for classification task. In this paper, we have proposed a method for classification of human cancer types by analysis of microarray gene expression data. We have used an ensemble feature selection algorithm for selecting subsets of 5, 10, 20 and 30 genes and applied random forest classifiers to obtain the classification accuracy and other performance parameters for comparison with existing solutions. We have been able to obtain 100% classification accuracy with just 5 genes on colon cancer data set with our algorithm.

Keywords – Feature Selection, Gene Expression Data, Machine Learning, Kernel PCA, Ensemble Techniques

1. INTRODUCTION

With the advancement in medical diagnostic technologies like DNA microarray technology [1], next generation sequencing etc. enormous volumes of genomic data has been generated. DNA microarray experiment produces the gene expression values of all thousands of genes in the genome of an organism being sequenced. So this data has very high number of dimensions and at the same time the number of observations or samples is very less.

This data can be analysed using computational analysis techniques like machine learning to understand various patterns in data. A study of such patterns can reveal an abnormality or a disease condition in the cell. This analysis can also reveal functional, regulatory relationships between different genes within a cell. It can also be used to study relationships between physiological functions and cellular interactions in an organism at the level of genes. After a microarray experiment, the raw data that is produced is pre-processed to perform background corrections, normalization, standardization, summarization [2, 3] etc. Once the preprocessing is over, the gene expression data sets are ready of analysis using machine learning techniques. The prominent tasks of machine learning that have been carried out by researchers on gene expression data sets are classification and clustering [4, 5, 6]. The tasks of classification and clustering are used to profile the gene expression data. Clustering aims at partitioning of genes into groups based on certain similarity measures in such a way that the genes which are related to each other by function or structure or effect are placed into same group and unrelated genes are placed in different groups. Classification is another important machine learning tasks that aims at labelling the activity of genes as normal or abnormal which can be used to diagnose a cell as normal or diseased. This can be used to classify patients according to health condition. Since gene expression data is a typical example of the data with very large number of dimensions and very small number of samples, the curse of dimensionality [7, 8] often figures in computational processing of gene expression data, in addition to the problem of 'large p and very small n', there is an acute class imbalance in such data besides the noise and missing values. From clinical research, we know that expression of very few genes is responsible for onset of diseases like cancer [9], therefore the task of identifying these few genes from expression of entire genome is an important task. These few genes, once identified, have the ability of correctly classifying patients as normal or diseased without the computational processing of entire genome. In this paper, the focus is to classify gene expression profiles as cancerous or non-cancerous. We have worked with colon cancer dataset [14]. Objective of a classification task is labelling of a gene in the given gene expression dataset as being relevant to diagnosis or the disease or not. Once we have found the genes not relevant to diagnosis, we drop them and retain the genes with most influence on diagnosis. This task of selecting the genes which are highly related to diagnosis is known as feature selection [10]. Depending upon the nature of machine learning task – supervised, unsupervised or semi-supervised, there are various methods for feature selection. In supervised feature selection, we have the information about diagnosis or labels of each sample, therefore the task of feature selection algorithm is to identify those genes whose expression values are highly correlated with value of label. Once the features have been selected, the classification task can begin.

Since the objective of classification task is to identify correct labels for the samples in gene expression dataset, the computational algorithm has to analyze each expression profile for its relevance to labels present during training of the classifier. This analysis on thousands of genes is compute intensive and the curse of dimensionality can influence the results. Therefore, the feature selection task precedes the classification task while processing gene expression datasets.

1.1. Background

The process of feature selection is a subset selection problem, i.e. the processing carried out to select a minimum number of most relevant and mutually unrelated genes which are sufficient to correctly classify a sample from gene expression dataset. The important task of differentiating normal human cell from a cancerous cell can be carried out by computationally observing difference in values of expression levels of genes in a sample taken from a person. It is much faster and efficient to observe just the relevant genes than it is to observe all twenty four thousand genes present in the genome. Thus it is important to find out the subset of genes that perform best with some classifier algorithm. The measure of being best if often classification accuracy, true positive rate, precision, recall, sensitivity and specificity

The process of feature selection on original data set with Y number of features is the process of selecting a minimal subset with X number of features, such that $X \subseteq Y$ and X minimizes the error in classification. [10,11,12] i.e., minimum number of samples are wrongly classified when basing the decision only on X. Error can be computed using a standard error function like mean squared error. Thus the process of feature selection is also a constrained optimization problem where we are trying to minimize classification error, in other words, we are trying to maximize classification accuracy. There can be constraints on the error function e.g., maximum permissible unit or error.

1.2 Advantages of Feature Selection

The advantages of feature selection are –

1. Removing the irrelevant features from the input dataset can lead to improvement in classification accuracy of the classifier.
2. Having only the relevant genes in the dataset can improve validation scores on test data.
3. This also increases generalizability of the classifier trained on this data.

4. Due to smaller dimensionality, the chance for overfitting is reduced.
5. Lower dimensionality of relevant feature set also reveals interesting aspects of the problem.
6. Lower dimensionality of input dataset, reduces the computation resources, time and memory required for processing.
7. Often, the classification task involve learning or search into the solution space, with less number of features, the search space is reduced.

Literature classifies feature selection methods as univariate or multivariate depending on whether each feature of the original dataset is considered individually to compute its relevance to class labels or a combination of features is considered at a time. While univariate feature selection considers genes individually, multivariate gene selection considers groups of genes at a time and therefore considers interdependencies and interrelations among genes.

Another way of classifying feature selection methods is as filter methods, wrapper methods, and embedded methods. This categorization is done based on the mechanism of validation used to identify relevance of a feature. Filter approaches exploit inherent properties of data for subset selection, these methods are independent of the predictor variables. The feature selection methods under the wrapper approach work on the principle of optimizing a predictor variables. The features that optimize value of predictor are retained in selected subset. The class of feature selection methods which fall under the category of embedded approach use a machine learning algorithm known as classifier as a part of the feature selection process. Each feature is evaluated by measuring its influence on the classification accuracy. The optimal subset of genes is one that maximizes the classification accuracy and at the same time the size of this subset should be minimum. Both wrapper and embedded methods have a drawback of overfitting and have high computational demands.

Feature selection methods which constitute filter methods use the information theoretic and statistical properties of data entropy, information gain, mutual information [9], correlation coefficient, consistency, fast correlation to identify the best genes. Correlation coefficient and fast correlation, consistency are multivariate filter methods as they act on multiple genes at a time. Both correlation and fast correlation methods compute the correlation coefficient of a gene or feature with the class label, the genes with highest values of correlation are preferred. However, we also want to select the genes which in addition to having highest correlation with label have least correlation with other genes in the feature set. We wish to retain most mutually unrelated genes that are most related to labels. A gene is considered highly relevant if it is highly correlated to the class of the sample and correctly predicts it and at the same time is uncorrelated to other genes in the input set of features.

Among important univariate methods of feature selection are mutual information [9] and information gain. In this method, the gain in information content is computed for each gene with respect to the label. Information gain is a measure of decrease of entropy due to inclusion of a feature. It is an important metric in training of decision tree based feature selectors and classifiers. For each feature, information gain can be computed by observing the difference between entropy of a dataset with or without the feature that is being evaluated. Mutual information is a related metric which is a measure of statistical inter-dependence of two variables. Mutual information indicates the amount of information that can be inferred about a feature from another feature. More the mutual information between two features, more is the redundancy.

Other prominent filter approach to feature selection is known as minimum redundancy maximum relevance. In this method, the genes which are most relevant to the class label and at the same time most dissimilar with other genes are selected in the subset. Many methods of feature selection perform the analysis of variates for differentiation of classes through identification of significant statistical differences between means of classes using metrics like sum of squares,

Fisher score, Laplacian score etc. Through calculation of these metrics the feature subsets are identified which maximize inter class dissimilarity measure and minimize within class dissimilarity measure and are able to maintain their locality.

Rest of this paper is organized as follows – Section II covers the survey of literature and the methods proposed for feature selection from gene expression data, Section III presents the proposed algorithm, in Section IV covers the results and discussion, Section V concludes the paper.

2. RELATED WORK

In this section, we cover the study of feature selection methods proposed in literature. The methods we present here are prominently applied to high dimensional gene expression data. These methods are primarily filter methods which are suitable for handling gene expression datasets.

In [1] authors demonstrated for the first time that accurate classification could be done by using computational approaches for analysis of gene expression datasets. The authors worked on Leukaemia data set and successfully classified it into two sub classes of leukaemia with very few relevant genes. The work in [2] presents the use of support vector machine as feature selection tool for the task of cancer classification from gene expression datasets. In [3], the researchers have employed regression technique called as least squares regression for identification of most relevant genes from the biological data for various classification tasks. Authors in [4], used a method called minimum redundancy maximum relevance as a feature selection approach and artificial neural networks as classifiers for classification tasks on biological data. Authors in [5], have employed evolutionary approach and fuzzy logic based method for feature selection. They have used Bee colony optimization combined with fuzzy logic for feature selection from gene expression data. Authors in [6], applied distributed correlation methods for feature selection from gene expression datasets for the classification problem. Paper [7] is a survey of prominent feature selection approaches for the task of classification of cancers using gene expression as input data. In [8], researchers have employed logistic regression as feature selection technique for selection of relevant genes for gene expression datasets. Authors in [9], applied a pipeline of mutual information and genetic algorithm to produce an ensemble algorithm for feature selection for cancer classification using gene expression profiles. In [10], the authors have used kernel SVM as s feature selection method on cancer gene expression data sets. In [11], authors used a hybrid algorithm consisting of an ensemble of cellular automata and an evolutionary algorithm called ant colony optimization for obtaining the optimal feature subset from microarray gene expression data. Authors in [12] compute best 'r' genes by dividing dataset into subsets. From each subset top genes are selected and finally all combined together to get the best overall genes of the dataset. The process is repeated till an optimal subset is obtained. Authors in [13], have applied an ensemble of Support vector machine and recursive feature elimination as a method for gene selection from gene expression datasets. In this study, we have experimented on the Colon cancer dataset [14], containing 60 samples, 2000 genes with two classes – normal and tumour.

3. PROPOSED METHOD

The proposed algorithm for classification is presented in this section. This method is an ensemble of mutual information and kernel PCA. The algorithm is as follows –

Algorithm 1: MIPCA Gene Selection

Input: I -Set of Gene Expression Data
Output: O- Subset of most relevant Genes
1. Compute the mutual information of all genes in I
2. Compute score of I with RFC with depth value of 4

3. Select top ranked 5, 10, 20 and 30 gene sets with highest MI values as X5, X10, X20, X30
4. Apply Kernel PCA to I and X5, X10, X20, X30 generating new Xs.
5. Apply RFC-4 to new Xs.
6. Compute the performance parameters.
7. Return the subset with maximum value of score

Algorithm 1: Gene selection Approach

Experimental set up consisted of a Windows 10 machine with Intel i7, 12 cores processor. Programming was done in Python 3. The dataset used is the Colon cancer dataset [14]. The dataset contains 60 samples, 2000 genes with two classes – normal and tumour.

We executed the experiment 20 times with each with full dataset, then with reduced data sets of lower number of features. We selected best 5, best 10, best 20 and best 30 genes using the algorithm presented above. The average values for all parameters were used for graphing and drawing conclusion. The classification was performed using Random Forest classifier with a depth of 4. During the feature selection phase, we used mutual information as first level of ranking of genes and kernel PCA as the second level of feature selection.

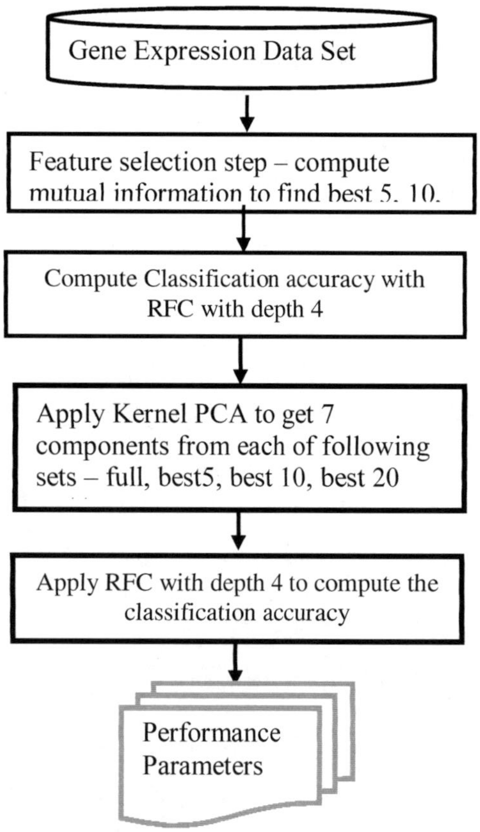

Fig.1. Schematic representation of the proposed scheme

Figure 1 shows the diagrammatic representation of the proposed algorithm. Each level shows the flow of data from top to performance parameters – classification accuracy and training time obtained at bottom end.

4. Results and Discussion

In the experiment, we found the pairwise mutual information in the features in full dataset [14]. The classification accuracy was noted without this computation. The top ranked 5, 10, 20 and 30 genes were noted and classification was performed using just the reduced feature sets, and corresponding performance noted. Then kernel PCA was applied on the full input set and the reduced sets and again the accuracy of classifier noted for comparison. The input data is standardized and normalized. Classification accuracy with the proposed method is 100% with a subset of 20 genes.

Number of Features	MIRFC	PCARFC	MIPCA-RFC	RFC
Full (2000 genes)	100	98	100	100
5	93	91	98	96
10	92	95	98	98
20	96	98	100	99
30	97	96	100	99

Table1. Comparison of Classification Accuracy

Table 1 shows that we get 100 % classification accuracy with full feature set as well as with best 10 features. Therefore these 10 features can be considered relevant to colon cancer.

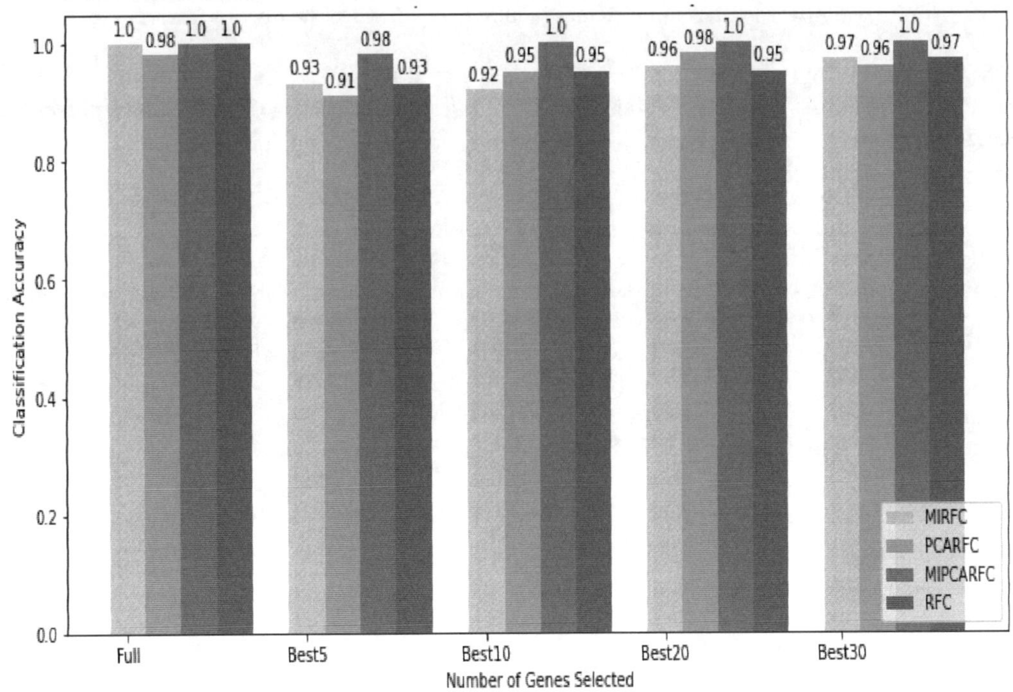

Fig. 2. Classification Accuracy Comparison

Figure 2 shows the comparison of classification accuracy obtained with original data set with 2000 features under four experimental conditions – first, using mutual information as feature selection algorithm and random forest as classifier, second, PCA as feature selector, RFC as classifier, third, proposed algorithm ensemble of MI and PCA as feature selector and RFC as classifier, fourth, using

RFC without explicit feature selection. The classification accuracy obtained with 30 selected genes is better than with 5, 10 or 20.

Table 1 shows the values of classification accuracy in percentages for the four experimental conditions mentioned above.

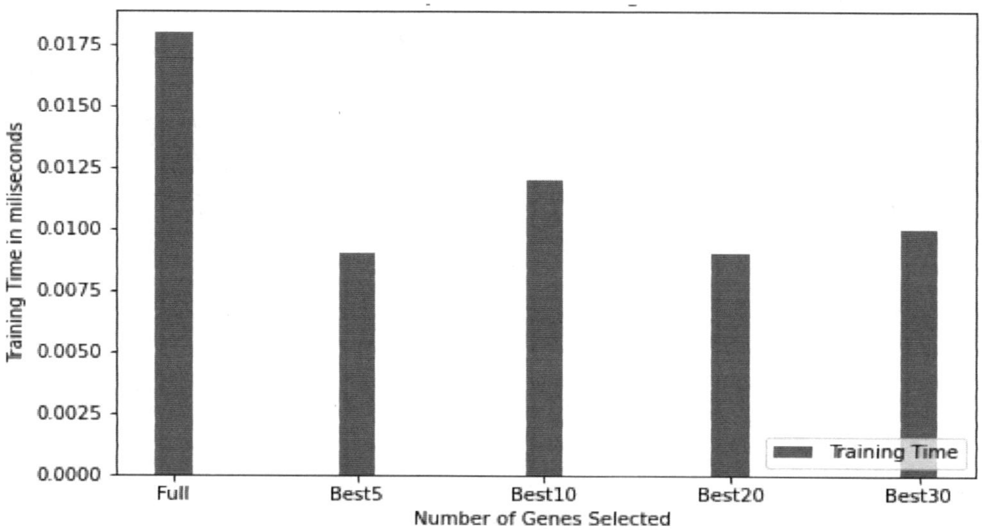

Fig. 3. Comparison of Training Times required with different number of feature

Figure 3 shows the training times required to fit the classifier with different sizes of feature set. It can be observed that the training times are highest for full feature set consisting of 2000 genes present in original data set.

5. CONCLUSION

Filter methods of feature selection prove to be most suitable for the feature selection from high dimensional gene expression data. These also are less compute intensive as compared to wrapper and embedded methods for this task. Among the classification algorithms, SVM and RFCs have been demonstrated to work very well with gene expression data. In this paper, we presented an ensemble technique for feature selection and RFC was used for classification of the data with and without feature selection. The algorithm identified 10 genes which could accurately classify the input dataset. Thus the minimal subset of genes has size 10 for accuracy of 100%. As a future work, the proposed algorithm has to be applied on other types of data and performance observed.

ACKNOWLEDGEMENT

Authors would like to thank Department of Science and Technology (DST), Government of India, for financially supporting this work under the scheme DST-ICPS 2019.

REFERENCES

[1] G T.R. Golub et. al., 1999, Molecular classification of cancer: Class discovery and class prediction by gene expression monitoring, *Science*, pp. 531–537.

[2] I. Guyon, J. Weston, S. Barnhill, V. Vapnik, 2002, Gene selection for cancer classification using support vector machines, *IEEE Transactions on Knowledge and Data Engineering*, vol. 25, no. 1, pp 1-14

[3] Lê Cao, K., Boitard, S. & Besse, P, 2011, Sparse PLS discriminant analysis: biologically relevant feature selection and graphical displays for multiclass problems, *BMC*

Bioinformatics, vol. 12, no. 253, https://doi.org/10.1186/1471-2105-12-253

[4] Masud Rana M., Ahmed K., 2020, Feature Selection and Biomedical Signal Classification Using Minimum Redundancy Maximum Relevance and Artificial Neural Network, *Proceedings of International Joint Conference on Computational Intelligence Algorithms for Intelligent Systems, Springer.*

[5] Nancy S.G., Saranya K., Rajasekar S., 2020, Neuro-Fuzzy Ant Bee Colony Based Feature Selection for Cancer Classification, *Springer Innovations in Communication and Computing. Springer, Cham 2020.*

[6] Alok Kumar Shukla, Diwakar Tripathi, 2020, Detecting biomarkers from microarray data using distributed correlation based gene selection, *Genes & Genomics.* https://doi.org /10.1007 /s13258 -020-00916-w

[7] Nada Almugren, Hala Alshamlana, 2019, Survey on Hybrid Feature Selection Methods in Microarray Gene Expression Data for Cancer Classification, *IEEE Access*, vol. 7, pp. 75833-44 10.1109/ACCESS.2019.2922987

[8] Zakariya Yahya Algamal, Muhammad Hisyam Lee, 2019, A two-stage sparse logistic regression for optimal gene selection in high-dimensional microarray data classification, *Advances in Data Analysis and Classification*, vol. 13, pp: 753–771, https://doi.org/10.1007/s11634-018-0334-1

[9] M. Jansi Rani, D. Devaraj, 2019, Two-Stage Hybrid Gene Selection Using Mutual Information and Genetic Algorithm for Cancer Data Classification, *Journal of Medical Systems*, vol. 43, no, 235, https://doi.org/10.1007/s10916-019-1372-8

[10] Medjahed SA, Saadi TA, Benyettou A, Ouali M, 2017, Kernel-based learning and feature selection analysis for cancer diagnosis, *Applied Soft Comput*ing, vol. 51, pp 39–48

[11] Fatemeh Vafaee Sharbaf, Sara Mosafer, Mohammad Hossein Moattar, 2016, A hybrid gene selection approach for microarray data classification using cellular learning automata and ant colony optimization, *Genomics*, vol. 107, pp.231–238, http://dx.doi.org/10.1016/j.ygeno.2016.05.001

[12] S. Miyano, S. Imoto, A. Sharma, 2012, A top-r feature selection algorithm for microarray gene expression data, *IEEE/ACM Transactions on Computational Biology and Bioinformatics*, vol. 9, no. 3, pp. 754–764

[13] P. Mundra, J. Rajapakse, 2010, SVM-RFE withmRMR filter for gene selection, *IEEE Transactions on Nano. Biosci*, vol. 9, no. 1, pp 31-37

[14] Alon U, Barkai N, Notterman DA, Gish K, Ybarra S, Mack D, Levine AJ, 1999, Broad patterns of gene expression revealed by clustering analysis of tumor and normal colon tissues probed by oligonucleotide arrays, *Proc Natl Acad Sci.*, vol. 96, pp. 6745-6750.A reference

IOP Conf. Series: Materials Science and Engineering 1074 (2021) 012005 doi:10.1088/1757-899X/1074/1/012005

Advanced Convolutional Neural Network Classification for Automatic Seizure Epilepsy Detection in EEG Signal

Venkata Ramana Mancha[1], Srinivasa Reddy E[2], Satyanarayana Ch[3]

[1]Assistant Professor, Gitam University, Visakhapatnam, A.P., India
[2]Professor, Nagarjuna University, Guntur, A.P., India
[3]Professor, JNTUK University, Kakinada, A.P., India

Abstract

Epilepsy is one of the irregular electro-physiological disorder appeared in human brain, which is characterized by tonic recurrent seizures, Electroencephalogram (EEG) is a sufficient test measure to maintain records with respects to electrical activity of brain and it is widely used in analysis and detection of electro epileptic seizures. Manual inspection of EEG signal extraction will take more time to process and it puts heavy complex on neurologists affects their performance. It is often difficult in identification of brain subtle but emergency changes in EEG wave forms by visual inspection based on research area for bio- engineers implement different types of methodologies for identification of such type of subtle. But all these algorithms/ methodologies don't perform efficient accuracy in classification of normal, ictal class instances. So that in this paper, we propose a novel system based on machine learning, which is single dimensional pyramidal ensemble convolutional neural network (1D-PECNN). Here ensemble means different parts of the signal are assigned to different models for efficient analysis of data. We also propose mathematical augmented approach for learning features. In 1D-PECNN model, system consist high amount of desirable and learnable parameters, in all cases proposed approach 1D-PECNN gives maximum accuracy (Approximately from 92%-99%) when compare to state-of-the methods.

Index Terms: Epilepsy, convolutional neural networks, visual inspection, electroencephalogram, epileptic seizures, non-linear analysis and feature extraction.

I. Introduction

Peoples who experienced seizure may influence epilepsy. Epilepsy is the average condition influencing around 65 million individuals worldwide.[1] According to the ongoing review about 2.3 million Americans were influenced by epilepsy. The general population who experienced seizure may loses their cognizant for quite a while causing modification in conduct and sensation. Epilepsy individual may have increasingly number of seizure types. Epileptic seizure has two sorts in particular incomplete (central) and summed up. The side effects for seizure were periodic blacking out spells and biting or flickering at unseemly time. The triggers for seizure are parchedness, photosensitivity, absence of rest, stress, and so forth. PNES, is the sort of non-epileptic seizure which is generally called pseudo seizures, are a moderately phenomenal confusion with a predominance of around 1 to 33 cases for every 100 000 and they represent 5– 20% of patients thought to have epilepsy [3]. There is potential for serious damage from the unfavorable reactions or teratogenicity of antiepileptic drugs endorsed to PNES patients [4], just as horribleness and mortality from intubation for delayed seizures [5].For recognizing and investigating epileptic seizure, EEG has set up. Electroencephalogram is the procedure in which seizure can be analyzed. The cathodes are utilized for acquiring the electrical signs from the human mind. EEG signals are commonly spoken to in high dimensional component space. It is hard to decipher EEG signals. For deciphering and breaking down high dimensional list of capabilities AI techniques are utilized. Specialists have proposed techniques for the recognition of seizures utilizing highlights separated from EEG motions by hand-designed systems.

A portion of the proposed techniques utilize ghastly (Tzallas et al., 2012) and worldly parts of data from EEG signals (Shoeb, 2009). An EEG flag contains lowfrequency highlights with long timeframe and high-recurrence highlights with brief timespan (Adeli et al., 2003) for example there is a sort of chain of command among highlights. Profound learning (DL) is a cutting edge ML approach which naturally encodes pecking order of highlights, which are not information subordinate and are adjusted to the information; it has indicated promising outcomes in my applications. Also, highlights extricated utilizing the DL models have appeared to be more discriminative and vigorous than hand-structured highlights (LeCun et al., 1995). So as to improve the exactness in the arrangement of epileptic and non-epileptic EEG signals, we propose a strategy dependent on DL.

We propose a novel system based on machine learning, which is single dimensional pyramidal ensemble convolutional neural network (1D-PECNN). We also propose mathematical augmented approach for learning high amount of features. Our proposed approach takes signals of EEG and fixed them into single data window with different attributes and they pass those instances into associated PECNN model. It outperforms the state of the methods for various issues concerning epilepsy discovery. The principle commitments of this investigation are: 1) information enlargement plans, 2) a framework dependent on a gathering of P-1D-CNN profound models for parallel just as ternary EEG flag grouping, 3) another methodology for organizing profound 1D-CNN model and 4) exhaustive assessment of the expansion plans and the profound models for identifying distinctive epilepsy cases.

II. BACKGROUND RELATED WORK

We present a concise audit of seizure-related wording, the seizure recognition writing, and the one-class SVM.

2.1. Seizure-Related Terminology

Seizure investigation alludes all things considered to calculations for seizure recognition, seizure expectation, and programmed center channel recognizable proof. These investigations are principally performed on the EEG. In this investigation, examinations were completed on the intracranial EEG (IEEG), which has significantly better spatial goals, higher flag to-clamor proportion, and more noteworthy transfer speed than scalp EEG. At the point when various channels are considered, the anode area that shows the most punctual proof of seizure movement is named the center channel. It is advantageous to depict sections of the EEG motions by their worldly nearness to seizure movement. The ictal period alludes to the time amid which a seizure happens. The interictal period is the time between progressive seizures. The unequivocal electrographic beginning (UEO) is characterized as the most punctual time that a seizure event is obvious to an epileptologist seeing an EEG without earlier information that a seizure pursues; the unequivocal clinical beginning (UCO) is the most punctual time that a seizure event is obvious by outwardly watching a patient. Seizure beginning in this paper is synonymous with UEO. It is significant that the UEO quite often goes before the UCO by a few seconds, and that numerous recently distributed papers characterized "seizure beginning" as the UCO.

2.2. Seizure Detection

Early endeavors to distinguish seizures started during the 1970s (Viglione, Ordon and Risch, 1970; Liss, 1973) and principally considered scalp EEG chronicles to identify the clinical (and less as often as possible) electrographic beginning of seizures. In 1990, Gotman revealed a strategy for mechanized seizure identification that accomplished 76% location precision at 1 Fp/hr for 293 seizures recorded from 49 patients (Gotman, 1990). In 1993, it was demonstrated that the brief timeframe mean Teager vitality could be utilized to identify seizures from electrocorticograms (Zaveri, Williams and Sackellares, 1993). Their identifier accomplished 100% recognition exactness on a 11-seizure database. In 1995, Qu and Gotman displayed an early seizure cautioning framework prepared on format EEG action that accomplished 100% location exactness at a mean recognition inertness of 9.35 seconds and false alert rate of 0.2 Fp/hr (Qu and Gotman, 1995). Comparative outcomes were additionally announced utilizing time-and recurrence area highlights grouped by a k-closest neighbor classifier (Qu and Gotman, 1997). In 1998, Osorio et al. guaranteed 100% identification affectability with a mean recognition idleness of 2.1 seconds utilizing a wavelet based measure called seizure power. They broke down a database of 125 patients, yet similar information were utilized for preparing and approval (Osorio, Frei and Wilkinson, 1998). The calculation was all the more widely dissected in 2002 utilizing disconnected electrocorticogram chronicles; once more, 100% affectability was accounted for, with identification latencies extending from 1.8 – 31.1 seconds (Osorio et al., 2002).

A few effective endeavors at seizure discovery utilizing fake neural system classifiers have been accounted for since 1996 (Khorasani and Weng, 1996; Webber et al., 1996; Gabor, 1998; Esteller, 2000). Assessment of 31 particular highlights (Esteller, 2000) demonstrated that fractal measurement, wavelet parcel vitality, and mean Teager vitality were particularly encouraging for seizure recognition. In 2001, Estelle announced an identifier dependent on the line length include that accomplished a mean discovery inertness of 4.1 seconds at a bogus caution rate of 0.051 Fp/hr (Esteller et al., 2001). An aggregate of 111 seizures (numerous subclinical) were utilized for approval. NeuroPace, Inc., in this manner announced a comparative locator dependent on this work accomplished 97% affectability at a mean discovery idleness of 5.01 seconds (Echauz et al., 2001). This finder was assessed on 1265 hours of IEEG information, yet was tuned heuristically in a patient-explicit way. The NeuroPace finder claims speak to the best in class in seizure recognition execution. Progressively total surveys of the seizure identification and forecast writing are accessible somewhere else (Litt and Echauz, 2002; Gardner, 2004).

III. PROPOSED IMPLEMENTATION

In this section, we describe the procedure of single dimensional pyramidal ensemble convolutional neural network (1D-PECNN). We describe the procedure of deep learning convolutional neural networks to train epileptic eye related data for feature extraction and the identify the relevant data relates to epileptic data.

Deep Convolutional Neural Networks: In this work, sequential procedure of the convolution neural networks on EEG data related to human. A point by point representation of the proposed recovery framework is appeared in Fig. 1. The hidden DCNN model intends to learn channel part by creating an increasingly unique portrayal of the information in each layer.

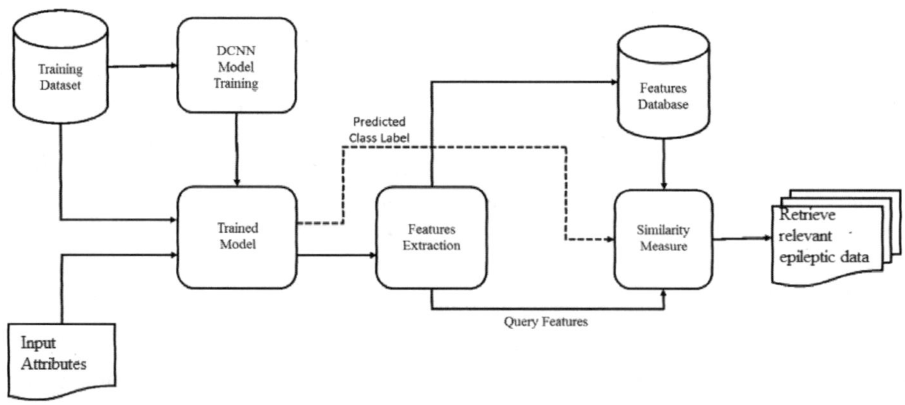

Figure 1 Proposed architecture implementation using convolutional neural networks.

In spite of its straightforward arithmetic, DCNN is as of now the most amazing asset in vision frameworks. The DCNN models by and large have three sorts of layers i.e., convolutional layers, pooling layers, and completely associated layers. The yield layer is commonly treated independently as a special layer and the model gets information tests at the information layer. Each convolutional layer produces highlight maps by convolving the part with information highlight maps. A pooling layer is intended to down example highlight maps created by the convolutional layers, which is frequently practiced by discovering nearby maxima in a neighborhood. Additionally, pooling gives translational invariance and in the then it lessens the quantity of neurons to be prepared in up and coming layers. In completely associated layers, every neuron has an increasingly denser association when contrasted with the convolutional layers. The piece of the DCNN before completely associated layers is known as feature extractor part and after that is known as classifier part. A de-followed portrayal of the structure utilized is exhibited in following subsections.

The model utilized for preparing comprised of eight layers, out of which five are convolutional layers and three are completely associated layers, as delineated in Fig. 2. The convolutional and completely connected layers are spoken to as CVL and FCL, where the subscript speaks to the layer number e.g., CVL 1 speaks to the first convolutional layer. The yield of last completely associated layer (FCL 3) has been bolstered to a soft-max capacity having 24 yields, which produce likelihood dispersions for each class mark. Subsequently, the probabilities vector of size 1 ×24 where every vector component relates to a class of dataset is gotten. The system acknowledges grayscale im-periods of measurement 224 ×224 as sources of info and not at all like the model presented in [11] utilize a lesser number of bits. The CVL 1 channels the info picture with 64 portions of size 11 ×11 with walk equivalent to 4 pixels. The walk is the separation between the focuses of open fields of neighborhood neurons in the portion map. The yield of the first convolutional layer is encouraged to a non-linearity and after that went through the spatial max pooling layer for abridging neighbor-ing neurons. Redressed straight unit (ReLU) [26] nonlinearity is ap-utilized to the yields of all convolutional and completely associated layers. This system with ReLUs has not just the capacity to get prepared a few times quicker than its proportional with tanh units [27] however it likewise permits to go further with evaporating angle issues.

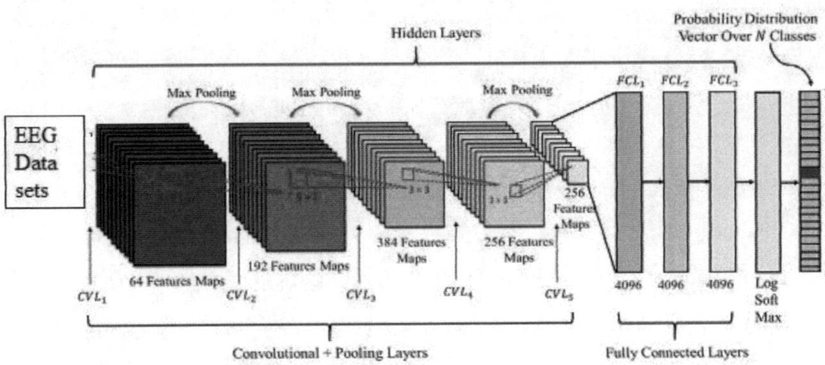

Figure 2 Deep convolution neural network classifier procedure to process .

The CVL 2 takes the yield of CVL 1 as information prepared by ReLU non-linearity and spatial max-pooling layers separately and channels it with 192 bits of size 5 ×5. The CVL 3 contains 384 parts of size 5 ×5 and it gets a contribution from the pooled yields of the second convolutional layers. Both the convolutional layers, CVL 4 and CVL 5 have 256 pieces of size 3 ×3. All completely associated layers have an equivalent number of neurons i.e., 4096.

Training Image Sequences:

Stochastic Gradient Descent (SGD) is commonly used calculation for preparing neural systems and it is productive learning with description straight under classifier a raised misfortune capacity like support vector machine (SVM). The two noteworthy favourable circumstances of utilizing SGD are effectiveness and straightforwardness in execution giving choices in tuning the system like various cycles, processing rate, rate rot, and so forth. A couple of inconveniences of SGD incorporate its requirement for hyper-parameters like various ages or emphasis optimized and regularized parameters. SGD update each parameter preparation test xi and mark yj. Eq. (1) is utilized for parameter update of the SGD

$$\theta = \theta - \eta . \Delta_{\theta} J(\theta; x^{i} . y^{j})$$

J be the objective function, will optimize the relations based on weights of each pixel.

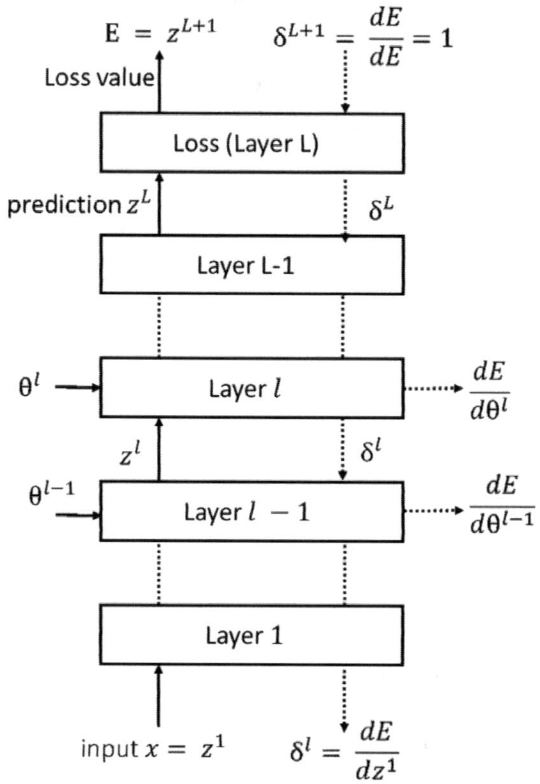

Figure 3.5 Description of different layers with different parameter sequences in training phase.

When the convolution model is effectively upgraded and prepared for grouping the medicinal pictures, pixel representations are extricated from last pixel to pixel formation in completer associated layer in pre-processing model i.e., from FCL 1 FCL 3 . For picture recovery task a locally settled highlights image data source for the entire preparing information is required. In this way, to make such highlights database, each picture x I from preparing set is feed sent to the prepared DCNN model for arrangement undertaking and afterward includes portrayal F 1 I , F 2 I , and F 3 I are extricated related to that particular picture from completely associated layers 1 3, individually. The F 1 I , speaks to a highlights database separated from FCL 1 and likewise F 2 I and F 3 I speaks to highlights databases removed from FCL 2 and FCL 3 , where I = 1 to P and P is equivalent to number of tests in preparing set. At whatever point an inquiry is defined, comparative pictures as that of question picture are recovered by looking at highlight portrayals extricated for inquiry picture features are representation using in Euclidean distance described as follows:

$$d(a,b) = \sqrt{\sum_{i=1}^{p} (a_i - b_i)^2}$$

As described in above, a, b, are the initial image data partitions separately which is shown in figure 3.5, and also describe the disseminating the pixel calculation of image from image data sources. Pictures have high image intensity or high similitude when contrasted with others is shown as recovery results to the client. At last, relative examination is performed for features portrayals separated from FCL 1, FCL 2, FCL 3 and convolution layer as far as recovery quality

3.1. Procedure to Explore Epilepsy in Convolutional Layers

Describe the procedure used in detection of epilepsy in human eye related EEG data sets.

Input: Human eye related EEG data with different attributes. Output: Epilepsy detection results (based on sequence of convolutional layers) Step 1: Import data relates to different attributes. Step 2: Convert attribute data into different inner signal values to convolutional layer for kernel feature extraction. Step 3: Forward non-linear activation layer to describe linear function based on convolutional layer with respect to no. of kernel attributes. Step 4: Calculate depth of kernel parameter based on fully connected layer communications. Step 5: Explore relevant features relates convolutional layer to extract epilepsy related data based on threshold at each attribute which consist epilepsy. Step 6: Predict simulated epilepsy results.

Algorithm 1 Step by step procedure to explore different statistical data evaluation in CNN.

Description for accessing relevant data from epileptic data, calculate the kernel parameter functionalities to describe connected layer communication in epileptic data evaluation.

3.2. Layers in Convolution

Convolutional layers operation in one-dimensional used to filter EEG signals to extract features. Convolution layer is generated by convoluting existing layers with respect to respective field (rf) and depth and is equivalent to number of features present in existing layer. Formally convoluting layer $A=\left\{a_{ij}:1\leq i\leq n,1\leq j\leq m\right\}$, where n is no. of channel in layer and m is the no. of units in channel with k[th] kernel $k^{i},l=1,2,.......,K$ each of the respective field and depth n convolutional layer $B=\left\{b_{ij}:1\leq i\leq q,1\leq j\leq K\right\}$ where

$$a_{ij}=\sum\nolimits_{d=1}^{q}\sum\nolimits_{e=1}^{k}w_{d,e}x_{i+d.j+e}$$

m is no. of units in each channel of layer, in convolutional layer, number of channels are integrated with no. of kernel specifications to extract discriminative features from input data as shown in figure 2. In proposed model, for each convolutional layer, it reduces pyramid structure based on no. of kernels in detection epilepsy.

IV. Performance Evaluation

We laid it on the line the explain of announcement, and the eventual story augmentation schemes. Then, we try evaluation measures, which have been secondhand to do justice to the stunt of the approaching system. After this, the training procedure has been elaborated. Finally, the of the first water data augmentation step by step diagram and

P-1D-CNN person to look up to have been latent by analyzing the results with diverse ways of data augmentation, and diverse 1D-CNN models.

4.1. Dataset Description

The informational collection utilized in this work was gained by an examination group at University of Bonn (Andrzejak et al., 2001) and have been broadly utilized for research on epilepsy identification. The EEG signals were recorded utilizing standard 10-20 terminal arrangement framework. The total information comprises of five sets (A to E), each containing 100 one-channel occurrences. Sets An and B comprises of EEG signals recorded from five solid volunteers while they were in a loose and wakeful state with eyes opened (An) and eyes shut (B), separately. Sets C, D, and E were recorded from five patients. EEG motions in set D were taken from the epileptogenic zone. Set C was recorded from the hippocampus arrangement of inverse side of the equator of the cerebrum. Sets C and D comprise of EEG signals estimated amid sans seizure interims (interictal), while, the EEG motions in Set E were recorded just amid seizure action (ictal) (Andrzejak et al., 2001). The detail is given in Table 1.

A	B	C	D	E
Non-Epileptic	Non-Epileptic	Epileptic	Epileptic	Epileptic
Eyes opened	Eyes closed	Interictal	Interictal	Ictal

Table 1 Data sets used for different attributes.

The number of instances collected in this dataset is not enough to train a deep model. Acquiring a large number of EEG signals for this problem is not practical and their labeling by expert neurologists is not an easy task. We need an augmentation scheme that can help us in increasing the amount of the data that is enough for training deep CNN model, which requires large training data for better generalization.

4.2. Performance Metrics

For assessment, we received 10-overlap cross approval for guaranteeing that the framework is tried over various varieties of information. The 100 signs for each class isolated into 10 overlap, each crease (10%), thusly, is kept for testing while the rest of the 9 folds (90% signs) are utilized for learning the model. The normal execution is determined for 10 folds. The execution was assessed utilizing understood execution measurements, for example, exactness, explicitness, affectability, accuracy, f-measure, and g-mean. A large portion of the cutting edge frameworks for epilepsy additionally utilize these measurements, the adjustment of these measurements for assessing our framework helps in reasonable examination with best in class frameworks. The meanings of these measurements are given underneath.

$$Accuracy(Acc) = \frac{TP + TN}{Total - Samples}$$

$$Specificity(Spe) = \frac{TN}{TN + TP}$$

$$Sensitivity(Sen) = \frac{TP}{FN + TP}$$

$$F - measure(F\_m) = \frac{2 * precision * sensitivity}{precision + sensitivity}$$

$$G - Mean(G\_m) = \sqrt{Specificity * Sensitivity}$$

where TP (true positives) is the quantity of strange cases (for example epileptic), which are anticipated as anomalous, FN (false negatives) is the quantity of strange cases, which are anticipated as should be expected, TN (genuine negatives) is the quantity of ordinary case that is anticipated as should be expected and FP (false positives) is the quantity of typical cases that are distinguished as anomalous by the framework.

4.3. Results

Table 2 shows our proposed accuracy values with respect to existing techniques with different value formats as follows:

Table 2. Different accuracy values with different attributes .

Accuracy				
Databases	Proposed Approach	Existing Previous [3]	Existing Previous [4]	Existing Previous [5]
Data Class 1	0.84	0.712	0.689	0.696
Data Class 2	0.736	0.51	0.708	0.562
Data Class 3	0.746	0.764	0.576	0.415
Data Class 4	0.832	0.604	0.484	0.423
Data Class 5	0.832	0.470	0.508	0.371

Figure 3 shows the accuracy presentation of proposed approach with different Elliptic Data databases.

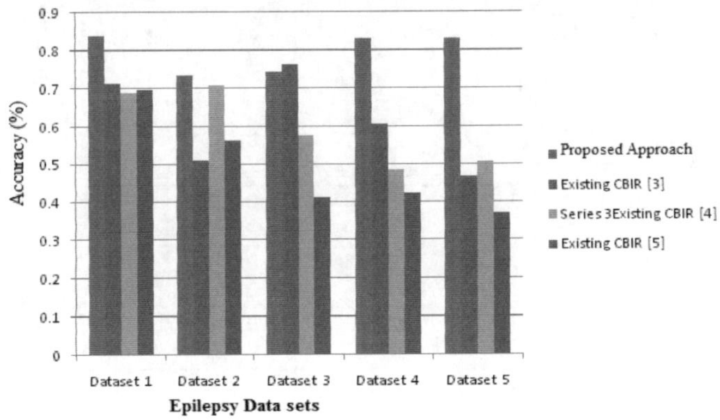

Figure 3. Performance evaluation of different accuracy values.

Table 3 follows precision representative values with different Epilepsy Data's in optimization of Epilepsy Data query retrieval.

Table 3. Different values with respect to precision.

Precision				
Epilepsy Data Achieves	Proposed Approach	Existin g System [3]	Existin g System [4]	Existing System [5]
Epilepsy Data Class 1	0.56	0.342	0.4245	0.352

Epilepsy Data Class 2	0.42	0.301	0.415	0.401
Epilepsy Data Class 3	0.472	0.399	0.28	0.338
Epilepsy Data Class 4	0.445	0.444	0.392	0.345
Epilepsy Data Class 5	0.472	0.333	0.335	0.345

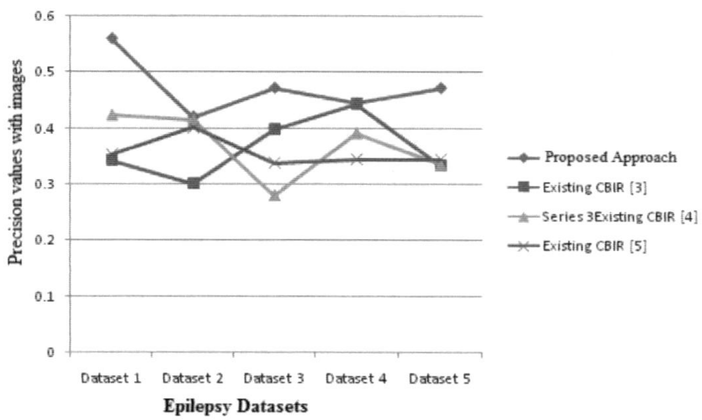

Figure 4. Performance evaluation of different attributes.

Figure 4 shows precision results of proposed approach with different representation values.

Table 4. Recall values with different Epilepsy Data databases.

	Recall			
Databases	Proposed Approach	Existing Previous [3]	Existing Previous [4]	Existing Previous [5]
Epilepsy Data Archive 1	0.61	0.412	0.501	0.427
Epilepsy Data Archive 2	0.552	0.346	0.496	0.468
Epilepsy Data Archive 3	0.598	0.486	0.346	0.367
Epilepsy Data Archive 4	0.61	0.462	0.454	0.329
Epilepsy Data Archive 5	0.61	0.376	0.396	0.356

Table 4 shows recall effective Epilepsy Data retrieval formation with relevant and irrelevant Epilepsy Data's from different Epilepsy Data datasets.

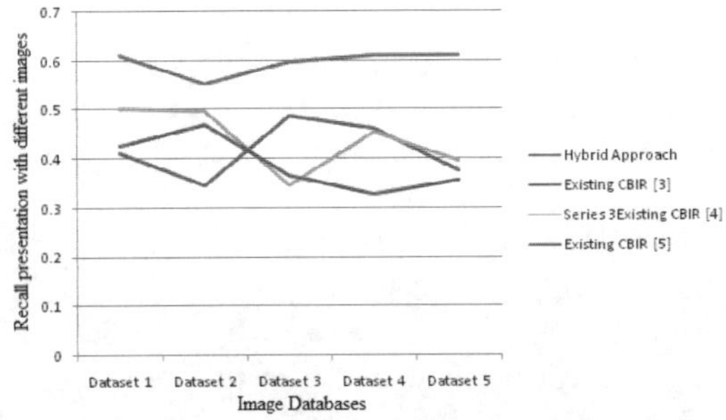

Figure 5. Epilepsy Data retrieval results with different attributes.

Figure 5 shows effective Epilepsy Data retrieval formations with different query Epileptic Data's from different Epilepsy Data datasets. In the all five different dataset recovery were done by querying five different pictures and for every dataset we get perfection, remember and F-measure principles is identified. The perfection and remember value evaluation between the suggested comprehensive methods functions removal centered on multi-objective marketing techniques demonstrates the efficiency of the suggested system is best over that of the current systems.

V. Conclusion

In this paper, a programmed framework for epilepsy identification has been proposed, which manages paired identification issues (epileptic versus non-epileptic or seizure versus non-seizure) and ternary recognition issue (ictal versus typical versus interracial). The proposed framework depends on profound realizing, which best is in class ML approach. The proposed framework gives extraordinary execution with less information and less parameters. It will help nervous system specialists in recognizing epilepsy, and will enormously decrease their weight and increment their proficiency. In practically every one of the cases concerning epilepsy recognition, the proposed framework gives an exactness of 99.1±0.9% on the University of Bonn dataset. The framework can be helpful for other comparative arrangement issues dependent on EEG mind signals. At present, the epilepsy discovery techniques recognize seizures after their event. In future, we will examine its convenience for identifying seizures preceding their event, which is a testing issue.

References

[1] Ihsan Ullah, Muhammad Hussain, Emad-ul-Haq Qazi, "An Automated System for Epilepsy Detection using EEG Brain Signalsbased on Deep Learning Approach", Paper presented at the Signal-Image Technology & Internet-Based Systems (SITIS), 2016.

[2] Acharya, U. R., Molinari, F., Sree, S. V., Chattopadhyay, S., Ng, K.-H., & Suri, J. S. (2012). Automated diagnosis of epileptic EEG using entropies. Bio Signal Processing and Control, 7(4), 401- 408.

[3] Cui, Z., Chen, W., & Chen, Y. (2016). Multi-scale convolutional neural networks for time series classification. arXiv preprint arXiv:1603.06995.

[4] Ince, T., Kiranyaz, S., Eren, L., Askar, M., & Gabbouj, M. (2016). Real-time motor fault detection by 1-D convolutional neural networks. IEEE Transactions on Industrial Electronics, 63(11), 7067-7075.

[5] Ioffe, S., & Szegedy, C. (2015). Batch normalization: Accelerating deep network training by reducing internal covariate shift. Paper presented at the International Conference on Machine Learning.

[6] Ji, S., Xu, W., Yang, M., & Yu, K. (2013). 3D convolutional neural networks for human action recognition. IEEE transactions on pattern analysis and machine intelligence, 35(1), 221-231.

[7] Kingma, D., & Ba, J. (2014). Adam: A method for stochastic optimization. arXiv preprint arXiv:1412.6980.

[8] Koffler, D., & Gotman, J. (1985). Automatic detection of spike-and-wave bursts in ambulatory EEG recordings. Electroencephalography and clinical neurophysiology, 61(2), 165-180.

[9] Zhang, T., Chen, W., & Li, M. (2017). AR based quadratic feature extraction in the VMD domain for the automated seizure detection of EEG using random forest classifier. Bio Signal Processing and Control, 31, 550-559.
[10] Tran, D., Bourdev, L., Fergus, R., Torresani, L., & Paluri, M. (2015). Learning spatiotemporal features with 3d convolutional networks. Paper presented at the Proceedings of the IEEE international conference on computer vision.
[11] Saab, M.E. and Gotman, J. (2005) A system to detect the onset of epileptic seizures in scalp EEG. Clinical Neurophysiology, 116, 427-442. doi:10.1016/j.clinph.2004.08.004
[12] Indiradevi, K.P., Elias, E., Sathidevi, P.S., Dinesh, S. and Radhakrishnan, K. (2008) A multi-level wavelet approach for automatic detection of epileptic spikes in the electroencephalogram. Computers in Biology and Medicine, 38, 805-816. doi:10.1016/j.compbiomed.2008.04.010
[13] Übeyli, E.D. (2009) Automatic detection of electroencephalographic changes using adaptive neuro-fuzzy inference system employing Lyapunov exponents. Expert Systems with Applications, 36, 9031-9038. doi:10.1016/j.eswa.2008.12.019
[14] Ghosh-Dastidar, S., Adeli, H. and Dadmehr, N. (2007) Mixed-band wavelet-chaos-neural network methodology for epilepsy and epileptic seizure detection. IEEE Transactions on Bio Engineering, 54, 1545-1551. doi:10.1109/TBME.2007.891945
[15] Kannathal, N., Choo, M.L., Rajendra, A.U. and Sadasivan, P.K. (2005) Entropies for detection of epilepsy in EEG. Computer Method and Programs in Biomedicine, 80, 187-194. doi:10.1016/j.cmpb.2005.06.012
[16] Ocak, H. (2008) Optimal classification of epileptic seizures in EEG using wavelet analysis and genetic algorithm. Signal Processing, 88, 1858-1867. doi:10.1016/j.sigpro.2008.01.026
[17] Srinivasan, V., Eswaran, C. and Sriraam, N. (2007) Approximate entropy-based epileptic EEG detection using artificial neural networks. IEEE Transactions on Information Technology in Biomedicine, 11, 288-295.
[18] T.Rajesh kumar, K.Geetha, R.Satheesh, S.Barkath Nisha, MRI Brain Image Segmentation using Fuzzy C Means Cluster Algorithm for Tumor Area Measurement, International Journal of Engineering Technology Science and Research, ISSN 2394 – 3386, Volume 4, Issue 9.
[19] V. Srinivasan, C. Eswaran, and A. N. Sriraam, "Artificial neural network based epileptic detection using time - domain and frequency-domain features", Journal of System, vol. 29, no. 6, (2005), pp. 647– 660.
[20] V. Srinivasan, C. Eswaran, and N. Sriraam, "Approximate entropy-based epileptic EEG detection using artificial neural networks," Transactions on Information Technology in Biomedicine, vol. 11, no. 3, (2007), pp. 288–295.
[21] A. S. Zandi, G. a Dumont, M. Javidan, R. Tafreshi, B. a MacLeod, C. R. Ries, and E. Puil, "A novel wavelet-based index to detect epileptic seizures using scalp EEG signals", Conf. Proc. Engineering in Medicine and Biology Society, vol. 2008, no. 2, (2008), pp. 919–922.
[22] G. Chen, "Automatic EEG seizure detection using dual-tree complex wavelet-Fourier features", Expert System Application, vol. 41, no. 5, (2014), pp. 2391–2394.
[23] A. Shoeb and J. Guttag, "Application of Machine Learning To Epileptic Seizure Detection", Proceedings of the 27th International Conference on Machine Learning , (2010) pp. 975–982.
[24] P. Bhatia and A. Sharma, "Different Techniques for Extracting Brain Signals for Human Machine Interface , a Review", Australian Journal of Information Technology and Communication, vol. 2, no. 2, (2015), pp. 31–34.
[25] S. Garg and R. Narvey, "Denoising and Feature Extraction of EEG Signal Using Wavelet Transform", International Journal of Engineering, Science & Technology, vol. 5, (2013), pp. 1249-1253.
[26] G. Kaushik, H. P. Sinha, and L. Dewan, "Bio Signals Analysis by Dwt Signal Denoising with Neural Networks", vol. 3, no. 1, (2013), pp. 1–18.
[27] I. Omerhodzic, S. Avdakovic, A. Nuhanovic, K. Dizdarevic, " Energy distribution of EEG signals: EEG signal wavelet-neural network classifier", arXiv preprint arXiv:1307.7897, 2013.

An Effective Analytics using Machine Learning Integrated Approaches for Diagnosis, Severity Estimation andPrediction of Heart Disease

N Satyanandam[1(a)], **Ch Satyanarayana**[2],

[1] Department of CSE, Bhoj Reddy Engineering College for Women, Hyderabad, Telangana, India;

[2] Department of CSE, JNTUK University College of Engineering, Kakinada, Andhra Pradesh, India;

[(a)]satya2606@gmail.com

Abstract. Heart Disease is one of the precarious issues in the medial domain whereby the number of cases is huge in the global scenario. The instances of heart transplants increase a lot every year even in the developed countries. Heart disorder identifies a number of heart disease disorders with the cardiac disorders including illnesses of the blood stream, coronary artery disease, cardiovascular rhythms and related cardiac defects. Cardiovascular diseasing commonly implies disorders requiring restricted or blocked blood pathways that can lead to cardiac attack, chest pressure (angina) or a stroke. Some cardiac disorders including key causes of heart disease are often regarded as influencing the cardiac muscle, valves and rhythms. Many cases of cardiac illness can be reversed or managed with a safe lifestyle. Heart Failure (HF) is not a disorder but a complex health illness. The rise in costs in health care, the increasing occurrence, the decreased quality of life, frequent hospitalizations and early mortality have turned HF into a global and Indian crisis and have emphasized the need to diagnose HF and determine its magnitude and efficacy. The principal goal of this research manuscript is to find a machine-based learning approach to diagnoses, intensity predictions and heart disease prediction. In this manuscript, we present the regression as well as ensemble learning based analysis on the benchmark dataset of Cleveland from UCI repository for the estimation of cardiac diseases. A further aspect is that 303 records are scheduled for this review. The knowledge from Cleveland consists of approximately 13 features and we have intended to group it in five groups in this report. The presented outcomes on the assorted algorithmic approaches are quite effective and ensemble-based approach is quite performance aware.

1. Introduction

In recent days, ICT applications are not confined to the sharing of data, cloud activity, social media, web servers and mobile applications. Informatics has influenced every field of social and corporate sector since last decade, including the health and the medical sciences [1]. Most health laboratories today have specialized computer-based instruments for the effective detection and selection of parameters for the human body, such as Magnetic Resonance Imaging (MRI), CT, Electroencephalography (EEG), ECG, ultrasound, mammography, Laparoscopy, X-Ray, and several more. It is having greater specificity in examining the human body that assists the surgeon or specialist in detecting the disease. Such approaches are used by the physicians to prescribe adequate care for the patients [2].

The software approaches and libraries are not only implemented in the vast testing devices of medical study but to test biological data obtained from diagnostic machines by computational approaches for multiple diseases [3][27]. The definition of bioinformatics is used in this sense to explain biological and medical data using analytical techniques and applications. These software suites use high-performance programming languages to process and analyze the biological data collection with the aims of discovering the conditions of the human body for successful implementations [4].

The high incidence of cardiac failure in adults is around 2%, and in patients older than 75 years of age it varies over 8%) [25]. Heart disease events in hospitals are very regular and 3–5 percent of hospital admissions are mostly focused on cases of heart disease [31]. In terms of clinical experience, it is a heart disease that triggers most hospital admissions [30]. The treatment for cardiac failure is substantially high and cannot be afforded by ordinary civilians and induces about 2% of overall health costs in many developing countries [27]. If we want to develop an efficient disease control plan, we need to carry out research on a wide spectrum of results, including disease diagnosis as soon as possible, severity-based evaluation and efficient prediction of adverse events[5][32].

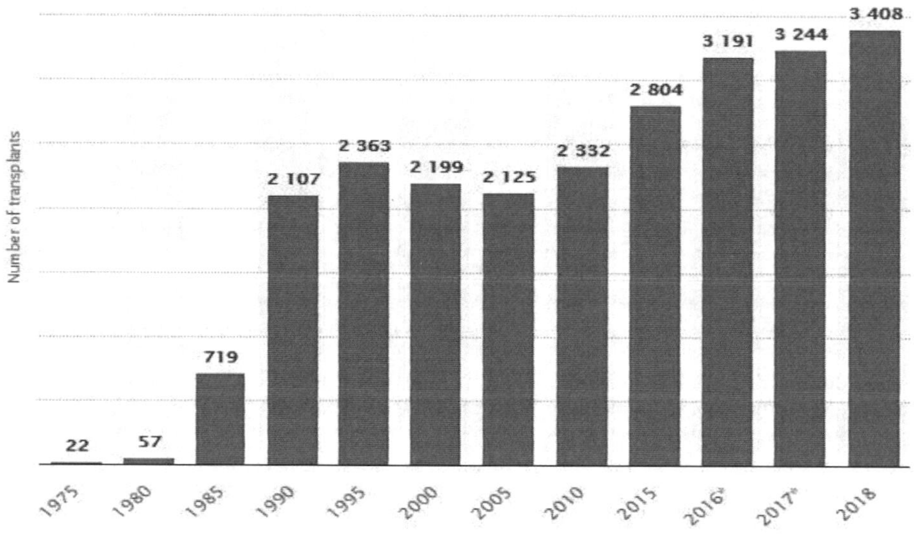

Figure 1 : Heart Transplants in USA (1975-2018 Source : Statista)

Artificial intelligence (AI) is the key domain of research that is implemented to replicate cognitive processes for human beings. It introduces a paradigm change into the health industry, assisted by improved access to health data and rapid advances in analytics [6]. Two big groups are primarily AI. Machine learning (ML) techniques that examine organized data such as graphic, genetic and electrophysiological data are part of the first group[24]. ML attempts to analyze the characteristics of the patient in medical procedures or predict the risk of the results of the illness. The second group involves Natural Language Processing (NLP) for the incorporation and refinement of organized medical data from unstructured data, such as medical notes and journals. The NLP procedures are aimed at translating text to structured data which are readable from computer and can be analyzed using ML techniques[7][29].

Machine learning should be viewed as an artificial intelligence subset involving the design and creation of algorithms, and functions like a container. These designing algorithms will improve the subject's actions on the basis of empirical evidence. Examples of observational results are used to demonstrate the interactions between variables identified. The emphasis of master learning is on automated learning and pattern recognition of the given examples and on making machines smart judgments or decisions[8].

In classification based predictive analysis [28], a sample is categorized in this protocol as one of many pre-defined groups, according to the methodology used to undertake testing of machine learning. The regression requires several computer tools and technique, which can be used successfully in the regression and classification process. For inputs and a single output variable to draw the final product, they use different input variables. Such examples of regression and classification activities are, Classification and Regression Vector Machines (SVM)[27], Naive classification bays, K-Nearest Neighbors (KNNs), Classification and Regression Tree (CART), Adaptive Regression Multivariate Splines (MARSplines) etc. [9].

2. Related Work and Review of Literature

New developments in science are structured to rely on a more precise and focused analysis of prediction models to assess the magnitude of cardiovascular disease based on clinical outcomes and computational methods [10].

The findings of Liu Q et al., work [11] suggested a conventional paradigm for Chinese medical practice based on a Bayesian approach for computer diagnostics. Study was also carried out in order to develop expert programmed for forecasting and diagnosing cardiac diseases in the context of general programming [12]. Assanelli et.al, [13] has recommended the use of ECG data for cardiac diseaseprediction.

In 1999 Carven [14] employed techniques such as text classification, together with the Naive Bayes algorithm classification and relational algorithms further. In 2001, Craven used the Hidden Markov models that are similar to the 2004 model of Rosario and Hearst, which primarily concentrated on object identification. In the sense of discrimination between two variables, namely diseases and medicines, Srinivasan et.al [15] used a history approach [16]. A lot of work has been performed on construction codes in order to extractposition.

In 2002 Feldman et al. used a rule-based method to isolate and explain interactions dependent on DNA, proteins, medications and diseases. In order to consider a number of biological and molecular relations, Friedman et al. intensify the construction of a rules-based structure by handcrafting a semantic grammar and collection of semanticconstraints [32][17]. Regulatory methods were commonly used in biomedical literature for the resolution of interaction extraction tasks. Either syntactic (POS) and syntactic (similarly) structures; or semantic knowledge in form of set patterns containing terms that cause a certain interaction is a key source of information used in this method

One of the disadvantages of using rule-based approaches is that more specialist commitment is needed than data-driven methodologies. The better regulatory systems are those with rules manually or semi-automatically constructed, manually refined and automatically removed. The fact that they yield decent outcomes for accuracy, though reclamation levels are also poor is a successful feature of rules-based systems. Syntactic rule-driven extraction schemes are dynamic frameworks driven on tools used for the allocation of POS tags or the extraction of syntactic parses. It is understood that such resources are not yet up to date in the biomedical literature, as they are for general English texts, so their success on phrases is not necessarily the greatest.

Polat et al. [18] suggested this definition and an AIRS based on blurry weighted pre- processing, which displays 92.59 per cent precision. The algorithms of SMO, Bagging, ANN and the use of Z-Alizadeh Sani data were suggested by Roohallah et.al [19], with 303 patients with 54 features. This analysis then showed that "Chest Pain and Age" had the greatest efficiency effect as the remaining characteristics and their reported accuracy was as high as 94.08%. ANN-ANFIS contrasting with MATLAB applications is dealt with in Abushariah et.al. [20].

The accuracy of the ANFIS 100% and ANN 90.74% training results have been determined. However, the precision of the experimental results has been obtained for ANN 87.04% and ANFIS 75.93%. Negar Ziasabounchi et.al [21] addressed ANFIS technique's precision. The UCI data with seven features is treated as an input in their analysis. The analysis findings indicate 92.30 percent accuracy. The accuracy obtained was 72.5% by five classes, but when analyzing it for two classes, which include patients and stable persons, the accuracy achieved was 90.57%.

3. Methodology andImplementation

Inferring the Cleveland data set can be described as a classification issue where cardiac disease data have been used. This problem is known to be a regression problem because we take any data part as vector, of which the time-series data are obtained. In fact, we hope that all components can be modified, so it can be considered a problem with regression in this experiment [23]. Traditional approach uses a node's regression predictor and takes little account of the association between time-nodes. Even with the exception of the first time we added goal to the feature set to boost predictionaccuracy.

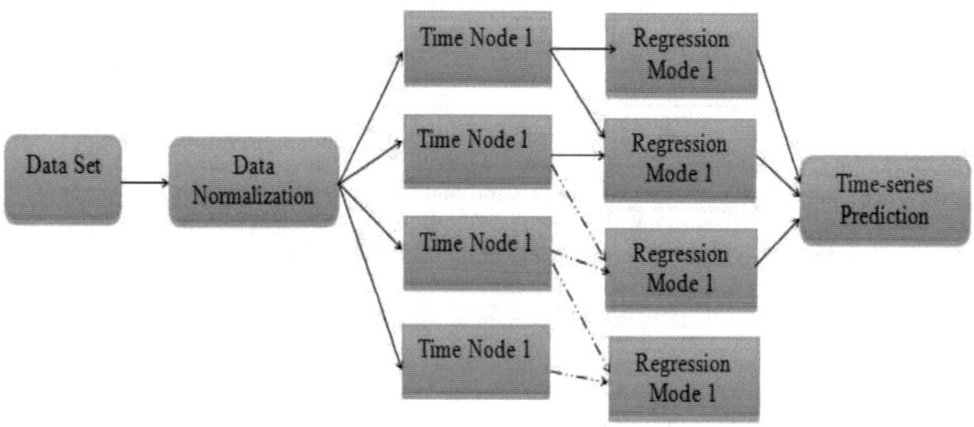

Figure 2: Proposed System Architecture

3.1 Polynomial Regression Based onMulti-Target

The role of multi-target polynomial regression is to induce data from the given polynomial equation, and it is capable of predicting the value of the multiple variables considered. As a simple platform for polynomial regression, based on multiple objectives, the principle of general linear model is used. The following is described in this strategy:

$$P = \beta_0 + \sum_{i=1}^{m} \beta_i \cdot T_i$$

$T_i = \prod_{j-1}^{n} X_j^{u_{i,j}}$ is denoted as terms..

In the above equation β_i, $i = 1, \cdots, m$ and β_0 are the constant vectors and they are no constants. The number of targets that has to be predicted is based upon the dimensions of these three vectors.

An example for this is given below,

$$P = (2,5,7) \cdot X_1^2 X_2 + (1,3,5) \cdot X_1 X_2^3 + (3,2,7).$$

IOP Conf. Series: Materials Science and Engineering 1074 (2021) 012006 doi:10.1088/1757-899X/1074/1/012006

This equation model is used in predicting the three targets brought under consideration and this is equivalent to the three single target polynomial equations given below,

i.e., P= (P_1, P_2, P_3),

where

$$P_1 = 2 \cdot X_1^2 X_2 + 1 \cdot X_1 X_2^3 + 3$$

$$P_2 = 5 \cdot X_1^2 X_2 + 3 \cdot X_1 X_2^3 + 2$$

$$P_3 = 7 \cdot X_1^2 X_2 + 5 \cdot X_1 X_2^3 + 7$$

Though the three equations P_1, P_2, and P_3 have same structure they have different coefficients for each target variable.

3.2 AlgorithmicApproach

Step 1: Input: Tr (x, y) and Ts procedure evaluation

Step 2: datasets Exit: Fault test E0

Step 3: Use regression algorithm to train the model L to the training set Tr with the initial node and measure the trainingerror.

Step 4: Move 2: Apply to the feature set the estimation value of the previous goal and use the regression algorithm to forecast the new aim and compute training errors.

Step 5: Iteratively execute step 2 before the final goal is specified. Combine and execute the test results with all the parameters.

Step 6: Close

The study provided statistics on Cleveland cardiac disease provides datasets at the following (https://archive.ics.uci.edu/ml/datasets/heart+disease).

This including 303 reports of 13 characters, as shown in Table. The results separated into 5 groups indicating loss of cardiac disease, and Grades 1 to 4 reflect a rise of heart disease incidence respectively. The features of these data are:

Table 1. Benchmark Dataset of Heart Disease for Prediction

Data Set Characteristics	Multivirat0065	Number od instances	303	Area	Life
Attribute Characteristics	Categorical, Integer, Real	Number of Attributes	75	Date Donated	01-07-1988
Associated Tasks	Classification	Missing Vales	Yes	Number of Web Hits	1309737

3.3 Benchmark Dataset for Analytics: Cleveland Heart Disease Dataset

Table 2 : Cleveland Heart Disease Dataset

Attribute	Description	Domain of Value
Age	Age in years	29 to 77
Sex	Sex	Male (1) Female (0)
Cp	Chest Pain type	TypicalAngina(1) Atypical Angina(2) Non-angina(3) Asymptomatic(4)
Trestbps	Resting blood sugar	94 to 200 mm Hg
Chol	Serum Cholesterol	126 to 564 mg/dl
Fbs	Fasting blood sugar	>120 mg/dl True (1) False(0)
Restecg	Resting ECG Result	Normal (0) ST-Twave Abnormality (1) LVhyper Trophy(2)
Thalach	Max heart Rate	71 to 202
Exang	Exercise include angina	Yes(1) No(0)
Old peak	ST depression induced by exercise relative to rest	0 to 6.2
Slope	slope of peak exercise ST segment	Up Sloping (1) Flat (2)
Ca	No.of. major vessels colored by fluoroscopy	Download sloping(3) 0-3
Thal	Defect type	Normal(3) Fixed Defect (6) Reversible defect (7)
Num	Heart Disease	0-4

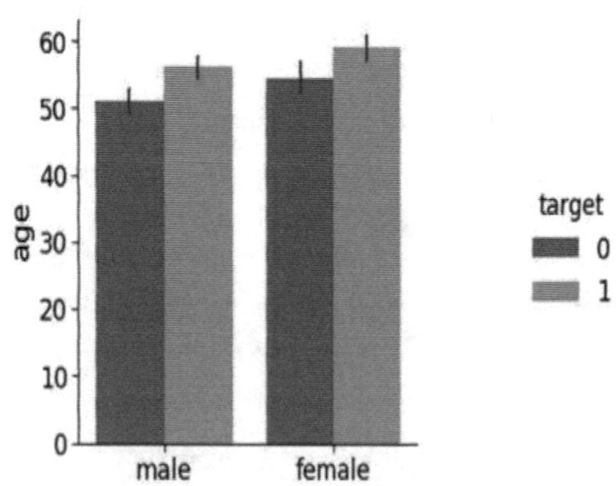

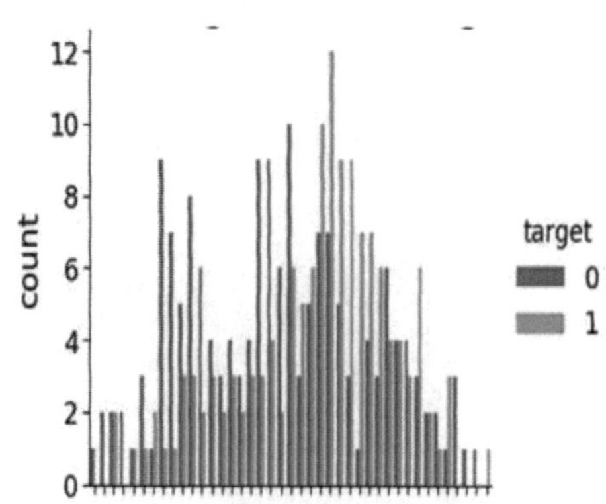

Figure 3: Distribution Patterns of Variation in Cleveland Benchmark Dataset

4. Experimental Results and EvaluationMetric

The estimation of results from the polynomial regression multi-target model results in an optimized and functional model. The classification methods are determined in conjunction with the following 1 to 4 formula with multiple Criteria, such as specificity, sensitivity, precision and accuracy to test. To calculate the indexes, we will use the uncertainty matrix. This matrix is a valuable tool to evaluate classification system output in the identification of data or evaluation in different groups. Ideally, the main diagonal of the matrix needs to contain a significant number of observe results, with the remaining matrix values being zero or equivalent tozero.

PP = Truly Classified Positive data labels
NF = Falsely Classified Negative data labels
PF = Falsely Classified positive data labels
NP = Truly Classified negative data labels

$$Specificity = NP /NP+NF \quad (1)$$
$$Sensitivity = PP /PP+PF \quad (2)$$
$$Precision = PP/PP+NF \quad (3)$$
$$Accuracy = PP+NP /PP+NP+NF +PF \quad (4)$$

Table shows the values of four indices for each class labels. The values have been calculated for C5.0 algorithm using the confusion matrix. The error rate or misclassification rate can also be calculated based on accuracy (formula 5) index.

$$Error\ Rate = 1\text{–}accuracy \quad (5)$$

Table 3: Evaluation Analysis

Category (Class) Label	Accuracy (%)	Specificity (%)	Precision (%)	Sensitivity (%)
0	79	65	77	72
1	78	67	76	75
2	78	68	78	76
3	77	66	78	78
4	78	68	79	77

The plot shows about the different categories specificity, sensitivity, precision and accuracy

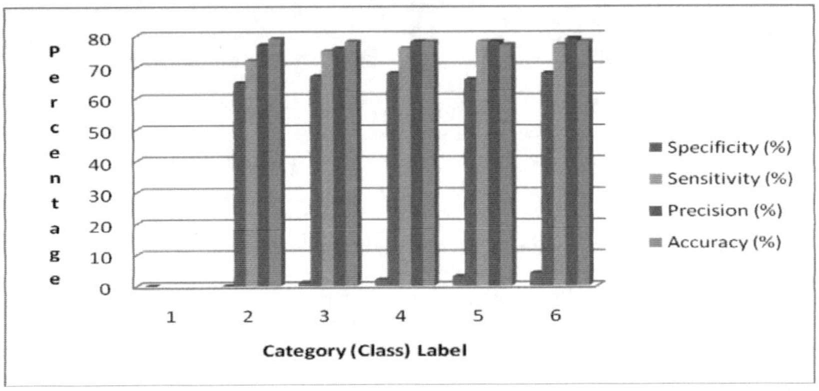

Figure 4: Assorted Patterns on Sensitivity, Specificity, Precision and Accuracy

IOP Conf. Series: Materials Science and Engineering 1074 (2021) 012006 doi:10.1088/1757-899X/1074/1/012006

The table shows the overall comparison of existing system with proposed system. AUC (Area Under Curve)-ROC is a performance metric for classification problems based on the different Threshold values. ROC is a likelihood curve and AUC measure the separability. In simple terms, we will learn from AUC-ROC metric how model can differentiate between classes. More AUC, better model, the better model is generated with the outputs

Table 4: Overall comparison of proposed system with existing system

Different algorithms	**Performance in Percentage**
Projected algorithm	92
Naïve Bayes Approach	90
Neural Network Based Implementation	91
CART Algorithm	83
ID3 Approach	65
J48 Approach	84

The cumulative performance level of proposed implementation with existing system is given

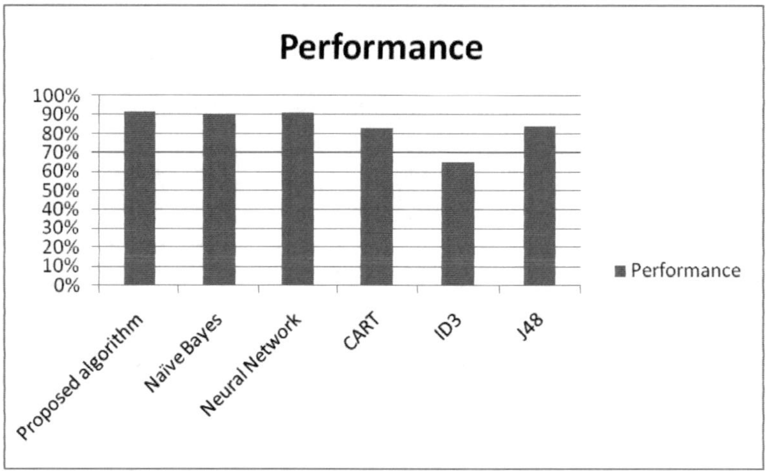

Figure 5: Overall performance

4.1 Integration with Assorted Ensemble Based for EffectualLearning

Different models of machine learning are used to build a hybrid ensemble learning models as weak learners in this role. These models may be Logistic Regression Model, Decision Tree, Support Vector Machine, K-Nearest Neighbor Model, Naive Bayes Model or other. The term hybrid is used here since a homogenous set of weak students is used in other ensemble models, but a heterogeneous selection of weak learners is used for thisassignment.

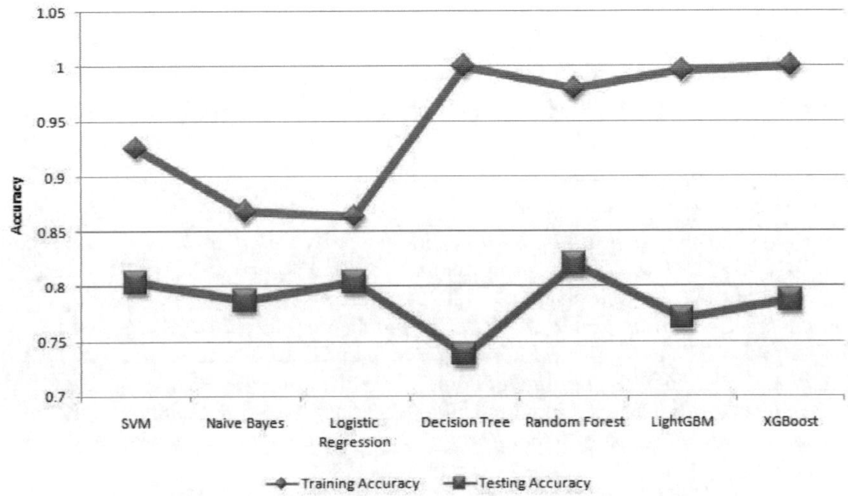

Figure :6 Evaluation process of Accuracy

Table 5: Models of Accuracy in Training and Testing

Models	**Training Accuracy**	**Testing Accuracy**
SVM	0.9256198347107438	0.8032786885245902
Naive Bayes	0.8677685950413223	0.7868852459016393
Logistic Regression	0.8636363636363636	0.8032786885245902
Decision Tree	1.0	0.7377049180327869
Random Forest	0.9793388429752066	0.819672131147541
LightGBM	0.9958677685950413	0.7704918032786885
XGBoos	1.0	0.7868852459016393

5. Conclusion

Bioinformatics based data analytics include the biological data's mathematical representations and are mostly used to provide statistical information through computational methods. Machine learning is a prominent domain of research for the bioinformatics and related segments which involves the development of algorithms to improve or create the actions involved in a given task by integrating new data in it. In the last few decades, the very large number of genetic sequence data generates large data banks which defy human researchers' ability to analyze and process these things effectively and without using computer methods. In addition to the ever more advanced application, other areas in medicine and biology are rapidly protected by machine learning.

For example, images from neuro-imaging methods such as CT and MRI are now being studied by machine-learning so that scientists can gain insight into the signs and characteristics of early diseases. This is especially helpful for brain and heart disease, as systems can dig through several thousands of findings in order to locate commonalities. Thus, the work of polynomial regression, i.e. the use of data in polynomial regression models, is discussed in this manuscript along with the tree-based approaches in association of ensemble learning. Polynomial structures have been commonly used in the past, but many people thought about it until the endorsement of machine learning. The presented work is having the integration patterns of assorted approaches on the base of ensemble learning so that the higher degree of accuracy can beachieved.

Acknowledgments

This research paper on studies An Effective Analytics using M L Integratedapproaches for Diagnosis, Severity Estimation and Prediction of Heart Disease is been supported by JNTUK College of Engineering, Kakinada, Andhra Pradesh, India.

References

[1] Rana, H. K., Akhtar, M. R., Islam, M. B., Ahmed, M. B., Lió, P., Huq, F., & Moni, M. A. (2020). *Machine Learning and Bioinformatics Models to identify Pathways that Mediate Influences of Welding Fumes on Cancer progression. ScientificReports, 10(1), 1-15.*

[2] Wood, A., Najarian, K., & Kahrobaei, D. (2020). *Homomorphic Encryption for Machine Learning in Medicine and Bioinformatics. ACM Computing Surveys (CSUR), 53(4),1-35.*

[3] Li, H., Tian, S., Li, Y., Fang, Q., Tan, R., Pan, Y., & Gao, X. (2020). *Modern Deep Learning inBioinformatics. Journal of molecular cellbiology.*

[4] Pikula, K., Zakharenko, A., Chaika, V., Kirichenko, K., Tsatsakis, A., & Golokhvast, K. (2020). *Risk assessments in nanotoxicology: Bioinformatics and computational approaches. Current Opinion inToxicology, 19, 1-6.*

[5] Nowbar, A. N., Gitto, M., Howard, J. P., Francis, D. P., & Al-Lamee, R. (2019). *Mortality from ischemic heart disease: Analysis of data from the World Health Organization and coronary artery disease risk factors From NCD Risk Factor Collaboration. Circulation: Cardiovascular Quality and Outcomes, 12(6),e005375.*

[6] Chicco, D., Heider, D., & Facchiano, A. (2020). *Artificial Intelligence Bioinformatics: Development and Application of Tools for Omics and Inter-Omics Studies. Frontiers in Genetics,11.*

[7] Chen, C., Hou, J., Tanner, J. J., & Cheng, J. (2020). *Bioinformatics methods for mass spectrometry-based proteomics data analysis. International Journal of Molecular Sciences, 21(8),2873.*

[8] Parvizpour, S., Pourseif, M. M., Razmara, J., Rafi, M. A., & Omidi, Y. (2020). *Epitope-based vaccine design: a comprehensive overview of bioinformatics approaches. Drug DiscoveryToday.*

[9] Liu, G., Zeng, H., Mueller, J., Carter, B., Wang, Z., Schilz, J., & Gifford, D. K. (2020). *Antibody complementarity determining region design using high-capacity machine learning.Bioinformatics, 36(7), 2126-2133.*

[10] I. H. Witten, E. Frank. (2005). *"Data Mining: Practical machine learning tools and techniques". 2nd Edition, Morgan Kaufmann, SanFrancisco.*

[11]Luo Y, Liu Q, Du CF, Long HX, Wang F, Zhu W, Zhou M, Xi J, Liu SD, Wen Y. (2014). *"Quality assessment for Chinese Meta-analysis on prevention and treatment of coronary heart disease". ChineseCirculation Journal. Dec, 29 (12): pp.979-982*

[12] J. Han and M. Kamber. (2001). *"Data Mining: Concepts and Techniques". San Francisco, MorganKauffmannPublishers*

[13] Deodato Assanelli, L. Cazzamalli, M. Stambini, M.L. Poeta. (1993). "*Correct diagnosis of chest pain by an integrated expert system", IEEE Xplore Conference: Computers inCardiology.*

[14]Jennifer G. Dy. (2004). *"Feature Selection for Unsupervised Learning", Journal of Machine Learning, pp.845-889.*

[15] Srinivasan, P., Rindflesch, T. (2002). "*Exploring text mining from Medline". In: Proceedingsof the AMIA Symposium.*

[16] M.A.Jayaram, Asha Gowda Karegowda. (2007). *"Integrating Decision Tree and ANN for Categorizationof Diabetics Data". International Conference on Computer Aided Engineering, December 13-15, 2007, IITMadras, Chennai,India.*

[17]Mark A. Hall, *"Correlation-based Feature Selection for Machine Learning", Department of Computer science, University of Waikato.*http://www.cs.waikato.ac.nz/ mhall/thesis.pdf.

[18]K. Polat and Salih Güneş, (2007). *"A hybrid approach to medical decision support systems: Combining feature selection, fuzzy weighted preprocessingand AIRS", computer methods and programs in biomedicine, Vol. 88, No. 2, pp.164-174.*

[19]R.Alizadehsani, J. Habibi, M. J. Hosseini, H. Mashayekhi, R. Boghrati, A. Ghandeharioun, B. Bahadorian and Z. A. Sani., (2013), *"A data mining approach for diagnosis of coronary artery disease", Computer methods and programs in biomedicine, Vol. 111, No. 1, pp.52-61.*

[20]M. A. M. Abushariah, A. A. M. Alqudah, O. Y. Adwan and R. M. M. Yousef, (2014). "*Automatic Heart Disease Diagnosis System Based on Artificial Neural Network (ANN) and Adaptive Neuro-Fuzzy InferenceSystems (ANFIS) Approaches." Journal of Software Engineering and Applications, 7, No.12.*

[21] N. Ziasabounchi and I. Askerzade, (2014*)."ANFIS Based Classification Model for Heart Disease Prediction", International Journal of Electrical & CS IJECS-IJENS, Vol. 14, No. 02,pp.7-12.*

[22] S. Bhatia, P. Prakash and G.N. Pillai, (2008). *"SVM based decision support system for heart disease classification with integer-coded genetic algorithm to select critical features", In Proceedings of the WorldCongress on Engineering and Computer Science, WCECS, San Francisco, USA, pp.22-24.*

[23] P. Palanikumar, S. Geofrin Shirly, S.Balakrishnan, (2015). *"An Effective Two Way Classification of Breast Cancer Images", International Journal of Applied EngineeringResearch, ISSN 0973-4562, Volume 10,Number 21, pp42472-42475.*

[24]Masetic Z, Subasi A. *Congestive heart failure detection using random forest classifier. Comput Methods Programs Biomed2016;130:54–64.*

[25]Yang G, Ren Y, Pan Q, Ning G, Gong S, Cai G, (2010), "*A heart failure diagnosis model based on support vector machine". 2010 3rd international conference on biomedical engineering and informatics (BMEI), vol. 3; 2010. p. 1105–8.*

[26]Zheng Y., Guo X., Qin J., Xiao S. (2015), *"Computer-assisted diagnosis for chronic heart failure by the analysis of theircardiac reserve and heart sound characteristics." Comput Methods Programs Biomed. 2015; I22: pp.372–383.*

[27] Balakrishnan S, K.Aravind, A. Jebaraj Ratnakumar, *A Novel Approach For Tumor Image Set Classification Based On Multi-Manifold Deep Metric Learning", International Journal of Pure and Applied Mathematics, Vol. 119, No. 10c, 2018,pp.553-562.*

[28]A.Jebaraj Rathnakumar, S.Balakrishnan, *"Machine Learning based Grape Leaf Disease Detection", Jour of Adv Researchin Dynamical & Control Systems, Vol. 10, 08-Special Issue, 2018. Pp.775-780.*

[29] N Satyanandam, Ch Satyanarayana, *"Heart Disease Detection Using Predictive Optimization Techniques", International Journal of Image, Graphics and Signal Processing(IJIGSP), Vol.11, No.9, pp. 18-24, 2019.DOI:10.5815/ijigsp.2019.09.02*

[30]N Satyanandam, Dr. Ch Satyanarayana, *A New Multilayer Perceptron Model to Detect Heart Disease Severity, International Journal of Scientific & Engineering Research, Volume 7, Issue 12, December-2016 ISSN2229-5518*

[31]N Satyanandam, Dr. Ch Satyanarayana, *Detection of Heart Disease Severity using A Novel Multilayer Perceptron Model: Validation through Major Datasets, Advances in Fuzzy Mathematics. ISSN 0973-533X Volume 12, Number 3 (2017), pp. 333-345 © Research IndiaPublications*

IOP Conf. Series: Materials Science and Engineering 1074 (2021) 012007 doi:10.1088/1757-899X/1074/1/012007

Sentiment Analysis using Neural Network and LSTM

Akana Chandra Mouli Venkata Srinivas[1,5], Ch.Satyanarayana[2,6], Ch.Divakar[3,7], Katikireddy Phani Sirisha[4,8]

[1]Professor & Head, Department of Computer Science, AMC Engineering College, Bengaluru, India
[2]Professor, Department of CSE, & Registrar, JNTUK, Kakinada, AP, India
[3]Professor, Department of CSE, SRKR Engineering College, Bhimavaram, AP,India
[4]Associate Professor, Department of Computer Science, AMC Engineering College, Bengaluru, India
[5] mouliac3@gmail.com
[6] chsatyanarayana@yahoo.com
[7] divakar_c@yahoo.com
[8] phanisirisha8@gmail.com

1.

Abstract. People put their opinions or views on various events happening in the society or world. Twitter is one of the best social networking sites where a huge amount of data generates on the daily basis. These data can be used to classify their tweets based on various sentiments attached to them. Numerous technologies are applied to analyse the sentiments of users. Sentiment analysis needs a very efficient method to manage long arrangement data and their drawn-out dependencies. In this paper, we have applied a deep learning technique to perform Twitter sentiment analysis. Simple Neural Network, Long Short-Term Memory (LSTM), and Convolutional Neural Network (CNN) methods are applied for the sentiment analysis and their performances are evaluated. The LSTM is the best among all proposed techniques with the highest accuracy of 87%. We have collected a Twitter dataset from Kaggle to perform our experiment. The future improvement of the proposed research should include REST APIs and web crawling-based solutions to get live tweets to perform real-time analytics. We have analysed 1.6 million tweets in our research work.

2. Introduction

Sentimental analysis is the textual mining applied to extract useful subjective information to understand the social sentiments about the brand which help in business improvement or quality control. The opinion about the product utility varies from person to person. Customer feedbacks are very important in terms of brand monitoring and product reputation. It is used to find negative, positive, and neutral emotions about the subject. Twitter is an American social networking service on which users put their posts, retweet and read them. It is one of the most popular platforms to express our opinions or comment on ongoing issues from all sources.

These days social media, blogs and networks produce huge information. People share their thoughts on various events and issues. This bulk information is used for opinion and decision making in a business. Twitter generates massive short tweets and because of that, the tweet size is reduced to 140 characters. Opinion and Sentiment Mining are exceptionally helpful to get the opinion from the users or to analyse their sentiments. Our main objective is to classify sentiments into different categories.

Recurrent Neural Network (RNN) theoretically claimed to maintain the long-term dependencies but practically fall as addressed by Bengio et al. [2]. Information correlation over a long period needed distance calculation between nodes. The distance calculation is done based on multiple products of Jacobian Matrix. This prompts all the more normally vanishing gradient and less continuous exploding gradients. To overcome this issue Hochreiter et al. [1] presented Long Short-Term Memory networks typically called LSTMs. It has shown an unbelievable performance in NLP and Image Processing applications.

LSTM is a kind of RNN equipped for learning long term conditions. The RNN is the combination of repeating neural network modules. Figure 1 shows a single layer RNN that depicts how RNN collaborates to infer information from each module.

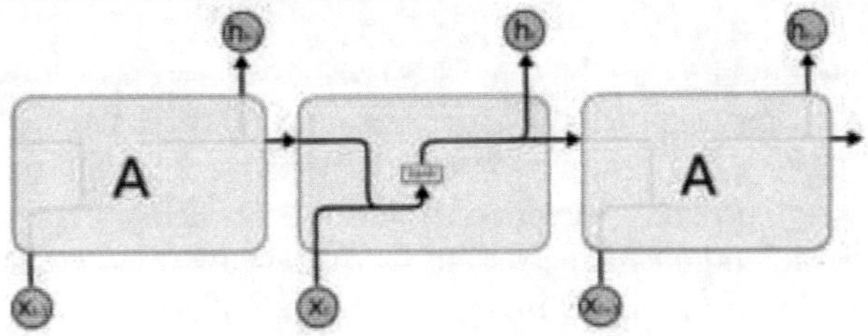

Figure 1: A single layer RNN

LSTM is a specialized RNN that has four network layers that interact uniquely. Equation (1) shows the outcome of the sigmoid function is.

$$f_t = \sigma(W_f . [h_{t-1}, x_t] + b_f) \quad (1)$$

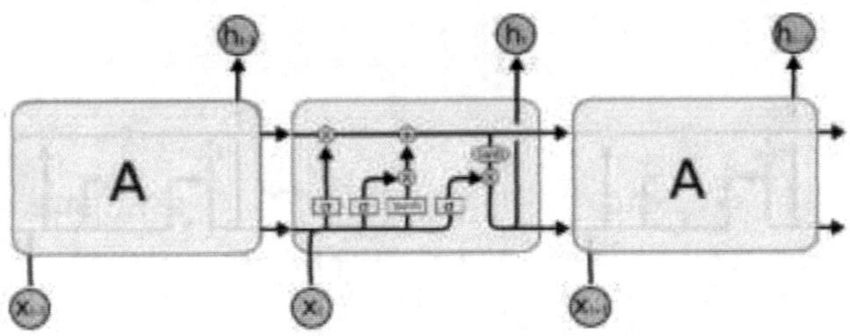

Figure 2: LSTM with interacting layers

Equation (2) represents the outcome of the cell state using the *tanh* function.

$$C_t = tanh(W_C . [h_{t-1}, x_t] + b_C) \quad (2)$$

These parameters are updated in each training step and stored. Mike Schuster et al. [3] portrayed a Bidirectional RNN that utilizes distinctive hidden layers, feed-forward to a similar output layer. The LSTMs have shown significant performance improvement in speech recognition, synthesis and entity recognition. The deep learning-based techniques can be applied to sentiment analysis [13].

In the research work we have used a LSTM and neural network-based approach to Tweeter sentiment analysis. We have taken Tweeter dataset for our experimental analysis.

3. Related work

This section deals with the related methods used for sentiment analysis. We have mainly concentrated on deep learning-based approach. Patel et. al. [4], have described a model using Recursive RNN to classify the movie reviews. The proposed system does sentiment analysis based on the review text.

Jan Deriu et. al. [6] have developed a Twitter sentiment classification using deep learning. In the proposed model they have used a Two-layer convolutional neural network. The main task is divided into 3 subtasks.

Anand et.al. [5] have posed a system to do the same task with CNN. In this approach, sentiment analysis is done using Natural Language Processing (NLP) as most text present on Twitter is in the form of emotions or opinions. The proposed model gives a comprehensive view of the emotions in terms of positive, negative, or neutral.

Fenna Miedema et al. [7] have proposed a system that used LSTM networks to develop a model that has to turn out to be most efficient among other approaches. Zainuddinet al. [8] have posed a sentiment classification approach using Support Vector Machines. This model lack accuracy in the case of long sentences obtained from a variety of resources.

Mathieu Cliché [9] has implemented a CNN and LSTM based approach that is trained on SemEval-2017 twitter dataset. This approach uses a large set of unlabelled dataset and pre-trained word embedding. This hybrid approach has shown better improvement in the accuracy level of the classification. The proposed approach included 5 steps – Reading CSV file (Twitter data), Pre-processing, Feature extraction, and classification. In the paper, the tasks are accomplished in two ways in the first approach. ML- based approach is applied to the dataset whereas in the second approach deep neural network-based methods are used.

The mode which is incorporating the LSTM approach shows improved performance as compared to other approaches. In the previous few years, the deep learning approaches have demonstrated critical improvement in the performance of NLP mechanisms. [10, 11-14].

4. Proposed system

In the present research we have used LSTM based approach to analyze the Twitter post. Our deep learning model is built using Keras Python libraries.

5. Methodology

4.1. Data Collection

Twitter sentiment dataset is collected from a Kaggle repository and analyzed for negative and positive sentiments. Figure 3 shows the negative sentiment in the red color bar whereas positive sentiments using a green color bar. Dataset is read using the panda library.

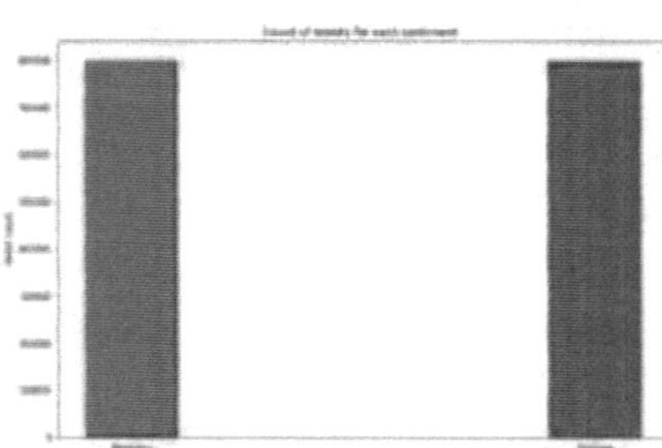

Figure 3: Sentiment classification

5.2. Data Pre-processing

The data preprocessing includes a set of operations to be performed on the dataset to extract sentiments. It cleans the tweets by removing stop words, spacing, or any special symbols. The step also lemmatizes and stems words. The processed data is further used for the model training and testing.

5.3. Training and testing with LSTM

The read dataset after cleaning is classified into training (80%) and test dataset (20%). The *Train_and_Test()* function is used to split the dataset. The trained model is tested further to check model efficiency.

6. **Experimental results**

6.1. Dataset Description

We have used 1.6 million tweets to form training and testing dataset. These datasets are cleaned for the further processing.

7. **Experimental setup**

The RNN based LSTM model is built using Keras and Tensorflow. Table 1 represents the network configuration parameters of the RNN-LSTM model.

Table 1: LSTM parameters

Parameters	Value
Batch-size	10
Epochs	128
Hidden layer-activation function	Relu
Output layer-activation function	Sigmoid
Optimizer learning-rate	Adam 0.001
Dropout	0.5
Loss-function	Binary cross-entropy

8. **Performance evaluation**

In this subsection, we are going to assess the performance of the applied approach and other related methods. Figure 4 shows the set of positive sentiments obtained from Twitter data.

Figure 4: Positive sentiments

Figure 5 shows the set of negative sentiments obtained from tweets.

Figure 5: Negative sentiments

The LSTM model classifies the sentiments with an accuracy of 85.04%. Figure 6 represents the accuracy and loss of our model in each iteration for 5 epochs.

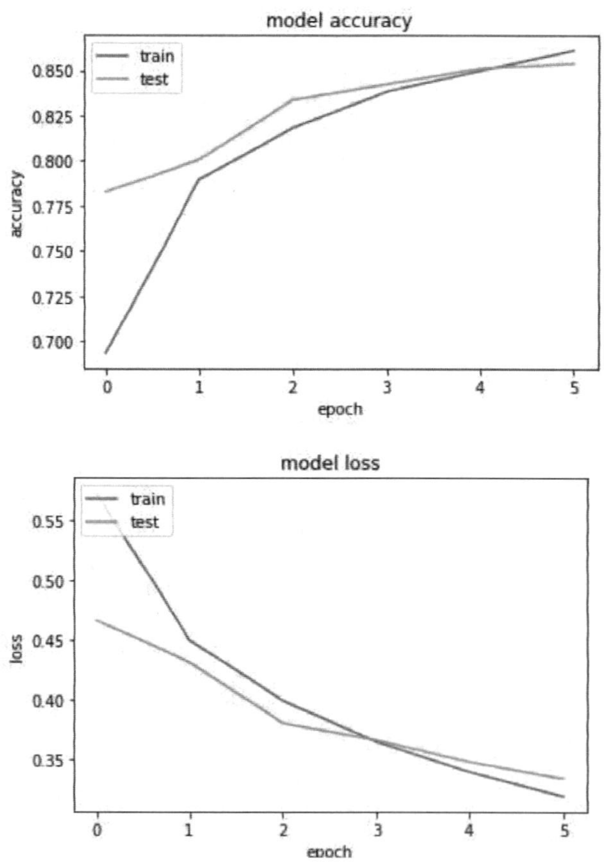

Figure 6. Model performance

The LSTM approach shows improved result compare to Convolutional Neural Network (CNN) and a simple neural network. Table 2 shows the performance comparison of the algorithms for 1.6 million tweets.

Table 2. Performance Matrix-I

Algorithm name	Training accuracy	Testing accuracy
LSTM	87.4	87.01
CNN	92	82
Simple Neural Network	85.3	74.7

Table 3 indicates the performance of the three different methods based on four parameters.

.

Table 3. Performance matrix-II

Methods	**Accuracy**	**Precision**	**Recall**	**F-measure**
LSTM	87	81	80	80
CNN	82	77	76	76
Simple Neural Network	81	76	75	74

9. Conclusion and future work

Our proposed research work addressed the need for RNN and LSTM for improved performance in Sentiment classification. In the paper, we have also addressed the different AI-based approaches and their performances. We have calculated the performance of the LSTM-RNN model. We have analysed 1.6 million tweets for our research to categorize them into Positive or negative sentiments.

The proposed approach has an 87% accuracy level which is best among all the approaches used in the current study. To conduct the experiment, we have collected a Twitter dataset from the Kaggle repository.

Although the proposed LSTM based approach is the best but to analyse real-time data and classify them into different emotions, we can use the REST API based web service approach. To solve scalability issues a big data platform can be chosen to analyse the huge number of tweets.

Acknowledgement

The author is thankful to all authors who have participated to conduct the experiment and research in the proposed work. We are thankful to AMC Engineering college for providing all resources and computing facilities to conduct this research.

References

[1] Hochreiter, Sepp & Schmidhuber and Jürgen Long Short-term Memory 1997 *Neural Computation* **9** 1735-80

[2] Bengio Y, Simard P and Frasconi P 1994 Learning long-term dependencies with gradient descent is difficult *IEEE Transactions on Neural Networks* **5** 2 157- 66

[3] Mike Schuster and Kuldip K 1997 Bidirectional recurrent neural networks *IEEE Transactions on Signal Processing* **45** 11 2673–2681

[4] Patel P, Patel D and Naik C 2021 Sentiment Analysis on Movie Review Using Deep Learning RNN Method

[5] Satapathy S, Zhang Y D, Bhateja V, Majhi R 2021 Intelligent Data Engineering and Analytics *J Advances in Intelligent Systems and Computing* **2** 1177

[6] Anand M, Sachin S 2017 *Proc. 5th Int. Conf. on Mining Intelligence and Knowledge Exploration: MIKE 2017* (Hyderabad, India) Sentiment Analysis of Tweets in Malayalam using Long Short-Term Memory Units and Convolutional Neural Nets

[7] Deriu, Jan & Cieliebak, Mark 2016 Sentiment Analysis using Convolutional Neural Networks with Multi-Task Training and Distant Supervision on Italian Tweets.

[8] Fenna Miedema and Sandjai Bhulai 2018 Sentiment Analysis with Long Short-Term Memory network Vrije Universiteit Amsterdam

[9] Zainuddin, Nurulhuda & Selamat, Ali 2014 Proc. Of 1st Int. Conf. on Computer, Communications, and Control Technology Sentiment analysis using Support Vector Machine pp 333-37

[10] Mathieu Cliché, *Proc. of the 11thInternational Workshop on Semantic Evaluations (SemEval-2017)* pp 573–80.

[11] Go A, Bhayani R and Huang L 2009 Twitter sentiment classification using distant supervision *CS224N Project Report Stanford* **1** 12

[12] Tomas Mikolov, Kai Chen, Greg Corrado and Jeffrey Dean 2013Efficient estimation of word representations in vector space arXiv preprint arXiv: 1301.3781.

[13] Danqi Chen and Christopher D Manning 2014 *Proceedings of the Empirical Methods in Natural Language Processing (EMNLP)* A fast and accurate dependency parser using neural networks pp 740–50

[14] Jeffrey Pennington, Richard Socher and Christopher Manning 2014 *Proceedings of the Empirical Methods in Natural Language Processing (EMNLP)* Glove: Global vectors forward representation pp 1532–43

Classification of Breast Cancer using Histology images: Handcrafted and Pre-Trained Features Based Approach

Jyoti Kundale[1]
Rajiv Gandhi Institute of Technology, India[1]
D.Y. Patil deemed to be University, Ramrao Adik Institute of Technology, India[1]
jyojadhav04@gmail.com[1]
Sudhir Dhage[2]
Sardar Patel Institute of Technology, India[2]
sudhir_dhage@spit.ac.in[2]

Abstract: Breast cancer has become a critical disease in women. The number of patients with breast cancer is quite high in India. It is of paramount importance to detect the disease in advance. Digital histopathology is one of the most advanced techniques for detection using machine learning. Artificial intelligence is going to be like a sunrise in the field of medicine. Deep neural networks have been successfully applied to the problem under consideration in the past. As, we know the feature extraction is one of the essential and crucial steps in case of classification. In this paper, we compare two approaches, first is feature extraction using traditional Handcrafted based and other is Transfer Learning based model (Pre-trained) for multiclass classification of Breast Cancer using Convolutional Neural Network (CNN) as a classifier. The models are trained using handcrafted features like Seeped Up Robust Features (SURF) and Dense Scale Invariant Feature Transform (DSIFT) techniques, later these extracted features are encoded by Locality Constrained Linear Coding method (LLC). In pre-trained model we have used VGG16, VGG19, ResNet50, GoogLeNet for feature extraction. The maximum accuracy for "SURF+CNN" is 92.88% for Handcrafted feature and in case of Pre-trained "GoogLeNet+ CNN" model gives 94%, both for 400X magnification factor.

Keywords: Histopathology Images, Locality Constrained Linear Coding, De-Novo Model, Pre-Trained Model, Handcrafted Feature, Convolutional Neural Network, Magnification Factor, Deep Learning

1. Introduction

Even though there have been outstanding advanced developments in the diagnosis and treatment of cancer, it has become a dangerous threat to human life. Cancer is a disease in which division of abnormal cell takes place in an uncontrollable fashion destroying other healthy body tissue finally resulting in a tumor. Breast cancer is a rampant and serious issue among females all over the world compared to the cancer of the lungs, brain or liver. Demographic aging and the adoption of bad daily routines by the people at large especially in the industrialized nations have become major contributing factors leading to an increase in the occurrence of disease-causing mortality rates to rise around the world by cancer. Breast cancer is leading in case of woman [1]. Around 9.6 million deaths occurred in 2018 due to cancer as reported by World Health Organization. In 2015, WHO reported 1.7 million (11.3%) victims of breast cancer and it is estimated that the number of new Breast cancer patients will grow by 70% in the next two decades. It is a well- known established fact that primary and exact diagnosis has a very significant role to play in the treatment thereby in- creasing the survival count of people suffering from Breast cancer by around 40% (World Health Organization) [2]. After the lung cancer Breast cancer is commonly found in now days in younger age group women. There are many reasons for Breast cancer like delay in child, alcohol consumption, hormonal changes, and obesity [3]. The foremost reason is being a woman and her day today life style. Now a day's women are very busy in her daily routine. She is managing her job as well as maintains social network also. She does not care about herself. In India, according to the International Agency for Research on Cancer, cancers of the lips, oral, lungs, neck and throat occur more in the males while females have to deal with breast, ovarian and cervical cancers. More females in the age group of 20 to 59 and in the age group of 59 plus die due to breast cancer. The trend is very disturbing as the number of patients is increasing in the younger age group more than in the older age group [4]. Hence there is a need for having a domain which will mainly focus on the early detection and cure of Breast Cancer among females. The main aim is to build systems that will detect and diagnose cancer an early stage so that proper treatment could be given to the patient. A complete cure of breast cancer is possible only if it is detected in an early stage. The rate of survival will improve by 95% if early detection of breast cancer is done for a patient. Thereby early detection can reduce the mortality rate automatically. Tumors in the breast are broadly divided into two types benign (i.e. ones that are not cancerous) and malignant (i.e. ones that are cancerous). Invasive and Non-Invasive also have subcategories which should be analyzed separately as every subcategory will need a unique course of cure. There are several ways for breast cancer detection like Ultrasound, Magnetic Resonance Imaging MRI, and Clinical examination, Mammograms, Biopsy and Histopathology [5]. Detection of cancer by the use of imaging is of prime significance because it can be commonly adapted and it is an effective and convenient way of detecting cancer when compared to other biopsy techniques such as Surgical (open) Biopsy (SOB), Vacuum Assisted Breast Biopsy (VABB), Core Needle Biopsy (CNB) and Fine Needle Aspiration (FNA) [6].

Breast Cancer for the female population has a huge bearing in the context of general public health; hence the urgent need to provide tools that will automatically classify Breast Cancer using digitized images from histological slides. High level expertise and more than one pathologist are required to give the final classification decision, which is a very tedious and time-consuming task. In the early stages of Breast cancer, misdiagnosis is a serious issue that could ultimately cost a precious human life.

There are various imaging modalities are present for detection of Breast cancer like Ultrasound, Magnetic Resonance Imaging (MRI), Digital Mammography (DM), Histopathology. Digital mammography images are efficient to store and Process, Cost effective and Works well with Computer Aided Device. Drawback of this Digital Mammography is it provides less spatial resolution and more preprocessing is required [7] [8]. MRI gives more details inner tissues and further it can be sent to biopsy. Disadvantage of MRI is not recommended to pregnant women and Image of MRI can be enhanced using some chemical agent which may cause allergy to patient [9]. Pros of Ultrasound imaging is not risky for pregnant women where images are captured in every angle in real time so reduces false negative. Patient is never exposed to radiation. Cons of Ultrasound Images are poor quality of images are not able to distinguished between Malignant and Benign tissue. Highly expertise are required to diagnose lump in Breast [8]. Images form Histology can be used as Whole Slide Imaging (WSI) or ROI (Region of Interest). Performance is better than other image modalities. Using histopathology, it is able to diagnose different types of cancer Images can be stored for long time for future analysis. Another option to histopathological diagnosis is computer- aided diagnosis (CAD) for breast cancer classification. Exact diagnosis can be done using the system of CAD as it is reasonable, easily acquired, quick and dependable. Drawback: Highly expertise and knowledgeable pathologist are required, as it is very time consuming and tedious task[10].

In the era of recent advances in technology, histopathological image analysis is used for the detection and diagnosis of breast cancer by most of the pathologist. Final classification is done labfelling by pathologist. It needs more concentrationa and very time-consuming task. Loss of attention may lead to misclassification of images. Automatic classification model is required to reduce mortality rate and to get proper treatment to patient.

The rest of the paper is organized as follows: The Background study of the proposed research. After that proposed methodology with two different feature extraction-based approaches basically handcrafted using SURF and DSIFT. In Experimental Analysis results are discussed with existing method followed by Conclusion and Future Scope is given.

2. Background

This section gives background study of various methods used by other researchers. The multi-class classification is done by Machine Learning method, De-Novo Model and Transfer Learning Based model. Based on pattern recognition and machine learning method handcrafted features are used for multiclass classification. Researcher have used ROI where features are selected from nuclei segmentation and used as for classification [11] [12]. In studies, it has been found that "texture" is used as feature for classification. There are various features descriptor like Local Binary Pattern (LBP), Scale Invariant feature transform (SIFT), Speeded UP Robust Transform (SURF). Gray–Level co-occurrence Matrix (GLCM), Histogram of Gradient (HOG) are used for feature extraction [13] [14]. Oriented fast and rotated brief method is later used which covered drawback of SIFT and SURF, which requires less computational power and gives better performance [15]. One class kernel principal component analysis (KPCA) model is used with different features for every image [16].

The proposed system makes use of handcrafted feature for multiclass classification, which will be helpful for pathologist. In past year, many researchers have been done using conventional convolutional model (De-Novo Model). De-Novo Model are model are developed from scratch. This type of model consists of convolutional layer, pooling layers. The variant of fully connected depends on the number of classes that has to be classified as output. Dabeer et al [17] used Nuclei structure, cytoplasmic quantity as feature extraction and compression as preprocessing technique for binary classification on BreakHis dataset. S. Margret et. has proposed filter and wrapper method for feature extraction for binary classification by using normalization as preprocessing [18].

CNN can be implemented using pre-trained model. There are various pre-trained models are available like GoogLeNet and AlexNet which are developed on non-medical images such as image-net dataset. The knowledge of these pre-trained networks is used and transfer to other medical images. On such type of pertained network parameters like weights, learning rate and layers of the networks are tuned is called parameter tuning. VGG16, VGG19, Inception and ResNet are already available pertained networks. Depends on the requirement of classification model fine-tuning of layers and weights are done. The authors [19] have used VGG16, ResNet pre-trained models and data augmentation as preprocessing techniques for multiclass and binary classification. In [20], binary classification is done by a pre-trained CNN (AlexNet) was used for binary classification using BreakHis Dataset. Mahboubeh Jannesari et al. have used ResNet V1150 and ResNetV1152 for classification of histopathological images. The classification is done for binary as well as multiclass of both benign and malignant. In preprocessing images are normalized, adjusted to specific to height and width, cropped means variation is done. They have use transfer learning techniques to improve performance of model using various version 1 through 4 and ResNet (V1 50, V1 101 and V1 152) frameworks. It is found that in all Cancer datasets examined, deep ResNet frameworks found to be resilient and accurate than Inception [21]. Amirreza Mahbod et al. has proposed the Fine-Tuned Deep Network Fusion model using ResNet 50 and ResNet 101 models. The two datasets have been used for experiment that is ICIAR 2018 grand challenge dataset and BioImaging 2015 challenge yielding for four classes' classification that is benign, normal, invasive carcinoma and in situ carcinoma. Preprocessing is one of the very essential steps before classification is done. In preprocessing normalization in order to reduce color variation and resizing is done using bi-cubic interpolation. Instead of using CNN from scratch single fine-tuned ResNet model gives better result [22]. Hafiz Mughees Ahmad et al. have proposed model based on transfer learning using AlexNet, GoogLeNet and ResNet for classification based on multiple cellular and nuclei con- figurations. Pre-processing of images are done using Macenku's method. Initially patch wise classification is done in order to reduce size of database. All of separate samples with 512 X 512 size. In next step, whole slide is categorized depending on majority voting. ResNet50 is more accurate for both sample and whole slide wise categorization [23]. In [9], the authors have used a novel deep network named as class structure–based deep convolutional neural network (CSDCNN) in which discerning features are learnt in a hierarchy from at the various level. The performance is compared with the LeNet and AlexNet. The accuracy they have got 92 to 95% for BreakHis Dataset for various magnification factor. Yao Guo et al. Have used hybrid

CNN model which is based on GoogLeNet model. Voting tactics and bagging strategies are used in order to lessen bugs and improve the performance. The hybrid CNN is used on the BACH gives Benign, Insitu, Invasive, Normal classes with accuracy 87.5% [24]. Zeya Wang et.al have proposed CNN with cascading loss for classification. There are four classes are defined like benign lesion, normal tissues, invasive carcinoma, in situ carcinoma. The patch wise classification takes place using VGG16 network. The loss classifies images into particular category effectively. Other important factor is for improve classification is global image pooling. Such type of structure helps to global information from high resolution feature map [25].

3. Proposed Methodology

This section gives idea about Handcrafted and Pre-trained model used as feature extractor for histopathology image dataset for multiclass classification using CNN as a classifier.

3.1 Handcrafted Feature Based Model

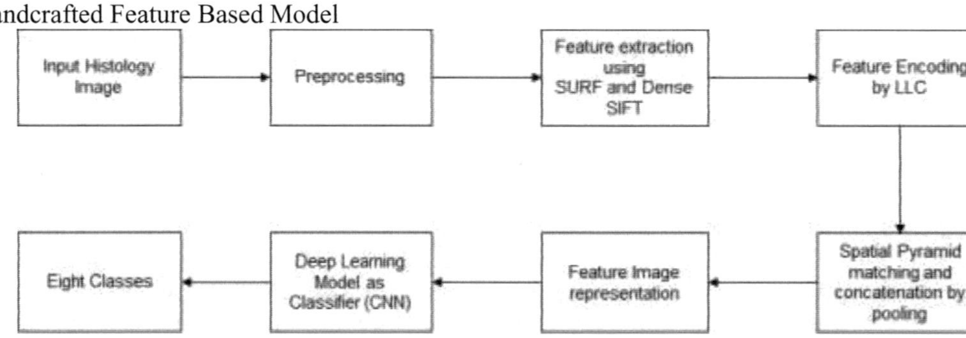

Figure 1. Magnification Independent Handcrafted Feature Based Classification Model

3.2 Pre-Trained Model as a feature Extractor Based Model

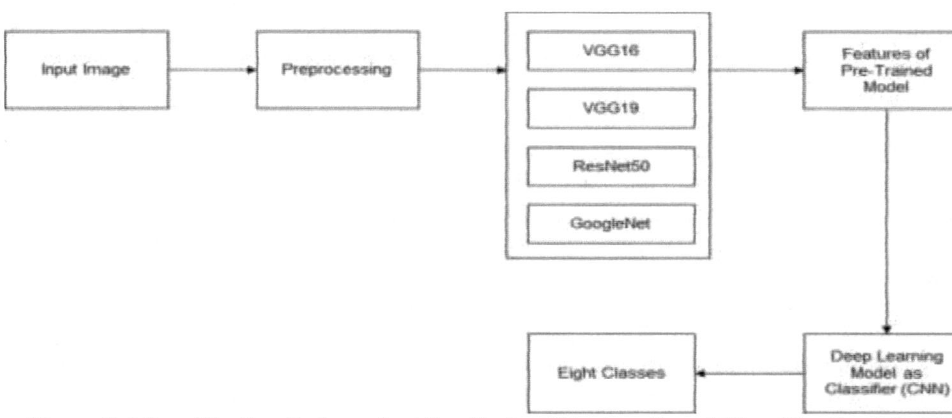

Figure 2. Magnification Independent Pre-Trained Feature Based Classification Model

For both the model Step1 (Input given), Step2 (Preprocessing) and Step4 (Convolutional Neural Network as classifier) are same. In Step3, two different feature extraction techniques are discussed.

Steps are follows:

Step 1: Input given as histopathology image dataset

Dataset from which images were taken as an input: BreakHis [26]
The BreakHis Dataset shown in Table 1 is used for experimental analysis. This dataset consists total 7909 images. This dataset contains benign and malignant images captured using a microscope. The benign categories include Tubular Adenoma (TA), Phyllodes Tumor (PT), Fibroadenoma (F), Adenosis (A). The malignant categories include Papillary Carcinoma (PC), Mucinous Carcinoma (MC), Lobular Carcinoma (LC) and Ductal Carcinoma (DC). The dataset contains images with various zooms specifically 40 times, 100 times, 200 times, 400 times.

Table 1. BreaKHis Dataset Details

Classes	Sub-classes	40X	100X	200X	400X	Total
Benign	Adenosis	114	113	111	106	444
	Fibroadenoma	253	260	264	237	1014
	Phyllodes Tumor	109	121	108	115	453
	Tabular Adenoma	149	150	140	130	569
Malignant	Papillary Carcinoma	145	142	135	138	560
	Ductal Carcinoma	864	903	896	788	3451
	Mucinous Carcinoma	205	222	196	169	792
	Lobular Carcinoma	156	170	163	137	626
Total		1995	2081	2013	1820	**7909**

Initially a pathologist starts extracting ROI at 40X magnification level and keeps zooming upto 400X magnification level to get more detailed information. It also provides detail classification of benign and malignant subclasses.

Step 2: Preprocessing

To avoid the problem of limited size of data and unbalanced data, augmentation technique is used as preprocessing. This type of preprocessing technique helps to reduce the over fitting issue in case of CAD system. In data augmentation techniques images are flipped, cropped, rotated, translated and interpolated in case of medical imaging. Here we have used, rotation and horizontal flip technique as data augmentation. It is possible to loss informative feature if other augmentation technique is used because histology images are rotation and reflection in nature. We have rotated images by 90,180, 270 degrees to increase the size of training dataset. In horizontal flip helps to double cardinality of images.

Step 3: Feature extraction

Feature extraction is most important step while doing classification. Because every feature carries important information. There may be morphological or texture feature which are related to nuclei or other tissue of breast. The given dataset consists of different magnification factors like 40X, 100X, 200X, 400X and each magnification shows various discriminative features which is helpful for multiclass classification.

a. Magnification Independent Handcrafted Feature Based Classification Model

IOP Conf. Series: Materials Science and Engineering 1074 (2021) 012008 doi:10.1088/1757-899X/1074/1/012008

In this, we have extracted features using Seeped Up Robust Feature (SURF) and Dense Scale Invariant Feature Transform (DSIFT) techniques. After extracting features encoding is done by local descriptor like locality constrained linear coding (LLC).

i. SURF: It is robust and faster than SIFT. There are three steps for SURF computation in first step given input is represented in scale space, in second step key points are detected by hessian matrix and orientation to that key point is assign. Finally, SURF descriptor is assigned to each detected key point. This feature descriptor generates 64-dimension feature vector. A rectangular area centered on the Interesting point is established and then the haar sum Wavelet reaction is performed on the point of interest into 4X4 sub region.

ii. DSIFT: It is opposite to SIFT technique. It gives 128 feature vectors by using histogram gradient method on 16X16 region. This descriptor uses grid in uniform way without identifying intersecting point. DSIFT gives more data/information in real time.

iii. Feature Encoding: While working with local descriptor it is necessary to do feature encoding to represent image for classification. Most of classifier accept input as fixed vector length. So, it is necessary to convert into fixed length vector to represent local descriptor into image. Here we are using locality constrained linear coding (LLC) for aggregating the local features to form entire image. LLC coding is variation of sparse coding where local coordinate of each descriptor combined using max pooling approach. LLC coding more efficient than Bag of words method in terms of its coding speed and representation.

b. Magnification Independent Pre-Trained Feature Based Classification Model

In 2012, AlexNet architecture is introduced for ImageNet Challenge having error rate of 16%. Later various variations of AlexNet with denser network are introduced. Here features are extracted using various variation of CNN model VGG16, VGG19, ResNet50, GoogLeNet (pre-trained) models. Each model has its own structure and gives variant vital feature used for classification.

i. VGG architecture: In 2014, Oxford Visual Geometry Group implemented VGG architectures to compete in the ILSVRC contest. There are two variants of VGG architecture one is VGG16 and VGG19. The key difference between two structures is number of convolutional layers. VGG16 possess 13 and VGG19 possess 19 convolutional layers. VGG19 is denser than VGG16. VGG architecture consists of five convolutional blocks. In first and second convolutional block consists of two convolutional layers with 64 and 128 filters each with 3X3 filter size. Third block of convolutional of VGG16 and VGG19 consists of three and four convolutional layers with 256 filters which are 3X3 in size. Fourth and fifth convolutional block also consists three and four convolutional layers with 256 filter in 3X3 size. These five convolutional blocks are separated by max pooling layer. VGG16 is 23 layers in depth with 138 million parameters and VGG19 is 26 layers in depth with 143 million parameters.

ii. ResNet: As we increase the number of layers in deep learning network then vanishing gradient become zero or too large. Other problem with that error rate is also increases between train and test dataset. In 2015, in order to avoid vanishing gradient problem Microsoft researcher has introduced new architecture called as Residual Network. This model uses a technique called s skip connections where it skips few layers and directly connects to output. There are four stages in ResNet50 architecture. In stage 1, three residual blocks with 3 layers each. To perform convolution operation in three layers of 1st block it uses 64, 64 and 128 kernel size with stride size is 2. The overall channel width gets doubled where size of the input decreased to half, when it switches from one stage to other.

iii. GoogLeNet: This architecture is discovered during ImageNet competition with rate of error is 6.65% in 2014.In this model different convolutional layer having different kernel size run in parallel with one max polling layer. Final output is combined is combined into single layer. Parallel structure helps to gives fast output without increasing the more depth. It consists of 23.8 million parameters with 159 in depth having filter sizes with 1X1, 3X3 and 5X5.

From each of these pre-trained model informative features are extracted. For VGG16 and VGG19 has output feature dimension size is 63X133X3. For Resnet50 and GoogLeNet, output Size is 126X266X3, 63X133X3 respectively.

Step 4: CNN as Classifier

Here, we have done classification using CNN as classifier is De-Novo Model, where the number of layers, size of filter and stride is decided at initial only. Such type of model is known as De-Novo Model. Performance of classifier can be checked by varying layer, filter size and activation function. CNN model from scratch gives better performance if it designs and trained efficiently as it can work on small size of datasets. The over fitting problem occurs if a greater number of levels are imposed on small set of datasets when training and testing is done. To optimize the overall performance, system requires huge number of computational resources and its very time-consuming task. We have experimented dataset by using convolution2dLayer with dimension (3,256), maxpooling2dLayer with stride size 2 and filter size 3X3. Relu as activation function is used as its give's better performance than other activation function like sigmoid, tanh in real time scenario. As we have done multiclass classification so that in fully connected layer, we have given total eight classes for classification. The final layer is the classification layer. This layer uses the probabilities returned by the softmax activation function for each input to assign it to one of the mutually exclusive classes.

4. Experimental Analysis

This section provides systematic approaches that represent experimental results with two proposed methods. Experiments are implanted in MATLAB 2018b with 8GB RAM. The various performance parameters are considered while evaluating the performance of the model like Accuracy, Precision, Recall, F1- Score. One of the metrics which is commonly used for model performance is confusion matrix which is plot of actual label vs predicated labels, based on following parameters False Positive (F_P): A sample without breast cancer i.e. negative is wrongly identified as positive. False Negative (F_N): A sample with breast cancer is wrongly identified as negative. True Positive (T_P): This indicates a woman with a cancer of the breast. True Negative (T_N): This indicates patient is negative in the test. There is other measure like specificity, sensitivity, precision, accuracy, AUC, F-Measure and volume under the ROC surface through which model performance can be measure.

- Accuracy: It indicates correctly classified total instances. It basically indicates that the number of patients is correctly predicted without cancer and the number of patients is correctly diagnosed with cancer.

$$A\_c = (T\_P + T\_N)/ (T\_P + T\_N + F\_P + F\_N)$$

IOP Conf. Series: Materials Science and Engineering 1074 (2021) 012008 doi:10.1088/1757-899X/1074/1/012008

- Sensitivity (S_n) or Recall (R_c) This test reveals just how many numbers of the overall positive cases are accurately predicted. The value of recall should be high as possible. It indicates that the number of patients having cancer correctly from the total number patient those having cancer. If the value of sensitivity is low patient hiving cancer are treated as normal which leads to threat to normal patient life.

  $S_n = T_P / (T_P + F_N)$

- Specificity (S_p) It indicate the number of patients not having cancer is classified as correctly as well as the number of patients is having cancer correctly. The value of specificity is should more but not more than Sensitivity (Sn) which has large impact in medical diagnosis.

  $S_p = T_N / (T_N + F_P)$

- Precision (Pr): This indicate that total number of patients for the positive predictions are accurately. This clearly depicts how many numbers of patient having cancer is predicted accurately. The value of recall and precision should more to diagnose patient correctly.

  $P_r = T_P / (T_P + F_P)$

- F1 Measure: This give combine impact of recall and precision. With the help of F-measure metric two models can be compare with high Sn and low Pr and vice versa.

  $F1 = (2 \cdot precision \cdot recall) / (precision + recall)$

4.1 Results Analysis: Features extracted using SURF and SIFT of existing methods and proposed method

This section gives result analysis based on extracting handcrafted features using SURF and SIFT and applying CNN as classifier for 40X, 100X, 200X and 400X magnification factor.

Table 2. Performance analysis for Magnification Independent Handcrafted Feature Based Classification Model

Feature + Classifier	Magnification Factor	Accuracy	Precision	Recall	F1-Score
SURF+CNN	40X	0.9214	0.6306	0.6092	0.6081
	100X	0.9267	0.6531	0.6409	0.6379
	200X	0.9228	0.6195	0.6270	0.6198
	400X	0.9288	0.6639	0.6440	0.6478
DSIFT+CNN	40X	0.9161	0.6075	0.6435	0.6231
	100X	0.9215	0.6317	0.6705	0.6485
	200X	0.9167	0.6013	0.6367	0.6170
	400X	0.9234	0.6416	0.6565	0.6456
DSIFT+CNN [27]	40X	60.58			
	100X	57.44			
	200X	70.00			

IOP Conf. Series: Materials Science and Engineering 1074 (2021) 012008 doi:10.1088/1757-899X/1074/1/012008

Feature + Classifier	Magnification Factor	Accuracy	Precision	Recall	F1-Score
	400X	46.96			
SURF+CNN [27]	40X	80.37			
	100X	63.84			
	200X	74.54			
	400X	54.70			
DSIFT + SVM [27]	40X	44.54			
	100X	51.68			
	200X	44.30			
	400X	35.54			
SURF+SVM [27]	40X	53.75			
	100X	44.30			
	200X	45.30			
	400X	40.88			

From Table 2, it is clear that features extracted using SURF + CNN get maximum accuracy that is 92.88 for multiclass classification.

4.2 Results Analysis: Features extracted using Pre-Trained model such as VGG16, VGG19, ResNet50, and GoogleNet of existing methods and proposed method

Table 3. Performance analysis for Magnification Independent Pre-Trained Feature Based Classification Model

Feature + Classifier	Magnification Factor	Accuracy	Precision	Recall	F1-Score
VGG16+CNN	40X	0.9247	0.6412	0.6057	0.6075
	100X	0.9286	0.6453	0.6450	0.6388
	200X	0.9246	0.6156	0.6240	0.6155
	400X	0.9301	0.6609	0.6316	0.6377
VGG19+CNN	40X	0.9183	0.6540	0.6178	0.6199
	100X	0.9249	0.6691	0.6622	0.6594
	200X	0.9199	0.6413	0.6491	0.6409
	400X	0.9261	0.6734	0.6590	0.6584
ResNet50+CNN	40X	0.9192	0.6397	0.6145	0.6108

Feature + Classifier	Magnification Factor	Accuracy	Precision	Recall	F1-Score
	100X	0.9256	0.6610	0.6509	0.6498
	200X	0.9223	0.6424	0.6360	0.6351
	400X	0.9266	0.6699	0.6534	0.6537
GoogLeNet+CNN	40X	0.9335	0.6028	0.5747	0.5724
	100X	0.9386	0.6202	0.6199	0.6136
	200X	0.9373	0.6043	0.6027	0.5995
	400X	0.9400	0.6342	0.6147	0.6166
CNN features + linear SVM [27]	40X	72.35			
	100X	67.68			
	200X	66.45			
	400X	64.95			
CNN features + RF [27]	40X	66.38			
	100X	65.12			
	200X	69.80			
	400X	67.96			
CSDCNN + augmented data [9]	40X	92.8 ± 2.1			
	100X	93.9 ± 1.9			
	200X	93.7 ± 2.2			
	400X	92.9 ± 1.8			
VGG19 + SVM (L, 1) [28]	40X	92.64	92.00	92.00	92.00
	100X	91.25	91.00	91.00	91.00
	200X	81.42	82.00	82.00	82.00
	400X	80.84	82.00	81.00	81.00
VGG16 + SVM (L, 1)[28]	40X	93.97	94.00	93.00	94.00
	100X	92.92	92.00	91.00	91.00
	200X	91.23	92.00	92.00	92.00
	400X	91.79	92.00	91.00	91.00

From Table 3, it is observed that accuracy for GoogLeNet+CNN is 94% for 400X magnification factor which more than another pre-trained model and handcrafted feature model.

4.3 Confusion matrix

a. Confusion matrix for SURF+CNN for 40X, 100X, 200X and 400X

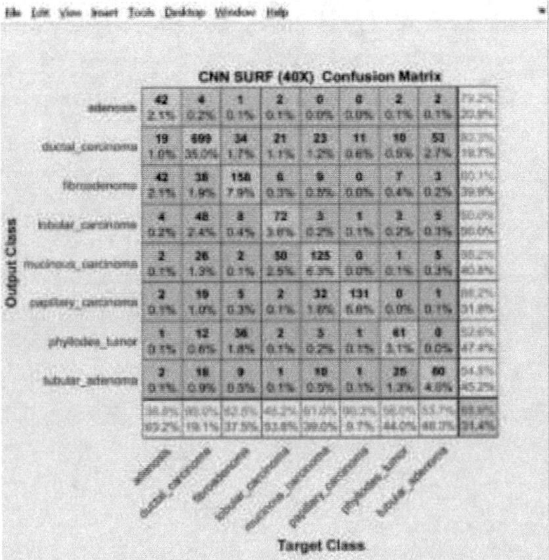

(i)

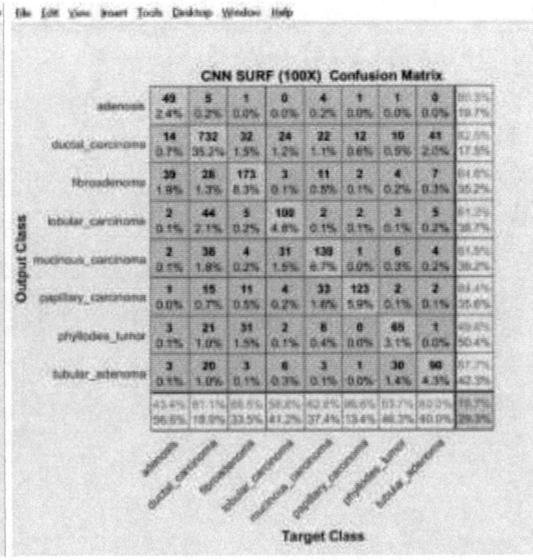

(ii)

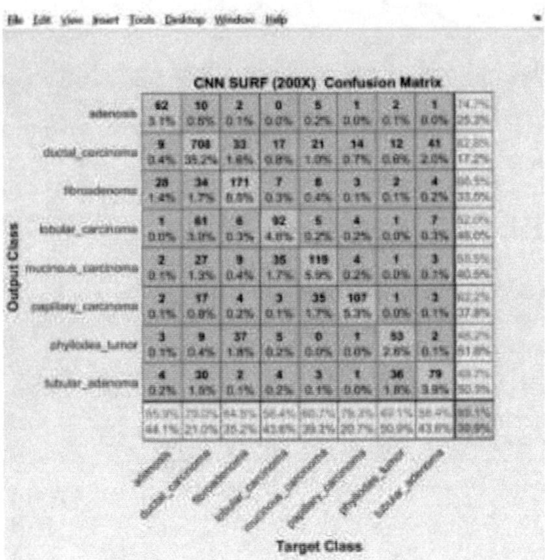

(iii)

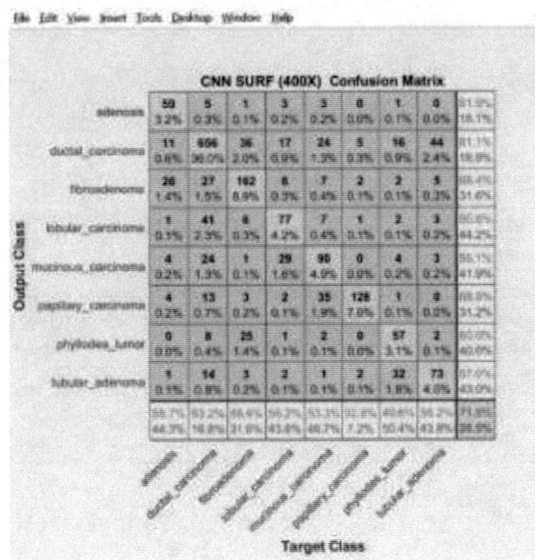

(iv)

b. Confusion matrix for SIFT+ CNN for 40X, 100X, 200X and 400X

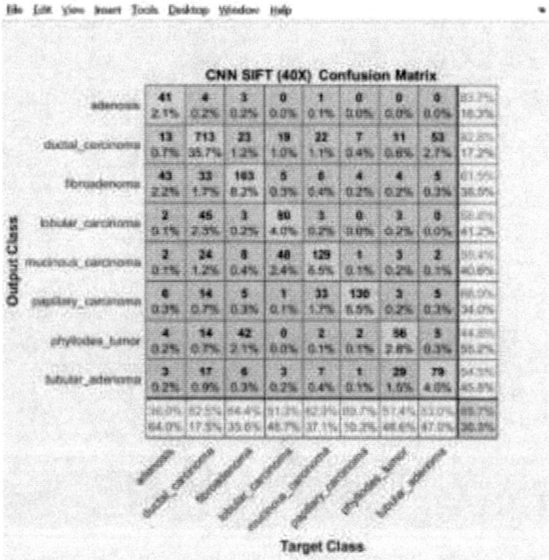

(i)

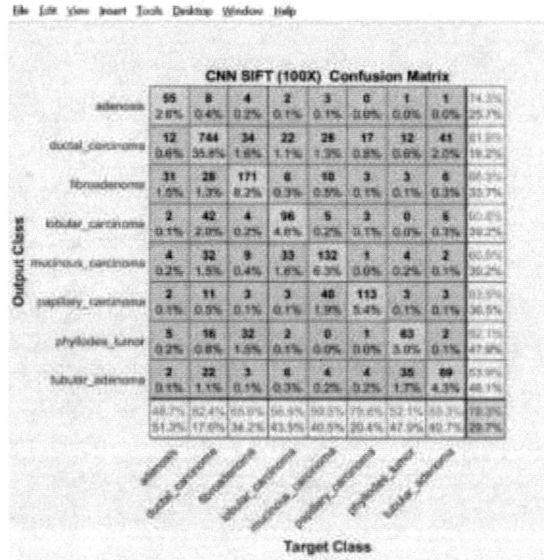

(ii)

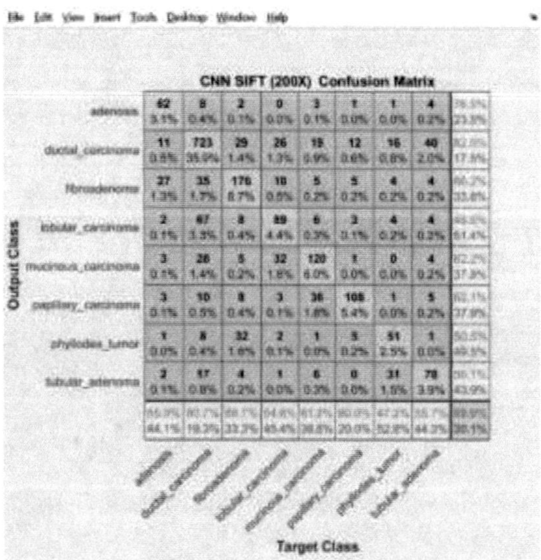

(iii)

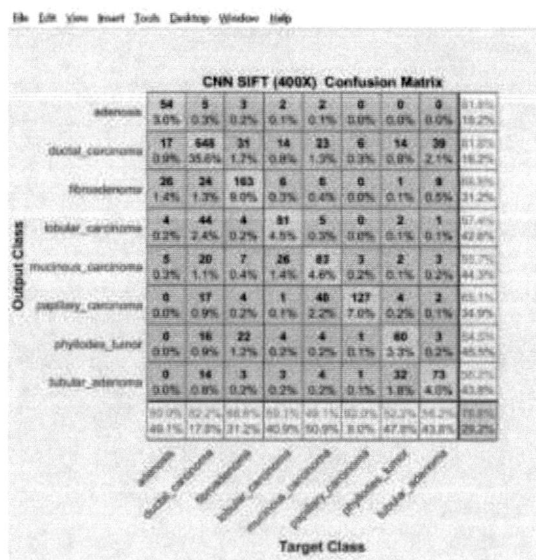

(iv)

c. Confusion matrix for VGG16 + CNN for 40X, 100X, 200X and 400X

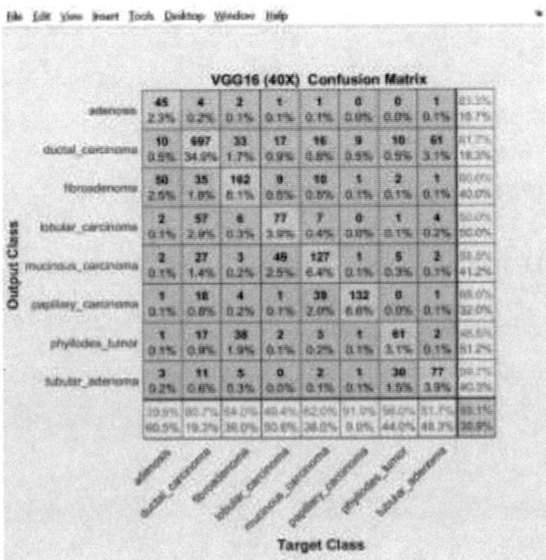

(i)

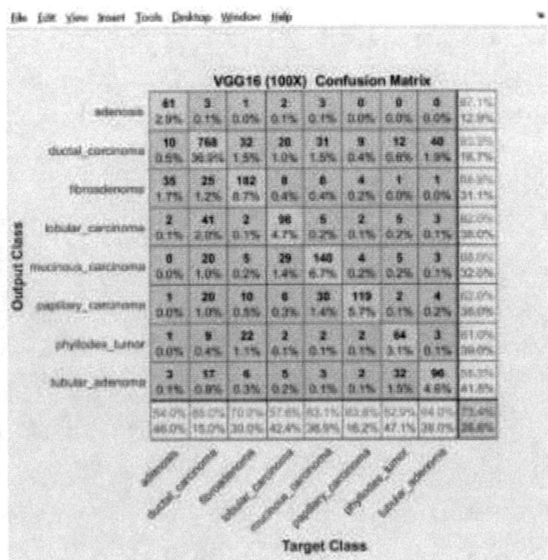

(ii)

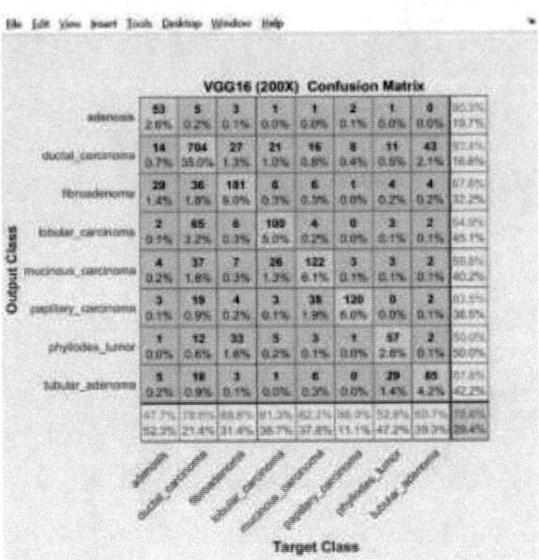

(iii)

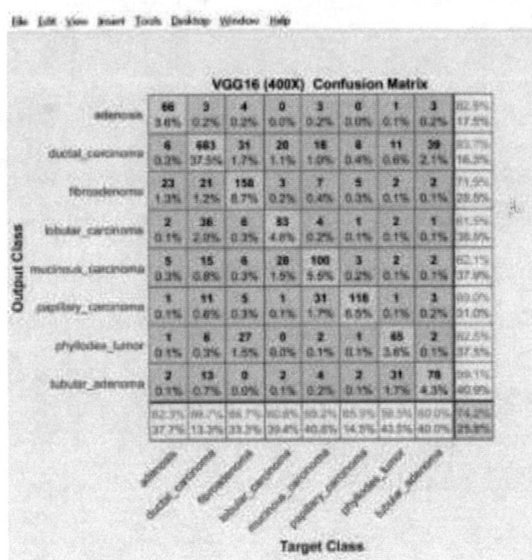

(iv)

d. Confusion matrix for VGG19 + CNN for 40X, 100X, 200X and 400X

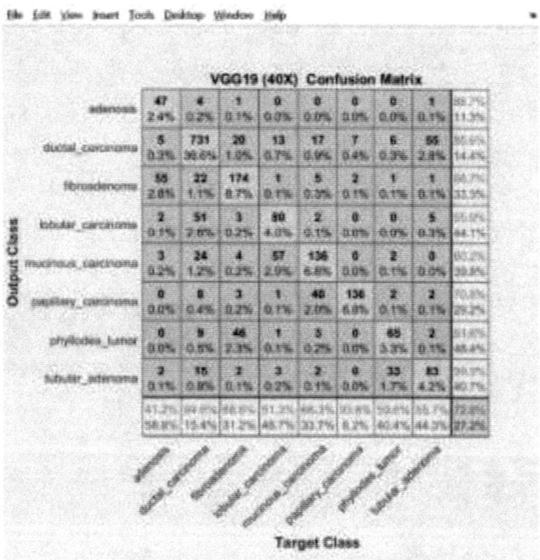

(i)

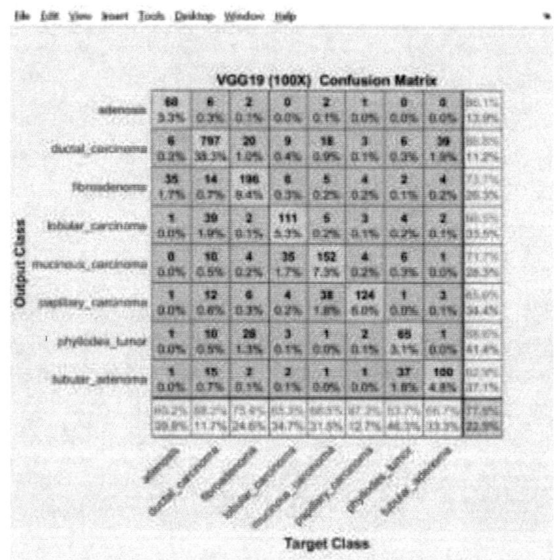

(ii)

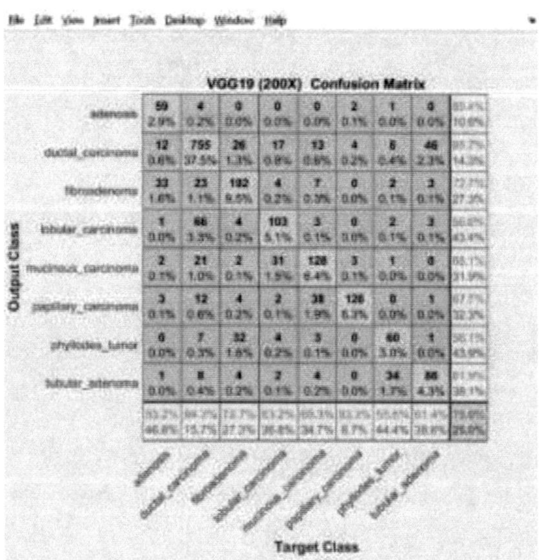

(iii)

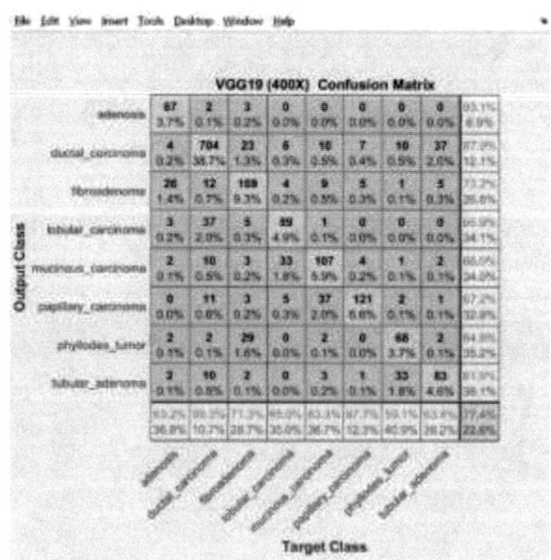

(iv)

e. Confusion matrix for ResNet50 + CNN for 40X, 100X, 200X and 400X

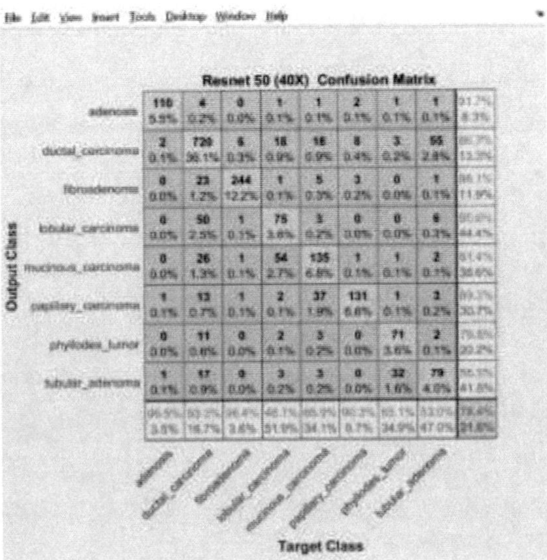

(i)

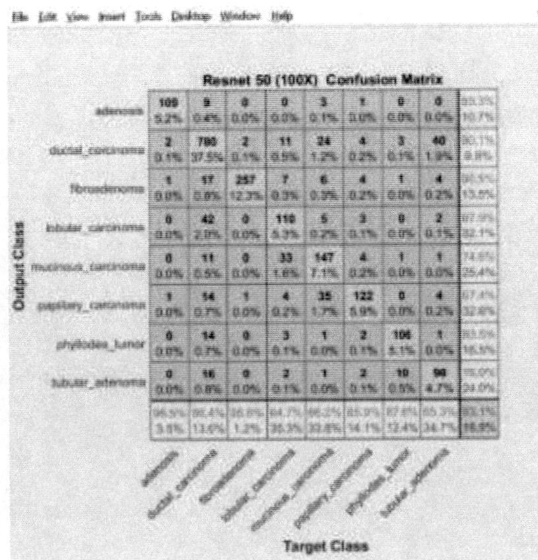

(ii)

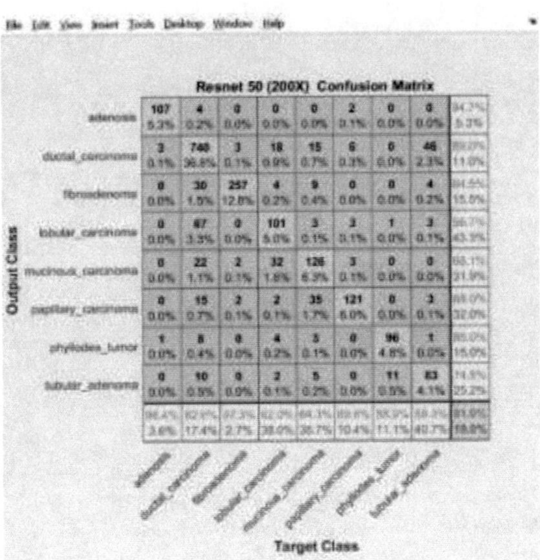

(iii)

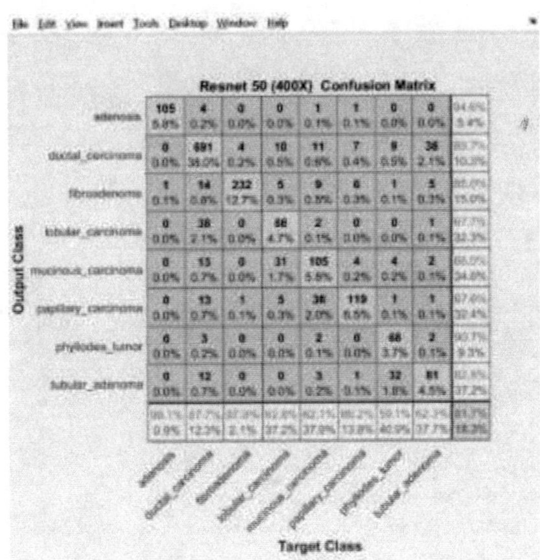

(iv)

f. Confusion matrix for GoogLeNet+CNN for 40X, 100X, 200X and 400X

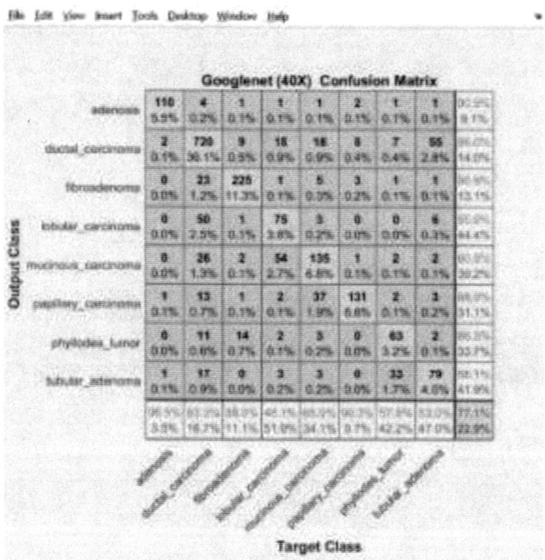

(i)

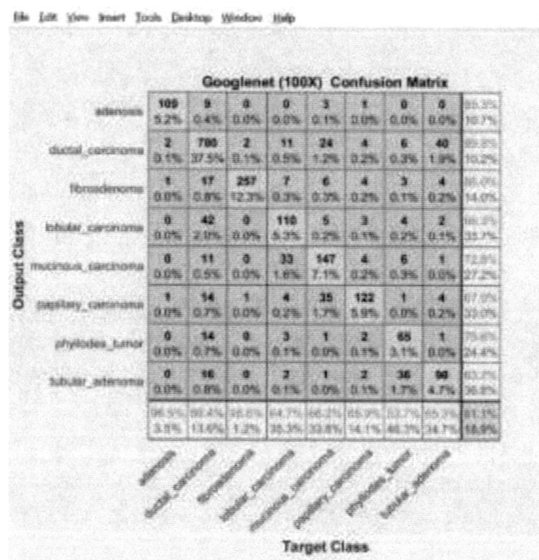

(ii)

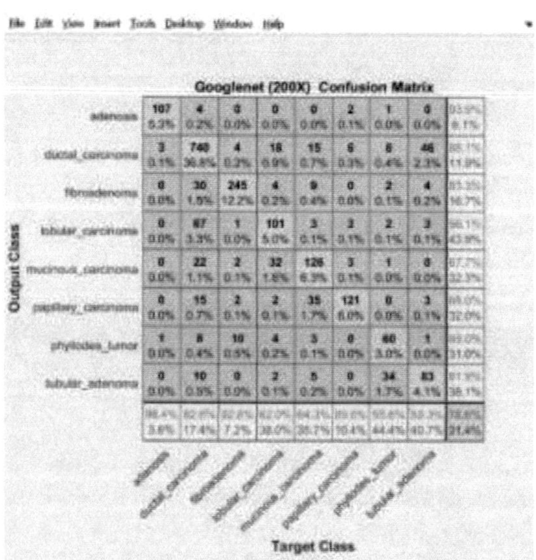

(iii)

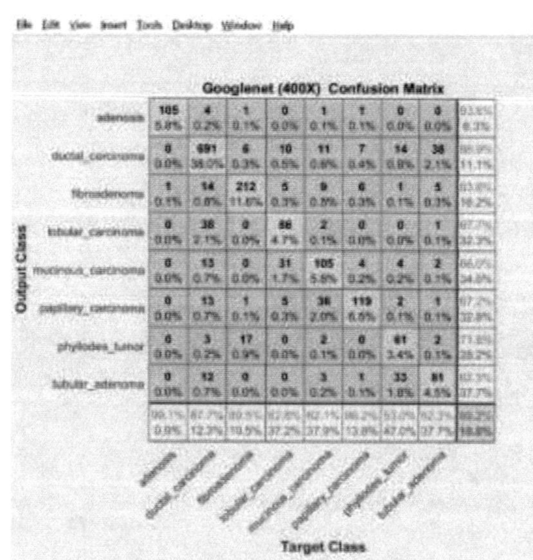

(iv)

5. Conclusion and Future Scope

In this paper, we have compared handcrafted features and pre-trained model for multiclass classification for breast cancer classification using deep learning model. It has been observed that pre-trained model (GoogLeNet) gives better performance i.e 94% accuracy than handcrafted feature-based model.

Experiments are done using CNN from scratch which gives good results than other conventional classifier for all magnification factors. In future other variants of pre-trained model can be used as a feature extractor as well as classifier. CNN with fine tuning of layer can be employed to improve the accuracy. The complexity can be increased with increase in magnification level. Computational efficiency of model can be improving by using parallel processing and cloud-based approach, as it will reduce the overall training time.

REFERENCES

[1] Bray F, Ferlay J, Soerjomataram I, Siegel RL, Torre LA, Jemal, 2018 Global cancer statistics 2018: GLOBOCAN estimates of incidence and mortality worldwide for 36 cancers in 185 countries. *CA Cancer J Clin*;**68**(6):394–424.
[2] Organization WHO. Available from: available: http://www.who.int/en/
[3] M. C. Chun. Breast Cancer: Symptoms, Risk Factors, and Treat-ment, Medical News Today Available from: https://www.medicalnewstoday.com/articles/37136.php.
[4] D. S. Shah. Statistics of breast cancer in india 2018. Available from: http://www.breastcancerindia.net/statistics/trends.html
[5] Altaf F, Islam SMS, Akhtar N, Janjua NK. 2019 Going Deep in Medical Image Analysis: Concepts, Methods, Challenges, and Future Directions. *IEEE Access*.**7**:99540–72.
[6] Murtaza G, Shuib L, Abdul Wahab AW, Mujtaba G, Mujtaba G, Nweke HF et al. 2020 Deep learning-based breast cancer classification through medical imaging modalities: state of the art and research challenges. *Artif Intell* Rev **53**(3):1655–720. Available from: https://doi.org/10.1007/s10462-019-09716-5
[7] Elmore JG, Jackson SL, Abraham L, Miglioretti DL, Carney PA, Geller BM, et al. 2009 Variability in interpretive performance at screening mammography and radiologists' characteristics associated with accuracy. *Radiology* **253**(3):641–51.
[8] Sadaf A, Crystal P, Scaranelo A, Helbich T. 2011 Performance of computer-aided detection applied to full-field digital mammography in detection of breast cancers. *Eur J Radiol.* **77**(3):457–61. Available from: http://dx.doi.org/10.1016/j.ejrad.2009.08.024
[9] Han Z, Wei B, Zheng Y, Yin Y, Li K, Li S. 2017 Breast Cancer Multi-classification from Histopathological Images with Structured Deep Learning Model. **Sci Rep**.7(1):1–10.
[10] Pantanowitz L, Farahani N, Parwani A. 2015 Whole slide imaging in pathology: advantages, limitations, and emerging perspectives. *Pathol Lab Med* Int 23.
[11] Kowal M, Filipczuk P, Obuchowicz A, Korbicz J, Monczak R. 2013 Computer-aided diagnosis of breast cancer based on fine needle biopsy microscopic images. *Comput Biol Med.*;**43**(10):1563–72.
[12] George YM, Zayed HH, Roushdy MI, Elbagoury BM. 2014 Remote computer-aided breast cancer detection and diagnosis system based on cytological images. *IEEE Syst J*.**8**(3):949–64.
[13] Alhindi TJ, Kalra S, Ng KH, Afrin A, Tizhoosh HR 2018 Comparing LBP, HOG and Deep Features for Classification of Histopathology Images. *Proc Int Jt Conf Neural Networks* ;2018-July:1–7.
[14] Belsare AD, Mushrif MM. 2015 Images using Texture Feature Analysis.
[15] Rublee E, Garage W, Park M. 2011 ORB : *an efficient alternative to SIFT or SURF* 2564–71.
[16] Zhang Y, Zhang B, Coenen F, Xiao J, Lu W. 2014 One-class kernel subspace ensemble for medical image classification. *EURASIP J Adv Signal Process* **2014**(1):1–13.
[17] Dabeer S, Khan MM, Islam S. 2019 Informatics in Medicine Unlocked Cancer diagnosis in histopathological image : CNN based approach. *Informatics Med Unlocked* 16(August):100231. Available from: https://doi.org/10.1016/j.imu.2019.100231
[18] Anouncia SM. *Knowledge Computing and Its Applications*. Vol. 1.
[19] Xie J, Liu R, Luttrell J, Zhang C. 2019 Deep learning based analysis of histopathological images of

breast cancer. *Front Genet.* **10**(FEB):1–19.

[20] Spanhol FA, Oliveira LS, Petitjean C, Heutte L. 2016 Breast Cancer Histopathological Image Classification using Convolutional Neural Networks I*nternational joint conference on neural networks (IJCNN) IEEE* Jul 24, pp. 2560-2567.

[21] Motlagh MH, Jannesari M, Aboulkheyr H, Khosravi P. 2018 Breast Cancer Histopathological Image Classification : A Deep Learning Approach.;1–8.

[22] Mahbod A, Ellinger I, Ecker R, Smedby Ö, 2018 Wang C. Breast Cancer Histological Image Classification Using Fine-Tuned Deep Network Fusion. I*n: Lecture Notes in Computer Science (including subseries Lecture Notes in Artificial Intelligence and Lecture Notes in Bioinformatics)* p. 754–62.

[23] Ahmad HM, Ghuffar S, Khurshid K. 2019 Classification of Breast Cancer Histology Images Using Transfer Learning. *In: Proceedings of 2019 16th International Bhurban Conference on Applied Sciences and Technology, IBCAST 2019.* p. 328–32.

[24] Guo Y, Dong H, Song F, Zhu C, Liu J. 2018 Breast Cancer Histology Image Classification Based on Deep Neural Networks Breast Cancer Histology Image Classification Based on Deep Neural *Springer International Publishing* Available from: http://dx.doi.org/10.1007/978-3-319-93000-**8**_94

[25] Wang Z, Dong N, Dai W, Rosario SD, Xing EP. 2018 Classification of Breast Cancer Histopathological Images using Convolutional Neural Networks with Hierarchical Loss and Global Pooling Vol. **2**. *Springer International Publishing* 745–753 p. Available from: http://dx.doi.org/10.1007/978-3-319-93000-8_84

[26] Spanhol FA, Oliveira LS, Petitjean C, Heutte L 2016. A Dataset for Breast Cancer Histopathological Image Classification. *IEEE Trans Biomed Eng.* **63**(7):1455–62.

[27] Bardou D, Zhang KUN, Ahmad SM. 2018 Classification of Breast Cancer Based on Histology Images Using Convolutional Neural Networks. *IEEE Access.* **6**:24680–93.

[28] Sharma S, Mehra R 2020. Conventional Machine Learning and Deep Learning Approach for Multi-Classification of Breast Cancer Histopathology Images—a Comparative Insight. *J Digit Imaging* **33**(3):632–54.

IOP Conf. Series: Materials Science and Engineering 1074 (2021) 012009 doi:10.1088/1757-899X/1074/1/012009

Ensemble Deep Learning for Brazil Currency Coin Prediction

[1]Naresh BabuMuppalaneni, [2]Ch. Prathima, [3]Akula Chandra Sekhar

[1]Assistant professor, Department of CSE, National Institute of Technology Silchar, Assam, India

nareshmuppalaneni@gmail.com

[2]Assistant Professor, Department of IT, Sree Vidyanikethan Engineering College, Tirupati, India

prathima.ch@vidyanikethan.edu

[3]Professor, Avanthi Institute of Engineering and Technology, Cherukupally(V), Vizianagaram District, India

chanduforu@gmail.com

Abstract. In this present fast growing environment the automatic coin reorganization and identification machines has a vital role in all financial allied fields. At present most of coin recognition techniques are depends on physical properties of the coin like length, width, weight etc. Whereas image processing techniques are based on extraction of colour of the coin, edge features of the coin and shape of the coin. For recognition and detection of Brazil currency we have designed Machine Learning, Deep Learning (DL) models in this paper. We have designed various a Machine Learning (ML) models with Convolutional Neural Network for identification and reorganization of Brazil currency. Brazil currency consisting of 5 Centavos, 10 Centavos, 25 Centavos etc. Each of which has different shapes and designs. Different deep learning models are designed and applied ensemble to find the better accuracy. The trained ensemble model is tested on various the datasets which consists of shifting of images, rotation and translated images.

1. Introduction

In this present fast growing environment the automatic coin reorganization and identification machines has a vital role in all financial allied fields. Coin detection/prediction has various applications. These techniques may used in automated toll gates, slot machines, vending machine, coin based certificate printing devices etc. There are 3 different types of traditional coin prediction techniques which use Image Processing, Electromagnetic and Mechanical techniques [1]. In image processing technique the coins are predicted based on the pattern of the coin. The image based coin prediction technique involves basically two steps, the first step is taking the picture of the coin and the second step is to compare it with the reference coin database. The dataset contains images of various coins taken in various angles. After capturing the image of coin, some image processing techniques like Edge detection, DCT, Segmentation, FFT, Gabor Wavelets image subtraction, ANN, SIFT, decision trees, etc. are used to process the image. In the Electromagnetic approach, the material properties to manufacture the coin are considered. So if two coins are made up of the same material then these techniques may fail. Mechanical coin prediction techniques usually based on the physical properties of the coin like weight, height and radius. However, if two coins have the similar physical attributes then this technique fails. Based onfeatures extracted, coins are predicted in various categories. For improving the classification accuracy, ensemble learning can be used. Ensemble Learning (EL) is a different Trending technique in Deep Learning (DL), which utilizes a group of Supervised and Unsupervised learning models to attain enhanced classification results than the fundamental learning algorithms alone. A. U. Tajane et al [1] developed Deep Learning model for recognition of Indian currency coins. Authors have developed own dataset of Indian currency. Authors have used Matlab and Alexnet model.

Chetan et al [2] proposed Side and Rotation invariant coin recognition technique on Indian coins. In their work, the authors have identified the coin in the given image using segmentation. The coin prediction is performed using the radius of coins and based on template matching by the rotating the image. It is observed that the accuracy is ~90%.

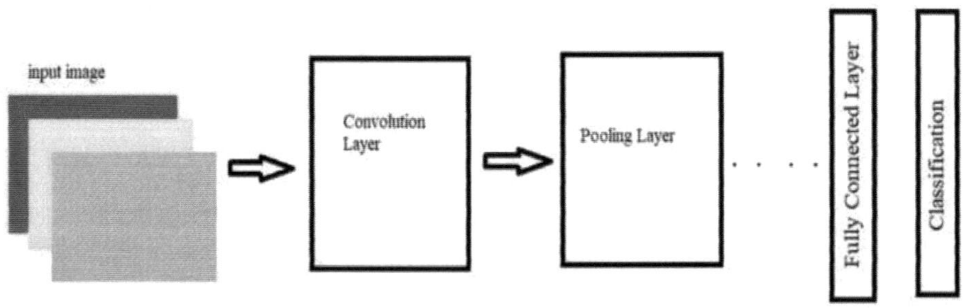

Figure 1: Skeleton Structure of CNN layers

Modi et al [3] proposed ANN based automatic coin recognition on Indian coins. In their work, the authors have generated feature vector of image of size 20x20 after preprocessing and pattern averaging the given input image. They have used 70 images of coins and generate 5040 images by different rotations and reported the accuracy ~97%.

Reisert et al [4] proposed a fast technique for coin recognition on CIS benchmark dataset. The authors have used inverse bilinear interpolation to address the small gradient changes in the input images. Further to improve the speed, FFT of feature function is precomputed. The accuracy they achieved is ~97%.

Veena et al [5] used Spatial Gray Level Dependency Matrix and Steerable Pyramid and Laws Mask to extract the features of input images. To increase the performance here we have proposed a new technique. This is a Deep learning based mechanism which uses the Convolutional Neural Network technique.

Nicola Capece et al [6] have implemented deep learning based coin recognition system for mobile devices. They have developed 5 different models for 5 different dataset of Euro coins. Some of the coins have reported the accuracy as 0% and some are reported as 100%. The authors have integrated the machine learning models with Android mobile App.

Imanol Schlag et al [7] developed deep learning model for prediction of ancient roman coins. They have categorize the coins as Very Fine, Fine, Extreme Fine based on the condition of the coins.

2. Methodology

In this present fast growing environment the automatic coin reorganization and identification machines has a vital role in all financial allied fields. Coin recognition is a challenging task. Coin detection/prediction has various applications. These techniques may used in automated toll gates, slot machines, vending machine etc. There are 3 different ways of traditional coin recognition techniques which uses Image Processing, Electromagnetic and Mechanical techniques. In this work for recognition of Brazil coin, we are experimenting with Convolutional Neural Network, a deep learning model. For classification problems, feature extraction is the challenging task. In the advent of Machine

Learning techniques we think coin recognition becomes easier, but feature extraction is a tough task. There are many ML Techniques for recognition of the coin; but how accurate our ML technique recognizing the coin depends on the features extracted with Machine Learning Model.
Convolutional Neural Networks (CNN) figure 1 is used to recognize and predict the Brazil coins. Using the handicraft features like color, dimensions for recognition of coin has not given high accuracy. CNN is being used for many applications like Character Recognition [11], traffic signal identification, object prediction etc., Hence we planned to consider the CNN in which the model itself extracts the features and helps in better feature extraction. CNN has multiple pooling layers, convolution layers, fully connected layers, activation functions. In the convolution layer the invariant features will generated by using the filter or kernel applied to the local neighbours. Then the invariant features are passed on to the next layer. The pooling layer is used to reduce the dimension of next convolution layer such that the number of parameters can reduced. The CNN architecture is represented in Figure 2.

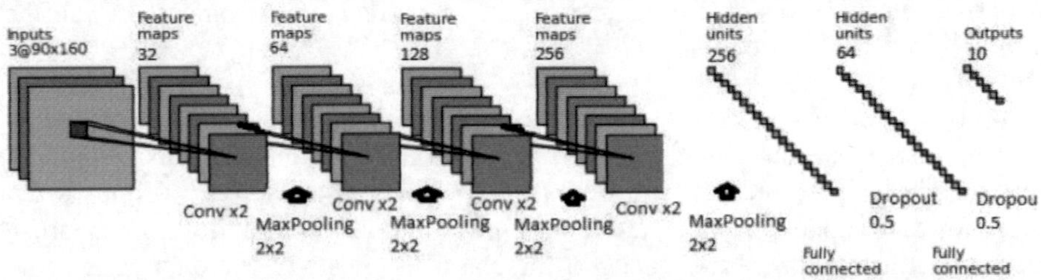

Figure 2: Feature extraction and pooling layers, Fully connected layer and classified outputs after voting (Dropouts) in CNN

2.1. Convolution Layer

CNN is like Neural Networks where all the neurons in one layer connects to neurons in the next layer. The input to the CNN is a two dimensional image. Unlike neural networks, the neurons in one layer may not connect with all other neurons in CNN. CNN has become fast and accurate because of sharing of weights between the two layers. The functioning of CNN is present in the Figure 5. The output of the convolution layer is dot product of image pixel values with kernel (weights). If the Kernel size is NxN and input image size is MxM then the final output after convolution with stride 1 is (M-N+1) * (M-N+1).

2.2. Pooling Layer

The importance of the pooling layer is to minimize the convolute image size to the next convolution layer also as to reduce the parameters to calculate in the training process. Pooling also helps in learning the transformations due to translation and rotation. Pooling layer has to include in between the two convolution layers. There are different ways of applying pooling like Max Pooling, Average Pooling as shown in fig 6 etc., The method for Max Pooling with size 2x2 is presented in the Figure 5. The filter chosen is 2x2, the output each 2x2 matrix of convoluted images is consider for applying pooling, it is maximum of each 2x2 matrix in the convoluted image. The input for pooling layer is output of convolution layer. In pooling layer if the Kernel size is NxN and input image size is MxM then the resultant output reduces by 1/N times.

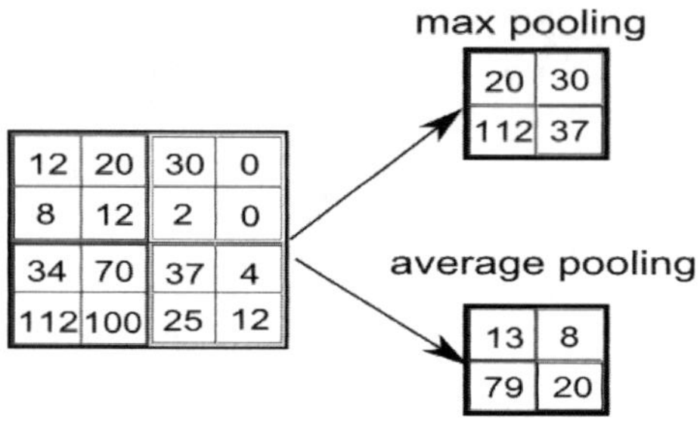

Figure 3: Max and Average Pooling

2.3. Fully Connected Layer

Fully Connected (FC) layer helps in classification. The key purpose of FC layer is learning with non linear function flattern. It means all the neurons in the preceding layer connected to all the neurons in the FC Layer.

2.4. Evaluation Criterion

The key goal of this paper is to recognize the Brazil coins through Deep learning Ensemble after classification scheme on the images acquired from multiparameters like, weight, size and adding multiple filters for each model and calculate the accuracy and loss of the model. A confusion matrix is a ML technique to calculate the performance of the classifier on set of test data where true values will be known as shown in fig 4. Let TP denote true positive, let FP denote false positive, let TN denote true negative, and let FN denote false negative.

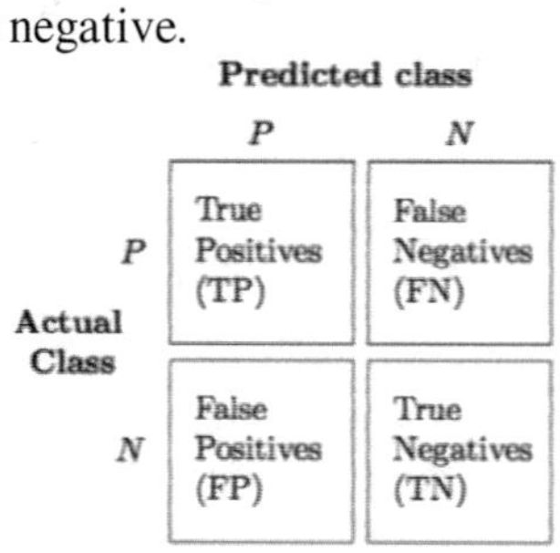

Figure 4: Confusion Matrix

2.5. Ensemble Learning

Ensemble Learning (EL) is a different Trending technique in Deep Learning (DL), which utilizes a group of Supervised and Unsupervised learning models to attain enhanced classification results than the fundamental learning algorithms alone. Ensemble Learning can be largely grouped into three types as follows:

• Bootstrap aggregating, also called bagging in fig 6, and proposed by Breiman [9] is a machine learning ensemble meta-algorithm intended to get better the stability and accuracy of machine learning algorithms used in regression and statistical classification. It also decreases the variance and assist in avoiding of overfitting. The random forest [10] algorithm is an instance of bagging, which groups a set of random decision trees to obtain high accuracy.

Boosting refers to a family of ensemble algorithms which transform a weak classifier to strong classifier. Boosting in fig: 7 is an ensemble method for improving the model predictions of any given learning algorithm. The idea of boosting is to train weak learners sequentially, each trying to correct its predecessor. The Adaboost still is the most broadly implementation of boosting.

• Bucket of models is an ensemble learning technique in which a selection algorithm of utilized to selects the best model in a group of various models. Mostly it approaches through cross-validation.

The architecture for ensemble model of deep learning is represented in Figure 5.

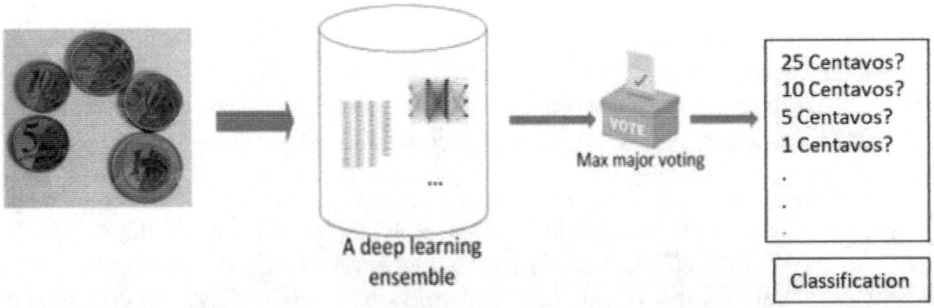

Figure 5: The Frame work of our Deep CNN Ensemble Model

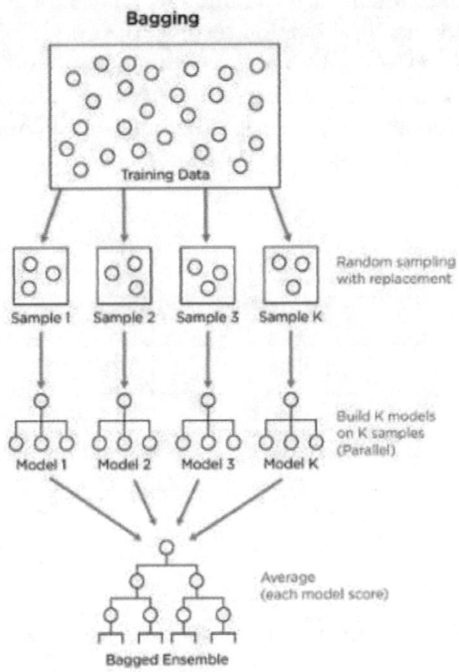

Figure 6: Bagging Ensemble

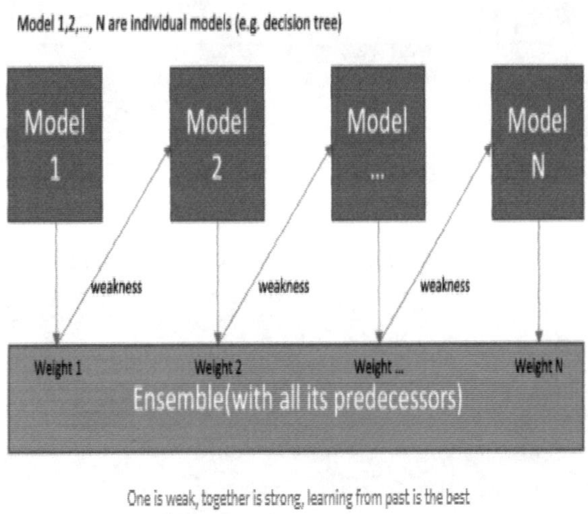

Figure 7: Boosting the weaker classifier

3. Results and Discussion

For recognition of Brazil coin, we employed Convolutional Neural Network, a deep learning model. In this paper we have used the dataset [12] of Brazil coins, which contains 3057. Brazil currency consisting of 5 Centavos, 10 Centavos, 25 Centavos etc.,. Each of which has different shapes and designs. To avoid the manual cropping of coin, we have used automated cropping technique with boundary detection which resulting clear and visible coins of 1662. We divide the dataset into train set of 1187 and test set of 475. We have designed four different CNN models by varying the filters, network dimensions. Training and Testing of Brazil coin images using CNN is performed on NVIDIA Quadro P2000 machine. We tested the Brazil coins on multiple models by using varying filters and network dimensions at layer of the model on feature extracted of the coin images. Results of various models are shown in Table-1 and in table 2 the Confusion Matrix of the ensemble model is presented. The confusion matrix represents the classification results. We got ~87.36% of accuracy which is more than the accuracy given by the models with handicrafts features. Model 1 is able to predict all 1 Real coins and the error rate more in classifying 10 centavos. In Model 2 and Model-3 the error rate is more and misclassifies many coins. Model-4 is able to predict all 1 Real coins and the error rate observed is less in classifying other coins. The architecture of four models and their accuracy, train, test loss is represented in figures 8-19.

Table 1: Accuracy of various models for prediction of Brazil coins

Model	**Accuracy**
Model-1	79.79%
Model-2	78.73%
Model-3	78.31%
Model-4	82.52%
Ensemble Model	**87.36%**

Table 2: Confusion matrix of the Ensemble Model

		Prediction				
		1 Real	50 centavos	25 centavos	10 centavos	5 centavos
Actual	1 Real	83	0	0	0	0
	50 centavos	0	75	1	0	7
	25 centavos	8	2	65	1	7
	10 centavos	0	0	9	64	10
	5 centavos	0	0	6	9	128

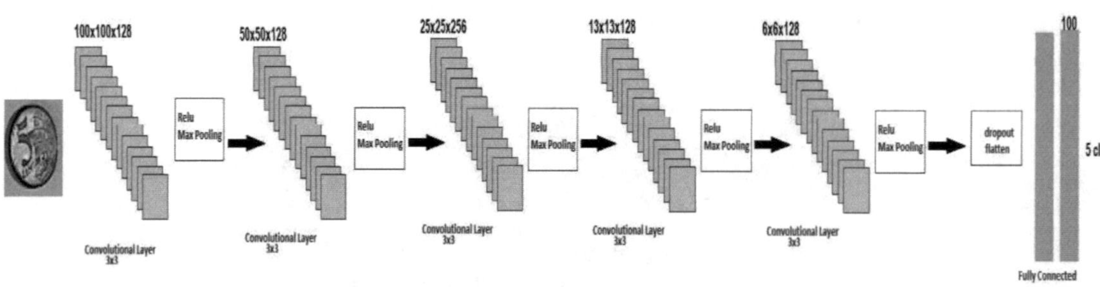

Figure 8: Model 1 Trained on data set

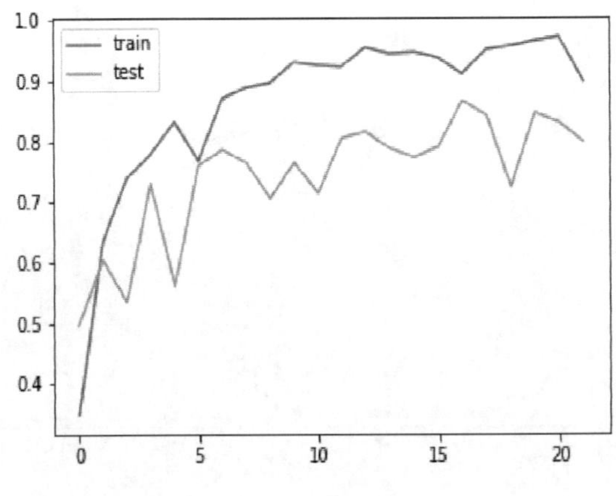

Figure 9: Accuracy of Model 1

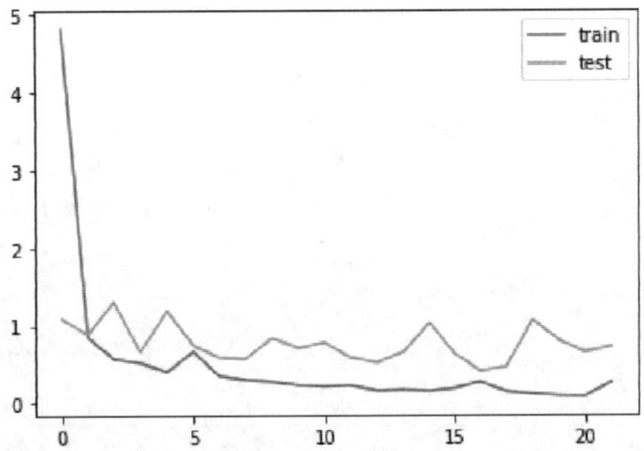

Figure 10: Loss for Model 1

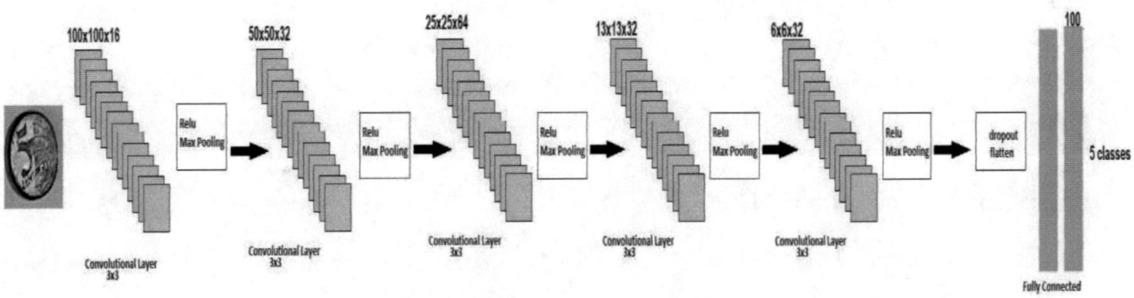

Figure 11: Model 2 Trained on data set

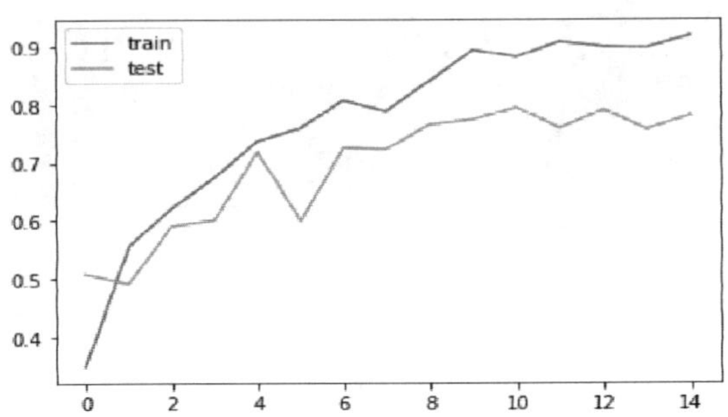

Figure 12: Accuracy of Model 2

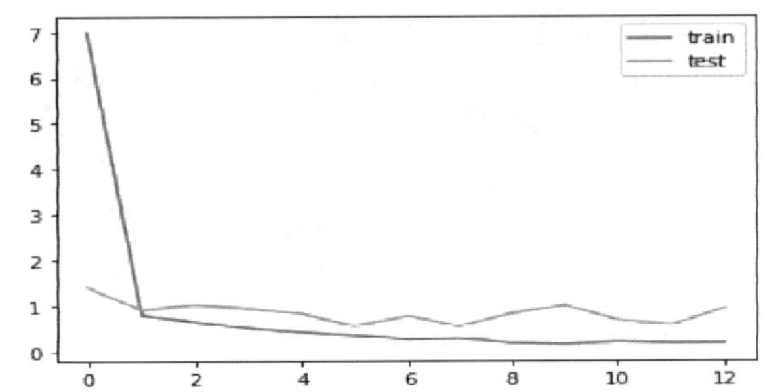

Figure 13: Loss for Model 2

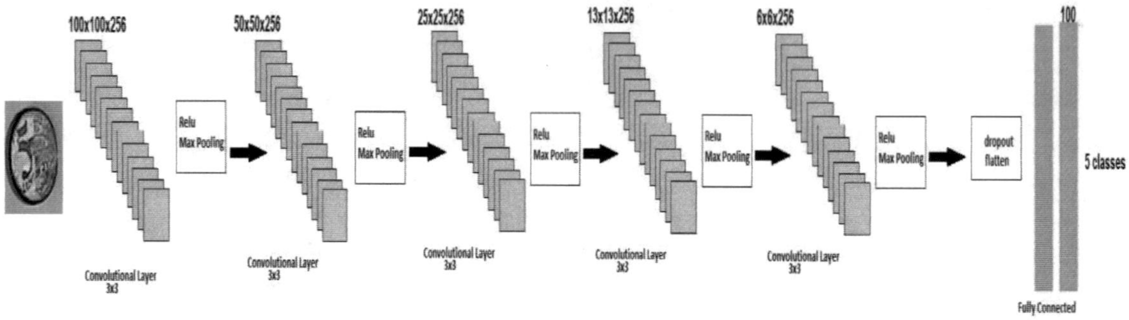

Figure 14: Model 3 Trained on data set

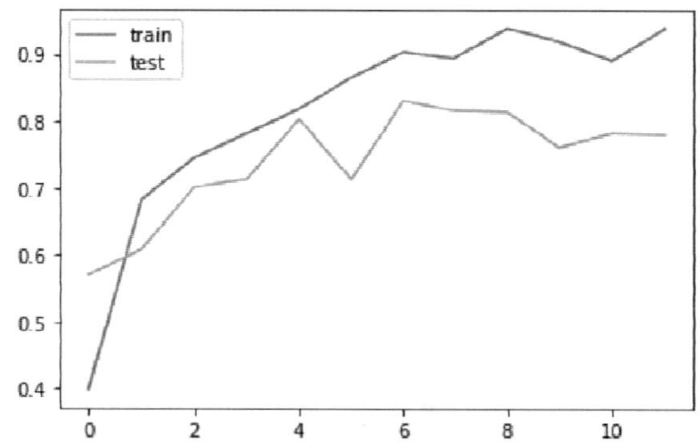

Figure 15: Accuracy of Model 3

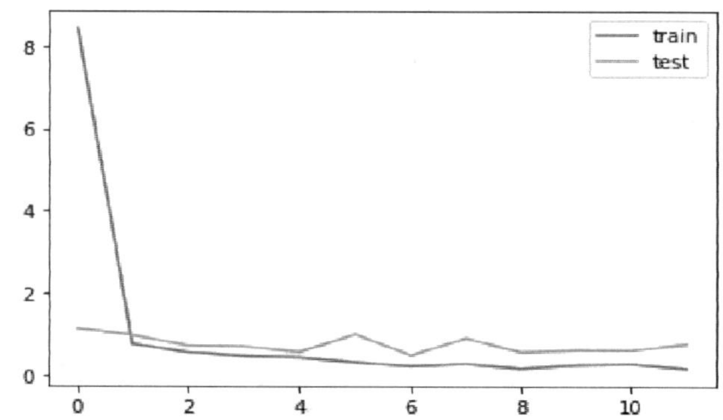

Figure 16: Loss for Model 3

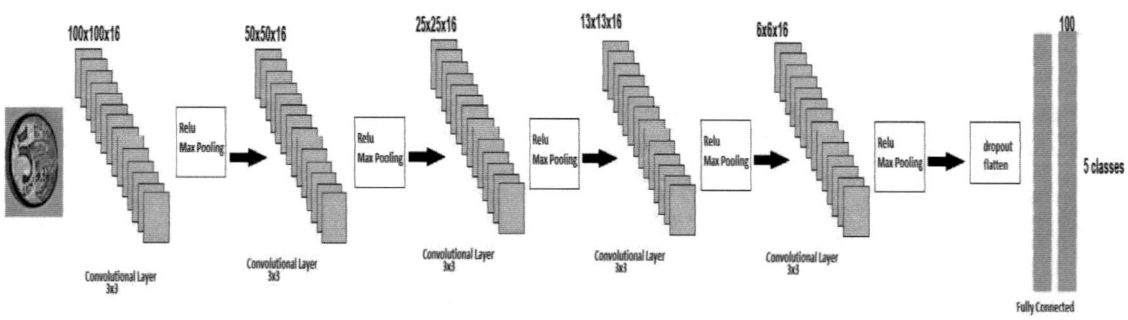

Figure 17: Model 4 Trained on data set

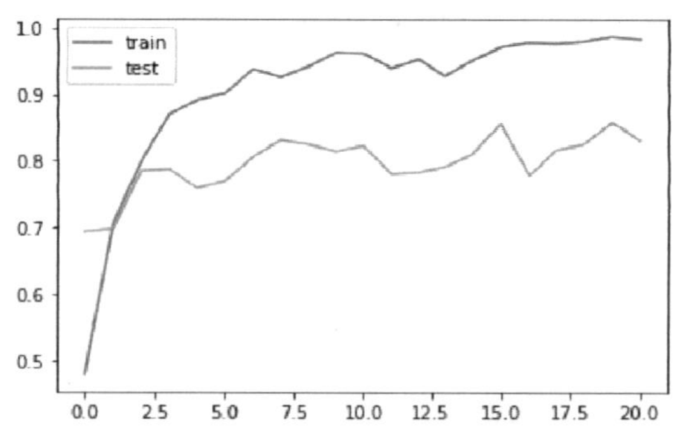

Figure 18. Accuracy of Model 4

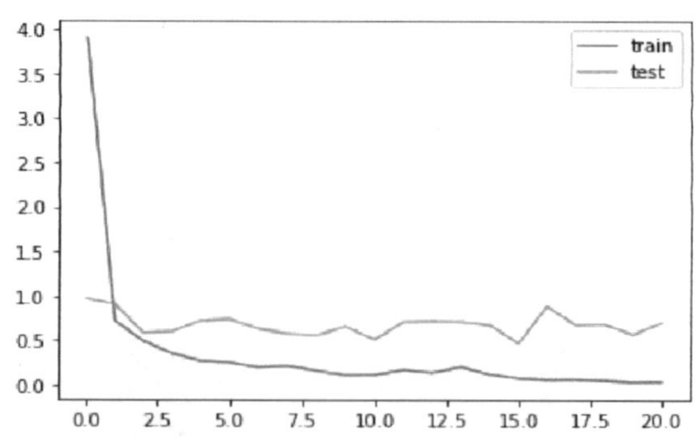

Figure 19. Loss for Model 4

IOP Conf. Series: Materials Science and Engineering 1074 (2021) 012009 doi:10.1088/1757-899X/1074/1/012009

4. Conclusion

Convolutional Neural Networks (CNN) is used to recognize and predict the Brazil coins. Using the handicraft features like color, dimensions for recognition of coin has not given high accuracy. In our work we consider Brazil currency coins of 5 Centavos, 10 Centavos, 25 Centavos etc. Each of which has different shapes and designs. For recognition of Brazil coin, we employed Convolutional Neural Network, a deep learning model. We have designed four different CNN models by varying the filters, network dimensions. Different deep learning models are designed and applied ensemble learning to find the better accuracy. Ensemble Learning (EL) is a different Trending technique in Deep Learning (DL), which utilizes a group of Supervised and Unsupervised learning models to attain enhanced classification results. We got ~87.36% of accuracy which is more than the accuracy given by the models with handicrafts features.

5. Future Scope

We will have to analyze why and where the prediction is failing. To increase the accuracy, in future we will adopt the LSTM and capsule network to handle the orientation.

Conflict of Interest: There is no conflict of interest.

6. References

1. A. U. Tajane, J. M. Patil, A. S. Shahane, P. A. Dhulekar, S. T. Gandhe and G. M. Phade, "Deep Learning Based Indian Currency Coin Recognition," 2018 International Conference On Advances in Communication and Computing Technology (ICACCT), Sangamner, 2018, pp. 130-134, doi: 10.1109/ICACCT.2018.8529467.
2. Chetan B.V., Vijaya P.A. (2013) A Robust Method of Image Based Coin Recognition. In: Kumar M. A., R. S., Kumar T. (eds) Proceedings of International Conference on Advances in Computing. Advances in Intelligent Systems and Computing, vol 174. Springer, New Delhi
3. Modi, Shatrughan, and Dr Bawa. "Automated coin recognition system using ANN." arXiv preprint arXiv:1312.6615 (2013).
4. Reisert, Marco, Olaf Ronneberger, and Hans Burkhardt. "A fast and reliable coin recognition system." Joint Pattern Recognition Symposium. Springer, Berlin, Heidelberg, 2007.
5. Veena H. N., Muruganandam M., and T. SenthilKumaran " Coin recognition using texture feature based on SPLM and SGLDM algorithm" AIP Conference Proceedings 2039, 020047 (2018); https://doi.org/10.1063/1.5079006
6. N. Capece, U. Erra and A. V. Ciliberto, "Implementation of a Coin Recognition System for Mobile Devices with Deep Learning," 2016 12th International Conference on Signal-Image Technology & Internet-Based Systems (SITIS), Naples, 2016, pp. 186-192, doi: 10.1109/SITIS.2016.37.
7. Schlag, Imanol, and OgnjenArandjelovic. "Ancient Roman coin recognition in the wild using deep learning based recognition of artistically depicted face profiles." Proceedings of the IEEE International Conference on Computer Vision Workshops. 2017.
8. Z.-H. Zhou, ''Ensemble learning,'' in Encyclopedia Biometrics, S. Z. Li and A. Jain, Eds. New York, NY, USA: Springer, 2015, pp. 411–416.
9. L. Breiman, ''Bagging predictors,'' Stat. Dept. Univ. California, Oakland, CA, USA, Tech. Rep. 421, 1994.
10. T. K. Ho, ''Random decision forests,'' in Proc. 3rd Int. Conf. Document Anal. Recognit., vol. 1. Aug. 1995, pp. 278–282.
11. N. B. Muppalaneni, "Handwritten Telugu Compound Character Prediction using Convolutional Neural Network," 2020 International Conference on Emerging Trends in Information Technology and Engineering (ic-ETITE), Vellore, India, 2020, pp. 1-4, doi: 10.1109/ic-ETITE47903.2020.349.
12. Luis Moneda, David Yonekura, ElloáGuedes, "Brazilian Coin Detection Dataset", IEEE Dataport, 2020. [Online]. Available: http://dx.doi.org/10.21227/kj4c-y809. Accessed: Jun. 30, 2020.

Visualizing UML's Sequence and Class Diagrams Using Graph-Based Clusters

Nakul Sharma

Department of Computer Science and Engineering

Koneru Lakshmaiah Education Foundation

Vaddeswaram,Guntur-522502, India

nakul777@gmail.com

Prasanth Yalla

Department of Computer Science and Engineering

Koneru Lakshmaiah Education Foundation

Vaddeswaram,Guntur-522502, India

prasanthyalla@kluniversity.in

Abstract. The paper discusses the creation of UML diagram based recommendation system using java and class files as the input. The existing systems do not make use of techniques available in text-mining for creating UML diagrams. The overall methodology makes use of keyphrase extraction, contextual similarity calculation, and graph-based clusters in creating UML diagrams. The existing systems survey of state-of-art UML diagram generation techniques and keyphrase extraction survey is also provided. A comparative analysis of the existing tools for generating UML-diagrams is also provided. The recommendation system generated is useful to maintenance engineers and software developers.

Keywords: Natural Language Processing, UML diagrams, Sequence Diagram, Source Code Analysis, Keyphrase Extraction, Clustering.

1. Introduction

UML diagram is considered as the industrial standard for providing visual representation in the design phase of SDLC. These diagrams are part of the Software Design Document which is a work product of the design phase. There are two categorizations of UML diagrams, namely the static or structural specification and the behavioral or dynamic specification [32].

In the current work, a literature review of UML diagram construction from text or source code is done. A comparative analysis of different methods used in UML diagram construction is also proposed. The authors also highlight UML diagrams as a recommendation system for developers as well as the maintenance engineer. The UML diagrams can give a visual insight into the effect of the system's changes as they are incorporated by the developer and the maintenance engineer [3]. In the existing literature, natural Language text is used in developing UML diagrams. These techniques however, cannot be applied to large documentations or source codes. This is because poor exploration of the contextual relationships among the source code and API documents.

Saroj and Monali proposed an unsupervised graph based keyword extraction method called Keyword Extraction using Collective Node Weight (KECNW) which determines the relevance of keyword by taking into cognizance various collective parameters. The authors also propose doing preprocessing and providing textual graph representation [63]. These systems are not suitable for large open source code projects because of high computational memory and time.

Aziz et. al. have proposed Extended ForUML which generates UML sequence & class diagram from Fortran source code. The technique applied is unsuitable for large open source projects since computational time & memory requirements are not met [69].

Mohammed et. al. proposed clustering technique by considering the knowledge of artifacts. However, authors proposed work cannot identify software architecture efficiently for large software projects [34].

In this paper, the authors propose a contextual similarity approach combined with cluster and graph creation. A multi-step approach involves keyphrase extraction, graph construction, clustering of related documents together and finally creation of UML class & sequence diagrams.

The rest of the paper is divided into the following sections: Section 2 discusses the related literature. Section 3 gives the proposed methodology while section 4 gives the result & discussion. Lastly, section 5 gives future scope & conclusion.

2. Literature Review

In the current section, a literature review based on UML diagrams generation, keyphrase extraction, and clustering are discussed. Table-1 gives a comparative analysis of research papers related to UML diagram construction.

Table 1.Disadvantages and advantages of the existing tools in generation UML diagrams.

Name of Tool Generated	Advantages of Tool	Disadvantages of Tool
Extended ForUML (2019)	1. Extended the ForUML system to include sequence diagrams generation 2. More efficient and useful in different scenarios.	Except for class and sequence diagram, other UML diagrams are not generated.
Automatic Builder of Class Diagram (ABCD) (2016)	The best properties of NLP techniques are employed in generating UML Class Diagrams.	1. The system lacks advanced mechanisms for checking repetitive information 2. Two synonyms were treated differently in the case of the class diagram.
RECAA (2015)	1. Provides an automated approach towards the development of the UML class diagram.	The UML diagrams such as activity, state-chart are not drawn.

	2. Provides feedback in case of a change in text.	
ForUML (2015)	1. ForUML can reverse engineer FORTRON code. 2. ForUML is able to convert FORTRON into UML code. 3. ForUML combines AgroUML in converting class diagram	1. Provide more relationships in form of dependency and realization. 2. Incorporation of other UML visualization tools such as Eclipse plugin
Class-Gen (2010)	Raw Evaluation and comparative analysis of the system (Class-Gen) developed was done.	Aggregation and inheritance not included in the system developed.
UMGAR (2008)	Automatic identification of OO Concepts	Not very efficient in generating state chart and class diagram
ER convertor (2008)	The tool used heuristics to convert English language	Does not propose a semantic heuristic
SENSE (2007)	1. Meta Model elements can be combined 2. Appropriate sematic encode	The application of pattern is difficult Tool does not study the syntactic heuristics
LIDA (2001)	Generates class diagrams	The tool is not fully automated.
CM-Builder (2000)	Generates class diagrams.	Manual refinement not yet stable.

Table 2Summary Of Literature Review on Developing UML Diagrams.

Title of Publication	Work Done/Conclusion of Paper	Future Work Purposed	References
Minimizing ambiguity in Natural Language Software Requirements Specification	Complete analysis of NL text to create UML Diagram artifacts	1. Improving accuracy by improving algorithm 2. Component and Deployment diagram not generated	[4]
From Natural Language Software Specifications to UML Class Models	Provide better accuracy for Conversion of NL to Class Models	OCL for natural language constraint	[5]
From User Requirements to UML Diagrams	Ontology and NLP techniques for developing class diagram. ANNIE Tool developed	1. Heuristics rules are not exhaustive 2. Algorithm for text analysis can be improved in order to generate complex UML Diagrams	[7]
Generating class models through controlled requirements	1. Overcoming ambiguity by making the language controlled 2. Generate effective, quality usecases (RAVEN) tool developed.	Benchmark for requirement validation is necessary part of future	[8]
UDRA: Reflecting Natural	RACE tool generated	Not	[10]

Language Text into UML Diagrams	for converting NL text to UML Diagram. The diagrams generated were class, object, deployment, package diagrams	mentioned	
Automatically generating object models from N.L. Analysis	Original problem statements are converted to 4w language. GOOAL tool made to complete the task of conversion	Not mentioned	[11]
A CBR approach to Text to Class Diagram Translation	REBUILDER UML developed for generating class diagram. Ontology used is domain ontologyfor conversion.	System should be evaluated by software engineers	[25]
A linguistic approach to the development of object-oriented system using the NL system LOLITA	OO Model extracted using NLP systems.	Implement the missing functionality & experiment with real life requirement documents	[23]
CM-Builder A Natural Language-based CASE Tool	A Case tool CM-Builder is used for undertaking conversion of text to UML Class diagram	1. Building Language corpus for real world applications 2. More grammar rules for the sentence types. 3. Discourse interpretation module improved 4. Extraction dynamic aspects from NL Text 5. Evaluation can improved	[19]
UCDA: Use Case Driven Development Assistant Tool for Class Model Generation	UCDA for generating use case diagram developed. Tool has advantages of reducing time to market, increasing productivity and helping novice developers	Automation can help software design effort and cost getting reduced	[26]
The MOVA Tool: A Rewriting Based UML modeling, measuring, validation tool	The MOVA tool is rewriting based UML tool developed with the aim of integrating SE software engineering curricula and	OCL expressions cannot be introduced with MOVA,so it is a future work	[27]

	the industrial software engineering process.		
Relative Extraction Methodology for class diagram generating using dependency graph	Relative Extraction methodology implemented Data structure's concepts such as BFS and DFS used for generating UML Diagrams	Multiplicity between classes constitute future work	[14]
Generating UML diagrams from natural language specifications	RAPID architecture given for the purpose of generating UML diagrams from NL Text	Not mentioned	[21]
From Requirements to UML Models & Back : How automatic processing of text can support requirements engineering	Aims for constructing requirement engineering specification with consistent models. RECAA tool that automates UML Diagram generation is developed.	Getting the stakeholder to undertake coding directly	[9]
Automatic Builder of Class Diagram (ABCD) : An application of UML generation from Functional requirements	Utilized statistical and pattern recognition techniques in generation of UML class diagrams.	1. System lacks advanced mechanism to deal with redundant information problem. 2. A deep NLP analysis can help in avoiding association method confusion.	[28]
Parsed Use Case description as a basis for Object Oriented Class Model generation	Created Class-Gen for creating UML Diagrams from natural language text	Potential exists in utilizing NLP tools for software development	[17]
Conceptual Modeling of Natural language functional requirements	Claims that automation can help save time of developers	None	[20]

There are several methodologies being used in developing UML diagrams. The most common diagrams which are developed are Use –case and class diagram[13[15][16][18] [30] . However little work has been done wrt using text, source code, API documentation for generating UML diagrams. In addition text mining techniques are not used extensively in generating UML diagrams.

Table 3.Summary Of Literature Review on Keyphrase Extraction

Reference number	Authors names	Title of Paper	Advantages of Proposed Work	Disadvantage of Proposed Work
34	G S. Mohammadi and H. Izadkhah	A new algorithm for software clustering considering the knowledge of dependency between artifacts in the source code	Similarity Index along with various hierarchical and search based approaches have been used	Authors proposed work cannot identify software architecture efficiently for large software projects.
35	V. U. Gómez, S. Ducasse and T. D'Hondta	Visually characterizing source code changes	Authors propose Visual representation of changes done on source code. text and visual representations are combined.	Sematic sequences are not addressed properly for large source code projects. Release masters are also not addressed
36	S. L. Abebe and P. Tonella	Extraction of domain concepts from the source code	Concepts of NLP ontologies mapped onto source code by the authors. Union of ontologies give better result.	None Mentioned
37	S. Bajracharya, J. Ossher and C. Lopes	Sourcerer: An infrastructure for large-scale collection and analysis of open-source code	The authors develop sourcerer which provides structural as well as textual medium for searching	Software Engineering Tools integration needed
38	Abilhoa, Willyan D., and Leandro N. De Castro.	A keyword extraction method from twitter messages represented as graphs.	Authors concluded that the graph developed using neighborhood edges provided better performance.	Other centrality measures can also be applied
39	Ziqi Zhang, Johann Petrak, Diana Maynard,	Adapted TextRank for Term Extraction: A Generic Method of Improving Automatic Term Extraction Algorithms	A generic method for term extraction proposed.	For pre-training in the seeding stage other techniques can also be used.
40.	Duari, Swagata, and Vasudha Bhatnagar.	Complex Network Based Supervised Keyword Extractor.	Text is considered as a complex network. The proposed methodology is domain ad language independent	Work can be extended to different Indian languages as well
41.	Wael Etaiwi, Arafat Awajan	Graph-based Arabic text semantic representation	Sematic relationships found for constructing AG, can be used in text summarization	AG for different Indian languages can be constructed.
42	Javad Rafiei-Asl, Ahmad Nickabadi	TSAKE: A topical and structural automatic keyphrase extractor	The main aim of keyphrase extraction is fulfilled	Macro & micro topic selection forms the future work.
43	Shansong Yang, Weiming Lu, Dezhi Yang, Xi Li, Chao Wu, Baogang Wei	KeyphraseDS: Automatic generation of survey by exploiting keyphrase information	Document summarization tool which overcomes the data sparsity problem.	Multiple layers ML/ DL algorithm can be applied
44	B. D. Farahani, S. O. Fatemi and M. Ghorbani	Automatic Keyphrase Extraction from Persian Scientific Documents Using Semantic Relations	Due to lack of linguistics resources in Persia author make use of thesaurus for conducting keyphrase extraction in Persian language	Keyphrase suggestion, conducting manual evaluation
46	Niraj Kumar, Kannan Srinathan, Vasudeva Varma	Towards Intelligent Text Mining Under Limited Linguistic Resources	All these techniques make use of graph for representing text.	Not suitable for large text documents.
47	Litvak, M., Last, M. & Kandel, A.	DegExt: a language-independent keyphrase	Performance is better than TextRank	Additional natural languages can be

		extractor		included
48	Kumar, Niraj, Kannan Srinathan, and Vasudeva Varma	A graph-based unsupervised N-gram filtration technique for automatic keyphrase extraction	Statistical feature extraction and co-location strength is used in keyphrase extraction	As Dbpedia & Wikipedia are evolving, the work can be extended to multiple domains and multilingual situations.
50	Yan Ying, Tan Qingping, Xie Qinzheng, Zeng Ping, Li Panpan	A Graph-based Approach of Automatic Keyphrase Extraction	Authors work is based on premise that word is essential if it appears in a sentence multiple times. Word & sentence relationship is unexplored.	Work can be extended to include full corpus of documents
57	Mahata, Debanjan	Key2vec: Automatic ranked keyphrase extraction from scientific articles using phrase embedding's	Authors develop a tool that uses phrase embedding for giving rank to scientific documents. Theme-weighted PageRank evaluates the candidates phrases.	The tool can be used in other domains such as news articles
58	Florescu, Corina, and Cornelia Caragea.	Positionrank: An unsupervised approach to keyphrase extraction from scholarly documents	Relative position of the word is used in ranking keywords for positionrank tool	Web-pages can be used for PositionRank.
59	Josiane Mothe, Faneva Ramiandrisoa, Michael Rasolomanana	Automatic keyphrase extraction using graph-based methods	Effect of word embedding on keyphrase extraction is studied.	None mentioned
60	Degl'Innocenti, Dante, Dario De Nart, and Carlo Tasso	A New Multi-lingual Knowledge-base Approach to Keyphrase Extraction for the Italian Language	Authors propose creating a multi-lingual tool for extracting keyphrases of English and Italian language	Keyphrase generation can be extended for other languages also.
61	Ravinuthala, V. V. M. K., and Satyananda Reddy Chinnam.	A keyword extraction approach for single document extractive summarization based on topic centrality	Authors use lexical relationships along with centrality measures for keyphrase extraction	None mentioned

The existing literature also has several survey papers on the topic of Keyphrase extraction [51-57]. Yakoob et. al proposed keyword searching in the cloud environment. The author gave the strategies for retrieving essential information from cloud records. The authors work is novel as keyword as well as keyphrase locations are assessed in cloud environment [49]. Selvaraj et. al. propose a new relationship for extracting essential words from a given document. Lexical association between words is used for finding relevant keywords. The authors propose an approach to vertex connectivity for identifying different centrality measures of keywords [62].

3. Proposed Methodology

In the existing systems, centrality measures have been applied to data structures and graph-based systems as well for clustering or for keyphrase extraction. Software Engineering research inculcates application from variety of domains [68-69].This trend of applicability is also see with NLP domain [68]. An extensive analysis of clustering algorithm is presented by the authors [45] [64] [67].
The authors proposed making use of contextual similarity in generating UML diagrams. This first step includes extracting the methods and fields from the source code and class files. Using the keyphrases extracted a Source code directed graph is constructed. The graph is used in calculating similarity index between two terms.

The proposed methodology is divided into following sub-sections:

3.1. Module-1 Probabilistic Weighted based contextual similarity measure for Source code and class files dependency graph [33]

The source code's metrics and class-file metrics are used in developing SDG graph. Here vertex V is source code methods and fields while the edges represent the weighted rank. The edge weights are calculated based on the probability of occurrence of methods and fields. The last step includes generating Contextual Source Code Dependency Graph similarity index [33].

3.2. Module-2 Contextual source code graph based clustering algorithm for class dependency diagram [33]

The similarity index computed in the previous module is now used in clustering related files together. The cluster of files indicates the relative relatedness as it exists according to similarity index. These clusters of files are used to construct the Class and Sequence diagram [33].

4. Results & Discussion

The project was tested on weka software. The project accepts three types of inputs, the first is directory location where the source code files are located. Figure-2 and Figure 3 shows the input provided for source code path and class path respectively.

Figure 2: Input for Source Code Path

Figure 3: Input for Class Path

The input of all the essential locations gets completed and categorization in form of tokens, for each word of the document gets started. As shown in figure-4, a snapshot of processing at tokenization stage gives information about different token in the file nameConsoleLogger.java.

Figure 4: Tokenization Stage of Project.

The source code and its documentation are scanned leading to the development of the graphs. The similarity measure is then used in the generation of clusters. These clusters are then used in the generation of UML's class diagram and sequence diagram.

IOP Conf. Series: Materials Science and Engineering 1074 (2021) 012010 doi:10.1088/1757-899X/1074/1/012010

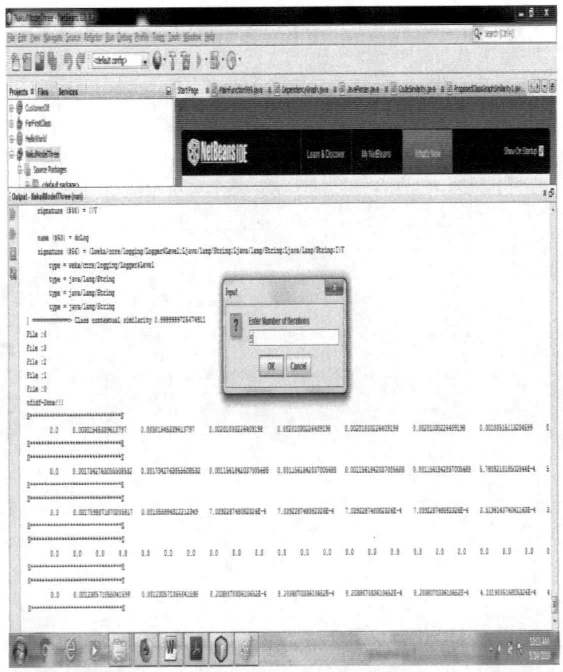

Figure 5: Accepting Input for Number of Iterations

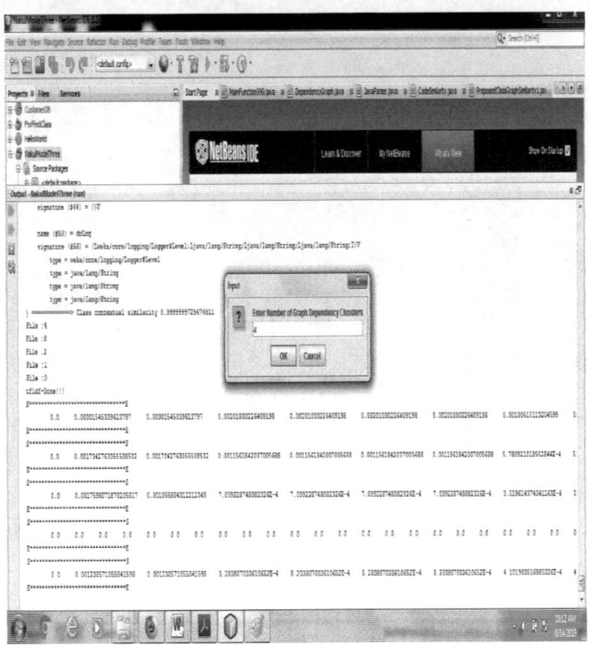

Figure 6: Accepting Input for Number of Clusters

IOP Conf. Series: Materials Science and Engineering 1074 (2021) 012010 doi:10.1088/1757-899X/1074/1/012010

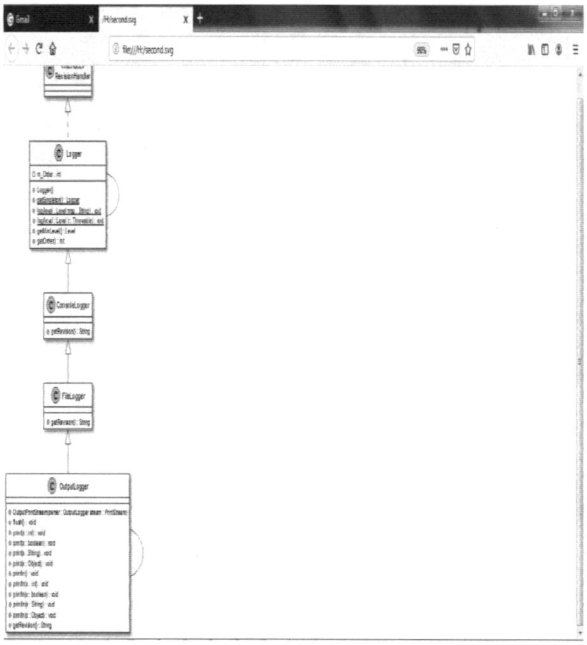

Figure 7: Class Diagram Generated

Figure 8: Sequence Diagram Generated

The current work can be seen as developing UML and sequence diagrams by making use of contextual similarity, clusters and keyphrase extraction from the source code and API documentation. The traceability link across the different artifacts within the SDLC holds the key to providing a better maintenance environment.

Table 4.Comparative Analysis of Different UML Tools Developed

Name of UML Tool	Techniques and Input files used for Conversion				
	NLP SOFTWARES	NLP AND Rules (Heuristics)	NLP AND XMI/XML	Source Code	API Documentation
Automatic Builder of Class Diagram (2016)	No	Yes	Yes	No	Yes
RECAA (2015)	Yes	Yes	No	No	Yes
CM-Builder (2000)	Yes	Yes	No	No	No
UMGAR (2008)	Yes	Yes	No	No	No
SENSE (2007)	Yes	Yes	No	No	No
ER convertor (2008)	No	Yes	No	No	No
LIDA (2001)	No	Yes	No	No	No
ForUML (2015)	Yes	Yes	Yes	Yes	No
Extended ForUML (2019)	Yes	Yes	Yes	Yes	No
Nakul Sharma et. al [33].	Yes	Yes	No	Yes	Yes

Table 5.Comparative Analysis of Proposed Work with Extended ForUML [69].

Sr. No.	Own Work	Extended ForUML [69]
1	Generate class and sequence diagram using API documents and source code as input	Generate sequence and class diagram using fortron code only.
2	Parsing of all java source code for names of methods and fields	Parsing of fortron source code done using Open Fortron Parser (OFP) available for Fortron Programming Language.
3	Creation of dependency graphs	Parsing of Fortron source code done using Open Fortron Parser (OFP) library.
4	Calculation of contextual similarity for the source code, class files and API documentation	Development of XMI representation using derived relationship between code chunks.

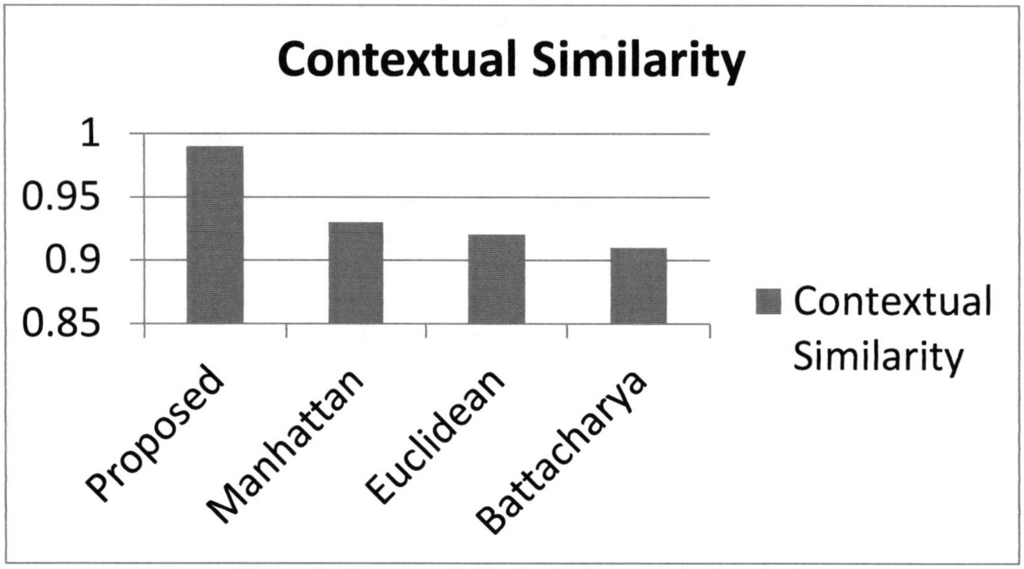

Figure 9: Contextual Similarity Proposed v/s existing systems

5. Conclusion & Future Scope

The paper discusses how UML diagram can be used as a tool for recommending most essential classes within a given set of project. A large-scale open source project cannot be assessed using the existing similarity measures. Hence, a new hybrid probabilistic model is proposed for large open-source projects. The future scope of current work includes:-

1. Construction of static and dynamic UML diagrams from Source Code, Natural Language artifacts.
2. Creating Recommendation System for different stakeholders within the SDLC.
3. Developing specialized ontologies for developers, coders, designers, UI experts to enable lesser time to market.
4. Developing traceability mechanism between the different software artifacts.
5. Centrality measures can be used to further recommendation system research.

References

[1] Pankaj Jalote. 2002. An Integrated Approach to Software Engineering, Second Edition, Narosa Publishing House.

[2] Bernd Bruegge, Allen H. Dutoit. 2012. Object Oriented Software Engineering Using UML, Patterns, and Java, Pearson Education, Tenth Edition ISBN: 978-81-775-8768-5.

[3] Dan Pilone, Neil Pitman. 2007. UML 2.0 In a Nutshell, O'Reilly Media, Third Edition. ISBN: 81-8404-002-4.

[4] A. Umber and I. S. Bajwa, "Minimizing ambiguity in natural language software requirements specification," *2011 Sixth International Conference on Digital Information Management*, Melbourn, QLD, 2011, pp. 102-107, doi: 10.1109/ICDIM.2011.6093363.

[5] Bajwa I.S., Choudhary M.A. (2012) From Natural Language Software Specifications to UML

Class Models. In: Zhang R., Zhang J., Zhang Z., Filipe J., Cordeiro J. (eds) Enterprise Information Systems. ICEIS 2011. Lecture Notes in Business Information Processing, vol 102. Springer, Berlin, Heidelberg. https://doi.org/10.1007/978-3-642-29958-2_15.

[6] Pushpak Bhattacharyya. 2012, Natural Language Processing: A perspective from Computation in Presence of Ambiguity, Resource Constraint and Multiliguality, In proceedings of the CSI Journal of Computing, 1, 2, (2012).

[7] Herchi, Hatem, and Wahiba Ben Abdessalem. "From user requirements to UML class diagram." *arXiv preprint arXiv:1211.0713* (2012).

[8] Giganto, Reynaldo. "Generating class models through controlled requirements." NZCSRSC-08, New Zealand Computer Science Research Student Conference Christchurch, New Zealand. 2008.

[9] Mathias Landhauber, Sven J. Korner and Walter F. Tichy. 2014, From Requirements to UML Models and Back: how automatic processing of text can support requirements engineering, In. Proc. Software Quality General, Springer, March 2014, Vol 22, Issue 1, pp 121-149.

[10] Ontology, Domain. "UDRA: Reflecting Natural Language Text in to UML Diagrams."

[11] H. G. Pérez-González, Jugal K. Kalita, Alberto Salvador Núñez Varela, and Richard S. Wiener. 2005. GOOAL: an educational object oriented analysis laboratory. In Companion to the 20th annual ACM SIGPLAN conference on Object-oriented programming, systems, languages, and applications (OOPSLA '05). Association for Computing Machinery, New York, NY, USA, 180–181. DOI:https://doi.org/10.1145/1094855.1094924

[12] Mencl V. 2004, Deriving Behavior Specifications from Textual Use Cases, In. Proc. of Workshop on Intelligent Technologies for Software Engineering (WITSE04, Sep 21, 2004, part of ASE 2004), Linz, Austria, ISBN 3-85403-180-7, pp. 331-341, Oesterreichische Computer Gesellschaft, September 2004.

[13] Imran Sarwar Bajwa and M. Abbas Choudhary. 2006, Natural Language Processing based automated system for UML diagrams generation In. Proc. 18th National Computer Conference (NCC-2006), Page No-1-6.

[14] Krishnan, Hema, and Philip Samuel. "Relative Extraction Methodology for class diagram generation using dependency graph." *2010 International Conference On Communication Control And Computing Technologies*. IEEE, 2010.

[15] Hina Afreen Imran Sarwar Bajwa and Behzad Bordbar, 2011. “SBVR2UML: A Challenging Tranformation”, In. Proc. of IEEE’s, 2011 Frontiers of Information Technology, Page No- 33-38.ISBN-978-0-7695-4625-4.

[16] Imran S. Bajwa, M. Asif Naeem, Riaz-Ul-Amin, and Dr. M. Abbas Choudhary. 2006, Speech Language Processing Interface for Object-Oriented Application Design using a Rule-Based Framework, In. Proc. Proceedings of 4th International Conference on Computer Applications, Rangoon, Myanmar, Feb 23-24.

[17] Elbendak, Mosa, Paul Vickers, and Nick Rossiter. "Parsed use case descriptions as a basis for object-oriented class model generation." *Journal of Systems and Software* 84.7 (2011): 1209-1223.

[18] Imran Sarwar Bajwa, Irfan Hyder, “UCD-Generator A LESSA Application for Use Case Design”, In. Proc. IEEE-International Conference on Information and Emerging Technologies, IEEE-ICIET, Pages 200-205 Karachi-Pakistan.

[19] Harmain, H. M., and Robert Gaizauskas. "Cm-builder: A natural language-based case tool for object-oriented analysis." *Automated Software Engineering* 10.2 (2003): 157-181.

[20] Sagar, Vidhu Bhala R. Vidya, and S. Abirami. "Conceptual modeling of natural language functional requirements." *Journal of Systems and Software* 88 (2014): 25-41.

[21] More, Priyanka, and Rashmi Phalnikar. "Generating UML diagrams from natural language specifications." *International Journal of Applied Information Systems* 1.8 (2012): 19-23.

[22] Fabian Friedrich, Jan Mendling and Frank Puhlmann. 2012, Process Model Generation from Natural Language Text, In Advanced Information Systems Engineering, Eds. Lecture Notes in

Computer Science. Springer Berlin Heidelberg, Berlin, Heidelberg, 482-496.
[23] Mich, Luisa, and Roberto Garigliano. "A linguistic approach to the development of object oriented systems using the nl system lolita." *International Symposium on Object-Oriented Methodologies and Systems*. Springer, Berlin, Heidelberg, 1994.
[24] Harmain, Harmain Mohamed, and R. Gaizauskas. "CM-Builder: an automated NL-based CASE tool." *Proceedings ASE 2000. Fifteenth IEEE International Conference on Automated Software Engineering*. IEEE, 2000.
[25] Harmain, H. M., and Robert Gaizauskas. "Cm-builder: A natural language-based case tool for object-oriented analysis." *Automated Software Engineering* 10.2 (2003): 157-181.
[26] Subramaniam, Kalaivani, et al. "UCDA: Use Case Driven Development Assistant Tool for Class Model Generation." *SEKE*. Vol. 7804. 2004.
[27] Clavel, Manuel, Marina Soledad Egea González, and Viviane Torres da Silva. "The MOVA tool: a rewriting-based UML modeling, measuring, and validation tool." *Actas de las XII Jornadas de Ingeniería del Software y Bases de Datos*. 2007.
[28] Ben Abdessalem Karaa, Wahiba, et al. "Automatic builder of class diagram (ABCD): an application of UML generation from functional requirements." *Software: Practice and Experience* 46.11 (2016): 1443-1458.
[29] Naeem M, Bajwa IS, Chaudhri AA, Ali S," A controlled Natural Language Interface to Class Models",13th International Conference on Enterprise Information Systems (ICEIS'11), China, pages 102 – 110, SciTePress, 2011.
[30] Omer Salih Dawood and Abd-El-Kader Sahraoui, "From Requirements Engineering to UML using Natural Language Processing – Survey Study", EJERS, European Journal of Engineering Research and Science, Vol. 2, No. 1, January 2017, 44-50
[31] "About the Unified Modeling Language Specification Version 2.5" , https://www.omg.org/spec/UML/2.5/About-UML/
[32] Hnatkowska, Bogumiła, Zbigniew Huzar, and Lech Tuzinkiewicz. "Data modeling with UML 2.0." *Proceedings of the 2005 conference on Software Engineering: Evolution and Emerging Technologies*. 2005.
[33] Nakul Sharma, Prasanth Yalla, "A Hybrid Weighted Probabilistic Based Source Code Graph Clustering Algorithm For Class Diagram And Sequence Diagram Visualization", International Journal of Scientific & Technology Research Volume 9, Issue 4, Pages 3124-3158.
[34] Mohammadi, Sina, and Habib Izadkhah. "A new algorithm for software clustering considering the knowledge of dependency between artifacts in the source code." *Information and Software Technology* 105 (2019): 252-256.
[35] V. U. Gómez, S. Ducasse and T. D'Hondta, Visually characterizing source code changes, Science of Computer Programming.
[36] S. L. Abebe and P. Tonella, Extraction of domain concepts from the source code, Science ofComputerProgramming98 (2015)680–706.
[37] S. Bajracharya, J. Ossher and C. Lopes, Sourcerer: An infrastructure for large-scale collection and analysis of open-source code, Science of Computer Programming 79 (2014) 241–259.
[38] Abilhoa, Willyan D., and Leandro N. De Castro. "A keyword extraction method from twitter messages represented as graphs." Applied Mathematics and Computation 240 (2014): 308-325.
[39] Ziqi Zhang, Johann Petrak, Diana Maynard, Adapted TextRank for Term Extraction: A Generic Method of Improving Automatic Term Extraction Algorithms, Procedia Computer Science, Volume 137, 2018, Pages 102-108, ISSN 1877-0509, https://doi.org/10.1016/j.procs.2018.09.010.
[40] Duari, Swagata, and Vasudha Bhatnagar. "Complex Network Based Supervised Keyword Extractor." Expert Systems with Applications 140 (2020): 112876.
[41] Wael Etaiwi, Arafat Awajan, Graph-based Arabic text semantic representation, Information Processing & Management, Volume 57, Issue 3, 2020, 102183, ISSN 0306-4573, https://doi.org/10.1016/j.ipm.2019.102183.

[42] Javad Rafiei-Asl, Ahmad Nickabadi, TSAKE: A topical and structural automatic keyphrase extractor, Applied Soft Computing, Volume 58, 2017, Pages 620-630, ISSN 1568-4946, https://doi.org/10.1016/j.asoc.2017.05.014.

[43] Shansong Yang, Weiming Lu, Dezhi Yang, Xi Li, Chao Wu, Baogang Wei, KeyphraseDS: Automatic generation of survey by exploiting keyphrase information, Neurocomputing, Volume 224, 2017, Pages 58-70, ISSN 0925-2312, https://doi.org/10.1016/j.neucom.2016.10.052.

[44] B. D. Farahani, S. O. Fatemi and M. Ghorbani, "Automatic Keyphrase Extraction from Persian Scientific Documents Using Semantic Relations," 2019 27th Iranian Conference on Electrical Engineering (ICEE), Yazd, Iran, 2019, pp. 1972-1978, doi: 10.1109/IranianCEE.2019.8786696.

[45] Kousar Nikhath, A., & Subrahmanyam, K. (2019). Feature selection, optimization and clustering strategies of text documents. International Journal of Electrical and Computer Engineering, 9(2), 1313-1320. doi:10.11591/ijece.v9i2.pp.1313-1320

[46] Niraj Kumar, Dr. Kannan Srinathan, Dr. Vasudeva Varma, Towards Intelligent Text Mining Under Limited Linguistic Resources. PhD Thesis, IIIT Hyderabad, http://hdl.handle.net/10603/120959.

[47] Litvak, M., Last, M. & Kandel, A. DegExt: a language-independent keyphrase extractor. J Ambient Intell Human Comput 4, 377–387 (2013). https://doi.org/10.1007/s12652-012-0109-z

[48] Kumar, Niraj, Kannan Srinathan, and Vasudeva Varma. "A graph-based unsupervised N-gram filtration technique for automatic keyphrase extraction." International Journal of Data Mining, Modeling and Management 8.2 (2016): 124-143. DOI: 10.1504/IJDMMM.2016.077198

[49] Yakoob, Sk. Dr. V. Krishna Reddy and C. Dastagiraiah, "Analysis of Keyword Searchable Methodologies in Encrypted Cloud Data." (2016), IJET.

[50] Yan Ying, Tan Qingping, Xie Qinzheng, Zeng Ping, Li Panpan, A Graph-based Approach of Automatic Keyphrase Extraction, Procedia Computer Science, Volume 107, 2017, Pages 248-, ISSN 1877-0509,https://doi.org/10.1016/j.procs.2017.03.087.

[51] S. Beliga, A. Meštrovic, S. Martincic-Ipšic, "An Overview of Graph-Based Keyword Extraction Methods and Approaches", Journal of Information and Organizational Sciences, VOL. 39, NO. 1 (2015), PP. 1-20.

[52] F. Bulgarov, C. Caragea, "A Comparison of Supervised Keyphrase Extraction Models", WWW 2015 Companion, May 18–22, 2015, Florence, Italy, ACM 978-1-4503-3473-0/15/05.http://dx.doi.org/10.1145/2740908.2742776.

[53] K. Saidul Hasan, V. Ng, "Automatic Keyphrase Extraction: A Survey of the State of the Art", Proceedings of the 52nd Annual Meeting of the Association for Computational Linguistics, Baltimore, Maryland, USA, June 23-25 2014, pages 1262–1273. http://acl2014.org/acl2014/P14-1/pdf/P14-1119.pdf

[54] T. Gupta, "Keyword Extraction: A Review", International Journal of Engineering Applied Sciences and Technology, 2017. Vol. 2, Issue 4, ISSN No. 2455-2143, Pages 215-220.

[55] S. Siddiqi, Aditi Sharan, "Keyword and Keyphrase Extraction Techniques: A Literature Review", International Journal of Computer Applications, Volume 109 – No. 2, January 2015. ISBN: 0975 – 8887.

[56] 56. Jose Mary, G., & Haritha, D. (2017). A survey on best keyword cover search. Journal of Advanced Research in Dynamical and Control Systems, 9(Special issue 14), 2217-2231. Retrieved from www.scopus.com

[57] Mahata, Debanjan, et al. "Key2vec: Automatic ranked keyphrase extraction from scientific articles using phrase embeddings." Proceedings of the 2018 Conference of the North American Chapter of the Association for Computational Linguistics: Human Language Technologies, Volume 2 (Short Papers). 2018.

[58] Florescu, Corina, and Cornelia Caragea. "Positionrank: An unsupervised approach to keyphrase extraction from scholarly documents." Proceedings of the 55th Annual Meeting of the

Association for Computational Linguistics (Volume 1: Long Papers). 2017.
[59] Josiane Mothe, Faneva Ramiandrisoa, Michael Rasolomanana. Automatic keyphrase extraction using graph-based methods. SAC 2018: The 33th ACM Symposium on Applied Computing, Apr 2018, Pau,France. pp.728-730. HAL Id: hal-02640988. https://hal.archives-ouvertes.fr/hal-02640988
[60] Degl'Innocenti, Dante, Dario De Nart, and Carlo Tasso. "A New Multi-lingual Knowledge-base Approach to Keyphrase Extraction for the Italian Language." KDIR. 2014.
[61] Ravinuthala, V. V. M. K., and Satyananda Reddy Chinnam. "A keyword extraction approach for single document extractive summarization based on topic centrality." International Journal of Intelligent Engineering Systems (2017).
[62] P. Selvaraj, V. K. Burugari, D. Sumathi, R. K. Nayak and R. Tripathy, "Ontology based Recommendation System for Domain Specific Seekers," 2019 Third International conference on I-SMAC (IoT in Social, Mobile, Analytics and Cloud) (I-SMAC), Palladam, India, 2019, pp. 341-345, doi: 10.1109/I-SMAC47947.2019.9032634.
[63] Biswas, Saroj Kr, Monali Bordoloi, and Jacob Shreya. "A graph based keyword extraction model using collective node weight." Expert Systems with Applications 97 (2018): 51-59.
[64] Vijay Bhaskar Reddy, Y., Reddy, L. S. S., & Sai Satya Naryana Reddy, S. (2017). Comparative study of density-based clustering algorithms. International Journal of Civil Engineering and Technology, 8(12), 763-767. Retrieved from www.scopus.com
[65] Kishore, P. V. V., Anantha Rao, G., Kiran Kumar, E., Teja Kiran Kumar, M., & Anil Kumar, D. (2018). Selfie sign language recognition with convolutional neural networks. International Journal of Intelligent Systems and Applications, 10(10), 63-71. doi:10.5815/ijisa.2018.10.07
[66] Prasada Rao, C., Siva Kumar, P., Rama Sree, S., & Devi, J. (2018). An agile effort estimation based on story points using machine learning techniques doi:10.1007/978-981-10-8228-3_20 Retrieved from www.scopus.com
[67] Ahad, A., Yalavarthi, S. B., & Hussain, M. A. (2018). Tweet data analysis using topical clustering. Journal of Advanced Research in Dynamical and Control Systems, 10(9 Special Issue), 632-636. Retrieved from www.scopus.com
[68] V MNSSVKR Gupta, CH.V. Phani Krishna , “Key Node Selection Network Analysis And Centrality Measurements On A Dataset Of Cancer Documents”, VOL. 14, NO. 5, MARCH 2019. ISSN 1819-6608.
[69] Aziz Nanthaamornphong, Anawat Leatongkam, "Extended ForUML for Automatic Generation of UML Sequence Diagrams from Object-Oriented Fortran" Hindawi, Scientific Programming, Volume 2019, https://doi.org/10.1155/2019/2542686

Detection of intruder using the improved dragonfly optimization algorithm

Nagaraju Devarakonda[1], S. Anandarao[2], Raviteja Kamarajugadda[3]
[1]Associate Professor at VIT University, E-mail: dnagaraj_dnr@yahoo.co.in
[2]Assistant Professor at Lakireddy Bali Reddy College of Engineering,
E-mail: sarvani@lbrce.ac.in
[3]Assistant Professor at Lakireddy Bali Reddy College of Engineering,
E-mail: ravimtcs11@gmail.com

Abstract

Now a days, we are completely dependent on the systems and also on the internet. Most of our valuable information is stored in the system or in the cloud. So, it's very necessary to protect our data and also our system from the outside intruders and also from the intruders already entered into our system. To protect our system, intrusion detection systems are constructed using many traditional techniques. Some of them are decision tree, KDD, data mining techniques, Artificial Neural Network. But there are many flaws in these systems. In this paper, we have used the improvised dragonfly optimization algorithm. Here the traditional dragonfly optimization algorithm is improved by adding the convergence and fitness function. This usage of improvised dragonfly optimization algorithm in detecting the intruders has given the best performance than the other methods. To justify the statement, we have shown the experimental study and also comparative study with other two optimization algorithm. Here first the detection rate is computed on the darknet 2020 dataset by using the traditional dragonfly algorithm, then the detection rate is again computed on the same dataset but with the improved the dragonfly optimization algorithm. Finally, in the comparative study it's very clear that the improved dragonfly optimization algorithm produced the accurate result than the other optimization algorithm.

Keywords: Dragonfly optimization algorithm, detection rate, Whale optimization algorithm, Host-based intrusion detection system, VM based Intrusion Detection System

1. Introduction

Our systems are protected from many malicious programs using intrusion detection systems. Building an efficient intrusion detection system is very important to protect our system/network from the many forms of malicious programs. There are four types of IDS, they are Network intrusion detection system (NIDS), Host-based intrusion detection system (HIDS), Perimeter Intrusion Detection System (PIDS), VM based Intrusion Detection System (VMIDS).

NIDS monitor the network, through which the system is connected, to check and to protect the system from outside unauthorized access. The HIDS always checks the system(host) for any abnormal behavior. If abnormal behavior is observed by HIDS, it immediately takes the necessary action.

The best and efficient IDS is very important to protect the data related to our business. By this efficient IDS system, organizations' clouds can be monitored for suspicious events. The combination of NIDS and HIDS gives the best results to identify the suspicious behavior. This can be improved by adding optimization algorithms. There are many optimization algorithms some of them are whale optimization algorithm, ant colony optimization algorithm, Bee optimization algorithm, firefly optimization algorithm, dragonfly optimization algorithm etc.

Conventionally many techniques are used to protect systems from malicious programs. Some of them are firewalls, anti-virus, signature-based malware detection, genetic network programming.Optimization techniques can bring-out best results. So, among many optimization techniques, the dragonfly optimization algorithm can be used to bring-out the efficient intrusion detection algorithm.

Here we are considering the search agent as the dragonfly and the prey as intruder or location of suspicious behavior in the system. By this technique we can detect the abnormal behavior in a very short time and can take the necessary action. Detecting the abnormal behavior in a short time, will result in less damage to the system resources.

To prove the efficiency of the system we have shown the experimental results.

2. Literature survey

The paper [1] shows the building of intrusion detection systems using the Decision tree, random forest, K-NN. This technique has given the best results in identifying IPV4 attacks but failed in identifying the IPV6 attacks. The paper [2] shows the detection system using clustering and KDD but this has resulted to High false positive Rate and high detection rate. The paper [3] shows the detection system using Data Mining and Machine Learning, but this technique has produced the result by taking a very long time. The paper [4] shows the detection system combination with Machine Learning, Ranking, Voronos clustering but it showed the delay in the Processing in the large scale of the network. The paper [5] shows the detection system combination with Common Path Mining but this technique is not able to show the best performance over large datasets and also capturing the data logs is also a difficult task. The paper[6] shows the detection system working using Artificial Neural Network (ANN), IDS. This paper concentrated on identifying the abnormal behavior in the android mobile. This technique produced the Accuracy of 85% and detection of 81%. The paper [7] has used Hybrid Cryptography to reduce the network overhead. This has shown better performance than the magnet but largely reduced the ratio of total packet received / Total packet sent.

The paper [8] has shown the hybrid whale optimization algorithm for efficient performance of intrusion detection systems. This paper concentrated on the network intrusion system. In the paper [9], Data Mining Classification Methods are used to build the efficient Intrusion Detection Systems, these methods were used. This has shown the model formed by the combination CART, Naïve Bayesian and Artificial Neural Network Model. This paper [10], shows the construction of an efficient intrusion detection system using Bayesian Network and the Hidden Markov Model (HMM). In this first, training of the model is done using model learning and then constructing the Bayesian Network and next initializing the HMM parameters using the bayesian network. In [11], discusses the usage of modified dragonfly optimization algorithms in picking the best features. This refined algorithm added the convergence function and a fitness function to the traditional algorithm. This refined algorithm is shown the best accuracy when compared with other algorithms. Now this refined algorithm can be used in our work for building the accurate intrusion detection system.

3. Modified DRAGONFLY OPTIMIZATION Algorithm

Three important stages are present in dragonfly optimization algorithms: separation, alignment and cohesion.

Formula for separation in the dragonfly optimization algorithm is as follows.

$$A_i = -\sum_{j=1}^{M} z - z_j \qquad (1)$$

In the above equation

z and z_j represents the position of dragonfly and the position of jth dragonfly respectively. The total count of dragonfly is given by M.

Formula for alignment in the dragonfly optimization algorithm is as follows.

$$B_i = \frac{\sum_{j=1}^{M} ve_j}{N} \qquad (2)$$

In the above equation ve_j represent the velocity of j-th neighboring dragonfly

Formula for cohesion in the dragonfly optimization algorithm is as follows.

$$C_i = \frac{\sum_{j=1} z_j}{M} - z \quad (3)$$

'z' and 'z_j' represents the position of dragonfly and the position of jth dragonfly respectively. The total count of dragonfly is given by M.

Mathematical representation to move the dragonfly nearer to the prey

$$D_i = z^+ - z \quad (4)$$

Here we can consider the intruder or location of the suspicious behaviors in the system in the place of the prey.

Mathematical representation to move away from the prey

$$E_i = z^- + z \quad (5)$$

To move the location of the dragonfly. Two vectors are used step vector (DZ) and position vector(Z). Step vector is calculated by the below formula. This step vector tells us the direction of movement of the dragonfly.

$$\Delta y_{t+1} = (aA_i + bB_i + cC_i + dD_i + eE_i) + w\Delta y_t \quad (6)$$

In the above equation, 'a' stands for separation weight, ' stands for separation of i-th search agent, 'b' stands for alignment weight, ' stands for alignment of i-th search agent, c stands for cohesion weight, 'stands for cohesion of i-th individual, 'd' stands for food source, ' stands for attraction to the food source, 'e' is enemy factor, ' stand for migration from the enemy(worst feature), w is the inertia weight, and t is the iteration counter.

The position vector can be calculated using below equation.

$$y = y_t + \Delta y_{t+1} \quad (7)$$

The convergence function is shown below. This is can be applied on the dragonfly algorithm[11]

$$y_i = e^{\{-[trial(i)-1].\frac{\ln 10}{D-1}\}} \quad (8)$$

Where, D - Dimension, trial- Number of inefficient searching iterations. trial(i) is a number and lies between1 ≤ trial(i) ≥D.

The fitness function used in this paper, to check the accuracy of the dragonfly algorithm is shown below

$$(x) = \sum_{i=1}^{d-1} [100(y_{i+1} - y_i^2)^2 + (y_i - 1)^2] \quad (9)$$

4. Results and Discussion

This paper has shown the usage of a modified dragonfly optimization algorithm in constructing the best intrusion detection algorithm. The intrusion detection system must be able to detect the abnormal behavior in the system. Some of the abnormal behaviors are: logging into the same system by trying a number of possible passwords, logging into the system with the correct password from various IP addresses, repeatedly opening the secure folder in the system without doing any work, and coping huge amounts of data from the system at a faster rate.

Here, to carry on the experiment we have used the darknet 2020 dataset to check the accuracy of the proposed method. In this paper, we have shown the detection rate of three algorithms: whale optimization algorithm, dragonfly optimization algorithm, improvised dragonfly optimization algorithm.

IOP Conf. Series: Materials Science and Engineering 1074 (2021) 012011 doi:10.1088/1757-899X/1074/1/012011

The steps are:

Step 1: The dataset (darknet 2020) is collected. The dataset (darknet 2020) is downloaded from, https://www.unb.ca/cic/datasets/index.html

Step 2: Traditional whale optimization algorithm (WOA) is applied on this dataset

Step 3: Compute the accuracy using the whale optimization algorithm

Step 4: Traditional dragonfly optimization algorithm (DOA) is applied on this same dataset

Step 5: Compute the accuracy using the Traditional dragonfly optimization algorithm

Step 6: Apply our improvised dragonfly optimization algorithm (IDOA)

Step 7: Compare the accuracy of improvised dragonfly optimization algorithm

The below table1 gives the description of the dataset used.

Table1. Description about the dataset

Dataset Name	Darknet 2020
Normal system calls	10932
Suspicious system calls	1858

The below table 2 shows the detection rate using the WOA, DOA, IDOA. Among these three algorithm IDOA has produced the highest detection rate.

Table 2. Comparison among the WOA,DOA,IDOA

Algorithm Used	**Detected Suspicious system calls**	**Detection rate**
WOA	1738	93.5%
DOA	1758	94.6%
IDOA	1800	96.9%

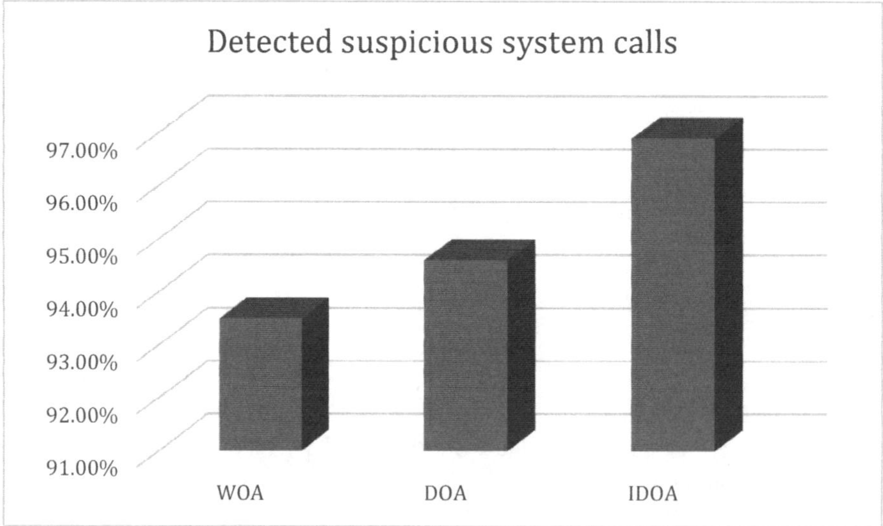

Fig.1. Shows the Detected Suspicious system calls by three algorithms

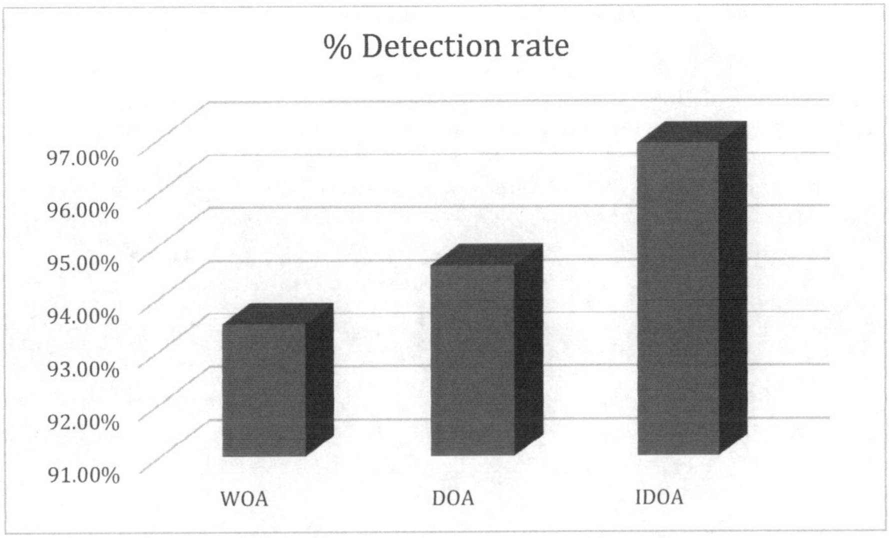

Fig.2. Shows the % detection rate by three algorithms

5. Conclusion

In this paper, the improvised dragonfly optimization algorithm is used in the construction of the intrusion detection system. The algorithm is improved by adding the convergence and fitness function. This technique of using improvised dragonfly optimization algorithms can perfectly detect the intruders with maximum accuracy than the other optimization algorithm. In this paper we also showed that our technique has given better performance than the traditional dragonfly and whale optimization algorithm.

Reference

1. Mohammed Anbar, Rosni Abdulah, Izan H. Hasbullah, YungWey Chong; Omar E. Elejla, "Comparative Performance Analysis of classification algorithm for Internal Intrusion Detection ", 2016 14th Annual Conference on Privacy Security and Trust (PCT), Dec 12-14,2016, Penang, Malaysia.
2. Weiwei Chen, Fangang Kong, Feng Mei, GuiginYuan, Bo Li, "a novel unsupervised Anamoly detection Approach for Intrusion Detection System", 2017 IEEE 3rd International Conference on big data security on cloud, May 16-18,2017, Zhejiang, China.
3. Anna L. Buczak, Erha n Guven, "A Survey of Data Mining and Machine Learning methods for cybersecurity intrusion detection", IEEE communication surveys and tutorials, vol. 18, Issue 2,2016.
4. Manoj s. Koli, Manik K. Chavan, "An Advanced method for detection of botnet traffic using Internal Intrusion Detection", 2017 International Conference on (ICICCT), March 10-11, 2017, Sangli, India
5. Shengyi Pan, Thomas Morris, Uttam Adhikari, "Developing a Hybrid Intrusion Detection System using Data Mining for power system", IEEE Transactions on, vol. 6, issues. 6, Nov. 2015.
6. Panagiotis I. Radogloa-Grammatikis; Panagiotis G. Sarigannidis, "Flow anamoly based Intrusion Detection System for Android Mobile Devices", 2017 6th International Conference on MOCAST, May 4-6, 2017, Kazani, Greece
7. Sharad Awatade, Shweta Joshi. "Improved EAACK: Develop Secure Intrusion Detection System for MANETS using hybrid cryptography", 2016 International Conference on computing communication control and automation (ICCUBEA), Aug 12-13, 2016, Maharashtra, India

8. R. Bilaiya and R. M. Sharma, "Intrusion detection System based on Hybrid Whale-Genetic Algorithm," *2018 Second International Conference on Inventive Communication and Computational Technologies (ICICCT)*, Coimbatore, 2018, pp. 822-825, doi: 10.1109/ICICCT.2018.8473082.
9. Srinivasulu, P., et al. "Classifying the network intrusion attacks using data mining classification methods and their performance comparison." *International Journal of Computer Science and Network Security* 9.6 (2009): 11-18.
10. Devarakonda, Nagaraju, et al. "Intrusion detection system using bayesian network and hidden markov model." Elsevier *Procedia Technology* 4 (2012): 506-514.
11. Devarakonda, Nagaraju, et al. "Unique Dragonfly Optimization Algorithm for Harvesting and Clustering the Key Features." *2019 IEEE 18th International Conference on Cognitive Informatics & Cognitive Computing (ICCI* CC)*. IEEE, 2019.

IOP Conf. Series: Materials Science and Engineering 1074 (2021) 012012 doi:10.1088/1757-899X/1074/1/012012

Automatic Vehicle Number Plate Recognition System Using Machine Learning

J M S V Ravi Kumar[1], B Sujatha[2] and N Leelavathi[3]
[1,2,3]Department of Computer Science and Engineering,
Godavari Institute of engineering and Technology (Autonomous), Rajahmundry, India

Abstract: The number plate recognition (NPR) system is one of the categories of smart transportation and detection mechanism (STDM). This is a combination of the technology in which the application enables the system to detect and automatically read the license id of number plate of vehicle from digitally captured images. Automatically capturing the license plate is the process of detecting and transforming the pixels data of a digital image into the plain text data or ASCII text of the number plate. Our project contains a method for the vehicle number plate recognition from the image using mathematical morphological operations (erosion, dilation). The main objective is to use and combine different morphological operations in such a way that the license plate of the certain vehicle can be detected and translatedeffectively. This is based on various operation such as image improvement, Gray scale transformation, Bilateral Filtering edge detection and getting the number plate from the picture of vehicle. After the completion of the above-mentioned steps, now the process of segmentation is being applied to detect the text present on number plate by making use ofmatching of template and OCR. This system is able to detect the license number accurately as well as quickly from the vehicle's picture.

*Keywords:*NPR, Vehicle Number Plate Recognition, Detection of Number Plate.

I.Introduction

The big combination of software things, under different perspectives of our current day globe, has led to the transformation of vehicles into conceptual resources in the field of datatechnologies [1]. Since the information applicationis meaningless without any information as the name implies, there is a huge requirement to redesignthe vehicle information among the information systems. It is also be achieved by external agents manually [2]or by special smartsystemwhichpermits detection andrecognition of vehicles by thelicense plates in real-time environments.

Between smart equipment, previously mentioned is made of the system of identification of the number plates of the vehicles [3]. Thismodel of vehicle number plate identification and recognition is used to identify the number plates and then make the detection of the plate that is to get the actual text from the picture and by using the modules of calculation, which use location algorithms [4], segmentation plate and character identification. Common usages are -

Car parking usage ,Electronic Toll Payment , Variety of vehicles in the road and Areas with less as well as more congestion

II.Literature Survey

Muhammad Tahir Qadrihas started the recognition the OCR techniques is used which is susceptible to misalignment and tovarious sizes [5]. The transformation achieved by himcan be used to advance the OCR identification from various size and angles in real-time. This programmed system for vehicle detection andrecognition system by making use of vehicle license number plate is exhibited [6]. A combination of image processing techniques of the model for detecting andrecognizing the vehicle from the databases stored in the system.

S. Kranthi, K. Pranathiinhave proposed that Automatic License/Number Plate Detection/Recognition (ANPR) is a method that detects a vehicle image and confirms their number of the license plate. ANPR is used in the presentation of stolen and crime suspectedvehicles [7]. ANPR can be easily used in various manners by using to identify it stolen vehicle on the highway or in parking lot or in traffic signals. This is all about the literature survey of the project "Automatic Vehicle Number Plate Recognition System Using Machine Learning" [8].

III. Existing System

Currently in many sectors vehicle is identified using the number plate[9] which is manually noted by a human which is a slow process.Thus, vehicle number plate recognition is an intensive manualprocess which can perhaps be automated using deep learning which forms the basis of this project[10].This is all about the existing system of the project"Automatic Vehicle Number Plate Recognition System Using Machine Learning".

IV. Drawbacks Of Existing System

The following are the drawbacks of the existing system in our project.

• More Manpower.

• More Expenditure.

• High Error Probability.

• Less Performance.

• More Dependency.

• Time Taking.

This is about the drawbacks of the existing system in the project"Automatic Vehicle Number Plate Recognition System Using Machine Learning" [11].

VProposed System

With the aim to connect all the smart vehicle number plate detection systems and to improve cost efficiency and reduce error. we are proposing a system which can be effectively reduce number of human resources and cost and detects vehicle number plates in a split of a second with no or negligible error.This system can be used for multiple purposed with a minimum setup cost for the required hardware [16].This is all about the proposed system of the project "Automatic Vehicle Number Plate Recognition System Using Machine Learning".

VI. Objectives Of Proposed System

The following are the objectives of the proposed system in the project.

- To make of Vehicle management easier.
- To reduce redundancy and error in number plate recognition.

IOP Conf. Series: Materials Science and Engineering 1074 (2021) 012012 doi:10.1088/1757-899X/1074/1/012012

- To generate analytical reports based on vehicle recognition.
- To minimize manpower required.
- To enable easier provision of remuneration to number plate detection and logging.
- Scalable
- Maintenance cost reduction.
- Effective.
- Affordable.
- Reduction of manpower.
- Human Error avoidance
- Secure
- Low cost
- Product excellence due to minimum requirements
- Optimized solution
- Increased reliability and faster reaction time compared to human workers

These are the objectives of the proposed system in the project "Automatic Vehicle Number Plate Recognition System Using Machine Learning".

VII. Algorithm

Step 1: Start
Step 2: Image Acquisition
Step 3: Image Desaturation
Step 4: Image Thresholding
Step 5: Optical Character Recognition
Step 6: Stop.
This is all about the algorithm in the project "Automatic Vehicle Number Plate Recognition System Using Machine Learning"[12].

VIII. Implementation

Loading and data pre-processing:
In this we train the model on the training data and validate it on the validation data. For this we need to create a validation set. This is done by creating a partition in the training set data[13].

IOP Conf. Series: Materials Science and Engineering 1074 (2021) 012012 doi:10.1088/1757-899X/1074/1/012012

Defining Model architecture:
We must define and elaborate how our model looksby answering questions like:
How many convolutional layers are needed?
What should be the function for activationapplied[14] in each layer?
How many hidden units do we want ineach layer?

Training the model:
To train a model, we need to:
Training the images and even map their corresponding true labels with validation images mapped with their corresponding true labels [15].

Estimation of performance:
Finally, we load the test data and go through the pre-processing step.
1. We at that point foresee the classes for these pictures utilizing the prepared model [16]. Picture Detection modules are:
2. *Image Acquisition*:
ANoIR enabled camera equipped for taking infrared photographs up to 3280×2464 pixels is utilized. This camera is likewise equipped for catching video at 1080p30, 720p60 and 640×480p90 goals which is an excellent video. To diminish the computational burden on system,[18]a camera arrangement to catch pictures with just 640×480 pixels can be used. Beforestarting to test the system with a real time/live image, a test image is preloaded to verify and test the algorithm as shown in below Figure 1.

Figure1: Input Image

3. *Image Desaturation:*
Right now, color picture is changed over into a grayscale picture by applying image desaturation.
This is done to decrease the intricacy of handling as it's simpler to register on grayscale picture than color picture. A color picture is changed over into grayscale advanced picture is a solitary example, conveying just the force data, and this can encourage the preparing far and away superior. OpenCV almost has the perfect capacity to change over color pictures to grayscale and the impact of this capacity is exhibited in Below Figure 2.

Figure 2: Desaturated Image

4. *Image Thresholding:*

Thresholding is nothing but the process of converting a grayscale image and generate a bi-level image which is easier to detect edges. It does not identify the main objects, but instead it separates them from the background to retrieve the information that we have to work with. At the time ofthresholding [18], it is very important to set the correct value of threshold which will best find a pixel as an object or a background. We use OpenCV's cv Threshold to achieve this, and the result is exhibited in below Figure 3.

Figure 3: Threshold Image

5. *Gaussian Filter:*

The Gaussian filterblurs the digital image by implementing an already existing Gaussian function. The Gaussian function is used to lowerunwanted detail, image noise. The visual effect of this Gaussian blurring technique is a smooth blur. The smooth haze of this Gaussian capacity obscuring looks like that of survey the picture through [17] a velvet or translucent screen.The Gaussian filter is regularly utilized as a pre-handling stage in computer vision calculations to upgrade picture structures at various scales. The impact of applying this capacity is exhibited in underneath Figure 4.

Figure 4:Gaussian Filtered Image

6.*Morphological Transformation:*

Morphological change examines geometrical structure inside a picture by testing it with little examples called structuring components. Therefore, a nonlinear image operator that is appropriate for investigating geometrical and topological structure is produced. The operator is applied to a picture so as to make certain highlights evident and recognize significant data from insignificant twists by lessening it to a skeleton. This procedure as a rule has four sorts of activities, which are extension, erosion, opening and shutting activity.

Figure 5 shows the impact of the picture after morphological change**.**

Figure 5: Image after morphological transformations

6. *Segmentation:*

The Image Segmentation is the way toward apportioning a picture into various parts which is then used to recognize objects or other pertinent data in advanced pictures. After morphological change, 2 fragments were distinguished as potential up-and-comers as delineated in underneath Figure. The fragments will at that point experience another significant procedure where the model will figure out which one has the best conceivable character. The fragment will at that point be segregated in a square formed picture, and it is tried with the goal that the longest rundown of potential character will be found and decided as the conceivable genuine number plate. The last picture with the most potential characters in it will at that point be picked to go for character acknowledgment in the following procedure. The impact of division on a picture is exhibited in underneath Figure 6.

Figure6:Segmented Image

7. Character Recognition:

In order todetect/recognize alphanumeric characters from the segmented image, the potential license plate is segmented later into individual set of characters before Optical Character Recognition testing as can be seen in below Figure 7.

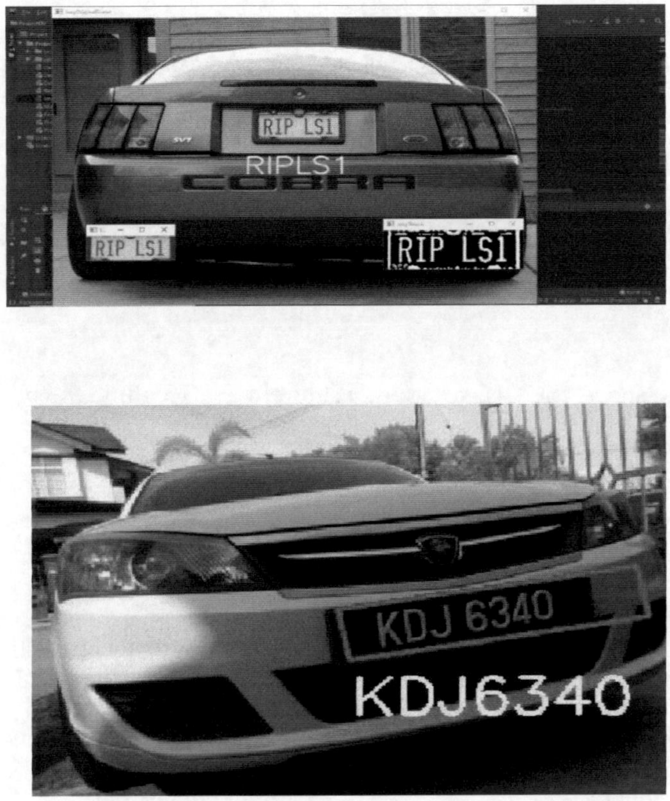

Figure 7: Final Image

IX. RESULT

The below figure8, shows python model training to detect and recognize license number plate. The below output is achieved by running the python program which trains the model using the training dataset and saves the model by using python libraries. Then a predictor file is run with a test image which is not included in the training dataset and gets recognized and for better visibility of detection the license plate area is marked by a red border.

Figure 8. Test model after training

Figure 9. GUI

Above figure 9, shows the GUI program written in .NET calling the python program and showing the results in a presentable format.

X. Conclusionsand Future Scope

This program can be used for variety of works where a license number is the primary way of recognizing a vehicle. Current application is completely usable for little uses as parking area or traffic line use.

In future we can include face acknowledgment and connection the entire framework with criminal database and recognize passing criminal outlaws.

XII. References

[1] Paper by J.M Mossi, L Sanchis, and A. Albiol, "The Detection of the Parked Vehicles Using the Spatiotemporal Maps," The IEEE Transactions on the Intelligent Transportation Systems, vol.12, no.4, pp. 1277-1291, 2011.
[2] Convolutional Neural Networks (CNNs) :http://cs231n.github.io/apr 2020

[3] Anton SatriaPrabuwono and Ariff Idris, "A Study of Car Park Control System Using Optical Character Recognition," in International Conference on Computer and Electrical Engineering, 2008, pp. 866-870.

[4] Automatic License Plate Detection & Recognition using deep learning https://github.com/GuiltyNeuron/ANPR

[5]WIKIPEDIA https://en.wikipedia.org/wiki/Automatic_number-plate_recognition/apr 2020

[6] Shiqiu Liu, Yiyu He,Zhenyu Xu, YifanZhu, and Han Huang, "The Chinese-style Plate Recognition Based on the Artificial Neural Network and Statistics," The Procedia Engineering, vol. 15, pp. 3556-3561, 2011.

[7]Kumar, Ravi, et al. "Reverse Engineering A Generic Software Exploration Environment Is Made Of Object Oriented Frame Work And Set Of Customizable Tools." International Journal of Advanced Research in Computer Science 2.5 (2011).

[8] Guoliang Fan and Xin Fan, "The Graphical Models for the Joint Segmentation and the Recognition of License Plate Characters," IEEE Signal Processing Letters, vol. 16, no. 1, pp. 10-13, 2009.

[9] You Shyang Chen and ChingHsue Cheng, “A Delphi-based rough sets fusion model for extracting payment rules of vehiclelicense tax in the government sector," Expert Systems with Applications, vol. 37, no. 3, pp. 2161-2174, 2010.

[10] Kumar, Ravi, et al. "System Testability Assessment and testing with Micro architectures." International Journal of Advanced Research in Computer Science 2.6 (2011).

[11] Paper by H. ErdincKocer and K KursatCevik named "Artificial neural networks-based vehicle license plate recognition," at Procedia Computer Science, vol no. 3, pp. 1033-1037, 2011.

[12]Estharakula, Suresh, and JMSV Ravi Kumar. "EBPH-MAC: Emergency Based Priority Hybrid Medium Access Control for Mobility Aware Cooperative WSN's In Indoor Industrial Monitoring." International Journal of Research 5.12 (2018): 1456-1465.

[13] ESTHARAKULA, SURESH, and DR JMSV RAVI KUMAR. "SECRBAC: Secure Data In The Clouds." International Journal of Research 5.15 (2018): 95-106.

[14] Kumar, J. M. S. V., et al. "Analyzing the Modern Tool-Supported UML-Based Static Reverse Engineering." International Journal of Advanced Research in Computer Science 3.4 (2012).

[15] Dara, Suresh, VaishnavaiPasari, NageswaraRaoBanoth, and JMSV Ravi Kumar. "Artificial Bee Colony Algorithm: A Survey and Recent Applications." International Journal of Pure and Applied Mathematics 120, no. 6 (2018): 313-321.

[16] ESTHARAKULA, SURESH, and DR JMSV RAVI KUMAR. "SECRBAC: Secure Data In The Clouds." International Journal of Research 5, no. 15 (2018): 95-106.

[17] RaviKumar JM, Rao KS. International Journal of Sciences & Applied Research.

[18] Parvathi, D. S. L., Leelavathi, N., Ravikumar, J. M. S. V., & Sujatha, B. (2020, July). Emotion Analysis Using Deep Learning. In 2020 International Conference on Electronics and Sustainable Communication Systems (ICESC) (pp. 593-598). IEEE.

IOP Conf. Series: Materials Science and Engineering 1074 (2021) 012013 doi:10.1088/1757-899X/1074/1/012013

Context-aware Multi-data Fusion to Achieve Finest Ride-share Schedule Using Big Data

Avinash Potluri[1] and N.V. Narendra Kumar[2]

IDRBT[1,2], Hyderabad

E-mail: potluri.avinash1@gmail.com[1]

Abstract. Enabling the taxi rides to share empty seats in their vehicles, to riders, is a win-win situation for both riders and drivers. Existing methods mainly target on minimizing the cost of travel for vehicles. In this work we focus on assigning a new rider to a vehicle considering the real-time context of both riders and drivers. The crux of the problem lies in analysing massive amounts of data in an instance of time to obtain the optimal ride schedule. We further study the scenario, to accommodate best compatible riders with minimal change in the route, to maximize the satisfaction of all possible entities in the business model in several aspects. Our system takes advantage from route recommendation and utility maximization approaches. We also inherit the benefits of collaborative filtering to provide best ride experience. The effectiveness of our approach is shown with experimentation on various datasets. Our work has significant importance, as it redefines the personalization preferences based on real-time context.

1. Introduction

From the traditional way of hailing a taxi service there has been a significant change because of the ridesharing benefits provided by several taxi-service brands in the industry. Ridesharing facility has gained huge interest in short time from all communities because of its one stop solution, providing eco-friendly and economical ride, reducing the traffic congestion. There can be a large scale adoption to this service once the autonomous vehicles come into place. Indeed, this has become the vision of several tech companies like Tesla, Waymo and Uber. With the estimate of having massive scale business these companies started to fine tune the utilization of available resources.

With the vast development of innovative technologies, individuals' every day life has a notable change. From online shopping to ride-sharing applications, there is a vast difference, compared to the traditional way of performing the same activity. In this paper, we address the practical problem in ridesharing, namely context-aware ridesharing, which assigns riders to the available vehicles considering the real-time context of both rider and driver. The objective is to maximize profit, minimize the detours and maximize the satisfaction of all possible entities in the business model. For instance, consider the following example, where we expect to have riders of similar context in a ride. Assume the capacity of the vehicle is three and there are two such vehicles and five distinct people waiting for a ride with approximately the same source and destination. Out of the five people three are going for an interview, where as the other two are going for a movie. Our work focuses on designing two ride schedules, where, in one ride schedule all the people going for interview are accommodated and in the other ride schedule the two people going for movie are accommodated.

The example shown above is completely different from the existing studies. We aim at improving the satisfaction of the users with best context relevant ride schedule. Given a set of users u1 , u2 , u3 , u4 , u5 , ...u_n and vehicles v 1 ,v 2 , ...v m the objective of our system is to recommend the association among users and vehicles, considering and assuring the context. The system has to return a best relevant setting in the available subspace. We drive around the issue of suggesting a viable intuitive solution for the ride-sharing e-business application. In this work, we utilize the available information to compute the end goal of improving the satisfaction of ride-sharing travelers. As figured, the key issue should be scheduling a route. With every request having a source and destination, existing works on ride-sharing has concentrated mainly on two viewpoints: accommodating all requests and minimizing the travel cost for the ride.

The following are the specific contributions in this paper:
– Develop an algorithm that recommends best path from all possible paths that can accommodate maximum number of passengers into the vehicle.
– We formulate a problem that recommends appropriate passenger into the vehicle based on real-time context.
– We overcome the computational bottleneck in computing while considering the contexts of both riders and drivers that redefines the personalization preferences.
– In addition we our method, maximizes the satisfaction of riders and drivers with minimal change in waiting time, route and cost.

The following is the organisation of the paper. In section 2 the overview of the problem and the need to explore this is emphasised. Section 3 deals with existing and proposed approaches. Section 4 shows the experimental results followed by conclusion and future work in section 5.

2. Overview of Problem Statement

The formulation of our problem comes from the following scenarios, Route Recommendation, Valid Scheduling for Dynamic Resources, Driver Interests, Similarity Matching and Pooling based on Context. We initially project the scope and need for formulating this problem. The overview of the problem with an example for each scenario is given below:

2.1 Route Recommendation

The ride sharing scenario where the shortest route is selected based on the initial rider, there is a chance of having a financial loss if the taxi does not find a customer in that route. The challenge is to recommend a route such that there is a maximum chance of accommodating the riders with minimum cost. This becomes more complex in the case where the riders should also have similar interests. The best suitable route is selected from the all possible routes available. Though we are greedy of getting the best route we need solve the issue in an instance of time. Consider figure 1, the nodes numbered 1 to 7 are nodes, where we have a ride request and a to h is cost incurred in that route. Assume the initial request is at 1 and the destination is at 7. Since, the shortest path is via 1, 4 and 7, the vehicle can accommodate only one more rider. But, if the route recommendation engine suggests another route i.e. either 1, 2, 3 and 7 or 1, 5, 6 and 7 we can accommodate three riders into the vehicle. Analysing all these possible routes for a ride is having significant role in maximizing the profit.

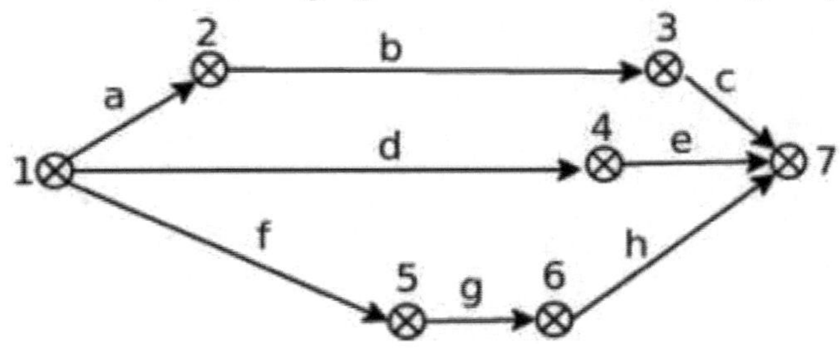

Fig. 1 Prossessing different routes

2.2 Valid Scheduling for Dynamic Resources

To accommodate the riders to a vehicle we need to compute. Our structure to store data for computing a ride schedule is as follows. We have start, maximum time to detour, schedule update and end. Firstly, the initial ride request, its location, start time and destination and end time. The maximum time to detour is given, where, we can update the current schedule without deviating much, the existing riders schedule. This is given as a schedule update to the driver. This depends on the number of available requests at that instance of time and the number of available vehicles.

Fig.2Schedule

2.3 Driver Interests

The selection of a vehicle and indeed the driver should match the interest of riders. For instance, if the language is common among the driver and the riders, the ease of clarifying address issues can be easily resolved. Similarly if the last ride taken by the driver is near his home, the satisfaction of the driver also increases. Our focus also lies in improving the satisfaction of riders as well as drivers. As shown in figure 3 D1 and D2 are the current locations of drivers with their vehicles. Let A, Band C be the riders requests. D1H and D2H are the homes of drivers. For D1, the schedule can be D1, A, C, D1H or D1, B, C, D1H and for D2, the schedule can be D2, A, C, D2H or D2, B, C, D2H. In contrast to the existing systems, which focus only on riders interests, our work also focus on drivers interest, and allocates the schedule which has minimum distance from their home from their last destination.

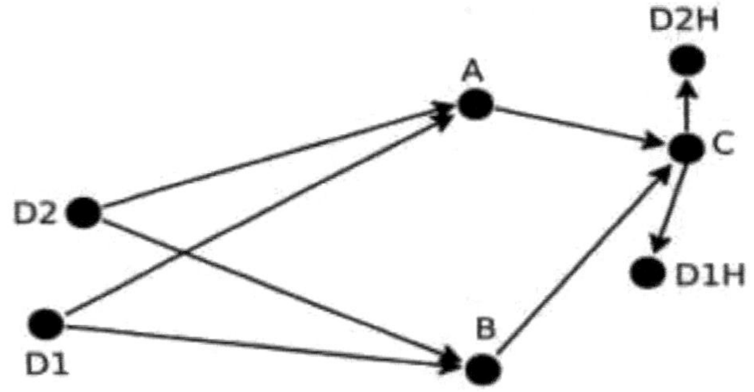

Fig. 3 Driver Scenario

2.4 Similarity Matching

The naive strategy to select riders in a vehicle is to select in the order in which the request is made in that route without exceeding the capacity of vehicle. But, to achieve maximum satisfaction from the riders we need to solve a bigger puzzle. With the constraint of effectively meeting the supply and demand we need to select the best matching riders in a vehicle. The matching can be starting with gender to the educational level. To evaluate the best match we can find the riders who have common friends. Similarity matching can be calculated based on the figure 4. Let A and B be two different riders, their common interests can be compared with a score for each feature in C1, C2 and soon CN .

From the available requests, we compute the score and the best matched score is used as one of the weights in deciding the riders in a vehicle.

A	B	C1, C2, .. CN

2.5 Pooling based on Context
To select the best compatible riders in a pool ride, the preferences of the users need to be known prior. Establishing correlation among the riders can guide us in selecting the appropriate riders. The preferences of users can be captured in several ways. Social media is one such place where the user interests can be highly captured. The huge availability of user interests like favorite color, interest in the kind of movies, favorite sports, favorite hero, favorite tourist place etc. in social media applications allows us to discriminate the riders. Though, the preferences obtained from social media benefits us, the optimal preference lies in the real-time context of the rider. We include the context-aware feature, which adds firmness in grouping the riders in a pool.

For instance, what can be the optimal selection..? whether selecting a group of users who are completely new to place or associating both new and old. By analysing the queries based on context-ware situations we can find common patterns. For example, the purpose of taking a ride can be going to a political meeting, hospital, mall, movie, competition, interviews, local trip, hotel, enjoy good weather, taking children to school and going to college etc. We can also benefit, if we observe the change of interests over time and contexts. Consider figure 5, there are five ride requests with approximately nearer source and destination. The context of visit for rider 1, 3 and 5 is interview and its same for all of them. The context of visit for rider 2 is movie and rider 4 is mall. As rider 2 and 4 is similar, the riders 2 and 4 has to be accommodated into the same vehicle. Riders 1, 3 and 5 need to be accommodated into same vehicle, which may improve the overall satisfaction of the system.

Fig. 5 Context Scenario

3 Approaches
As discussed above, we need to respond to the requests of the riders. We consider a set of ride requests and set of vehicles avaliable in an instance of time. Before the context based rider allocation representation is given, we discuss some naive approaches from the literature. We cover naive rider assignment approach, greedy approach and context-aware approach.

3.1 Naive Rider Assignment Approach
There are several approaches for allocating rider to a vehicle. In general, the rider allocation system, for a given request raised from a user the system designs a schedule. This is based on the source,

destination and the instance of time. Usually this will be the shortest path. The path that needs to be traversed to pick-up and drop-off is assigned and shared to the driver and the rider. Similarly, for all riders. If the system gets other rider requests in the initial scheduled route, the system reschedules the route without deviating much from the initial schedule and accommodates the other riders. The algorithm shown in 1 describes the naive rider assignment approach. We assume there are a set of ride requests and set of vehicles avaliable in that instance of time.

Algorithm 1 Naive Rider Assignment
1: Input : rider r i and Vehicle v j
2: Output : A schedule for rider and vehicle
3: T s : = { T s | s = 1, 2, 3, ..., s } where each T s is a valid schedule
4: T t : = { T t | t = 1, 2, 3, ..., t } where each T t is a valid schedule at an instance of time t from riders source and destination.
5: Selection of rider and update the schedule by traversing all possible schedules.

3.2 Greedy Approach
In the process of improving the naive approach of assigning a rider to a vehicle, a greedy approach has been developed. In this approach we tend to have a parameter tuning option, where, we can tune the system based on our interests. If our interest is to maximize the profit, we can tune the system accordingly. If our interest is to minimize the time, the system can be tuned accordingly. We can even select the best compatible customers into the vehicle. If we are greedy to match the interests of the users, we obtain the social connections from social networking sites and find the common interests of the riders. We find the best score among the riders combination and design a schedule. This can also be done based on the common language they speak. For instance we design a schedule with A, B and C with driver d1 and D, E and F with driver d2 considering the common language of drivers and riders. The system benefits the parameter to which we are greedy. In practice we can change to suitable parameter to which we are greedy, as the set of ride requests and set of vehicles avaliable in that instance of time change dynamically.

Algorithm 2 Greedy Approach
1: Input : available rider requests R and available vehicles V
2: Output : A schedule for rider and vehicle
3: T t : = { T t | t = 1, 2, 3, ..., t } where each T t is a valid schedule at an instance of time t from riders source and destination.
4: Retrieve all possible vehicles { V i | x = 1, 2, 3, ..., v }
5: For each vehicle V i , find the best combination of riders { R j | j = 1, 2, 3, ..., r } such that the cost is minimized.
6: For the selected rider vehicle combination, update the schedule S

3.3 Context-aware Solution - Our Approach
The proposed approach is a context based solution for assigning rider to a vehicle. This approach overcomes the drawbacks of traditional methods, which focus on maximization of profit, minimization of detour etc. In contrast to these techniques our method groups the riders in a single schedule considering the real-time context. The input to our method is set of riders who raised request of a ride and set of vehicles available in that instance of time. v z is the vehicle available, r x is the rider available, ru u is a possible route from the given source to destination and c s is the context of rider / driver. Initially a request is considered and based on the source and destination, a nearest vehicle is assigned. This is an initial schedule. Now, based on the available pending requests which are near to the initial source and destination, all possible routes are analyzed and a new route is rescheduled without having much deviation from the initial schedule. Let the difference be α time, β is the distance. Based on the context of rider in the vehicle, each of the other requests are given a score.

From the all possible routes, a route which has the chance of accomidating maximum number of riders is selected. Similarly, the similarity scores are also calculated based on the common interests of the riders and drivers. Each of these parameters is given a weightage. Let the weightage be w a , w b , w c . These weights can be tuned accordingly based on the requirement of the system. Once the wights are finalized, the scores are sorted and the schedules are designed. These schedules consists of the riders, who have similar context as well as interests. This guarantees the increase in the overall satisfaction of the system. Finally, a fine tuned route schedule is given to the riders, calculated from all possible combinations, such that the satisfaction score is maximized.

Algorithm 3 Rider-Route Assignment
1: Input : rider r x and vehicle v z
2: Output : Finest schedule for rider and vehicle
3: V z : = { V z | z = 1, 2, 3, ..., z } where v z is the vehicle available # V z at an instance of time t .
4: R x : = { R x | x = 1, 2, 3, ..., x } where r x is the rider available # R x at an instance of time t.
5: RU u : = { RU u | x = 1, 2, 3, ..., u } where ru u is a possible route # RU u from source to destination.
6: C s : = { C s | s = 1, 2, 3, ..., d } where c s is the context of rider / driver r x # R x
7: Selection of Route - RU such that the maximum capacity of vehicle is equalized.
8: Compute the score of satisfaction among the riders R x 1 , R x 2 ,..., R x x in vehicle V z combinedscore:: R(x 1) ./ R(x 2) ./ R(x 3) ./ ... ./ R(x x)
9: Fine tuning the combination of routes and riders such that the satisfaction score is maximized.

In our approach, we use the intelligence of Collaborative Filtering to find the best match. From the origin in 18 th century, till today there has been extensive research in improving the accuracy and performance of Collaborative Filtering techniques. The taxonomy of autoencoders using Maximum-likelihood estimation in a regular autoencoder evolves from:

$$\mathcal{L}_\beta(x_u;\theta,\phi) \equiv E_{q_\phi(z_u|x_u)}[\log p_\theta(x_u|z_u)] - \beta.KL(q_\phi(z_u|x_u)||p(z_u))$$

$$\theta^{AE},\phi^{AE} = argmax \textstyle\sum_u \log p_\theta(x_u|g_\phi(x_u))$$

In this scenario, our focus in not on optimizing the Collaborative Filtering technique, but rather we used one of the best available Collaborative Filtering techniques [Liang et al.(2018)Liang, Krishnan, Hoffman, and Jebara]. The route recommendation is inferred from [Huang et al.(2014)Huang, Bastani, Jin, and Wang]. The combination of Variational Autoencoders for Collaborative Filtering and route recommendation with context information improves the overall satisfaction of the system.

4 Experiments

We used synthetic data to show our context-aware approach. We consider a set of ride requests and set of vehicles available in an instance of time. The user interests are categorized into different types and are considered for finding similarities.

The graphs shown below are the satisfaction scores obtained for the synthetic datasets in various settings. The improvement in the satisfaction of the users highlight the importance of context.

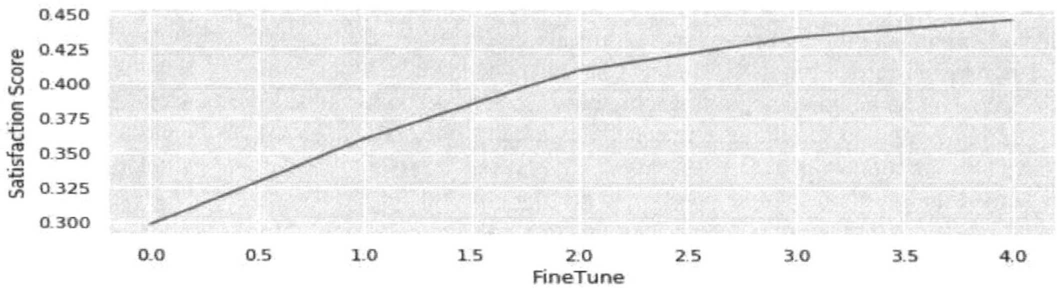

Fig. 6 Satisfaction Score

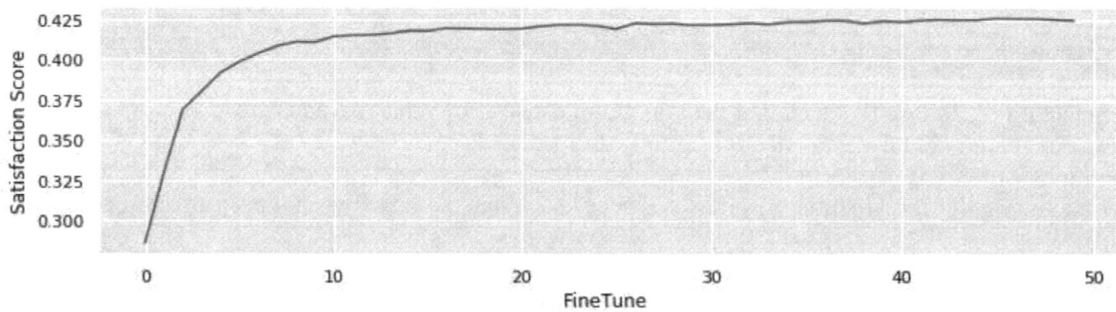

Fig. 7 Satisfaction Score

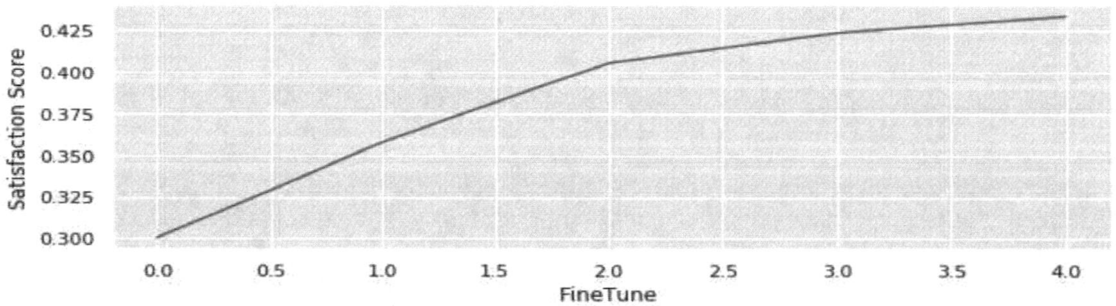

Fig. 8 Satisfaction Score

5 Conclusion

In this paper, we develop a Context-aware Multi-data Fusion approach using collaborative filtering and route recommendation system. Our approach enables to go beyond the traditional models capacity which minimizes cost. We empirically show through experimentation that, considering real-time context improves the satisfaction of users in the system. This paper best suits the current ride-sharing needs and in future when there is a large scale adoption to the autonomous vehicles. Our context-aware approach recommends the best compatible riders in a vehicle, capturing the real-time context, similarity and common interests with minimal change in time and route. In future one can explore the best way to capture the information since the real-time context and route recommendation are completely data-driven.

Deep Recurrent Q Reinforcement Learning model to Predict the Alzheimer Disease using Smart Home Sensor Data

Dhanusha.C[1] , A.V.Senthil Kumar[2]

[1]Research Scholar,Hindusthan College of Arts and Science,Coimbatore.
[2]Professor,Department of MCA , Hindusthan College of Arts and Science,Coimbatore.
E-Mail : kunjukannan020817@gmail.com

Abstract. Early detection of alzheimer disease is an essential process of the elderly persons when they are with the mild cognitive impairments. This work adapts the smart home based alzheimer disease detection by recording the daily activities of the resident's equipped with the sensor devices in their appliances as smart home. Using sensor data based clinical assessment of alzheimer patient is very complex as it is uncertain to predict such vague monitoring and understanding its deep characteristics. To acquire the deep characteristics of sensor dataset unlike the existing conventional supervised learning paradigm, this paper constructs an optimized self-learning model known as reinforcement learning process. This proposed model Deep Recurrent Q Learning based Reinforcement Model (DRQLRM) Process comprised of two stages they are pretraining using recurrent neural network to overcome the overfitting and fine tuning of the parameters involved in deep neural network and Deep Q learning process is used as reinforcement approach which uses the agents rewards and penalties to determine the optimal agents and their experience to predict the unknown pattern in the uncertain condition in effective way. The proposed model produced better result in CASA smart home test bed for investigating the presence of alzheimer
Keywords: Alzheimer, Smart home, Sensor data, Deep Q learning, Reinforcement, Recurrent Neural Network

1. Introduction

Activity Learning (AL) has the ability to define about the models which can identify human actions and the role of smart environments with the objective of helping [1]. In recent research development usage of smart home-based health care applications are started increasing to provide high quality of caring services especially for elderly persons [2]. This paper focuses on using smart home sensor systems to collect the daily activities of the elderly persons by sensing with the help of sensors appliances and it is recorded. The collected sensor data is used to predict the presence of alzheimer by monitoring the resident's daily behavior in terms of Activity of Daily Living (ADL), without disturbing them and if any changes in their order of activities it can be effectively identified at the early stage of alzheimer's by developing deep learning models.

In this paper the advantage of using Deep reinforcement learning model is adapted as it has the capability to handle voluminous data and enhanced evaluation measure to discover, appraise and acknowledge widespread data procedures for Alzheimer Disease Detection.

Few important aspects to be considered for constructing deep reinforcement learning are

- With restricted sample data, understanding the patterns and its underlying basic structure

- Objective functions have to be reviewed with the persistent illustration of events
  - The proposed framework needs the ability to consistently achieve better results even in dynamic environment as the alzheimer disease prediction is done daily activities of aged peoples using sensor data collected from smart home.

2 Related Work

Prafulla et al [3] in their work stated that it is feasible to conduct alzheimer prediction process using smart home sensor data and activity behavior of resident's. The model relies only on the statistical features of the collect information and it is less reliable in uncertain environment. Tanveer et al [4] reviewed numerous works of alzheimer detection by extracting the features, selecting potential ones and classification. Most of the works are related to the supervised learning models. Geert Litjens et al [5] in their work reported that using statistical and deep learning method can be used to handle the uncertainness very effectively. Novak et al [6] developed an abnormality based self-organizing map which investigates regular activities of elderly adults to detect cognitive impairments. Gray et al [7] constructed the random forest classifier for multimodal classification to detect the Alzheimer disease by analyzing MRI and PET data. Ane Alberdi et al [8] designed a smote boost algorithm to handle the misbalancing during feature selection to detect changes in their activities due to cognitive impairments and advise them to meet medical experts at the earliest.

3 Problem Statement

The existing supervised learning models developed for alzheimer disease detection performs the process of prediction by utilizing the labels defined for each instance used as input during the training time. But in this proposed work reinforcement learning solves the problem on its own instead of receiving explicit instruct to perform its task. But while using Q learning model [10] as reinforcement it is very difficult to distinguish and investigate the alzheimer disease detection. The Deep Q learning is empowered by Recurrent Neural Network which performs pre-training to gain historical expirence and apply for new patients disease detection. Thus, the objective of this proposed framework is to construct an effective self-reinforcing model which deeply analyse the daily activities of elderly person's and discover their cognitive

4 Reinforcement Learning

Reinforcement is a type of artificial intelligence which have the behaviour of dynamic programming characteristics which constructs and offers training to the algorithm by exploiting a policy of reward and penalty [9]. Unlike machine learning models, reinforcement learning solves the problem on its own instead of receiving explicit instruct to perform its task. Markov Choice Procedure is considered for learning process which embrace the performance of RL. By collaborating and interacting with the atmosphere, RL functions as an agent during its learning process. This agent will either rewarded or penalized based on their correct or wrong actions. RL as an agent itself learns without the assistance of humans by increasing its rewards and restraining its penalties.

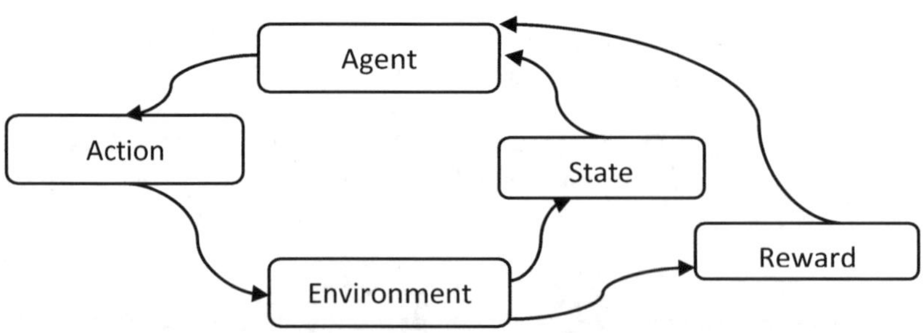

Figure 1. Workflow of the Reinforcement Learning (RL) Process.

The workflow of the reinforcement learning process is depicted in the figure 1. Consider an agent who is in the state 'st 'and desides to conduct an action 'ac'. The agent receives a reward RW (st, ac) and trabsfers to another new state st' when an action is completed. To accomplish mapping function between state and action is done based on a policy, it is used in each state to discover specific action to be done by an agent. Discovering an optimal policy π^* is the main objective of an agent in its lifetime. İt amplifies the whole reduced reward as formulated below as

$$\pi^*(\mathrm{st}) = \underset{ac \in Agent}{\operatorname{argmax}} \gamma \sum\nolimits_{st' \in State} pl_{stac}(st', ac)V^*(st', ac)$$

where $V^*(st', ac)$ is a value function that is used for estamiting state–action for evaluating the expected reward based on π. The utmost optimized function is accomplished by applying greedy optimal policy, evaluated by finding an agent which obtains highest reward of all other states. The function for optimality is denoted in the equation

$$V^*(st', ac) = \mathrm{RW(st, ac)} + \max_{ac \in Agent} \gamma \sum_{st' \in State} pl_{stac}(st', ac)V^*(st', ac)$$

Henceforth, reinforcement learning agent detects the presence of alzheimer through the interactions. It maximizes value of rewards by defining policy of best bellman and adopting and dynamic programming is used for value function.

5 Q-LEARNING Model

The Q learning model is one of the most important progress in RL, it evaluates which action the agent have to take next related to function of action value. It determines the existing value in a definite state after that it creates specification in the concern state. It assesses single state action value by developing temporal variance control algorithm as a policy[1]. The Deep Q function considers present state 'st' as well as action 'ac' as input and generates needed reward for the action of concern state. Before initializing the environment, Q function arbitrarily fixes a value. In later stages, after analyzing the environment it updates the value by providing optimal value for the action ac at the state 'st'. Meanwhile, agent iteratively perform sequence of action until it reaches its maximum reward.

6. Methodology: Deep Recursive influenced Q Reinforcement Learning Model for Prediction of Alzheimer Disease

The proposed Deep Recursive influenced Q Reinforcement Learning which considers forecasting of alzheimer disease as regression problem which can be fixed with supervised learning paradigm. This work used smart home-based behavior data to predict the presence of alzheimer by considering aged people's activities of daily living with the help of smart home test bests and its corresponding activity performance feature is treated as input parameters. While using RL the efficnecy of attaining knowledge about alzheimer detecting agents is identified by their otbained rewards so far. This consequence in unbalanced response releated to agents whic acclimatize their efficacy together with classifcaiton using supervised learning. Because the agents won't identified from available instances or records which are considered as input values are not competently known in the process of learning. Those factors impose that agent needs to be more effective in recognizing alzheimer patient's characterisitcs in deep. Thus, this work adapts the deep Q learning model which designs a predicting environment related to the input parameters which translates

IOP Conf. Series: Materials Science and Engineering 1074 (2021) 012014 doi:10.1088/1757-899X/1074/1/012014

conventional learning process to reinforcement model.
Let us consider the environment as an alzheimer prediction process, it includes specific feature combinations as parametric and constraint value that helps to predict alzheimer disease and individual combination comprised of isntance set along with labels. While an agent begins its play it discovers Activities of Daily Learning (ADL) datasets value of parameters invovled in the actions to achieve its rewards. Each near by or more approximate predicted value of the target (presence/ absence) of alzheimer, the agent attains either positive or negative reward based on its action related to its state at a specific time. The agent recieves an aggregate score when it ends its entier process.

To handle the huge information of Activities of Daily Learning collected from smart homes sensor devices, this work used the Recursive Neural Network (RNN) as stack of layers for Deep Q network (DQN) agent by utilizing the parameters of RNN and mapping finction is applied between RNN output to Deep Q values.
Let xt represents input of trainign dataset at time t, the hidden state is signifed as Ht at time t. Present xt input and hidden layer state Ht_{-1} which is a previous one fixes Ht and Ot denotes present layer's output at time t. The expected output Yt of training data and observed output Ot discover the error rate at time t is denoted as L. The weight distributed by RNN are signified as u, v and w. The Activation function of hidden layers is represented by F and b1 and b2. are the threshold across RNN.

The state vlaue of hidden layer at time t (i.e) Ht is formulated as

$$Ht = f(u * xt + w* Ht\text{-}1+ b1)$$

The output Ot predicted at time 't' is formulated as

$$Ot = f(v*ht+b2)$$

The error (L) occurred in RNN during time 't 'is equated as

$$Lt = Ot - Ht$$

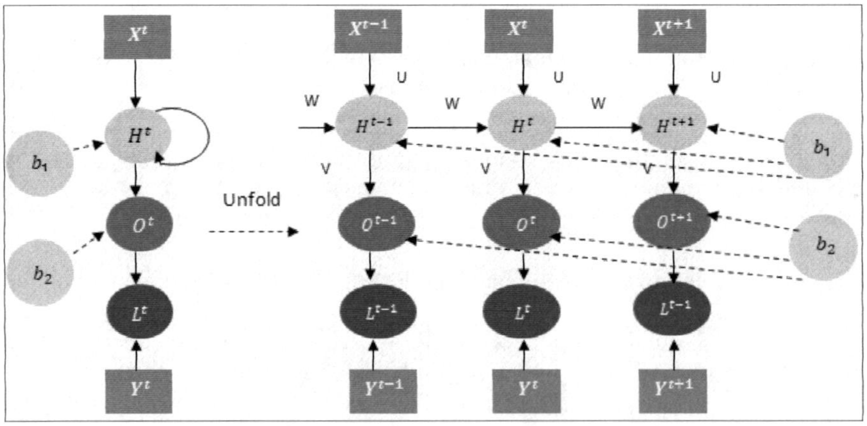

Figure 2. Recurrent Neural Network.

The RNN vital characteristics is that it can efficiently identifies the alzheimer presence is achieved by actual features repesentaiton as self-learning process among layer to layer and the constrain of sparse size which resolves problem of overfitting. In this work RNN is designed with a single input layer followed by three hidden layers and finally with one output Q layer. All the RNN layer is comprised of activation function known as ReLU and regularization L1 is utilized. This consequence absolute values of the data parameters to be penalized in the network while the data volume is high. Prior to training the model, an enormous state- action pairs space which consequeses to instability among correlation of data. During training phase to ensure non-divergence experience replay is used from the agents experience and it is stored in replay memory which holds the infroamtion of the concern state its action and reward obtained for that environment of the current time stamp along with next time stamp's state.

Primarily, at time 't', replay of experiece 'rp' will save experience of agent and consequence to a group of particular expereience set. An experience of individual Et at a time 't' is expressed as

$$Et = (stt,\ act, rpt, st_{t+1})$$

And the memory at time t is defined as

$$Memt = \{E1, \ldots Et)$$

To eliminate the divergence among parameters replay experience is an prominent idea which enables the agents to identify their experience during process of learning. The DRQN's training process comprised of two stages, in the first stage RNN is used for pre-training and in the second step DQN agent training is done. Based on ε Greedy policy [12] the agents choose and execute an action with arbitrary probability ε. Here, the chance of occurence (probability) 1-ε selects concern action denoting q value which is maximum. Deep Q learning is optimized by adapting stochastic gradient learning. The weights assigned to the network is done by data invovled in training process.

The algorithm of Deep Recursive influenced Q Reinforcement Learning Model for alzheimer prediction is shown as follows.

Input: CASA dataset
Output: Alzheimer prediction
Procedure
Begin

Stage 1: RNN for pre-training the network model

- Assign the memory capacity of replay as N
- Assign the RNN with arbitrary weights wti
- For i = 1 to n
- Train ith Hidden layer Hi
- Parameters of Hi is saved
- End for
- Set Q action value with hidden layer's parameters such as weight and bias
- Set parameter of Q' (target action value) same as the parameters invovled in Q.

Stage 2: DQN agent Training process

- For event 1 to m
- Set the observation series by genearting predicted patient arbitrarily
- For t = 1 to M
- Choose a arbitary action 'act' with chance of possibility ε
- Do action 'act' to recieve reward 'rwt'
- Arbitrarily produce next state st(t+1)
- Memory (Mem) is saved as (stt, act , rpt , stt+1)

IOP Conf. Series: Materials Science and Engineering 1074 (2021) 012014 doi:10.1088/1757-899X/1074/1/012014

- Owing to network parameter θ gradient descend is applied (rwt, - Q(stt ; act; θ)2)
- Assign Q' = Q
- End for
  - End for

7 Results and Discussions

The Effectiveness of Deep Recurrent Q Learning based Reinforcement Model (DRQLRM) model for alzheimer disease detection using smart home-based behavior data is analyzed by evaluation the model using various evaluation metric. The DRQLRM is implemented using python code and it is compared with three supervised learning models they are Random forest Tree (RF), Conventional Artificial Neural Network (ANN) and Deep Neural Network (DNN). The dataset used for predicting alzheimer is collected from CASA [13] composed of Activities of Daily living (ADL) of 400 elder peoples using smart home testbeds. It holds the information of 1BHK single resident apartments fortified with sensors in door, light, temperature, water and motion. The evaluation metrics used for determining the efficiency of four different prediction models are shown in the table 1 and they are RMSE, MSE and Accuracy.

Table 1. Performance analysis

Prediction Models	RMSE	MSE	Accuracy
DRQLRM	0.02	0.12	0.98
DNN	0.08	0.27	0.91
ANN	0.10	0.38	0.86
Random forest	0.71	0.8	0.79

The figure 3 shows the performance of the proposed DRQLRM produce less RMSE value compared to DNN, ANN and RF. This is because the reinforcement model learns through its experience of agents and its action to forecast the symptoms of the alzheimer based on activities of daily learning of the older peoples. The other algorithms worked as a conventional supervised learning model which uses the labels during training phase.

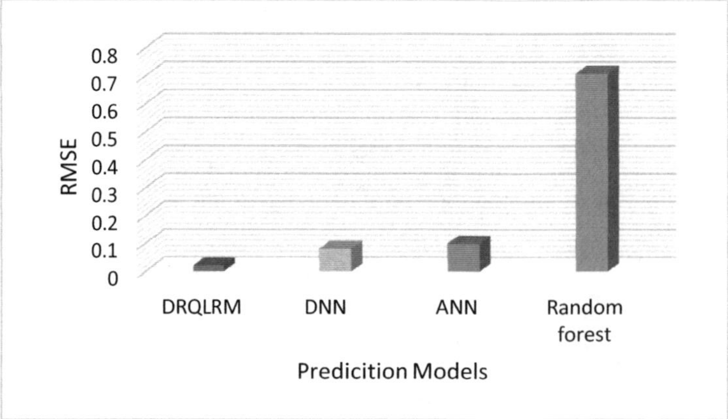

Figure 3. Comparative analysis based on RMSE.

The MSE value obtained by DRQLRM, DNN, ANN and RF is depicted in the figure 4. The results proved the efficiency of the reinforced model of Deep Q learning which enhances the learning process of the agents by applying greedy algorithm-based policy for providing rewards and penalties for their right actions and wrong actions respectively for alzheimer disease detection.

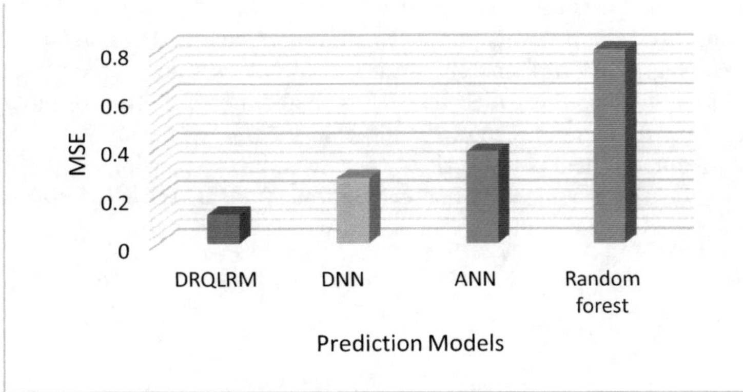

Figure 4. Comparative analysis based on MSE.

The figure 5 illustrates accuracy value obtained by four prediction models in alzheimer using smart home testbeds. The results revealed that by understanding the depth characteristic of activities of daily learning by the elderly peoples, the earlier stage of alzheimer can be more accurately predicted by the proposed DRQLRM by applying Recursive neural network during pre-training phase and applying the Q values to evaluate the expected output and controlling the overfitting of the parameters. Thus, the existing models provide less accuracy because of the vagueness in determining the presence of alzheimer using smart home datasets.

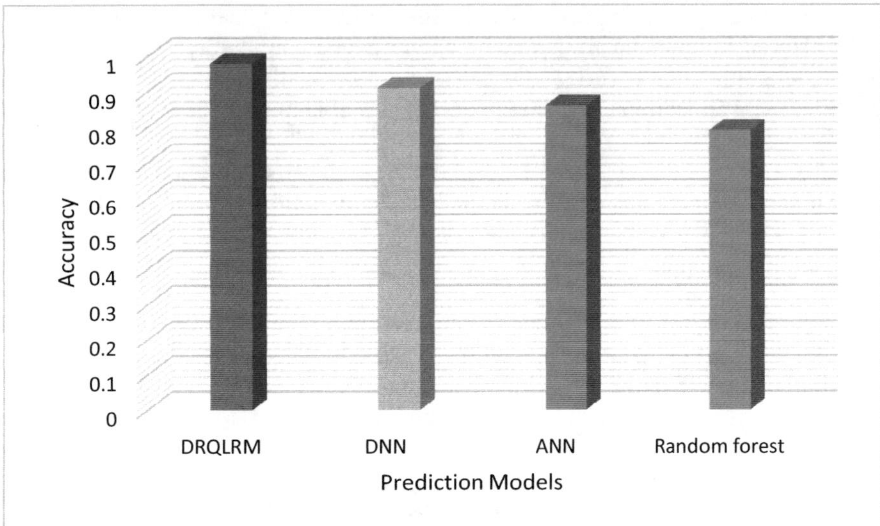

Figure 5. Comparative analysis based on Accuracy.

8 Conclusion

This paper constructs the reinforcement model by adapting deep Q learning for predicting the presence of alzheimer among elderly persons by examining the activities of their daily learning process. The sensor data are collected and used for investigation, this reinforcement model empowers its learning process by adapting the recurrent neural network and Q learning model is used to overcome the problem of overfitting and parameter fine tuning. It predicts the presence of alzheimer based on the irregularities of their daily activities. It self learns with the agents reward and penalty obtained during its training phase without using the class labels and recurrent neural network is used to gain the previous experience during pre-training phase. The simulation results provides detailed evaluation of the DRQLRM for alzheimer disease detection compared to the other deep learning and conventional supervised models.

References

1 Bravo J, Fuentes L, de Ipina, D.L. Theme issue: Ubiquitous computing and ambient intelligence. Pers. Ubiquitous Comput. 2011, 15, 315–316.
2 Rashidi, P.; Mihailidis, A. A Survey on Ambient Assisted Living Tools for Older Adults. IEEE J. Biomed. Health Inform. 2013, 17, 579–590.
3 Prafulla Nath Dawadi, Diane Joyce Cook, Maureen Schmitter-Edgecombe, Automated Clinical Assessment from Smart home-based Behavior Data, IEEE J Biomed Health Inform. 2016 July, 20(4): 1188–1194.
4 Tanveer M. Richhariya, Bharat, Khan, Riyaj Rashid, A.H. Prasad, Mukesh Khanna, Pritee Lin, Chin-Teng, (2020). Machine Learning Techniques for the Diagnosis of Alzheimer's Disease: A Review. ACM Transactions on Multimedia Computing, Communications and Applications.
5 16. 35. 10.1145/3344998.
6 Geert Litjens, Thijs Kooi, Babak Ehteshami Bejnordi, Arnaud Arindra Adiyoso Setio, Francesco Ciompi, Mohsen Ghafoorian, Jeroen Awm Van Der Laak, Bram Van Ginneken, and Clara I Sánchez. 2017. A survey on deep learning in medical image analysis. Medical image analysis 42 (2017), 60–88.
7 Novak, M.; Binas, M.; and Jakab, F. 2012. Unobtrusive anomaly detection in presence of elderly in a smart-home environment. In ELEKTRO, 2012, 341–344. IEEE
8 Gray, K.R.: Machine learning for image-based classification of Alzheimer's disease, Ph.D. thesis, Imperial College London (2012)
9 Ane Alberdi, Alyssa Weakley, Maureen Schmitter-Edgecombe, Diane J. Cook, AsierAztiria, Adrian Basarab, Maitane Barrenechea, Smart home-based prediction of multi-domain symptoms related to Alzheimer's Disease, IEEE Journal of Biomedical and Health Informatics PP(99):1-1, 2018
10 M. L. Littman, ``Value-function reinforcement learning in Markov games," Cognit. Syst. Res., vol. 2, no. 1, pp. 55_66, Apr. 2001.
11 T. Mannucci, E.-J. van Kampen, C. de Visser, and Q. Chu, "Safe exploration algorithms for reinforcement learning controllers," IEEE transactions on neural networks and learning systems, vol. 29, no. 4, pp. 1069–1081, 2018.
12 C. A. Merck and S. Kleinberg, "Causal explanation under indeterminism: A sampling approach." in AAAI, 2016, pp. 1037–1043
13 T. Mannucci, E.-J. van Kampen, C. de Visser, and Q. Chu, "Safe exploration algorithms for reinforcement learning controllers," IEEE transactions on neural networks and learning systems, vol. 29, no. 4, pp. 1069–1081, 2018. [325] C. A. Merck and S. Kleinberg, "Causal explanation under

indeterminism: A sampling approach." in AAAI, 2016, pp. 1037–1043
14 D. Cook. Learning setting-generalized activity models for smart spaces. IEEE Intelligent Systems, 2011
15 Dhanusha C, .A.V Senthil Kumar. "Intelligent Intuitionistic Fuzzy with Elephant Swarm Behaviour Based Rule Pruning for Early Detection of Alzheimer in Heterogeneous Multidomain Datasets" 'International Journal of Recent Technology and Engineering (IJRTE)', ISSN: 2277-3878, Volume-8 Issue-4, November 2019. Page No.: 9291-9298.Scopous Indexed
16 Dhanusha.c,Dr.A.V Senthilkumar, Enriched Neutrosophic Clustering with Knowledge of chaotic Crow Search Algorithm for Alzheimer Detection in Diverse Multidomain Environment "International Journal of Scientific & Technology Research(IJSTR)",ISSN:2277-8616,Volume-9 Issue-4,April 2020 Edition. Page No:474-481,Scopus Indexed.
17 Dhanusha.c,Dr.A.V Senthilkumar,Dr.Ismail Bin Musirin, Boosted Model of LSTM-RNN for Alzheimer Disease Prediction at their Early Stages" International Journal of Advanced Science and Technology(IJAST) Vol. 29, No. 3, (2020), pp. 14097 – 14108".

An Efficient Machine Learning Techniques Based on IoT for Effective Load Forecasting

K.Rameshwaraiah[1] Sankarpu Sateesh Kumar[2] K.Srinivasa Babu[3] T.Madhu[4]
[*1,3]Professor Computer Science and Engineering
[*2,4]Assistant Professor Computer Science and Engineering
[*1,3,4]Nalla Narasimha Reddy Educational Society's Group of Institutions
[*2]RGUKT – Srikakuram, A.P

Abstract— Internet of Things (IoT) networks (IoT) are computer networks which have an acute IT protection problem and in particular a computer attack detection problem. In order to solve this issue, the paper recommends the combination of machine learning approaches and concurrent data processing. The framework is developed and a new approach to the combination of the main classifiers intended for attacks on IoT networks. In which the accuracy ratio to the training time is the integral measure of efficacy, the problem classification statement is developed. We recommend the use of the data processing and multi-threaded mode provided by Spark to accelerate the speed of training and testing. In addition, a technique is suggested for preprocessing data set, which results in a large reduction in the sample length. An experimental examination of the proposed approach reveals that the precision of IoT networks' attack detection is 100 percent and the processing speed of data sets increases proportionally to the number of parallel threads.

Keywords: Machine learning, parallel processing, classifier design and evaluation

I. Introduction

The concept of the Internet of Things (IoT) is now being used increasingly in many areas of computer and technology engineering. Hospitals, residential administration, automation, traffic control, transport by rail and air, energy usage, development, etc .. There is a clear growth of IoT networks. IoT networks have key characteristics that separate them from many traditional computing networks. They link traditional computer network nodes, electronic "things," all kinds of devices and networks (WiFi, sensors, cell networks, and other networks) into a common global computer network. They are therefore linked to different types of networks and networks. This ensures the interchange of products and conventional computing equipment. Other characteristics of IoT networks include the high heterogeneousness of its components, geographic distribution and a very complex physical and logical structure with multiple interface points, low network node computation capacities and high data flow and transmission volumes. The features of IoT networks are very acute in security. The use of proven information protection resources which are adequate for conventional computer networks in IoT networks becomes ineffective. The key factors are the limited processing power and imperfection of IoT 's applications and the lengthy periods with which protection applications is upgraded.

The protection of current IoT networks is thus one of the key guidelines for improving in-service security technologies at present. Software bugs and their early fixes generate a threat to irregular behaviour on IoT computers. This suggests that modern and more efficient approaches for detecting and countermeasuring such attacks are desperately needed.

The use of machine learning techniques is one of the popular modern approaches for identifying suspicious events and attacks within computer networks.

The paper's contribution is the following

- It provides an overview of the issue regarding combined use of machine learning techniques and concurrent data processing to detect attacks.
- System architecture that enables anomalous events to be observed on IOT devices depending on the

IOP Conf. Series: Materials Science and Engineering 1074 (2021) 012015 doi:10.1088/1757-899X/1074/1/012015

solution suggested. Compared with the classificator mixture system, a weighted composition of classifications at the last stage of the curriculum allows the possibility of simultaneous training for the simple classifiers and increased the accuracy performance.A new method of generating a training sampleformachine learning algorithms, focused on the usage of parallel computing, is presented. Compared to [8], the implementation of this method is also per- formed for the local mode.

- The analysis discussed in this paper aims to measure the accuracy rate from the perspective of binary attack detection on IoT Networks based on the estimate of a certain class of potential aggressions. Based on the analysis of results obtained, conclusions are drawn regarding the possibilities of the approach proposed for further development..

II. RELATED WORK

In the research community the use of machine learning and parallel data technology is widespread. Despite this, a limited amount of work is carried out in the framework of IoT to address the difficulties of concurrent data processing frameworks, e.g. DryadLINQ, Hadoop, Spark and MPI.

The DryadLINQ platform is a Microsoft platform that offers parallel and distributed software creation tools. This application uses specific expressions that convert a dataset. The execution of the programme is structured as an acyclic graph, with vertices as processes and borders as interprocess channels.

The Map Minimize principle is introduced by Hadoop. Data analysis is done in two steps accordingly. The first stage (map) involves the isolation and propagation of the input data between the device nodes. The second stage (reduce) requires simultaneous collection of data and aggregation on these nodes.

Spark is a computational platform for resilient and concurrent distributed data sets (RDDs). Data RDD is grouped into blocks so that we can control them on many devices independently. One of the principal properties of the RDD is that if one computer fails, the independent partitions can be recovered. This provides the cluster based on Spark of fault tolerance. In comparison to Hadoop, Spark gives the ability to store memory results and slow computational methods. This helps you to gain the speed at which iterative algorithms are performed, particularly in machine learning algorithms.

MPI (Message Forwarding Interface) is another method for creating parallel structures. The MPI's basic unit is the message. Two methods are used in messages for the contact between processes: transfer and receipt. Because of the low levels of parallel processes, a high data processing speed is given. The popularity of this technology when designing supercomputers is largely decided.

In several academic papers, the challenge of designing parallel methods of processing data is solved and extended to different fields. In order to detect the malicious malware in a mobile application network, Shamily, Bauckhage and Alpcan in particular are suggesting implementations of the Distributed Support Vector Machine (SVM). To this reason, the distributed algorithm has been developed which underlies the SVM setup to resolve the quadratic binary classification problem. Scherbakov et al. recommend a web-server Apache log review framework. The architecture of the device consists of three stages, including interpretation, business logic and data layers. Kim and Yu suggest introducing the CEP framework, which is designed for bus traffic control and integrates Earthquake, Esper and Hadoop. Similar equipment is also used in the study of medical results.

There are many mechanisms that cross the topics of machine learning and concurrent data processing. TensorFlow, DistBelief, MXNet, and Piccolo [24, 25] are thus oriented not to an overall distributed computing paradigm applied on the Hadoop and Spark systems, but to solving machine learning issues under distributed computing conditions. One common feature of these systems is that they store and

control the mutual state using parameter servers. Shared state is expected for multiple ma-chines to scal model training. In addition, in a single dataflow diagram TensorFlow describes all machine learning tasks with parameters. This architecture varies from parallel systems in two ways for general purposes in order to boost machine learning performance. Next, the model maintains many runs on overlapping subdiagrams of the overall diagram. Application for IoT authentication, malware identification and access management regarded the numerous methods of supervised, unregulated learning.
Work has been provided on deep learning approaches to detect cyber threats. The proposed method consists of several steps: (1) the main analysis part is applied; (2) the neural network is pretrained using a restricted Boltzman machine; (3) the deep neural network training; (4) the output signal is generated based on a Softmax regression. Approaches are also systematically studied to develop parallel data processing structures. However, the use of machine learning techniques is improperly evaluated in these applications to ensure the security of IoT users. This distance is being eliminated by the solution suggested in this article.

III. METHODOLOGY &FRAMEWORK

Dataset Description:

We have used dataset for experiments «Detection of IoT Botnet Attacks». There are records in this data collection reflecting network stream vectors amongst nine commercial IoT devices. Two botnets have created anomalous network traffic: Mirai and BASHLITE.

The dataset comprises 7009270 documents separated into a set of classes: a range of attack classes and a benevolent class of traffic. The following classes comprise a selection of attack classes: ack_Mirai,scan_Mirai, syn_Mirai, udp_Mirai, udpplain_Mirai, combo_BASHLITE, junk_BASHLITE,scan_BASHLITE,tcp_BASHLITE, and udp_BASHLITE. We found these attacks to be the most common for IoT and the most traditional to validate the approach to attack detection based on computer and Big Data approaches.

The format of the display of documents is CSV: 115 fields (network stream features) are separated by a comma for each record. Each field is not empty and has a valued form that allows the use of this data set in classification and clustering.

Data Training:

We also developed the method for creating training samples to improve the precision of the attack classification. Duplicate data is omitted in the first place. And the data, which are weakly correlated, are retrieved. The degree that things are identical is determined by the Pearson correlation coefficient.

Data Loading and Sample Formation:

The records are put in 11 csv files for each computer. Per file is equal to one of eleven grades. The Python programming language and Spark DataFrame API have been chosen to allow for efficient storing and processing of csv files.
The SparkSession entry point has been used to build the dataframe object. The object DataFrameReader is used to load data from external storage into the data structure (e.g. object-relative databases, file systems, key-value stores). The API helps you to load data from files with formats CSV, JSON, PARQUET, TEXT, JDBC.

Model Training, Testing and Evaluation:

The experiments investigated a limited number of common classificator types:decision tree (DT), Random

IOP Conf. Series: Materials Science and Engineering 1074 (2021) 012015 doi:10.1088/1757-899X/1074/1/012015

Forest, DNN, support vector machine (SVM) and Extreme Machine Learning(EML).The MLlib library implements these models with their learning algorithms. There are a vast variety of data analytical functions obtained in this collection, using machine learning and mathematical techniques. These features are designed for distributed mode execution..

IV. EXPERIMENTAL RESULTS

In this paper author is evaluating performance of various classical algorithms such as SVM, Random Forest and Naive Bayes etc. to detect attacks on network using IDS datasets such KDD, NSL but this classical algorithms unable to predict dynamic (if attacker introduce new attacks with changes in attack parameter) attacks and need to be trained in advance to detect such attacks to overcome from this problem author has evaluate performance of Deep Neural Network (DNN) algorithm with dynamic attack signatures and detection accuracy of DNN shown to be better compare to all classical algorithms.

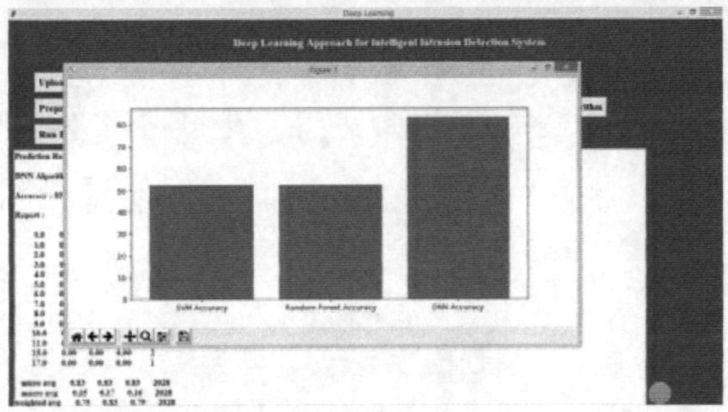

In above graph x-axis represents algorithm name and y-axis represents accuracy and DNN is more accurate technique.

We use parallel processing techniques for effective online load forecasting, which takes less time compared to regular loading The graph above shows that parallel processing takes less time than normal processing.

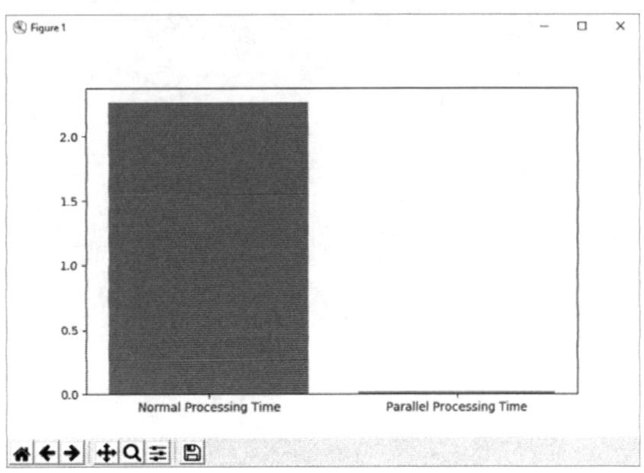

V. CONCLUSION

A new approach to detect attacks on IoT devices based on machine learning and parallel data processing was proposed in the paper. The layout of simple classifiers for attack detection in IoT networks was calculated on the basis of the study of cutting-edge implementations of machine learning techniques and parallel data processing for the solution of computer security problems. This involves SVM, RF, DNN and EML Based on simultaneous data processing; a classification issue statement was developed in which the fundamental efficacy measure was the accuracy-to-time ratio for preparation and research. An experimental test of the solution proposed found that the precision and speed of attack detection in the IoT Network was substantially increased. The sensitivity is approximately 100% and during the identification the velocity increases due to the number of parallel threads. The use of methods for integrating simple classifiers will improve the accuracy of the basic classifiers shown.

REFERENCES

[1] D. Evans, "The Internet of Things How the Next Evolution of the Internet IsChanging Everything," CISCO white paper, 2011.
[2] Zh.J. Shi and H. Yan, "Software Implementations of Elliptic Curve Cryptography," *International Journal of Network Security*, vol. 7, no. 1, pp. 141–150, July2008.
[3] M. Yassine and A. Ezzati, "Towards an Efficient Datagram Transport Layer Security for Constrained Applications in Internet of Things," *International Review on Computers and Software*, vol. 11, no. 7, pp. 611- 621, 2016, doi:10.15866/irecos.v11i7.9438.
[4] G. Apruzzese, M. Colajanni, L. Ferretti, and A. Guido, "On the Effectiveness of Machine and Deep Learning for Cyber Security", in *Proc.of the 10th International Conference on Cyber Conflict (CyCon)*, pp. 371– 390,2018,doi:10.23919/CYCON.2018.8405026.
[5] Th.Th. Nguyen and V.J. Reddi, "Deep Reinforcement Learning for Cyber Security", CoRR, http://arxiv.org/abs/1906.05799,2019.
[6] D.S. Berman, A.L. Buczak, J.S. Chavis, and Ch.L. Corbett, "A Survey of Deep Learning Methods for Cyber Security", *Information*, vol. 10, no. 4, 122, https://www.mdpi.com/2078-2489/10/4/122, 2019, doi: 10.3390/info10040122.
[7] M. Usman, M.A. Jan, X. He, and J. Chen, "A Survey on Representation Learning Efforts in Cyber security Domain", *ACM Comput. Surv.* vol. 52, no. 6, Article 111, 28 pages, October 2019, doi: 10.1145/3331174.
[8] I. Kotenko, I. Saenko, A. Kushnerivich, and A. Branitskiy, "Attack detection in IoT critical infrastructures: a machine learning and big data processing approach", in *Proc. of the 27th Euromicro International Conference on Parallel, Distributed and Network-Based Processing (PDP)*, pp.340-347,2019,doi:10.1109/EMPDP.2019.8671571.
[9] Kotenko I.V., Saenko I.B., Kushnerevich A.G., "Architecture of the Parallel Big Data Processing System for Security Monitoring of Inter- net of Things Networks", SPIIRAS Proceedings. Issue 4(59). pp.5-30, 2018, doi:10.15622/sp.59.1.

Fast Hybrid Adaboost Binary Classifier For Brain Tumor Classification

[1]S.Jayaprada [2]G.JayaLakshmi [3]L.KanyaKumari

[1]Associate Professor, Department of Computer Science & Engineering, Gudlavalleru Engineering College, Gudlavalleru, Andhra Pradesh, India. jayaprada79@yahoo.com
[2]Assistant Professor, Department of Information Technology, VR Siddhartha Engineering College, Vijayawada, Andhra Pradesh, India. jaya1123@gmail.com
[3]Assistant Professor, Department of Information Technology, Andhra Loyola Institute of Engineering and Technology, Vijayawada, Andhra Pradesh, India. kanyabtech@yahoo.com

ABSTRACT

The brain tumor is a dangerous disease and its characterization is a difficult undertaking for radiologists in light of the heterogeneous idea of the tumor cells. Computer-Aided tasks and its implementation of the current models with their frameworks would suffice the design metrics to recognize and relate the different tumors including brain though the process with scanning of the brain would emphasize on MRI. The models with support vector machine-nearest neighbor, naïve bayes analysis are obtained aren't enough to produce the different scenario at each set of layers and its correlation values of the performance observed. We propose a design model with fast Adaboost binary classifier, ensemble approach with fast boosting algorithm with the pre-trained model's dataset to analyze the different problem which proposes a strategy for multiple application scenarios of feature extraction to provide a classification model for different tumors in the brain to improve different performance parametric values for each trained and test sets with the prediction algorithm with each design formulation.

Keywords:

Tumor, Brain-tumor, Classification, Adaboost, ensemble,CNN,RCNN

1. INTRODUCTION

Medical Advancements in the field of biomedical aspects have defeated different sicknesses over the most recent couple of years yet peoples are as yet, experiencing disease because of its unusual changes in the real-time world. This tumor related to the brain is as yet a critical issue for mankind. The tumor in the brain for this disease has been genuine and most extreme rising afflictions as of till known scenario. In the USA, right around 23,000 patients were distinguished mind tumor malignant growth in 2015 [1].In one more report of disease markers [2], this ailment is consistent in the two grown-ups and youngsters. Roughly 80,000fresh instances of essential cerebrum tumors were accounted for in 2018 [3]. Meningioma spoke to 36.3% (29,320), Gliomas 26.5%(21,200), Pituitary tumors spoke to almost 16.2% (13,210)and the remaining of the cases had a place with different kinds of cerebrum tumor, for example, Malignant, Medulloblastoma, and Lymphomas. The chief reasons for such ailment are disease related sickness with dismalness. Successful treatment of this illness is significant which depends on its convenient and precise recognition. Presently the treatment of the tumors and its examination has become more costly and hectic states to identify each and differentiate the tumors. So their regenerative and active scenario would affect the different systems utilized to implement such drastic measures. Unpredicted nature of the changes in cell generation and its accumulation as a cluster results in formation of a clad known as tumor. Along these lines, cerebrum tumors are an autonomous and unpredictable proliferation of synapses [4], [5].

2. MOTIVATION

Boosting is an ensemble method that tries to create a strong classifier from a number of weak classifiers. The machine learning technology that discovered to answer problem like that if the data is partial, or the chosen representation is over-fitted with the related data, the representation cannot illustrate the exact performance is ensemble approaches. Ensemble approaches are mainly modeled into three types, specifically, boosting stacking, and bagging. Bagging is a procedure of permitting redundancy as arbitrarily classifying training statistics to a forecast form in the procedure of knowledge. Boosting is alike to bagging, but it reprocesses the data that the previous representation cannot categorize fine in the learning procedure by given that weight. Stacking is an ensemble method that is as well identified as great learning. It utilizes different forms, such as random forest, k-nearest neighbors, and support vector machine as support learners to make innovative data based on expected results. Next this novel data is used for one more analytical model, which is called meta-learner, to in conclusion obtain the predicted assessment. Low-level processes such as thresholding, edge detection, and morphological techniques, are speedy and be able to be used for brain tumor segmentation. But, the routine of these methods extremely depends scheduled the apparent difference in the intensities among tumor and non-tumor areas. We propose a fast hybrid binary classifier for this. The phrase hybrid is used here as, in other ensemble models, a identical set of weak learners is used excluding here, a various different set of weak learners is used. The hybrid ensemble learning form, assembled by these weak learning models, is functional in the job of brain images tumor, on-tumor classification.

3. LITERATURE REVIEW

AI approaches have been widely utilized in different spaces including clinical diagnostics and precautionary medication. A predetermined number of studies, in any case, contain focused on determination of cerebrum tumor particularly utilizing attractive reverberation imaging (MRI). Generally ML techniques train with test customary ML calculations on MRI information. As of late, a portion of the methodologies have utilized DL for the determination of cerebrum tumor.

Many studies in the literature used single classifier techniques for brain tumor classification. Many studies declared that classifier ensembles suggest improved performance, compared through single classifiers [11].The ensemble of classifiers has been used for the data mining domain [12].

S. Mary Praveena and Ila Vennila.[3] proposed a structure that utilized an arrangement called tri-design.With scenario of different data set from the Neural net observations different scenario of the tumors especially related to brain are investigates and analyzed. The creators additionally considered information expansion methods to acquire results' precision. This exploration accomplished an exactness of 98.69% by utilizing engineering for upgrading grouping and identification.

Subsequently, these authors W. Yang Q. Feng M. Yu Z. Lu Y. GAO Y. Xu Et Al.[4] improvise a three dimensional structure modeling of CNN that naturally reviews glioma from customary multi-arrangement scanning of each observed tumor part by characterizing a locale of intrigue consequently. The trained model and its performance accuracy was turnout to be 89.5 %, notwithstanding the exactness of anticipating tumor that was accomplished at 92.98% dependent on areas of interests.

Sugandha Agarwal, O.P. Singh and Deepak Nagaria., [6] additionally received the idea of profound exchange for arrangement of pictures and utilized a similar information source talked about in [5]. The design model and its functionaries were classified and represented with test and train scenario of each set of case utilized. To characterize patients requirements and their identification of the tumor with each section of the images from the patients are observed with the values of 98%.

Convolve nets with as a CN network model, by B. Cheng, M. Liu, H. I. Suk, D. Shen and D. Zhang., [8] characterized cerebrum tumors. The investigation improved the degree of precision of the design and its features with patients were at 86.50% by imparting the different component layers with each set of parametric criteria to enhance the design with 64 levels by getting a variety the guides of Caps Net at some layer implemented with convolution.

Mr. Manish Saraswat and Mr. Ajay Kumar Sharma., [9] the design model with layers connecting with convolution and rectification have characterized with different specific scenario of difference observed from the results which are actuated with the value of 98.5% on preparing and 84.19% for approval. The investigation accomplished a precision proportion of 91.9% by utilizing the previously mentioned framework arrangement with back propagation NN.

Sushma Laxman Wakchaure and Anil Khandekar., [10] built up an engineering dependent on CNN for highlights extraction. To improvise a specific structure and its importance we have different learning layers and its arrangement with 3X3 model. The investigation professed to accomplish a precision of 81% that was additionally upgraded to highlight arrangement model of CNN dependent on ELM (outrageous model for to learn different machine algorithms represented).

4. DATASET DESCRIPTION:

The dataset used is taken from the Kaggle. The dataset have a total of 253 MRI images. Out of them, 155 are marked "yes", which shows that there is a tumor and the remaining 98 are marked "no", which says that there is no tumor.

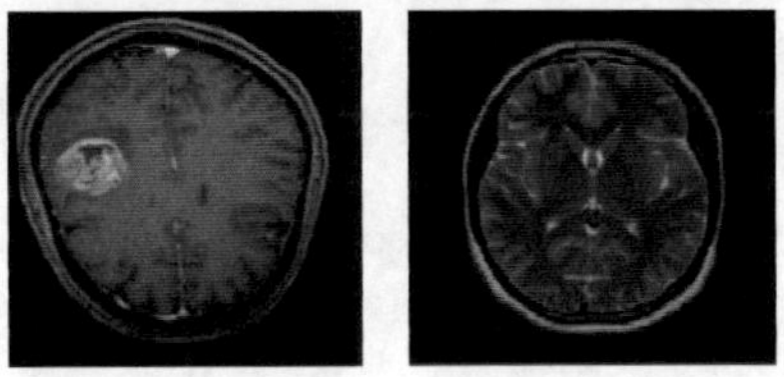

MRI with a tumor MRI without a tumor

Figure 1: Sample image of an MRI

The above figure 1 shows the Tumor in the left side of the brain and in the second image, there is no tumor formation.

Implementation of Existing models: CNN and mask RCNN:

CNN:

The CNN Model is designed as a neural network using *Keras* library with different convolutional and pooling layers.

Steps:

1) Pre-Training the CNN model: for this we create a "***train generator***" and "***validation generator***" to store split images of the training and the test sets into two classes (yes &no).

2) Training the CNN model: we fit the image data to the trained neural network for around 150 "***epochs***" and at the end of the 100th epoch, it was observed that the trained CNN model has validation accuracy of 70.85% which denotes that model can correctly classify about 71% of the test set images as Tumor/Non-tumor.

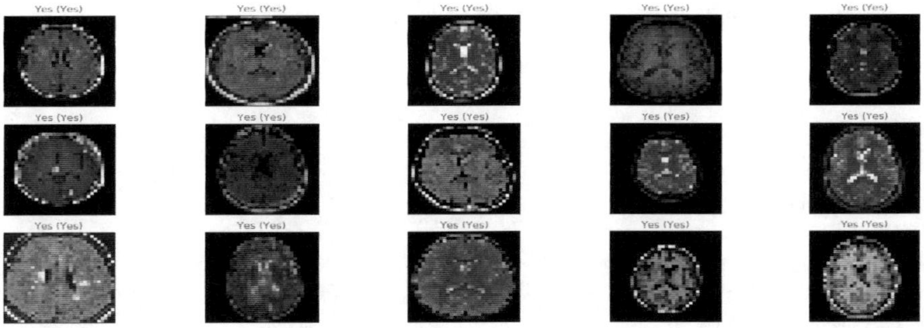

Figure 2: Results of Brain Tumor classification using CNN

After the training, the “accuracy” and “loss” of both “train generator” and “validation generator” for all the 100 epochs(iterations) are plotted.

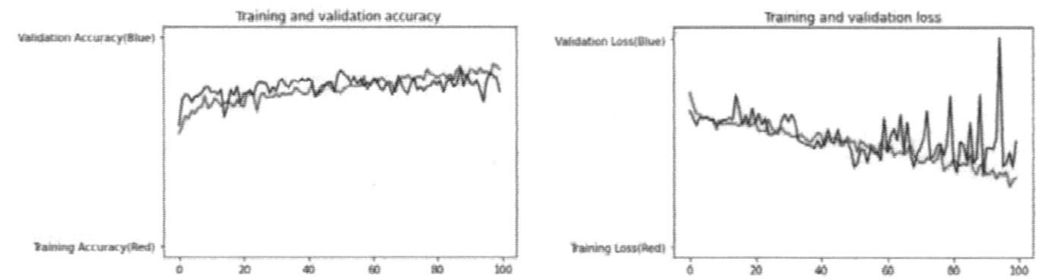

Figure 3:Accuracy and Loss graph

Mask RCNN :

The mask RCNN model is used to construct an definite detector to spot out on the position of the tumor on the MRI scan. For this implementation annotations are added for scans with VGG Image Annotator (VIA).And dataset is divided into 4 sets training, test, validation and annotations. Mask R-CNN to make an actual detector which will point out on the position of the tumor on the scan.

Steps:

1) Configuration for training on the brain tumor dataset.
2) Get the x, y coordinates of points of the polygons that build up the draw of each object illustration. These are stored in the contour attributes.
3) Generate occurrence masks for an image .A Boolean array of shape [height, width, instance count] with one mask per instance.
4) Training and validation the dataset.

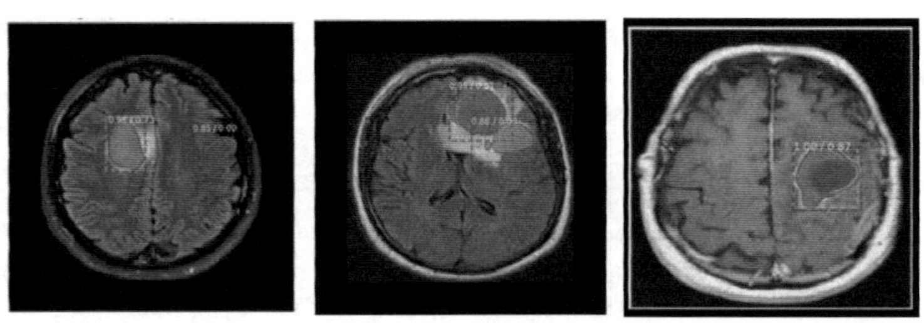

Figure 4: Results of Brain Tumor classification using RCNN

Region-based convolutional neural network shows accuracy results 83% for the same dataset which have been used for CNN

5. PROPOSED MODEL

Fast Adaboost binary classifier:

The novel extensions of the Fast Adaboost binary classifier is to improve tumor classification rate. The key idea is to present the boosting algorithm with a synopsis of the training dataset such that the learning algorithm can rapidly and competently prepare the classifier.

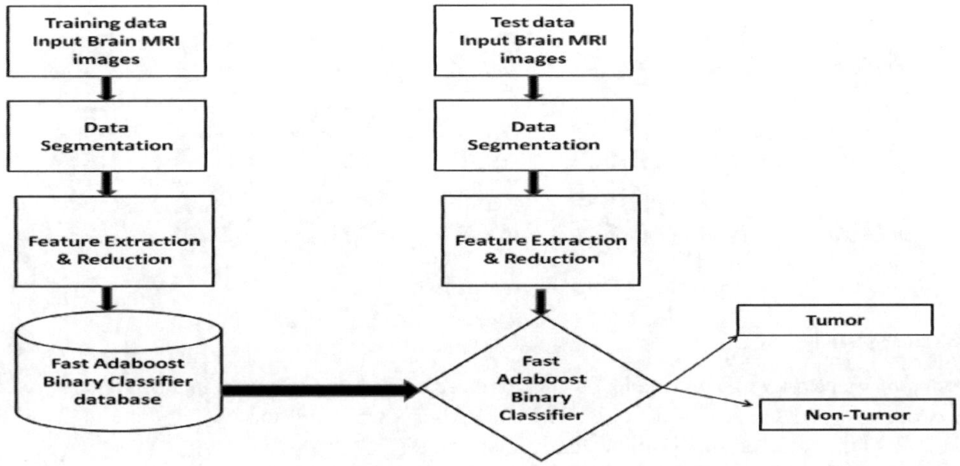

Figure 5: Proposed Algorithm fast Adaboost binary classifier block diagram

According to literature review, it was originate that brain tumor classification is extremely important as elevated accuracy is needed when human life is concerned. As Adaboost learning algorithm is believed to be computationally capable but training is time intense in image processing and classification, we propose to build classifier model based on boosting ensemble methods which selects a reduced set of training points, thus reducing large computational difficulty for training and rising the speed of the training progression.

If any classifier is given unlabeled image, it cannot classify that image with the knowledge obtained through the training process. So we have to train the classifier with a set of images with labels.

The same dataset considered for CNN and RNN is considered for proposed algorithm implementation.

a) Data Augmentation:

Clearly dataset indicate that we have data imbalance i.e., number of observations per class is not uniformly distributed. This data imbalance is solved using data augmentation because this is very important part in medicine as there would be very less number of unhealthy patients than the number of healthy patients in many instances. We performed image enhancements like flip & rotate. Data augmentation transformed the dataset into more number of images i.e., increases the size of the training set.

Flip and Rotation:

Random flipping creates a mirror likeness of an original image beside one (or more) chosen axis This operation helped AHM, especially benefitting background tumor information i.e., with respect to their location within the brain which would be or else complex for example having brain tumors positioned only in the left/ right hemisphere. This process is followed through right exclamation to suit the original image size.

After data augmentation ,the dataset contains almost equal number of **1085** images of "yes" class and **979** images of "no" class.

This data set now divided keen on training and test data sets. The training dataset contains 80% (1651 images) and test set contains 20% (413 images).

Removed any colors in the image to convert it to the gray scale by CvtColor function (python)

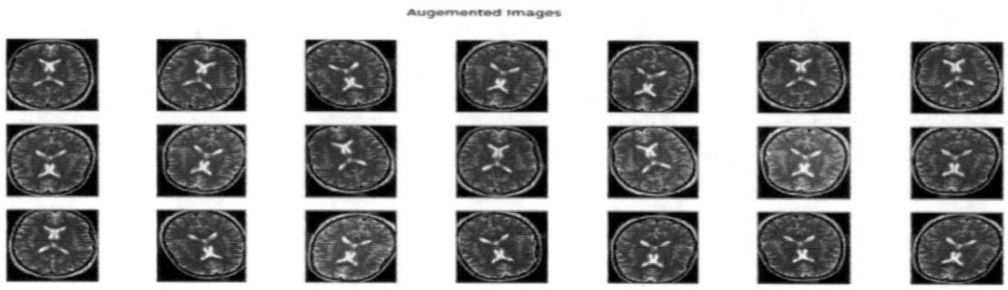

Figure 6: Augmented Images

b) Data Pre-processing:

Pre-processing of image prior to segmentation is vital for correct discovery of tumor. In this phase, we execute noise and artifacts decrease and sharpening of edges. There are slight chances of noise being there in the MRI images. Thus the key job of the preprocessing is to file edges in the image.

Filters be used to remove the noise from the image as preprocessing step. Here we used median filter for reducing the intensity difference between one pixel and its neighbor's pixels. All images are preprocessed using median filter, edge detection, binarization using thresholding.

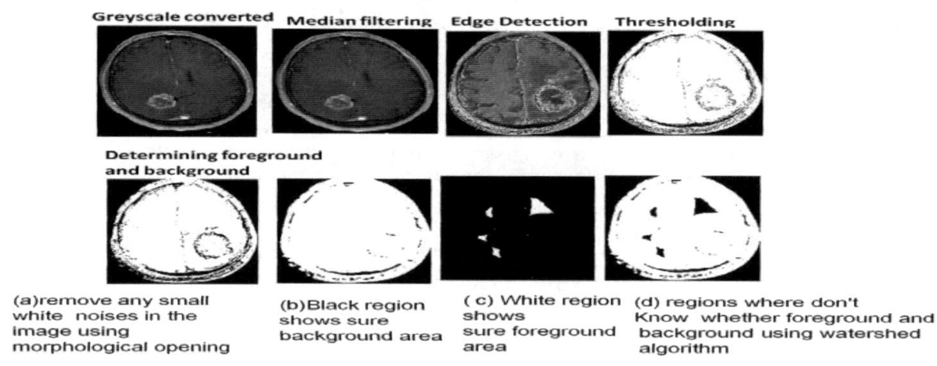

Figure 7: Preprocessing steps

c) Segmentation:

Segmentation is the procedure of dividing the image into smaller objects. So that objects in the similar cluster as alike as possible and objects in dissimilar clusters are as different. The segmentation procedure involves binarization using a thresholding technique, enhancing the image using morphological transformation, segmentation and edge detection.

The preprocess image is segmented by means of thresholding and tumor is identified. The white part in the image shows the tumor region in an MRI.

IOP Conf. Series: Materials Science and Engineering 1074 (2021) 012016 doi:10.1088/1757-899X/1074/1/012016

Figure 8: Tumor identification in segmentation process

Feature Extraction:

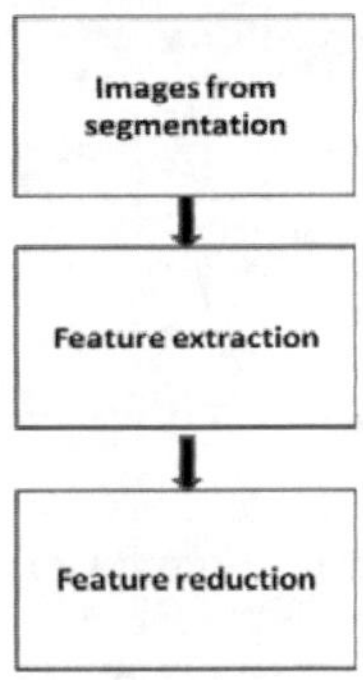

Figure 9:Process of features extraction.

GLCM (texture-based features) is used to extract features from images as we need to perform a binary classification of them using a fast Adaboost binary classifier which needs these features to get trained on. After obtaining the segmented images from the above process, we extract GLCM features and store them. GLCM indicates for Gray-Level Co-occurrence Matrix. Texture Analysis Using the Gray-Level Co- Occurrence Matrix (GLCM) is a numerical system of exploratory surface that considers the spatial link of pixels. Here feature describe as properties in class region, it have every shape feature.

Image	Mean	Varianc	Standa	Entrop	Skewn	Kurtosi	Contra	Energy	ASM	Homog	Dissimi	Correla	PSNR	SSIM	MSE	DC	Target
Image1	23.45	2539	50.39	0.651	1.984	5.421	181.5	0.782	0.611	0.847	2.765	0.969	97.97	0.777	0.171	0.304	1
Image2	4.398	834.9	28.89	0.954	6.495	43.35	76.75	0.973	0.946	0.981	0.549	0.96	110.3	0.978	0.01	0.839	1
Image3	3.244	642.1	25.34	0.966	7.773	61.76	81.75	0.98	0.961	0.985	0.54	0.944	112.3	0.985	0.006	0.85	1
Image4	8.511	1126	33.56	0.869	3.763	15.11	362.3	0.922	0.85	0.949	2.766	0.859	102	0.881	0.068	0	0
Image5	21	2235	47.28	0.685	1.936	4.722	312.4	0.804	0.647	0.88	3.007	0.939	97.64	0.766	0.185	0	0
Image6	0	0	0	1			0	1	1	1	0	1	inf		0		0
Image7	11.35	999	31.61	0.761	2.534	7.395	303.9	0.854	0.73	0.902	3.441	0.866	99.21	0.795	0.129	0	0
Image8	0.405	68.38	8.269	0.995	20.39	416.8	17.79	0.997	0.994	0.998	0.114	0.886	111.4	0.985	0.008	0.41	1
Image9	5.956	937.4	30.62	0.927	5.015	26.15	57.23	0.957	0.916	0.974	0.508	0.973	111.4	0.982	0.008	0.914	1
Image10	6.184	895.2	29.92	0.917	4.707	23.17	73.71	0.951	0.905	0.969	0.686	0.964	113.1	0.984	0.005	0.945	1
Image11	0.261	52.28	7.231	0.997	27.72	769.5	18.02	0.998	0.997	0.999	0.092	0.849	67.18	0.983	0.012	0.192	1
Image12	0.122	25.52	5.052	0.999	41.5	1723	18.23	0.999	0.998	0.999	0.088	0.687	63.96	0.967	0.026	0.048	1
Image13	4.742	750.2	27.39	0.94	5.781	35.14	51.38	0.965	0.931	0.977	0.479	0.97	70.67	0.986	0.006	0.923	1
Image14	2.32	395.3	19.88	0.972	8.539	74.27	38.27	0.984	0.968	0.988	0.306	0.958	64.9	0.968	0.021	0.593	1
Image15	2.869	365.3	19.11	0.953	6.707	47.02	49.72	0.972	0.945	0.981	0.535	0.94	71.42	0.985	0.005	0.905	1
Image16	17.81	1220	34.92	0.61	1.69	4.153	107.3	0.753	0.567	0.837	2.359	0.961	96.58	0.72	0.236	0	0

Figure 10: Result of shape and texture feature

Feature selection and extract has been used for 17 features from Tumor image array.

1) Low Variance Filter to remove features that have low variance data sample. These Features are: Solidity, Euler number, Extent.

2) Normalize the rest of features to create feature values Convergent as likely using Standard deviation & Mean.

IOP Conf. Series: Materials Science and Engineering 1074 (2021) 012016 doi:10.1088/1757-899X/1074/1/012016

3) Lastly, PCA was used to perform linear dimensionality reduction.

The output obtained proves that only three features: homogeneity, correlation, dissimilarity will have an effect on or have the best effect in classification problem.

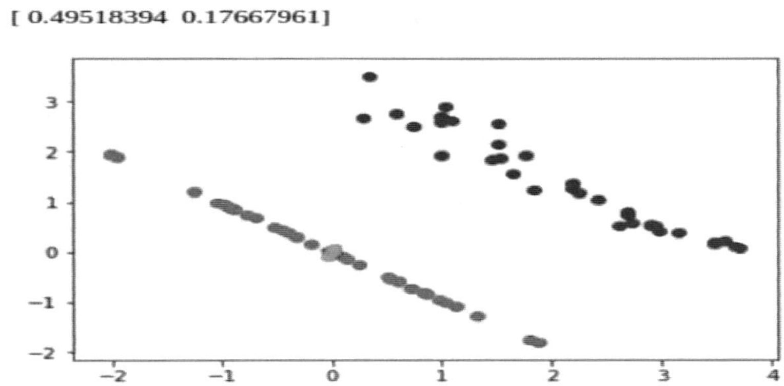

Figure 11: PCA output for 14 input features

As we know set of several models functioning together on a only one set is called an ensemble. The technique is called Ensemble Learning. It is a great deal and more useful to use all different models rather than any one. To execute fast adaboost binary classifier we mixed different classifiers where the class which has been predicted generally by the weak learners resolve be the ultimate class forecast of the ensemble form.

Test accuracy 0.904048
Training duration 0.06604sec

Table 3 : Results of CNN ,RCNN and start-of-art techniques

ALGORITHMS	ACCURACY for 150 iterations (%)
PROPOSED APPROACH FAHM (Fast Adaboost Hybrid Model)	90.4%
CNN	68%
RCNN	73%
Abiwinanda et al. CNN [13].	84.19%

DISCUSSION

1. Figures from 2-7 representing the CNN and RCNN model of comprising an accuracy of 64 % and 73% with 150 epochs.
2. Similarly our proposed design with 150 epochs have been modeled with 160 cycles and learning rate of 0.043. Hence the total learning rate is 100-160 *0.043 = 90.4 %.
3. Classification of the different parameters related to brain tumor are implemented and observed in figure 6 with confusion matrix.
4. Each calculated values and its representation are tabulated as mentioned in table 3.

CONCLUSION

The detection, segmentation and uncovering of infecting region in brain tumor MRI images are a dull and time consuming job. The dissimilar anatomy construction of human body can be seen by an image processing notion. We have taken heterogeneous collection of weak learners (machine learning models:Logistic Regression Model, K-

Nearest Neighbor Model, Support Vector Machine, and Naive Bayes Model) to construct our hybrid ensemble learning model, applied in the task of classification of brain MRI images. Our proposed system has accomplished the accuracy of 90.4% accuracy utilizing the given dataset.

In future, we will implement algorithms with more accuracy even for single image and also for large datsets, and identification of tumor types using artificial intelligence.

REFERENCES

[1] Neelum Noreen, Sellappan Palaniappan, Abdul Qayyum, Iftikhar Ahmad,Muhammad Imran and Muhammad Shoaib, 2019 "A deep learning model based on concatenation approach for brain tumor classification and identification" , *IEEE Access* ,Volume **8.**

[2] Q. T. Ostrom H. Gittleman J. Xu C. Kromer Y. Wolinsky C. Kruchko et al. 2016 "Cbtrus statistical report: primary brain and other central nervous system tumors diagnosed in the united states in 2009–2013" ,*Neuro-oncology* vol. **18**.

[3] S. Mary Praveena and Ila Vennila,2010 "Optimization Fusion Approach for Image Segmentation using K-Means Algorithm", International Journal of Computer Applications, **2** 7.

[4] W. Yang Q. Feng M. Yu Z. Lu Y. Gao Y. Xu et al. 2012 "Content-based retrieval of brain tumor in contrast-enhanced mri images using tumor margin information and learned distance metric" *Medical physics*, **39** 11.

[5] M Monica Subashini and Sarat Kumar Sahoo, 2013 "Brain MR Image Segmentation for Tumour Detection using Artificial Neural Networks" , *International Journal of Engineering and Technology* **5** 2.

[6] Sugandha Agarwal, O.P. Singh and Deepak Nagaria, 2017 "Analysis and Comparison of Wavelet Transforms For Denoising MRI Image", *Biomedical & Pharmacology Journal*, **10** 2.

[7] H. Chen, Y. Zheng, J.-H. Park, P.-A. Heng and S. K. Zhou, 2016 "Iterative multi-domain regularized deep learning for anatomical structure detection and segmentation from ultrasound images", *International Conference on Medical Image Computing and Computer-Assisted Intervention.*

[8] B. Cheng, M. Liu, H. I. Suk, D. Shen and D. Zhang, 2015 "Multimodal manifold-regularized transfer learning for MCI conversion prediction", *Brain imaging and Behavior*, **9** 4.

[9] Mr. Manish Saraswat and Mr. Ajay Kumar Sharma, 2013 "Genetic Algorithm for optimization using MATLAB", *International Journal of Advanced Research in Computer Science,* **4** 3

[10] Sushma Laxman Wakchaure and Anil Khandekar, 2015 "visualization of 3d view of detected brain tumor and calculation of its volume", *International Journal of Technical Research and Applications*, **3** 6.

[11] Zolfaghar R.2012, "Diagnosis of Diabetes in Female Population of Pima Indian Heritage with Ensemble of BP Neural Network and SVM" *Int. J. Comput. Eng. Manag.* **15** 4.

[12] Bashir,S. Qamar,U. Khan,F, 2016,"IntelliHealth: Amedical decision support application using a novel weighted multi-layer classifier ensemble framework", *Journal of Biomedical Informatics* **59**.

[13] Abiwinanda N., Hanif M., Hesaputra S.T., Handayani A., Mengko T.R.2018, "Brain Tumor Classification Using Convolutional Neural Network" *Proceedings of the World Congress on Medical Physics and Biomedical Engineering*; Prague, Czech Republic. 3–8 pp. 183–189.

[14] S. Vasavi S. Jayaprada V. Srinivasa Rao, 2011"Extracting semantically similar frequent patterns using ontologies" *International Conference on Swarm, Evolutionary, and Memetic Computing,* Vol. 2, pp. 157-165.

[15] G.Siva Nageswarao,M.Kiran Babu ,T. Srinivasarao, 2019 " Efficient scheduling Measures for improving real time systems using prediction process" *Journal of Computational and Theoretical NanoScience*, **16** 5.

SAMS: Smart Agriculture Management System Using Emerging Technologies IoT, AI -A Study

B. Srinivas Kumar[1], S.G. Santhi[2], K. Kranthi Kumar[3]

1.Research Scholar 2. Associate Professor 3. Assistant Professor
1,2 Department of Computer Science and Engineering, Annamalai University.
3.Department of Information Technology, Gudlavalleru Engineering College, Gudlavalleru.
Email: bskgec@gmail.com, kk97976@gmail.com

Abstract. Farming or Agriculture is the science and art of cultivating plants. It is the vital enrichment in the growth of human living. It creates food and the other related things by farming with domestic animals and will enable people to live long and healthy life. There are two agricultural categories, one is traditional Agriculture and the second one is Modern Agriculture. In Traditional Agriculture Farmers had faced many difficulties to Farm, but with the modern Agriculture Farmers can be able to see many improvements in Framing due to the agronomy, use of agricultural chemicals such as pesticides and fertilizers, Plant breeding and technological growth , these enrichments in Farming would yield good crop.(IoT)Internet of Things very often called as a system it can able to connect real physical Things, Implanted sensors, Embedded software's along with different technologies to connect and to transfer data amongst various diverse devices over the Internet. It is the hopeful expertise or technology which gives accurate and most reliable solutions on the way to too many domains. There is an extensive set of wide range of real time applications of IoT devices such as Industrial, Consumer, Commercial, Infrastructure may made IoT is the one of the best Technology to see enhancements in many domains. Solutions which are given by IoT would reduce human efforts in Farming and may provide sound of crop. In this paper a small survey will be presented to see how Internet of Things and its equivalent or other represent technologies AI, Cloud Computing would be useful to provide SMART AGRICULTURE along with Modern agricultural improvements will also be presented.

Keywords: Smart, Technology, IoT, AI, Agriculture.

I. Introduction

The Internet of Things (IoT), likewise recognized as the web of everything or the Mechanical net, is another innovation worldview abstract as a worldwide organization of machines and gadgets suited interfacing with each other. The IoT is perceived together of the most areas of future innovation and is increasing Brobdingnagian thought from a large scope of comes [1]. Applications of IoT has originated in many areas, it integrates several heterogeneous technologies [2]. Seeable of IoT real time devices or physical things have become autonomous [3]. Present police investigation scented by Wireless detector Network (WSN) advances cuts across various zones of the newest existing. It Can have the capability to quantify, derive and realize environmental indicators, after complex etiologies of traditional properties to metropolitan conditions. The enlargement of those gadgets during a transfer effectual organization makes the web of Things (IoT), whereby sensors and actuators combine systematically with the climate around U.S., and also the information is shared across stages to create up a typical operating image. Invigorated by the proceeded with change of partner variety of engaging far off advances, for instance, RFID marks and embedded locator and component centers, the IoT has wandered out of its soonest organizes and is that the accompanying reformist development in unique web the online the net into an extremely consolidated Future web [4]. Within the time of the IoT, a massive live of police investigation gadgets gathers, and in addition produce totally different tangible info when your time for an honest scope of fields and applications. Upheld the temperament of the device, these gadgets can finish in

Brobdingnagian or quick/constant info streams. Applying investigation over such information streams to

Fig-1 Crop Growth

get new information, anticipate future experiences, Associate in Nursing choose management decisions may well be a pressing cycle that creates IoT a commendable worldview for organizations, and a high personal satisfaction rising innovation [5]. IoT connects heterogeneous devices [6]. Transportable Distributed computing is another innovation that indicates to a framework wherever each info storing and data in making ready work outside the cellular phone. Another current innovation is net of Things. Net of Things is another innovation that is filling quickly within the field of broadcast communications. All the additional expressly, IoT connected with remote broadcast communications. It is the basic objective of the affiliation and collaboration among things and things that sent through the remote organizations is to satisfy the target set to them as a consolidated element. Also, there's a fast improvement of the 2 innovations, Distributed computing and net of Things, respect the sector of remote correspondences [7]. Voluminous measures of data are delivered, since the previous decade because the cutting down of the net of things gadgets increments. However, such information isn't valuable while not fact-finding force. Numerous monumental info, IoT, and investigation arrangements have authorized people to induce vital understanding into vast info created by IoT gadgets [8]. Creating agriculture as good is Associate in Nursing rising thought [9]. Agriculture improvement is dependable in temperature change [10], if climate is monitored through IoT Technology it might be higher to ascertain sensible ends up in the good Framing. Commercial enterprise is taken into account because the premise of life for the human species because it is the primary wellspring of food grains and different crude materials. It assumes essential perform within the development of nation's economy. It in addition provides vast enough work occasions to the people. Development in geographic area is prime for the advancement of monetary state of the state. Sadly, varied ranchers truly utilize the customary techniques for cultivating that brings concerning low yielding of harvests and organic merchandise. Yet, anywhere mechanization had been dead and people had been supplanted by programmed hardware, the yield has been improved. After there's got to execute current science and innovation within the commercial enterprise space for increasing the yield. An oversized portion of the papers means that the use of remote sensing element network that gathers the knowledge from numerous forms of sensors and later on send it to primary employee utilizing remote convention. The gathered info. gives the information concerning numerous natural elements that in goes assists with checking the framework. Checking natural components is not ample and complete arrangement to boost the yield of the harvests. There are variety of various components that influence the profitableness

Fig-2 Farming Vs IoT

to extraordinary degree. These variables incorporate assault of creepy crawlies and irritations which might be forced by showering the harvest with legitimate bug spray what is additional, pesticides. Also, assault of untamed creatures and fowls at the purpose once the yield grows up. there's to boot edibleness of burglaries at the purpose once yield is at the section of collection. Even once reaping, ranchers likewise face problems away of reaping crop. During this manner, to provide answers for all such problems, it's vital to make coordinated framework which can upset all parts influencing the profitableness in every stage like; development, collection what is additional, post collection storage. This paper consequently proposes a framework that is useful in checking the sector info in addition as dominant the sector activities which provides the ability. The paper targets creating a farm sensible utilizing motorization and IoT advancements. The that includes things to see of this paper incorporates keen GPS based mostly far-flung controlled automaton to perform undertakings like; weeding, showering, wetness detection, flying creature and creature surprising, keeping caution, and then forth Besides, it incorporates sensible water system with keen management addicted to constant field info. Thirdly, keen storeroom the board that incorporates; temperature repairs, humidity support and felony discovery within the storeroom. Dominant of each one in every of these activities are going to be through any far shrewd contraption or computer related to internet and therefore, the activities are going to be performed by interfacing sensors, Wi-Fi or ZigBee modules, camera and actuators with miniature regulator and raspberry pi. Farming assumes indispensable perform within the advancement of rural nation. In Asian nation concerning seventieth of world depends on cultivating and thirty-third of the country's capital comes from cultivating. Problems regarding factory farm are regularly frustrating the advancement of the state. The most declare this issue is sensible factory farm by modernizing this customary method for farming. Consequently, the enterprise targets creating farming shrewd utilizing motorization and IoT innovations. The that includes

IOP Conf. Series: Materials Science and Engineering 1074 (2021) 012017 doi:10.1088/1757-899X/1074/1/012017

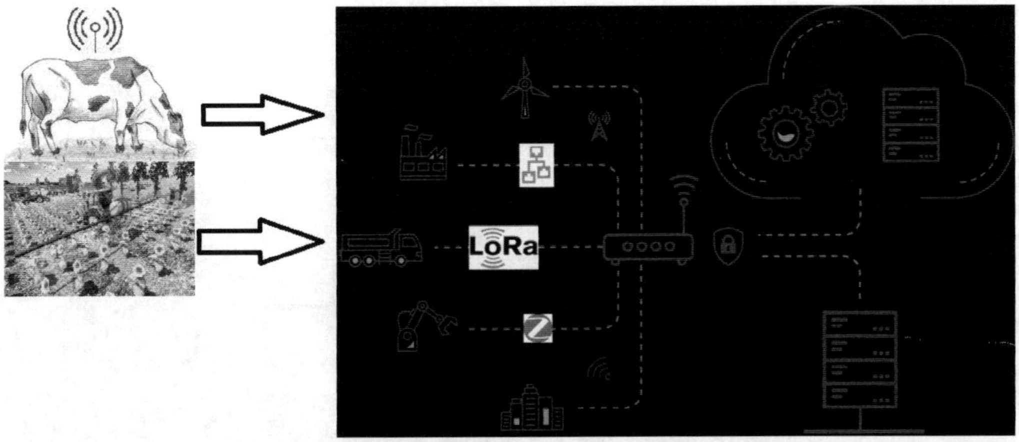

Fig-3 Interaction of various things in Smart Farming

II. Emerging Technological Applications for Smart Agriculture

Smart Agriculture includes a good vary of applications most of that ends up in numerus applications, a number of the sensible technologies square measure shown in below table. Technology has brought the intense influence on Agriculture. current insights uncover that the worldwide world goes to gain nine.6 billion by 2050. Also, to require care of this monumental world, the farming business is restricted to receive the online of Things. Among the difficulties like extraordinary climate conditions, environmental condition changes, ecological impact, IoT is annihilating these difficulties and serving to U.S. to satisfy the necessity for a lot of food. in the course of the globe, mechanical advancements, as an example, farm trucks and gatherers occurred and brought into the husbandry activities within the late twentieth century. Also, the farming Business depends smartly on inventive thoughts because of the systematically developing interest for food.

Sno	Technology	Remarks
1	IOT	Internet of Things
2	Big Data	Big Data
3	AI	Artificial Intelligence
4	ML	Machine Learning
5	DL	Deep Learning
6	BCT	Block Chain Technology
7	MPC	Mobile Phone Communication

Table -1 Smart Technology

Smart Agricultural systems application are listed here:

There are many application areas for Smart Agriculture, the main goal of this system is to reduce human involvement and to make system autonomous. Emerging technologies like what were mentioned in the above are very often used by many researchers. Automation in any area would give better results and it aims to improve system work and at the same time it provides good results ,in this paper are also when it is suggested Technological Farming it yields good crop, it warns farmers if any bad climate, it informs farmers if any food and other devices theft.

Role of AI and IoT in Agriculture:

A) Drones: Drones helps to monitor crop and to take necessary actions to improve the growth of the crop.

B) Weather Forecasting or Climate Estimation: Most of the Farming depends on weather, monitoring weather reports and giving weather report results to Farmers may give good improvement in the growth of Farming.

C) Machines/Robots: Robots are useful to reduce manual work and they play vital role to produce crop with in time.

D) Image Analysis: It will be useful to do some research on Farming

IOP Conf. Series: Materials Science and Engineering 1074 (2021) 012017 doi:10.1088/1757-899X/1074/1/012017

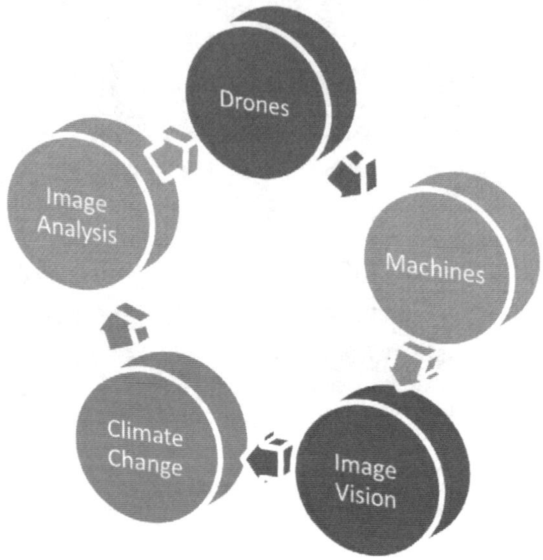

Fig-4 Role of Smart Technology

According to global share market IoT projects reports have been displayed here,

Sno	Segment	Percentage
1	Smart City	20
2	Smart Education	15
3	Smart Energy	13
4	Connected car	13
5	Industry	22
6	Smart Agriculture	6
7	Smart Health	5
8	Smart Retail	4

Table -2 Smart City

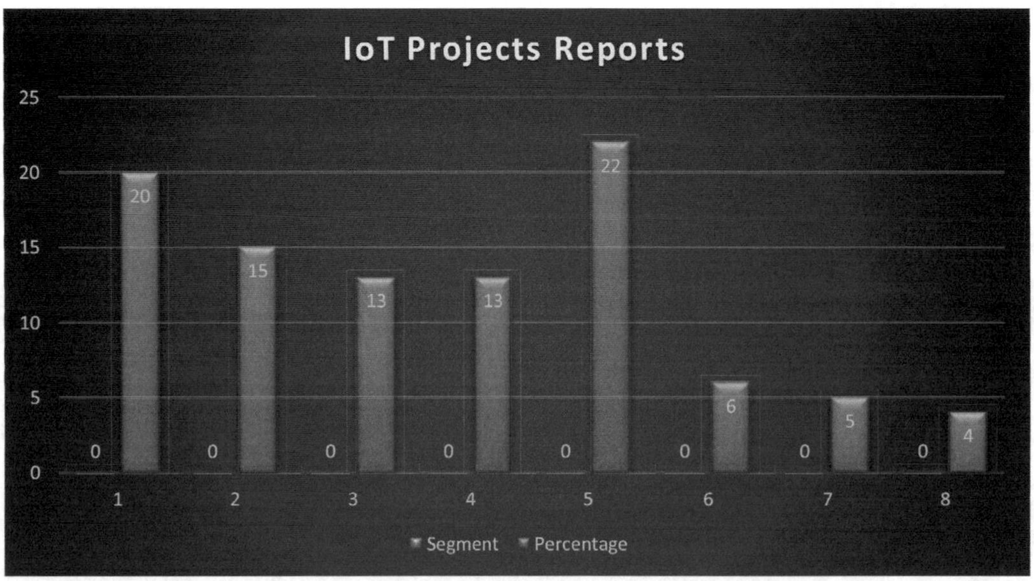

III.Security of Smart Agriculture

Security is a common problem in IoT. Some of the security and privacy of the smart agriculture are discussed here,

1. Data Authentication.
2. Data Reliability.
3. Data Integrity.
4. Data Leakage.
5. Data Theft.

To address the above said problems the system has to follow Some specific Cryptographic Algorithms.

IV.Smart Agriculture Issues

1. Internet Connection Issue.
2. Hardware Installation is Difficult.
3. Cloud Connection Disturbances

V. Conclusion

The rural space is of imperative significance for the district. it's surfing a cycle of get to a laissez-faire economy, with vital changes within the social, legitimate, auxiliary, helpful and graciously set-ups, almost like the case with all completely different areas of the economy. good cultivating is filling in significance due to the combo of the extending worldwide people, the increasing interest for higher harvest yield, the requirement to utilize characteristic assets proficiently, the rising use and complexness of knowledge and correspondence innovation and therefore the increasing demand for atmosphere keen. This paper has presented some study how emerging technology created a change in Agriculture.

References

[1]. Lee, I., & Lee, K. (2015). The Internet of Things (IoT): Applications, investments, and challenges for enterprises. *Business Horizons*, *58*(4), 431-440.

[2]. Huh, S., Cho, S., & Kim, S. (2017, February). Managing IoT devices using blockchain platform. In *2017 19th international conference on advanced communication technology (ICACT)* (pp. 464-467). IEEE.

[3]. Reyna, A., Martín, C., Chen, J., Soler, E., & Díaz, M. (2018). On blockchain and its integration with IoT. Challenges and opportunities. *Future generation computer systems*, *88*, 173-190.

[4]. Gubbi, J., Buyya, R., Marusic, S., & Palaniswami, M. (2013). Internet of Things (IoT): A vision, architectural elements, and future directions. *Future generation computer systems*, *29*(7), 1645-1660.

[5]. Mohammadi, M., Al-Fuqaha, A., Sorour, S., & Guizani, M. (2018). Deep learning for IoT big data and streaming analytics: A survey. *IEEE Communications Surveys & Tutorials*, *20*(4), 2923-2960.

[6]. Mekki, K., Bajic, E., Chaxel, F., & Meyer, F. (2019). A comparative study of LPWAN technologies for large-scale IoT deployment. *ICT express*, *5*(1), 1-7.

[7]. Stergiou, C., Psannis, K. E., Kim, B. G., & Gupta, B. (2018). Secure integration of IoT and cloud computing. *Future Generation Computer Systems*, *78*, 964-975.

[8]. Marjani, M., Nasaruddin, F., Gani, A., Karim, A., Hashem, I. A. T., Siddiqa, A., & Yaqoob, I. (2017). Big IoT data analytics: architecture, opportunities, and open research challenges. *IEEE Access*, *5*, 5247-5261.

[9]. Prathibha, S. R., Hongal, A., & Jyothi, M. P. (2017, March). IoT based monitoring system in smart agriculture. In *2017 international conference on recent advances in electronics and communication technology (ICRAECT)* (pp. 81-84). IEEE.

[10]. Lipper, L., Thornton, P., Campbell, B. M., Baedeker, T., Braimoh, A., Bwalya, M., ... & Hottle, R. (2014). Climate-smart agriculture for food security. *Nature climate change*, *4*(12), 1068-1072.

Analyzing and Predicting Covid-19 epidemic using Machine Learning Techniques

[1]A.Lakshmanarao, [2]Ch.Seshadri Rao, [2]Sridevi Gadde

[1]Department of IT, Aditya Engineering College, Surampalem,A.P
[2] Department of CSE, Raghu Engineering College, Visakhapatnam, A.P, India

Email: laxman1216@gmail.com, seshadrirao.chinta@raghuenggcollege.in, Sridevi.gadde@raghuenggcollege.in

Abstract. The entire world is suffering from a novel disease called covid-19 caused by a coronavirus since 2019. The main reason for the seriousness of the disease is the lack of efficient legitimate medication or vaccine. The World Health Organization (WHO) suggested several precautions to regulate the spread of disease and to reduce the contamination thereby reducing deaths. In this paper, we analysed the covid-19 dataset available in Kaggle. The previous contributions from several authors of similar work focused on covid-19 datasets having a limited number of samples. Our paper used the dataset updated till November 15th 2020. Three different aspects are considered mainly in this paper, namely the number of confirmed cases, number of recovered cases, and number of death cases. All the aspects are analysed in a daily and weekly manner. We applied linear regression, polynomial regression, and holt's method to predict the future number of confirmed, recovered, and death cases. This analysis is useful for the health sectors and frontline workers to help reduce the contamination caused by this disease.

1. Introduction

Covid-19, a disease caused by coronavirus ringing dangerous alarms from 2019 worldwide. The number of cases at the end of January 2020 is 581, out of which 571 from china, but the number of cases worldwide in November 2020 is 58,900,547. The number of cases in India is 9,177,840[1] now. This disease spreads in several ways, but the main cause for spreading is the virus affected patient's droplets produced through cough or sneezing. If a person is having close contact with a covid-19 patient, then there more chances of getting covid-19. The virus survival time varies from place to place. It depends on the type of surface, humidity, temperature, etc. The recovery period for covid-19 affected people is 14 days. So, if a person is tested as a covid-19 positive case, then that person must be in quarantine for 14 days. As research is going on for the covid-19 vaccine, all governments are suggesting the people take precautions to stop the spreading of the corona virus. The general symptoms of covid-19 are cold, fever, cough, and shortness of breath. Using a face mask and sanitizers is suggested as a primary measure. Although the Russian government released a vac-cine named 'Sputnik', many of the countries are not accepting its complete solution for covid-19. World Health Organization (WHO) declares covid-19 as a pandemic [2]. SARS (Severe Acute Respiratory

Syndrome-2003) are MERS (Middle East Respiratory Syndrome-2012) are also another similar virus disease causes thousands of deaths.

2.Literature Survey

Machine Learning coined by Arthur Samuel is not a new concept. Although it was introduced in the 1950s, the real usage of machine learning boosts from the last decade. Today machine learning is applied to every field and the health industry is not an exclusion. S Dhamodharavadhani [3] et.al proposed Statistical Neural Network Models for COVID-19 Mortality Rate Prediction in India. They con-ducted different experiments on two different COVID-19 datasets and achieved good RMSE values. M. Rubaiyat Hossain Mondal[4] et.al applied polynomial regression for finding a number of cases. They also proposed classification algorithms Multilayer Perceptron, Logistic Regression, XGBoost for classification of covid-19 patients. LamiaaA. Amar [5] et.al analysed four months of covid-19 data from February 2020 to June 2020 in Egypt. After analysing the dataset with polynomial regression, they concluded that the epidemic of the virus in Egypt would not end in the near future. Lodangi Nagakrishna [6] et.al forecasted covid-19 cases using regression analysis. They have done experiments with 7 months of data in India (jan2020 to July 2020) and concluded that there is no chance of reducing cases quickly. Rajan Gupta [8] et.al applied the ARIMA model for predictive analysis of the covid-19 dataset and concluded that the growth rate is high worldwide (until march-2020) but there is a chance of exponential growth. Ekta Gambhir [9] et.al proposed polynomial regression, support vector regression for analysing covied-19 data from January 2020 to June 2020. Kolla Bhanu Prakash [10] proposed various classification algorithms like decision tree classifier., random forest classifier, logistic regression, naïve Bayes classifier and concluded that the age group of 20 to 50 are more affected by covid-19. Bathula Preetham Kumar Reddy [11] applied Linear Regression model to predict future assumptions of covid-19 disease. Li Yan[12] et.al proposed XGBoost machine learning based model for the prediction of mortality rates of covid-19 patients and achieved an accuracy of 90%. Behrouz Pirouz[13] et.al applied binary classification techniques to identify the key factors causing covid-19. With their experiments they have shown that humidity and average temperature are having the highest impact on the disease. G.D. Barmparis [15] et.al proposed Gaussian spreading hypothesis for estimating infection horizon of covid-19. With their model, they predicted daily infections successfully.

3. Research Methodology

The covid-19 dataset is downloaded from Kaggle [14]. Later, we analysed data in different perspectives and then we applied regression models. The proposed method is shown in Figure-1.

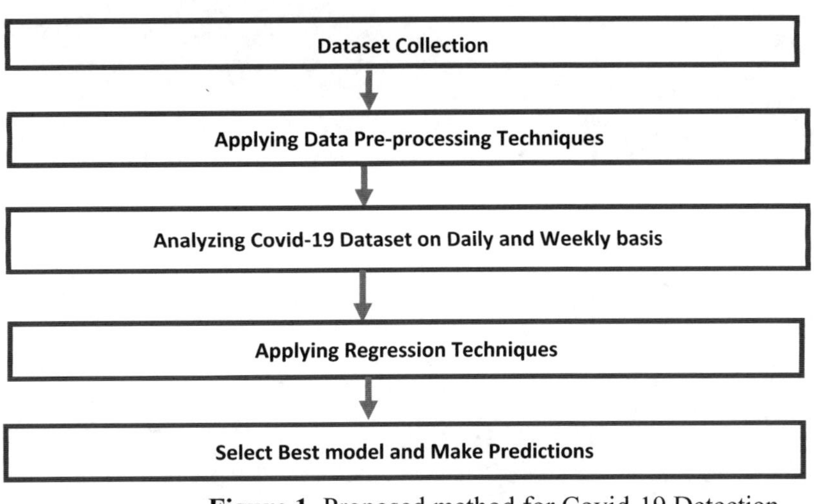

Figure 1. Proposed method for Covid-19 Detection

IOP Conf. Series: Materials Science and Engineering 1074 (2021) 012018 doi:10.1088/1757-899X/1074/1/012018

The dataset was collected from Kaggle repository. The dataset contains 8 columns namely Date, Province/State, country, last updated date, confirmed cases, recovered cases, death cases. There are missing values in the column 'Province/State'. We calculated total number of confirmed cases, death cases, recovered cases, active cases and closed cases from the given data. This information is tabulated in Table-1.

Table 1.DATASET DETAILS

Total no. of confirmed	Total no. of recovered	Total no. of death cases	Total no. of active cases	Total no. of closed cases
54370186	34955148	1317139	18097899	36272287

3.1. Distribution plots

From the table-1, the active cases distribution plot and closed cases distribution plots are drawn and they are shown in Figure-2.

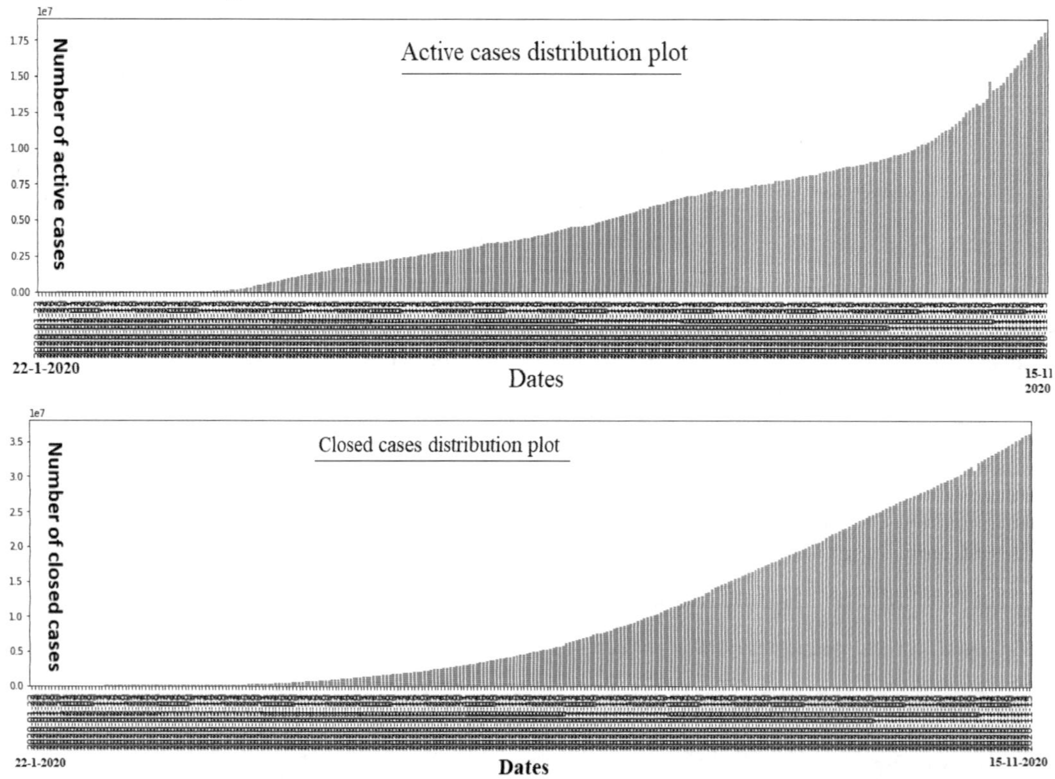

Figure 2. Distribution plots of active and closed cases

3.2. Weekly progress of confirmed, recovered, death cases

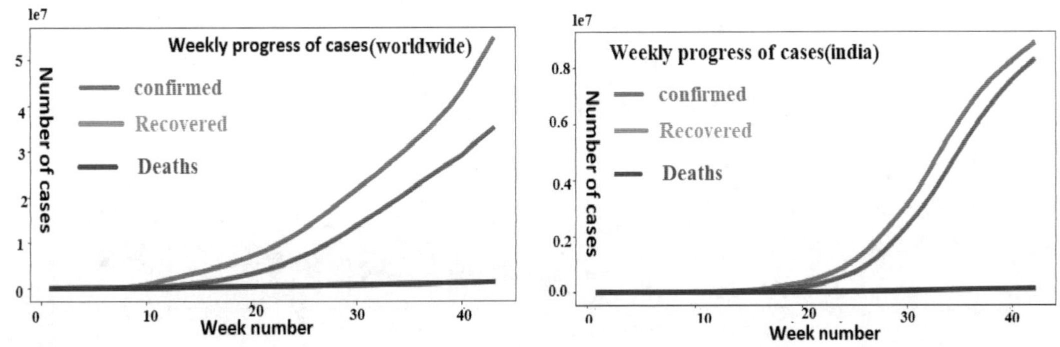

Figure 3. Weekly progress of cases worldwide, only in India

Weekly progress of confirmed, recovered, death cases are shown in Figure-3. From figure-3, it is observed that death cases are increasing in a similar manner from week-1 to week-43, but the number of confirmed cases are raising high. Weekly increase in number confirmed and death cases are shown in Figure-4.

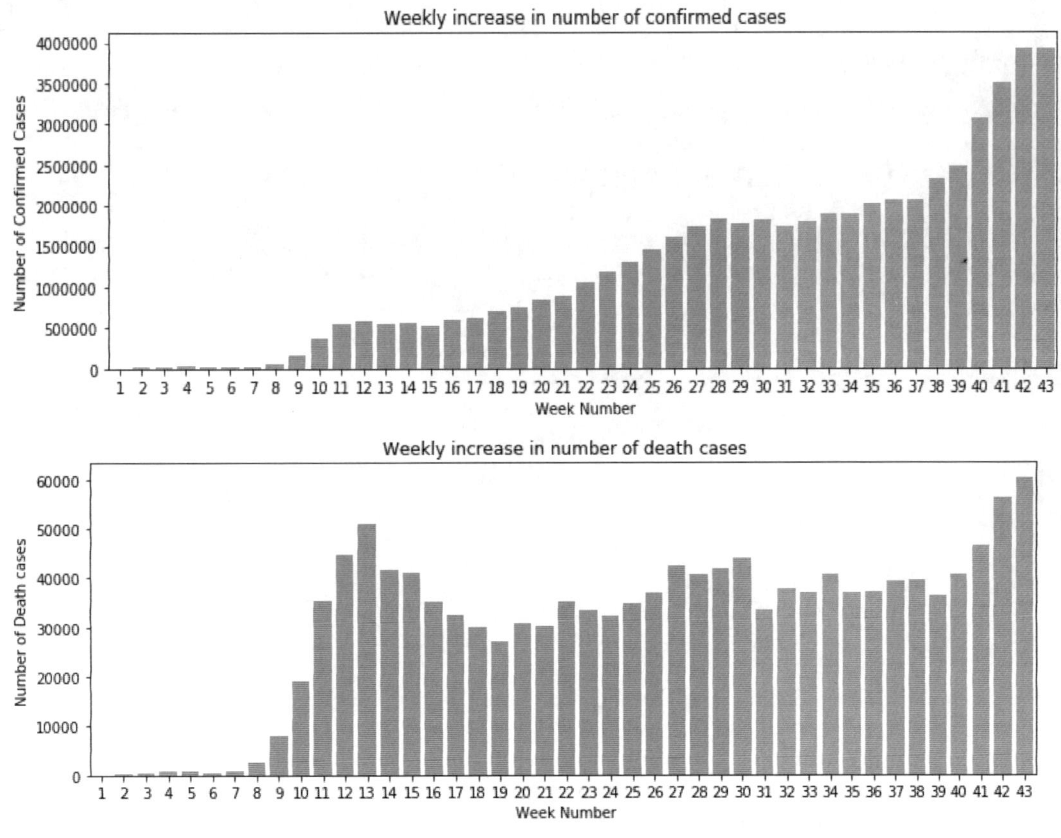

Figure 4. Weekly increase in number of confirmed and death cases

From Fig-4, average increase of confirmed, recovered, death cases are observed:
Average increase in number of confirmed cases everyday: 181838.0
Average increase in number of recovered cases everyday 116907.0
Average increase in number of death cases everyday 4405.0

IOP Conf. Series: Materials Science and Engineering 1074 (2021) 012018 doi:10.1088/1757-899X/1074/1/012018

3.3. Daily progress of confirmed, recovered, death cases

Daily progress of covid-19 cases is depicted in Figure-5. From Figure-5, it is observed that the gap between confirmed cases and recovered cases is increased from October 2020.

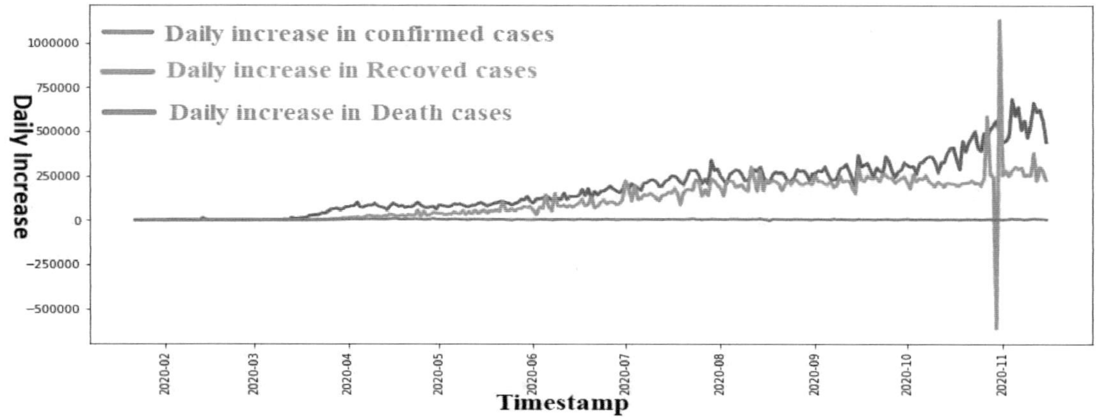

Figure 5. Daily progress of cases

3.4. Country wise Mortality rate

An important chrematistic of any dangerous disease is morality rate. It indicates how many deaths happened due to disease. Health authorities uses mortality data for evaluating their priorities/measures for control of disease. The morality data of top 15 countries are shown in Figure-6.

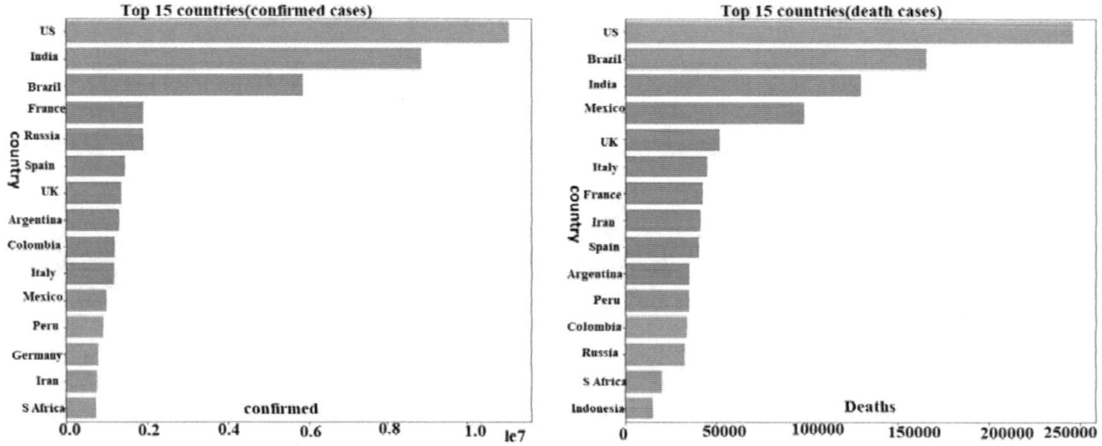

Figure 6. Confirmed and death cases for top 15 countries

4. Experimentation and Results

We applied regression algorithms for the prediction of confirmed cases, death cases, and recovered cases. We applied regression techniques in two different ways. In the first method, we considered day number as an independent variable and number of cases(confirmed/recovered/deaths) as a dependant variable. In the second method, we prepared a dataset structure where the previous 20 days' cas-es are used for predicting the number of cases on a particular day. In both cases we divided the dataset into training and testing sets in 80%,20% ratio.

4.1. Appling Regression techniques with day number and number of cases

4.1.1. Applying Linear Regression

Linear Regression is a simple technique where dependent feature(y)is estimated from independent feature(x). Here number of cases is dependent feature and time is independent feature. But linear regression cannot find nonlinear relationship between features.

4.1.2. Applying Polynomial Regression

Polynomial regression is used to find nonlinear relationship between features.In polynomial regression, Polynomial features are generated by raising current features to an exponent. If X is a feature, then we raise it to X power 2, X power 3, X power 4 etc...

4.1.3. Applying Holts method

The method is an exponential smoothing method. This applicable to the time series, value, trend, and seasonality data. It forecasts the data with trend.

4.1.4. Results of Linear Regression, Polynomial Regression, Holts method

Table 2. RMSE comparison of regression algorithms

Algorithm	RMSE(Confirmed)	RMSE(Recovered)	RMSE(Deaths)
Linear Regression	14810771	11386638	114504
Polynomial Regression	12965349	4420199	3728451
Holts method	3070855	616482	32717

From Table-2, it is easily observed that holt's method done better estimation, when compared to linear and polynomial regression.

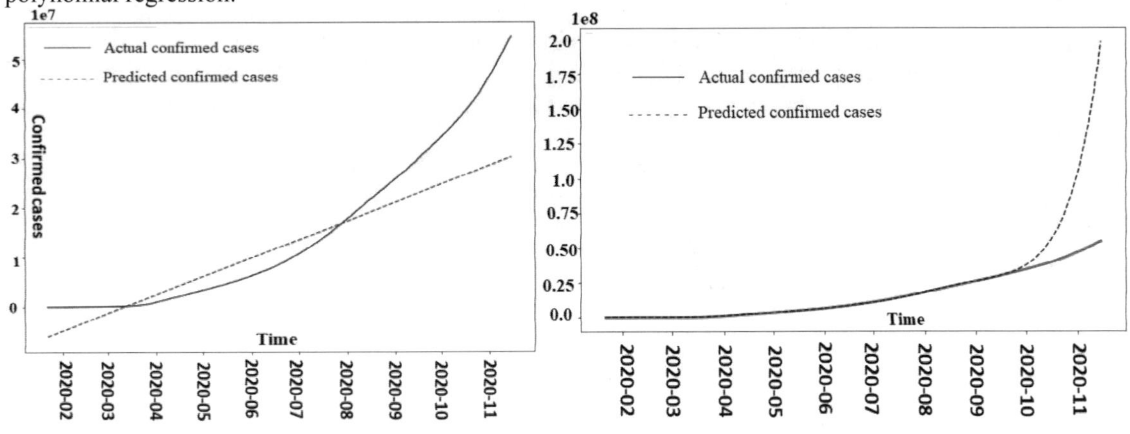

Figure 7. Linear regression, polynomial regression plot for confirmed cases (world-wide)

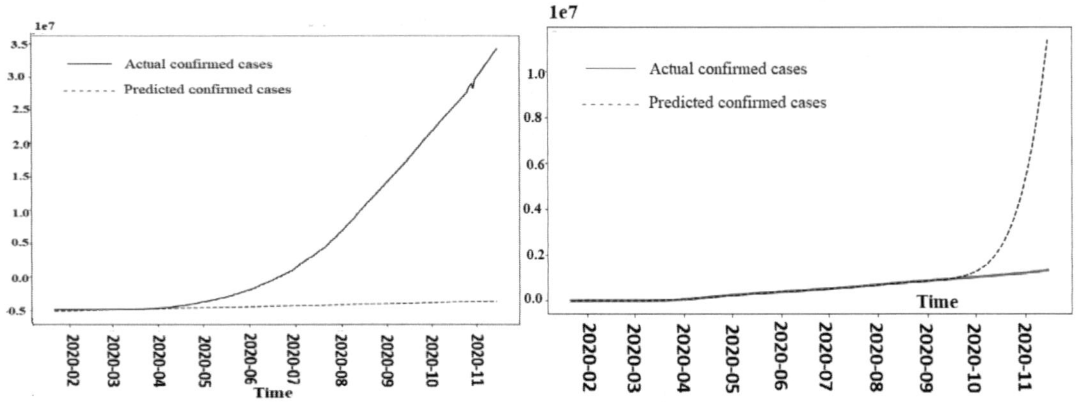

Figure 8. Linear regression, polynomial regression plot for death cases (world-wide)

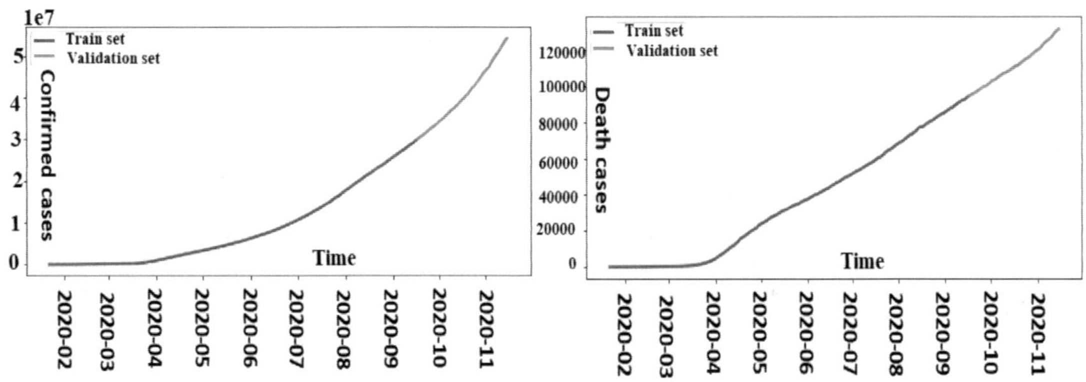

Figure 9. Holts method plot for Confirmed, death cases (world-wide)

4.2 Appling Regression techniques with previous days cases as predictor

We created a dataset with previous 15 days cases as independent variables and current cases as dependant variable. Here any particular cases on a day are predicted from its previous 15 days cases. After preparing dataset with this structure, we applied linear regression.

Table 6. Linear Regression technique with previous 10 cases as predictor

	Confirmed cases		Recovered cases		Death cases	
	RMSE	R2	RMSE	R2	RMSE	R2
Linear Regression	41465	0.99	211899	0.97	1129	0.97

5. Conclusion

In this paper, we analysed the number of confirmed cases, the number of re-covered cases, and the number of death cases on a daily, weekly basis in the covid-19 dataset. We applied regression techniques in two different approaches, the first one used the number of days as a predictor and the second approach used the previous days' cases as a predictor to predict the future number of con-firmed, recovered, death cases. Our experimental results on the dataset shown a good R-squared value of 0.99.

References

[1] https://covid19.who.int/table

[2] https://www.who.int/emergencies/diseases/novel-coronavirus-2019

[3] S Dhamodharavadhani, R Rathipriya and Jyotir Moy Chatterjee2020COVID-19 Mortality Rate Prediction for India Using Statistical Neural Network Models*Frontiers in Public Health*Volume-8, Article-441

[4] M. Rubaiyat Hossain Mondal, Subrato Bharati, Prajoy Podder 2020 Data analytics for novel coronavirus disease ELESEVIER- *Informatics in Medicine Unlocked 20 (2020) 100374* https://doi.org/10.1016/j.imu.2020.100374

[5] Lamiaa A. Amar, Ashraf A. Taha, Marwa Y. Mohamed 2020 Prediction of the final size for COVID-19 epidemic using machine learning: A case study of Egypt*Infectious Disease Modelling, August 2020, ELSEVIER*, https://doi.org/10.1016/j.idm.2020.08.008

[6] Lodangi Nagakrishna, Dr. Vijay Thawani, Kunda Gharpure 2020 COVID-19 Forecasting Using Regression Analysis *VIMS Health Science Journal* Volume 7 - Issue 3 - September 2020, DOI: 10.46858/vimshsj.7307

[7] Rajan Gupta, Gaurav Pandey,Poonam Chaudhary, Saibal K. Pal2020 Machine Learning Models for Government to Predict COVID-19 Outbreak *Digital Government: Research and Practice*volume 1, No. 4, Article 26

[8] Gupta, R, Pal, S. K 2020 Trend Analysis and Forecasting of COVID-19 outbreak in India*medRxiv*march-2020, https://doi.org/10.1101/2020.03.26.20044511

[9] Ekta Gambhir, Ritika Jain, Uma Tomer, Alankrit Gupta 2020 Regression Analysis of COVID-19 using Machine Learning Algorithms *International Conference on Smart Electronics and Communication (ICOSEC)IEEE Explore*pp 65–71

[10] Kolla Bhanu Prakash, S. Sagar Imambi, Mohammed Ismail, TPavan Kumar, YVR Naga Pawan 2020Analysis, Prediction and Evaluation of COVID-19 Datasets using Machine Learning Algorithms*International Journal of Emerging Trends in Engineering Research* volume-8(May)

[11] Bathula Preetham Kumar Reddy 2020Prediction of Covid-19 Using Regression Analysis *International Research Journal of Engineering and Technology (IRJET),* Volume: 07 Issue:06

[12] Li Yan ,Hai-Tao Zhang, Jorge Goncalves 2020 An interpretable mortality prediction model for COVID-19 patients *Nature Machine Intelligence* Volume: 02 pp 283-288

[13] Behrouz Pirouz, Sina Shaffiee Haghshenas, Sami Shaffiee Haghshenas,Patrizia Piro 2020 Investigating a Serious Challenge in the Sustainable Development Process: Analysis of Confirmed cases of COVID-19 (New Type of Coronavirus) Through a Binary Classification Using Artificial Intelligence and Regression Analysis*Sustainability 2020,* 12, 2427

[14] https://www.kaggle.com/sudalairajkumar/novel-corona-virus-2019-dataset

[15] G.D. Barmparis, G.P. Tsironis 2020 Estimating the infection horizon of COVID-19 in eight countries with a data-driven approach *Chaos, Solitons and Fractals 135 (2020)* 109842, 2020 Elsevier

Timely and Accurately Predict Rainfall by using Ensemble Predictive Models

[1]Veera Ankalu.Vuyyuru, [2]Giduturi.Apparao , [3]S. Anuradha

[1]Research Scholar, CSE Department, GITAM Deemed To be University,Visakhapatnam,AP,India.

[2]Professor,CSE Department,GITAM Deemed To be University,Visakhapatnam,AP,India.

[3]Assistant Professor,CSE Department,GITAM Deemed To be University,Visakhapatnam,AP,India,

[1]veeraankalu14@gmail.com, [2]apparao.giduturi@gitam.edu, [3]asesetti@gitam.edu

Abstract. Weather and rainfall are important factors for human life. By depends upon the rainfall agriculture, horticulture harvesting, and goods transportation. These all are goods and supply chain processes. If in supply chain process breaks any point, Ultimately, the farmer will get the loss. Timely predicting rainfall helps the farmers and agriculture and horticulture stock manage to maintain people require India's coastal area. For these reasons, this paper proposes the Ensemble Models (Catboost, Boost). Most of the authors are working on rainfall prediction using statistical models. Using statistical models to analyze and predict a huge amount of data is very difficult, depending upon the features. But using Ensemble models is likely to boost up the elements, and apply the classification to prediction makes it easy. This paper discusses and Compares the statistical decision tree model with ensemble models to find out the difference between the characteristics of algorithms and how they impact the timely predict the rainfall.

1 Introduction

The weather conditions around the world are changing quickly and continuously. Right predictions are central in today's everyday life. From agriculture to industry, from travel to regular transport, we are heavily dependent on weather forecasting. As the world is suffering from continuous climate change and its side effects, it is very important to forecast weather without error to ensure simple and smooth mobility and healthy day-to-day operations. [1].

In India, 30-50 percent of wheat yield variation depends on climate variability, and yields are especially susceptible to adverse climate conditions. Seasonal rainfall and water scarcity, particularly during spring, are the main causes of India's different crops. [2].

In this regard need to collect the data from the metrology department of rainfall to apply to the statistical and ensemble models. Here need to know the statistical model and ensemble model.

The method of applied statistic analysis to the data set is statistical modeling. A gathering (or mathematical model) of observed data is a statistical model. When data analysts use different statistical models for their analysis, the information can be more strategically understood and interpreted. This practice helps them classify connexions between variables, make projections of the future datasets, and visualize rather than select the raw data[3].

An analyst must accumulate or restore data in a folder, cloud, social network or plain-excellence file before a statistical model is used. Analysts would need to have a sound interpretation of the data system and administration, including the processing, storage, and retention of data. Many employed in this sector should also express their enthusiasm for facts and proof and recognise the basics of data exploitation [4]. Once the time has come to analyze the data, analysts can choose to use a range of statistical models. According to Mello, the most common techniques fall into the following two groups: supervised learning, including regression and classification models, and unsupervised learning, including clustering algorithms and association laws[3][5].

The Ensemble Model integrates multiple models, which have been designed using the same or different algorithms, to create a single model for use. Created with methods such as bagging boosters and Bayesian model averages, the ensembles seem to be incredibly complex yet tend to surpass their component models with new results. [3][7].

If uncertainty is measured by feature rather than structure, that location of the razor is restored rather than how it appears. The sophistication of a black-book modelling procedure calculates by re-sampling a GDF system rather than, say, counting parameters. A series of experiments on a two-dimensional decision tree topic have shown that the bagging of many trees has less GDF difficulty than a single tree part, avoiding the paradox of ensembles' generalisation. [3]..

2 Literature Survey

To study environmental changes on the earth depends upon the different attributes or parameters. These things technical termed as a feature of the object. Here the item is nothing but a thing that required hidden patterns and classified to determine the future. So in this section study, the rainfall predictions apply the statistical models and choose the models' way and accuracy.

Ahmad et al. undertake research as a research study (1) developing satellite remote-sensing estimates of maize acreage in Pakistan's typical maize area, (2) defining a statistical-empirical maize yield forecasting or (3) finally evaluating the impact of temperature on the inter-annual variability of maize yield over a decade. Classification models were validated around the region of the sample by 200 randomly chosen field check points. The effects of maize mapping work utilises Landsat's multi-timoral uniform vegetation difference index (NDVI) and surface soil (LST) data to assess interannual maize yields. The yield predictors were selected through a primary variable screening and fed into the least absolute shrinkage and regression model (LASSO) [7][13].

Chul-Min Ko et al. The research aimed to enhance the predictive precision of intense rainfall, using a machine learning technique to predict the hydrological effect. In this analysis, machine learning with the XGBoost technique uses to correct the quantitative precipitation forecast (QPF) given by the Korea Meteorological Administration (KMA) for the production of a hydrological quantitative precipitation forecast (HQPF) for flood modelling [9][14].

Diez-Sierra et al Eight mathematical and machine learning methods were used to estimate long-term daily precipitation in semi-arid climates (Tenerife, Spain). Cross-validation to recreate at 17 gauges 36 years of regular rainfall results. It's independent for each indicator. Compared to the observed data, the reconstructed sequence chooses the optimum hyperparameters for each model unit. The model efficiency evaluation utilises multiple rainfall rate and regular, weekly and annual aggregate scales. [10][20].

Elder et al. Analyze a modelling community of models trained by various approaches or algorithms to generate the final predictions. We will see in this chapter that clusters will surpass single-algorithm models. This discovery, though, appears to refute Occam Razor's theory that the most basic answer is usually the correct one. We would discuss this obvious contradiction and prove that such uncertainty is a positive thing .

Lazriet al. Discuss the precipitation forecasts of MSG (Meteosat Second Generation) using a multi-classifying machine-based learning model. Learn and test a multi-class model by comparing MSG satellite data and radar data. The outcomes of the classification calculated six certitude coefficients. Tier 2 classification is accomplished by the Random Forest Classifier (RF2), taking as input parameters the trust coefficients. Six types of precipitation strength gain exceptionally high intensity of precipitation, mild to high intensity of precipitation, mean heavy precipitation, low to medium intensity of precipitation, low intensity of precipitation and no précipitation. The comparisons between the findings of the multi-classifier and the results from independently used classification models indicate a substantial increase in classification accuracy [11].

Lei Xu et al. Discuss seasonal precipitation predictions in regional or local areas to help direct farming and urban water resource management. The North American Multi-Model Ensemble (NMME) is a seasonal forecasting method offering global precipitation forecasts. Spatial analysis reveals that the corrected NMME precipitation forecasts display the best South China ability with an average RMSE of about 30 mm, whereas the worst in Central and Southwest China with an 80 mm RMSE. Despite the correction, uncertainties of summer seasonal precipitation forecasts and severe wet cases are still high. WSVM and WRF methods can serve as an important tool in bias correction of NMME precipitation forecasts[12].

Some of the authors also discuss the statistical model and ensemble models and predict the model. The following sections are data set description ensemble models and heuristics of ensemble models algorithm flow, and mechanism and results discuss.

3 Rainfall Data Set Description

In Indian atmospheric conditions, rainfall is significant. Most of India depends upon agriculture and horticulture. Suppose the rain will be a help in the growth of agriculture productivity. In the Indian Economy, a significant part is the Agriculture and Horticulture. So for this reason, the thesis mainly focused on the rainfall predictions system. So In the system need to predict the rainfall range. The system requires data. This section discusses mostly the dataset's data set and features—the dataset collecting from the Indian Metrology data(IMD) data source. And data Set contains around 150000 records of data it is approximately 1960 to 2019 years of the monthly rainfall data of the Vishakhapatnam region, and it includes attributes like Location, Minimum Temperature (Degrees), Maximum Temperature (Degrees), Rainfall (Milli Meters), Evaporation, Sunshine, Industry, Relative Humidity, Pressure, Cloud Coverage.

The Sample Data set will consist mainly of numerical and categorical data. This particular data required to apply to a model is complicated. Why because this specific data needs to generalize and remove the noise. Sections, subsections and subsubsections

3.1.1 shows the sample data set with a description of the data set attributes.

Location	*MinTemp*	*MaxTemp*	*Rainfall*	*Evaporation*	*Sunshine*	*WindGustSpeed*	*relative humidity*	*Pressure*	*Cloud*	*RainToday*	*rain tomorrow*
VSKP	13.4	22.9	0.6	5.8	10.6	44	71	1007.7	8	No	No
VSKP	7.4	25.1	0	5.8	6	44	44	1010.6	NA	No	No
VSKP	12.9	25.7	0	4	9.4	46	38	1007.6	NA	No	No
VSKP	9.2	28	0	3.2	10.4	24	45	1017.6	NA	No	No
VSKP	17.5	32.3	1	4.8	5.1	41	82	1010.8	7	No	No
VSKP	14.6	29.7	0.2	2.6	5	56	55	1009.2	NA	No	No
VSKP	14.3	25	0	3.2	9.5	50	49	1009.6	1	No	No
VSKP	7.7	26.7	0	7.2	9.7	35	48	1013.4	NA	No	No
VSKP	9.7	31.9	0	8.8	11	80	42	1008.9	NA	No	Yes
VSKP	13.1	30.1	1.4	7	8.3	28	58	1007	NA	Yes	No
VSKP	13.4	30.4	0	8	11.9	30	48	1011.8	NA	No	Yes
VSKP	15.9	21.7	2.2	6	5.9	31	89	1010.5	8	Yes	Yes

The above data, the table consists of the details of the attributes are mention below
MinTemp& MaxTemp: Mintemp define as minimum temperature, and max temp defines as maximum temperature, and both measure on a scale of the degree in Celsius.
Rainfall: The quantity of rain which is measure in mm for a day.
Evaporation: It is a measure in the form of a millimeter. Evaporation calculation per day
Sunshine: The number of hours of bright daylight sunlight.
WindGustSpeed The velocity (km / h) in the 24 hours to midnight of the highest wind gust. RelHumid: It measures the relative moisture on the scale of the percentage. It calculates in the form of relative humidity in the air.
Cloud:
A cloud-obscured fraction of the sky, measured in "oktas," is a unit of eighths. It records how the cloud covers many eighths of the sky. A 0 measure indicates an apparent atmosphere while an eight suggests that it is completely overcast.
Rain Today Integer:
One if precipitation (mm) in the 24 hours to 9 am exceeds 1mm, otherwise 0.

4 Ensemble Models Are Applied Dataset

In this section, discuss the Ensemble model's functionalities flow of the algorithms on rainfall datasets. Here two algorithms taken for the predictions, one is CATBOOST and XGBOOST.

4.1 System Architecture

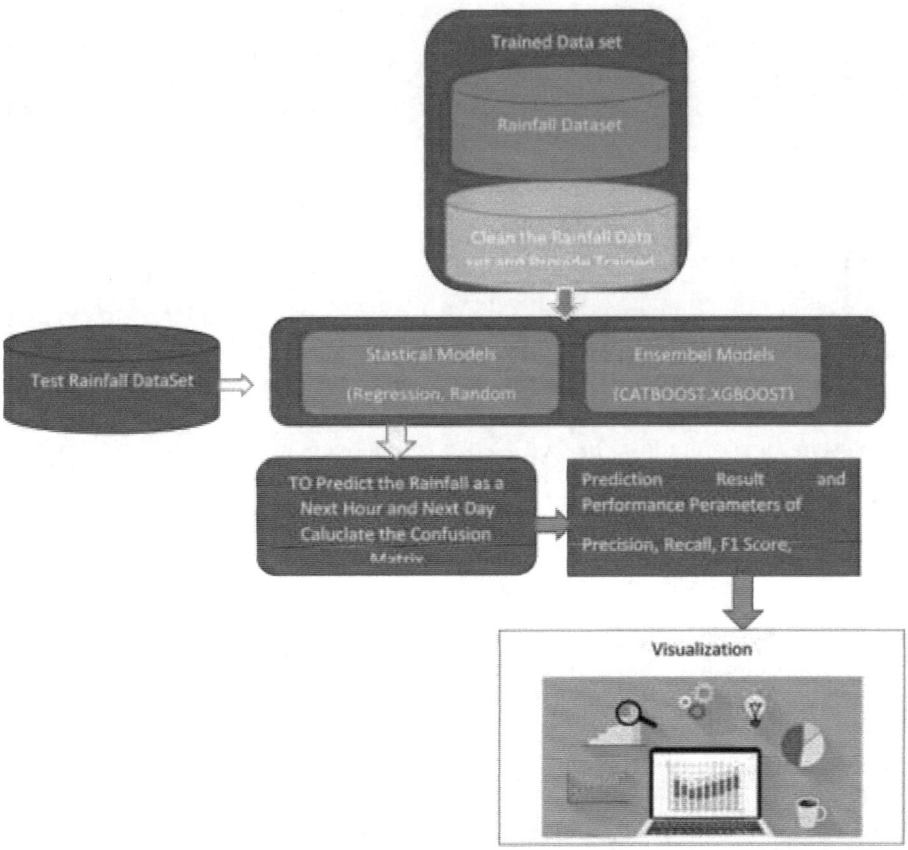

Figure 4.1.1System Architecture

4.2 CATBOOST

CatBoost is a new decision boosting gradient (GBDT) algorithm that can manage categorical features correctly. This algorithm varies on the following aspects from conventional GBDT algorithms [15].

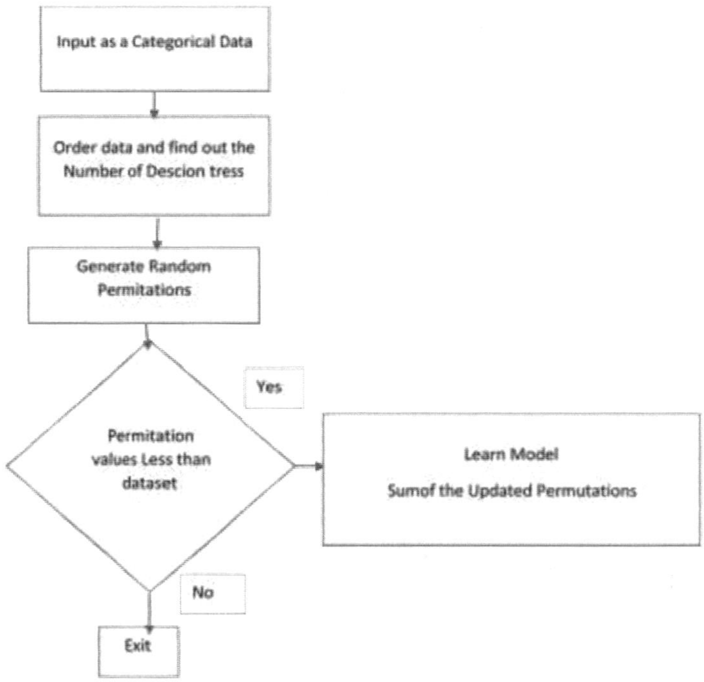

Figure4.2.1.1CATBOOST Flow Diagram

4.2.1Functionalities of CATBOOST

It works with categorical features rather than preprocessing during training. CatBoost requires the whole testing dataset to be included. Goal statistics (TS) are a highly effective means of handling categorical functionality and minimal data loss. In fact, CatBoost conducts a random permutation for each event, computing an average mark value with, for instance, the same category value in a permutation before the specified category value [16]. Role variations. All the categorical characteristics are new. CatBoost uses a gullible approach to perceive the combinations when constructing a new tree break. No mix thinks about a first break, but in the second and subsequent split data sets, Catboost integrates both presets and categorical functions. Both divisions selected from the tree are treated as a two-value category and merged. Categorical features are developing without discrimination. By using the TS process, the distribution can differ from the initial distribution to convert categorical characteristics into numerical values. The variance of this distribution would trigger the answer to depart, which for standard GBDT approaches is an inevitable problem and establishes a new methodology for solving the gradient variation by theoretical analysis; Quick scorer. Speedy scorer. CatBoostutilises forgotten trees as base predictors for the same dividing parameters in the entire stage of the tree. These trees are balanced and less prone to be overcrowded. In ignorance trees, each leaf index encodes as a binary vector with a length equivalent to the depth of the branch. This principle uses CatBoost model assessor model predictions as all binaries use float, statistic and single-hot encoded capabilities. [17]. These moves are nothing but CATBOOST algorithms to test the rainfall data predictions. The XGBOOST algorithm is another Ensemble type algorithm, as follows in the flow diagram and in functionality [18].

4.3 XGBOOST

XGBoost was initially proposed in 2011 and continuously refined and improved in many study follow-ups studies by Tianqi Chen and Carlos Guestrin. The paradigm is a method of learning based on the Boosting Tree model. For traditional boosting trees [20], only the first derivative details are used. Due to the rest of the former n-1 trees, dispersed teaching is not easy when training the nth tree. XGBoost conducts Taylor's second-order extension of the loss feature and can utilize CPU multithreading to perform parallel loading. XGBoost employs a variety of techniques to avoid overfitting. [22].

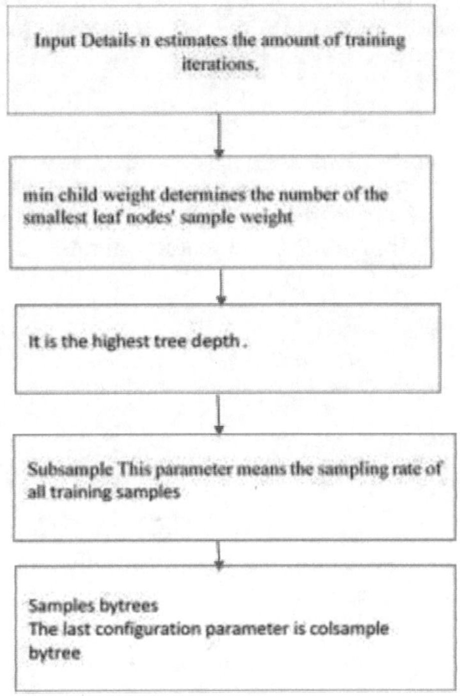

Figure 4.3.1.1XGBOOST Algorithm Functionalities as a Flow Diagram

4.3.1 XGBOOST Functionalities

In this experiment, the regression model XGBoost trains with each target gene separately, with an input mark of 943 genes, suggesting that the input dimension was 943, which is very high. However some XGBoost techniques to prevent overpasses will mitigate overpasses and increase predictive regression accuracy[21][23].

The XGBoost design of the experiment has been adapted to guarantee optimal model performance:

1New estimators n estimators evaluate the amount of training iterations. Too limited estimators can overlook fitness and leave the model incomplete. The estimator, though is typically too large and not secure, so it overflows.

2. Crete's Min Weight

The weight of the smallest leaf nodes is determined above by the weight of the smallest boy, to avoid excess.
3. Limit Profiling
It's the forest's full distance. The larger the tree heart, the more complicated the tree model and the better the ability, but the model is much simpler to overwrite.
4. Under-review
This parameter means the sampling rate of all samples.
5. Example (s)
The last parameter setup is a tree coal sample. When each tree forms, it is the sampling rate. The sampling rate of the hallmark gene for this mission is equal.
6. Rate for learning
The learning rate is an essential parameter for several XGBoost-related algorithms. It significantly affects the performance of the model. Each move's weight reduces to make the model longer lasting.

5 Experimental Results

The above section discusses how predictive ensemble models are working their functionalities. These ensemble models practically working is required. In this section, algorithms implemented using python and using the library's sklearn, NumPy, and matplot libraries implement the system. Python to implement the algorithms like Logistic Regression, Decision Tree, Random Forest, CATBOOST, XGBOOST[24].

5.1 DATA SET PREPROCESSING

Table 5.1.1.1 shows the sample data set as a training data set with attributes.

Location	*MinTemp*	*MaxTemp*	*Rainfall*	*Evaporation*	*Sunshine*	*WindGust Speed*	*relative humidity*	*Pressure*	*Cloud*	*RainToday*	*rain tomorrow*
VSKP	13.4	22.9	0.6	5.8	10.6	44	71	1007.7	8	No	No
VSKP	7.4	25.1	0	5.8	6	44	44	1010.6	NA	No	No
VSKP	12.9	25.7	0	4	9.4	46	38	1007.6	NA	No	No
VSKP	9.2	28	0	3.2	10.4	24	45	1017.6	NA	No	No
VSKP	17.5	32.3	1	4.8	5.1	41	82	1010.8	7	No	No
VSKP	14.6	29.7	0.2	2.6	5	56	55	1009.2	NA	No	No
VSKP	14.3	25	0	3.2	9.5	50	49	1009.6	1	No	No
VSKP	7.7	26.7	0	7.2	9.7	35	48	1013.4	NA	No	No
VSKP	9.7	31.9	0	8.8	11	80	42	1008.9	NA	No	Yes
VSKP	13.1	30.1	1.4	7	8.3	28	58	1007	NA	Yes	No
VSKP	13.4	30.4	0	8	11.9	30	48	1011.8	NA	No	Yes
VSKP	15.9	21.7	2.2	6	5.9	31	89	1010.5	8	Yes	Yes

Table 5.1.1.2 shows the sample data set as a training data set with attributes.

Min temp	*MaxTemp*	*Rainfall*	*Evaporation*	*Sunshine*	*Wind Gust Speed*	*relative humidity*	*Pressure*	*Cloud*
13.4	22.9	0.6	5.8	10.6	44	71	1007.7	8
7.4	25.1	0	5.8	6	44	44	1010.6	6
12.9	25.7	0	4	9.4	46	38	1007.6	5
9.2	28	0	3.2	10.4	24	45	1017.6	6
17.5	32.3	1	4.8	5.1	41	82	1010.8	7
14.6	29.7	0.2	2.6	5	56	55	1009.2	7
14.3	25	0	3.2	9.5	50	49	1009.6	1
7.7	26.7	0	7.2	9.7	35	48	1013.4	5
9.7	31.9	0	8.8	11	80	42	1008.9	1
13.1	30.1	1.4	7	8.3	28	58	1007	9
13.4	30.4	0	8	11.9	30	48	1011.8	4
15.9	21.7	2.2	6	5.9	31	89	1010.5	1

The above tables discuss the training data table in the training data table consists of the attributes with the labels. Second, without a data table as a test data table after getting the data set to apply, the pre-processing techniques analyze the data.

5.1.1 DATA EXPLORATION

Range Index: 94773 entries, 0 to 94772

5.1.1.1 Data columns (total of 24 columns):

Sno	*Column*	*Non*	*#NAME?*	*Dtype*
---	------	----	----------	-----
0	Date	94773	non-null	object
1	Location	94773	non-null	object
2	MinTemp	94313	non-null	float64
3	MaxTemp	94554	non-null	float64
4	Rainfall	93718	non-null	float64
5	Evaporation	52579	non-null	float64
6	Sunshine	45783	non-null	float64
7	WindGustDir	89081	non-null	object
8	WindGustSpeed	89089	non-null	float64
9	WindDir9am	87119	non-null	object
10	WindDir3pm	92116	non-null	object
11	WindSpeed9am	93615	non-null	float64
12	WindSpeed3pm	93014	non-null	float64
13	Humidity9am	93482	non-null	float64
14	Humidity3pm	93067	non-null	float64
15	Pressure9am	85072	non-null	float64
16	Pressure3pm	85119	non-null	float64
17	Cloud9am	58619	non-null	float64
18	Cloud3pm	57476	non-null	float64
19	Temp9am	93982	non-null	float64

IOP Conf. Series: Materials Science and Engineering 1074 (2021) 012019 doi:10.1088/1757-899X/1074/1/012019

20	Temp3pm	93514	non-null	float64
21	RainToday	93717	non-null	object
22	RISK_MM	94772	non-null	float64
23	RainTomorrow	94772	non-null	object

The above represents the data exploration consists of the Column name object types represent a data set as trained data set representation. It consists of 94773 entries of records and 24 columns of the dataset.

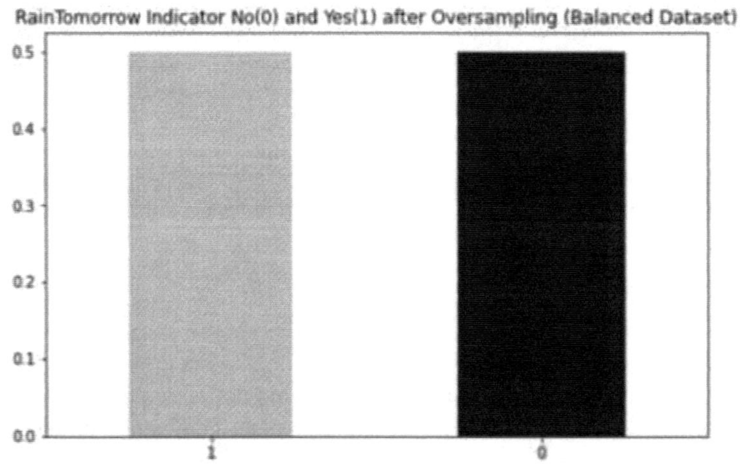

Figure 5.1.1.1 Rain Tomorrow Indicator No(0) and Yes(1) after Oversampling (Balanced Dataset

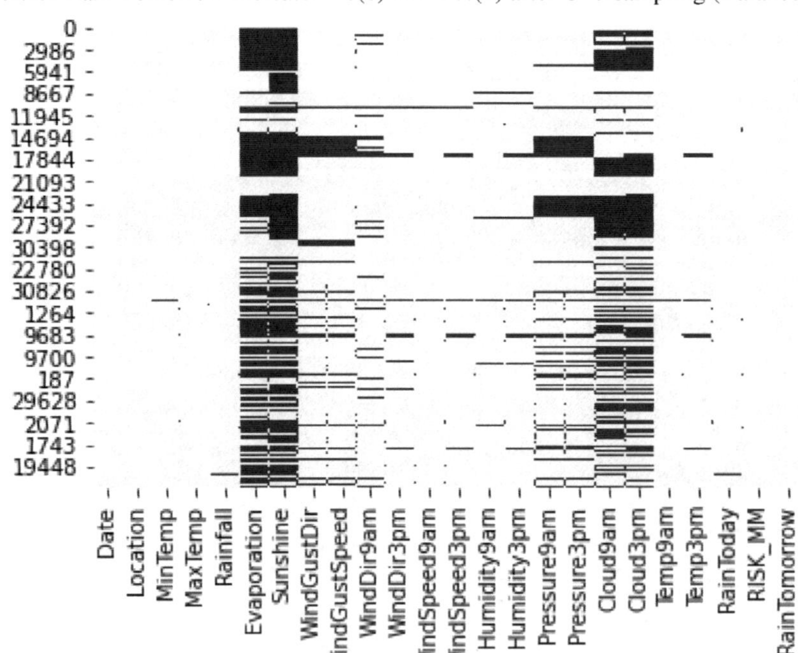

Figure 5.1.1.2 Heat Map for the Trained data set

The Below Table shows how to evaluate the MICE package (Multiple Imputation by Chained Equations). Afterward, we will detect outliers using the Inter-Quartile Range and remove them to get the final working data set.

Table 5.1.1. 3 the Inter-Quartile Range and remove them to get the final working data set

Date	**1495**
Location	6
MinTemp	8.4
MaxTemp	7.8
Rainfall	2.6
Evaporation	3.6
Sunshine	6.001871
WindGustDir	7
WindGustSped	18
WindDir9am	8
WindDir3pm	6
WindSpeed9am	13
WindSpeed3pm	11.53649
Humidity9am	24
Humidity3pm	32
Pressure9am	6.9
Pressure3pm	6.9
Cloud9am	4.039764
Cloud3pm	3.943048
Temp9am	7.4
Temp3pm	7.4
rain today	1
RISK_MM	5.8
rain tomorrow	1

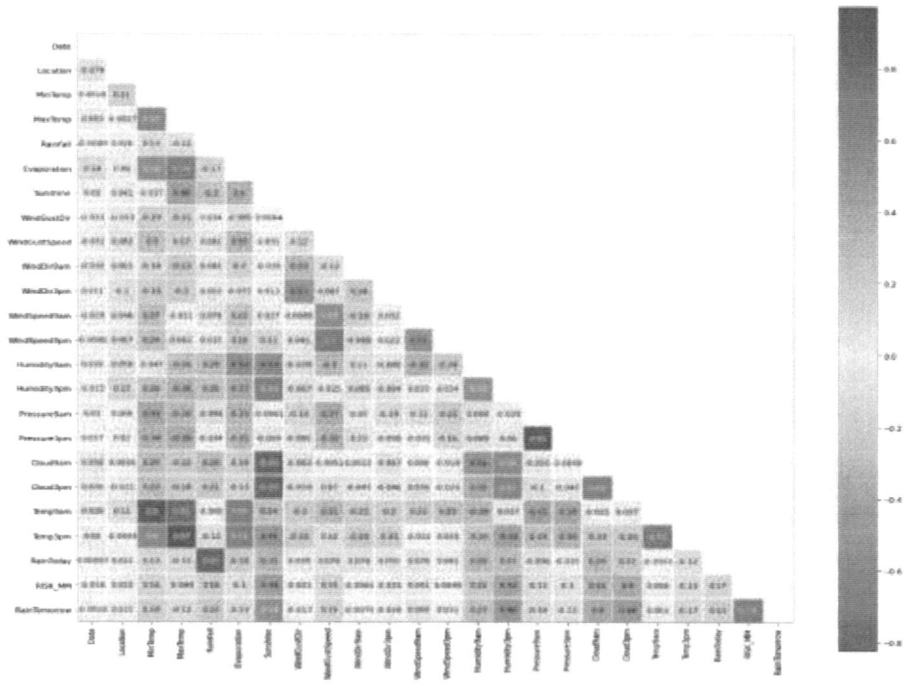

Figure 5.1.3 Correlation Heat Map between Attributes

Figure 5.1.14 Attributes Maps

5.2 FEATURE SELECTION

Feature selection and feature enhance mechanisms to apply to the rainfall data set.

Date Location MinTemp MaxTemp Rainfall Evaporation Sunshine WindGustDir WindGustSpeed WindDir9am WindDir3pm WindSpeed9am WindSpeed3pm Humidity9am Humidity3pm Pressure9am Pressure3pm Cloud9am Cloud3pm Temp9am Temp3pm RainToday RISK_MM

RainTomorrow

0	0.115284	0.03125 0.571429		0.518987	0.435873
	0.565373	0.468635	0.866667	0.521127	0.866667
	0.933333	0.516057	0.615385	0.539969	0.212121
	0.223214	0.277286	0.858482	0.409095	0.515957
	0.533627	0.174811	0.000000	0.0	
1	0.115575	0.03125 0.400000		0.574684	0.373192
	0.561494	0.727678	0.933333	0.521127	0.400000
	1.000000	0.108527	0.564103	0.236236	0.242424
	0.309524	0.297935	0.278259	0.223735	0.523936
	0.600252	0.174811	0.000000	0.0	
2	0.115866	0.03125 0.557143		0.589873	0.373192
	0.690262	0.775584	1.000000	0.549296	0.866667
	1.000000	0.490587	0.666667	0.168740	0.292929
	0.220238	0.324484	0.277728	0.167728	0.625000
	0.570937	0.174811	0.000000	0.0	
3	0.116157	0.03125 0.451429		0.648101	0.373192
	0.579307	0.756129	0.266667	0.239437	0.600000
	0.000000	0.286821	0.230769	0.247486	0.151515
	0.517857	0.445428	0.237588	0.173953	0.547872
	0.658881	0.174811	0.076923	0.0	
4	0.116448	0.03125 0.688571		0.756962	0.477660
	0.637367	0.384110	0.866667	0.478873	0.066667
	0.466667	0.184939	0.512821	0.663712	0.323232
	0.315476	0.244838	0.769158	0.670914	0.539894
	0.744161	0.174811	0.015385	0.0	

5.3 LOGISTIC REGRESSION ALGORITHMS IMPLEMENTATION

Table 5.3.1 Shows the Logistic Regression Algorithm Performance applies to the rainfall data set

Parameters	***Performances***
Accuracy	0.77
ROC Area under Curve	0.75
Cohen's Kappa	0.51
Time Taken	1.37

Table 5.3.2 Shows the Logistic Regression Algorithm Confusion Matrix apply to the rainfall data set

	precision	***recall***	***f1-score***	***support***
0	0.79388	0.85497	0.82329	15866
1	0.73959	0.6498	0.69179	10057
accuracy			0.77537	25923
macro avg	0.76673	0.75238	0.75754	25923
weighted avg	0.77282	0.77537	0.77228	25923

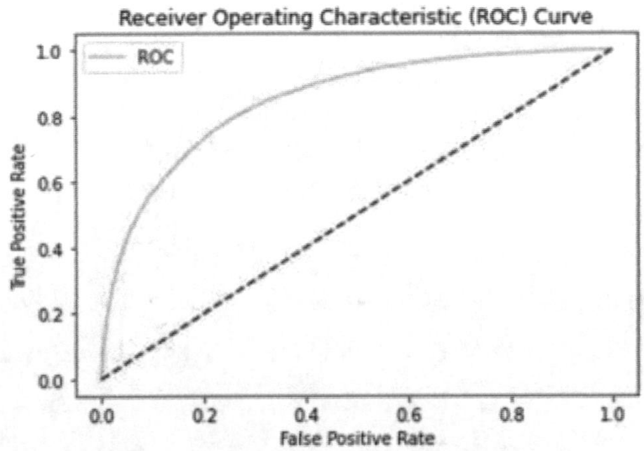

Figure 5.3.1 Logistic ROC Curve

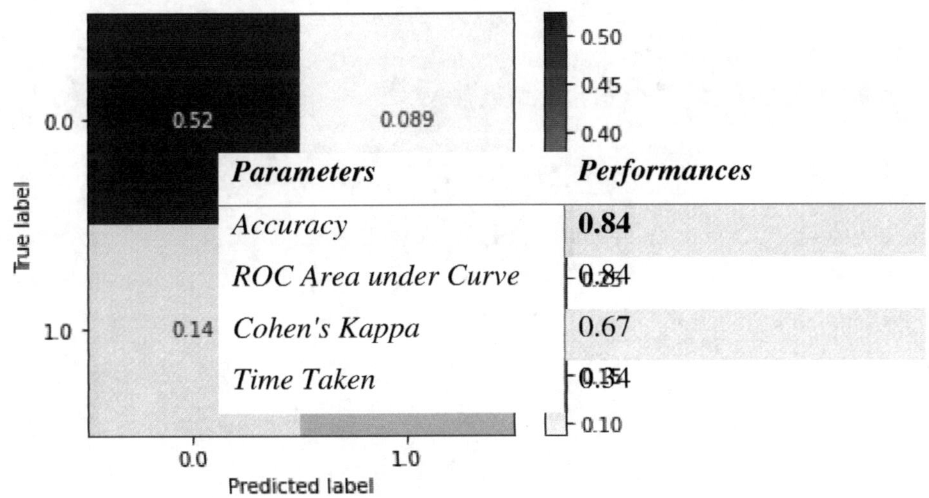

Parameters	*Performances*
Accuracy	**0.84**
ROC Area under Curve	0.84
Cohen's Kappa	0.67
Time Taken	0.54

Figure 5.3.2 Confusion Matrix for Logistic

5.4 DECISION TREE

Table 5.4.1 Shows the Decision Tree Algorithm Performance applies to the rainfall data set

IOP Conf. Series: Materials Science and Engineering 1074 (2021) 012019 doi:10.1088/1757-899X/1074/1/012019

Table 5.4.2 Shows the Decision Tree Algorithm Confusion Matrix apply to the rainfall data set

	precision	*recall*	*f1-score*	*support*
0	0.88726	0.85466	0.87065	15866
1	0.78327	0.82868	0.80533	10057
accuracy			0.84458	25923
macro avg	0.83527	0.84167	0.83799	25923
weighted avg	0.84692	0.84458	0.84531	25923

IOP Conf. Series: Materials Science and Engineering 1074 (2021) 012019 doi:10.1088/1757-899X/1074/1/012019

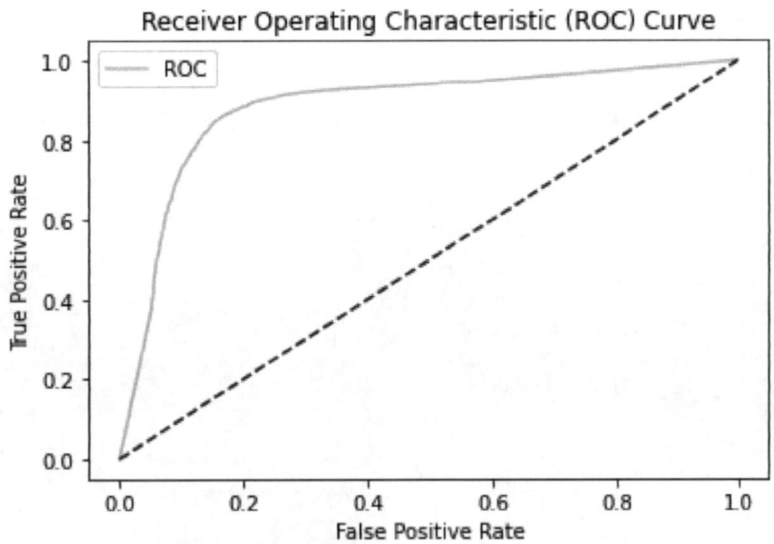

Figure 5.4.1 Decision Trees ROC Curve

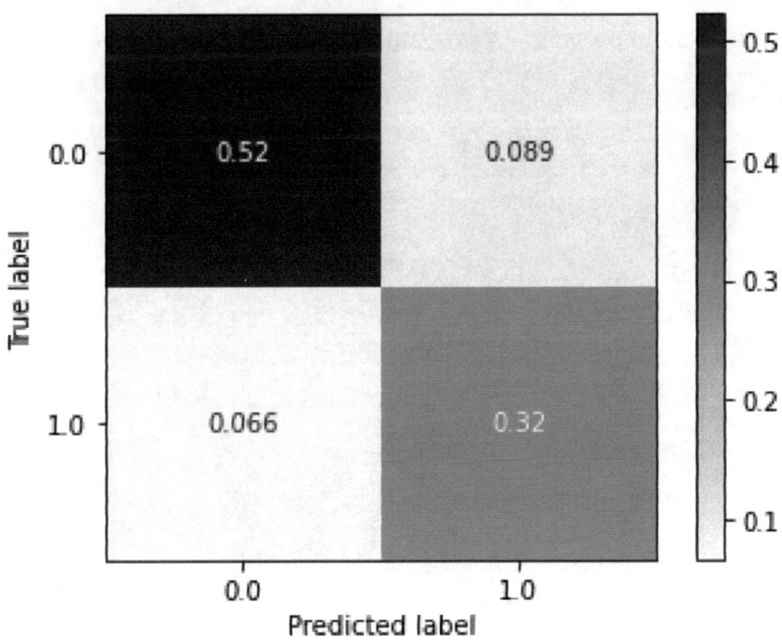

Figure 5.4.2 Confusion Matrix for Logistic

5.5 RANDOM FOREST

Table 5.5.1Shows the RANDOM FOREST Algorithm Performance applies to the rainfall data set

Parameters	***Performances***
Accuracy	0.92
ROC Area under Curve	0.92
Cohen's Kappa	0.84
Time Taken	20.0

Table 5.5.1 Shows the RANDOM FOREST Algorithm confusion Matrix apply to the rainfall data set

	precision	***recall***	***f1-score***	***support***
0	0.94714	0.93502	0.94104	15866
1	0.89951	0.91767	0.90850	10057
accuracy			0.92829	25923
macro avg	0.92332	0.92634	0.92477	25923
weighted avg	0.92866	0.92829	0.92842	25923

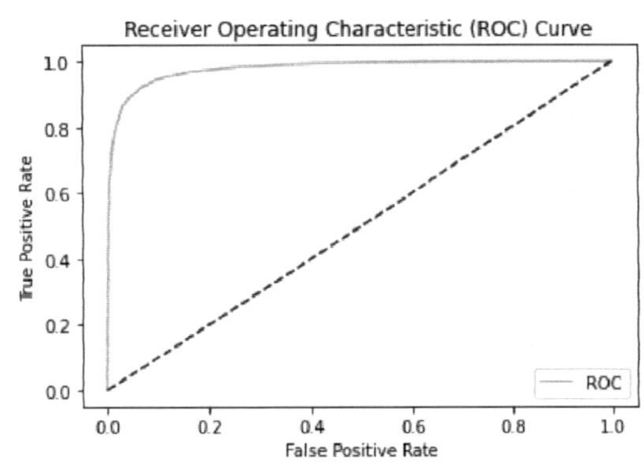

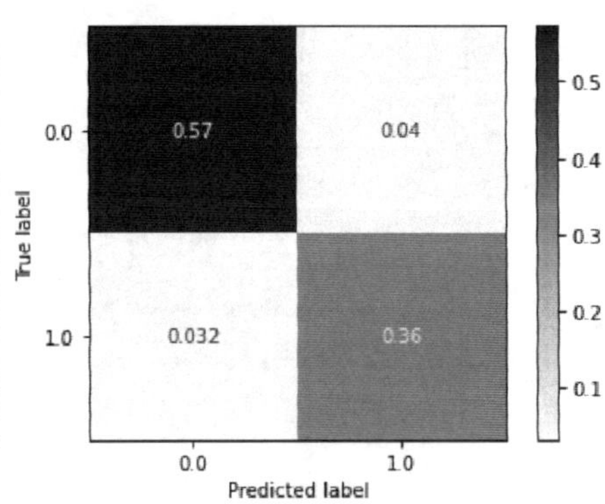

Figure 5.5.1 Decision Trees ROC Curve**Figure 5.5.2**Confusion Matrix for Decision Trees

IOP Conf. Series: Materials Science and Engineering 1074 (2021) 012019 doi:10.1088/1757-899X/1074/1/012019

5.6 XGBOOST

Table 5.6.1 Shows the Performance applies to the XGBOOST Algorithm rainfall data set

Parameters	***Performances***
Accuracy	**0.95**
ROC Area under Curve	**0.95**
Cohen's Kappa	**0.91**
Time Taken	**105.97**

Table 5.6.2 Shows the XGBOOSTAlgorithm confusion Matrix apply to the rainfall data set

	precision	***recall***	***f1-score***	***support***
0	0.98714	0.94743	0.96368	8541
1	0.92310	0.97100	0.94644	5551
accuracy			0.95671	14092
macro avg	0.95180	0.95921	0.95506	14092
weighted avg	0.95789	0.95671	0.95689	14092

Parameters	*Performances*
Accuracy	**0.943**
ROC Area under Curve	**0.923**

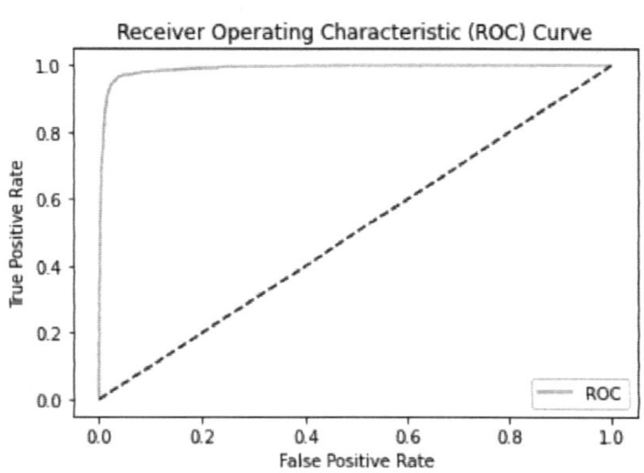

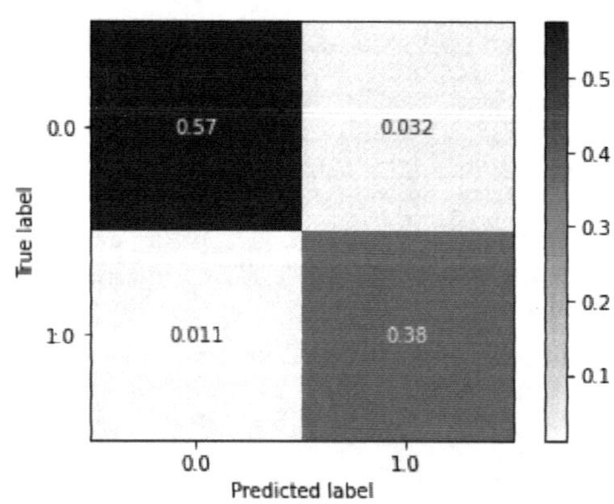

Figure 5.6.1 XGBOOST ROC Curve**Figure 5.6.2** Confusion Matrix for XGBOOST

5.7 CATBOOST

Table 5.7.1 Shows the CATBOOST Algorithm Performance applies to the rainfall data set

Cohen's Kappa	**0.90**
Time Taken	**100.5**

Table 5.6.2 Shows the CATBOOSTAlgorithm confusion Matrix apply to the rainfall data set

	precision	*recall*	*f1-score*	*support*
0	0.97714	0.93743	0.97368	8541
1	0.91310	0.97100	0.94644	5551
accuracy			0.95671	14092
macro avg	0.94180	0.94921	0.95506	14092
weighted avg	0.94689	0.946671	0.95689	14092

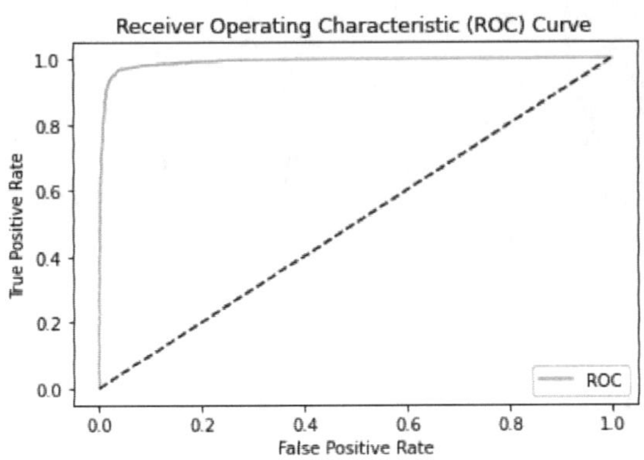

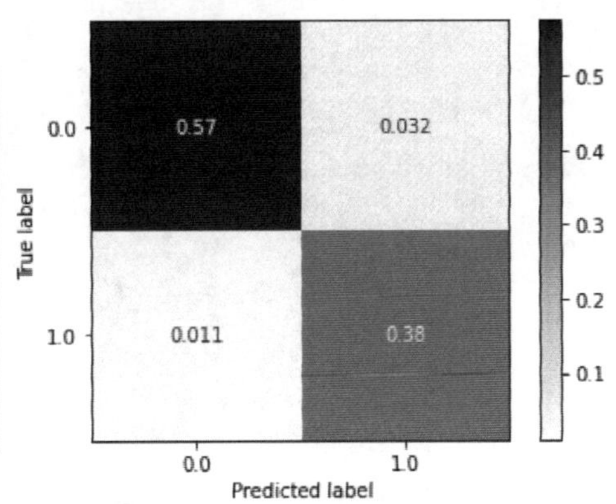

Figure 5.6.1 CATBOOST ROC Curve**Figure 5.6.2** Confusion Matrix for CATBOOST

5.8 MODEL COMPARISONS

IOP Conf. Series: Materials Science and Engineering 1074 (2021) 012019 doi:10.1088/1757-899X/1074/1/012019

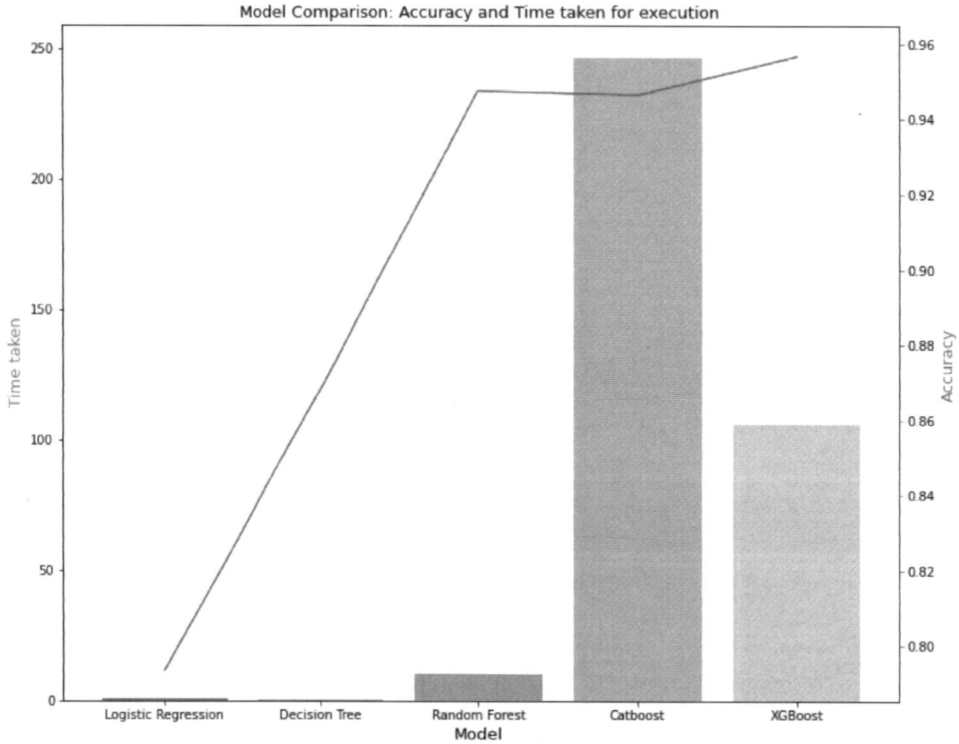

Figure 5.8.1 Barcharts Compare the Models Accuracy of all the algorithms

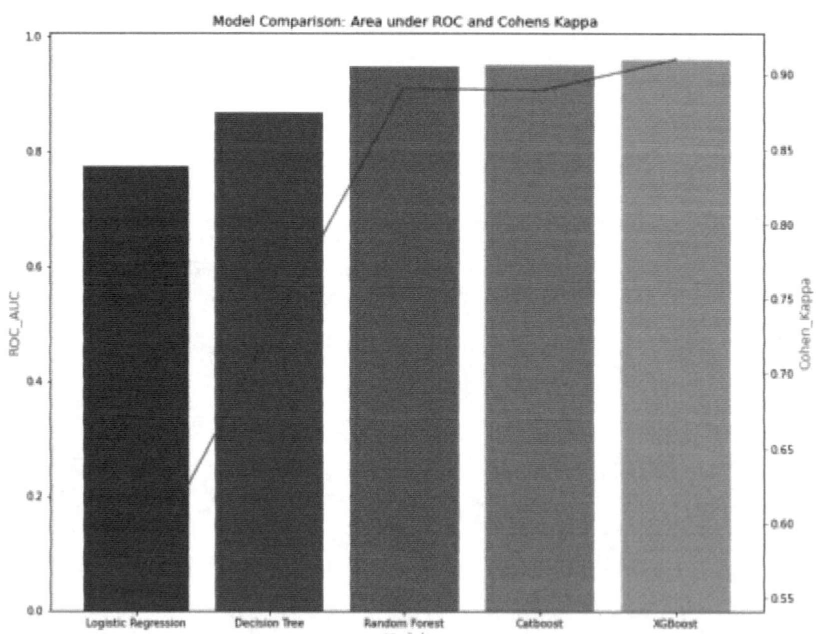

Figure 5.8.2 Bar charts Compare the Models ROC Curve and kappa Accuracy of all the algorithms

6. Conclusion

In this paper, the objective is to find out the accurate predictions. These days weather and changes and rainfall information are very important. By dependence on rainfall, horticulture, and agriculture, they are farming very crucial. In this regard, to find out the accurate rain, rain is possible next hour or the next day. This type of weather scenario consists of a huge amount of data. This historical data is very important for trained data, So here machine learning algorithms trained this huge amount of data to analyze and predict heuristic algorithms. This work mainly discusses the statistical algorithms logistic, random forest and decision tree algorithms, and Ensemble algorithms CATBOOST and XGBOOST algorithms apply the rainfall prediction using these algorithms and find out every algorithm's performances. To find out accurately predicting the rainfall to compare all models. In this context, XGBOOST is the best performing algorithm to compare with the remaining algorithms.

References

1. Jakaria, A. H. M., Hossain, M. M., & Rahman, M. A. (2020). Smart weather forecasting using machine learning: a case study in tennessee. arXiv preprint arXiv:2008.10789.

2. Wang, B., Feng, P., Waters, C., Cleverly, J., Li Liu, D., & Yu, Q. (2020). Quantifying the impacts of pre-occurred ENSO signals on wheat yield variation using machine learning in Australia. Agricultural and Forest Meteorology, 291, 108043.

3. John Elder. The Apparent Paradox of Complexity in Ensemble Modeling, 291, 108043.

4. Tufaner, F., &Özbeyaz, A. (2020). Estimation and easy calculation of the Palmer Drought Severity Index from the meteorological data by using the advanced machine learning algorithms. Environmental Monitoring and Assessment, 192(9), 1-14.

5. Das, A., Khan, M. N., & Ahmed, M. M. (2020). Detecting lane change maneuvers using SHRP2 naturalistic driving data: a comparative study machine learning techniques. Accident Analysis & Prevention, 142, 105578.

6. Fullhart, A. T., Nearing, M. A., McGehee, R. P., &Weltz, M. A. (2020). Temporally downscaling a precipitation intensity factor for soil erosion modeling using the NOAA-ASOS weather station network. CATENA, 194, 104709.

7. Shamshirband, S., Hashemi, S., Salimi, H., Samadianfard, S., Asadi, E., Shadkani, S., ... & Chau, K. W. (2020). Predicting standardized streamflow index for hydrological drought using machine learning models. Engineering Applications of Computational Fluid Mechanics, 14(1), 339-350.

8. Ahmad, I., Singh, A., Fahad, M., & Waqas, M. M. (2020). Remote sensing-based framework to predict and assess the interannual variability of maize yields in Pakistan using Landsat imagery. Computers and Electronics in Agriculture, 178, 105732.

9. Ko, C. M., Jeong, Y. Y., Lee, Y. M., & Kim, B. S. (2020). The Development of a Quantitative Precipitation Forecast Correction Technique Based on Machine Learning for Hydrological Applications. Atmosphere, 11(1), 111.

10. Diez-Sierra, J., & del Jesus, M. (2020). Long-term rainfall prediction using atmospheric synoptic patterns in semi-arid climates with statistical and machine learning methods. Journal of Hydrology, 124789.
11. Huang, G., Wu, L., Ma, X., Zhang, W., Fan, J., Yu, X., ... & Zhou, H. (2019). Evaluation of CatBoost method for prediction of reference evapotranspiration in humid regions. Journal of Hydrology, 574, 1029-1041.
12. Li, Y., & Qin, Y. (2019). The Response of Net Primary Production to Climate Change: A Case Study in the 400 mm Annual Precipitation Fluctuation Zone in China. International Journal of Environmental Research and Public Health, 16(9), 1497.
13. Kuradusenge, M., Kumaran, S., &Zennaro, M. (2020). Rainfall-Induced Landslide Prediction Using Machine Learning Models: The Case of Ngororero District, Rwanda. International Journal of Environmental Research and Public Health, 17(11), 4147.

14. Jakariya, M., Alam, M. S., Rahman, M. A., Ahmed, S., Elahi, M. L., Khan, A. M. S., ... & Ali, M. S. (2020). Assessing climate-induced agricultural vulnerable coastal communities of Bangladesh using machine learning techniques. Science of the Total Environment, 742, 140255.
15. Jakariya, M., Alam, M. S., Rahman, M. A., Ahmed, S., Elahi, M. L., Khan, A. M. S., ... & Ali, M. S. (2020). Assessing climate-induced agricultural vulnerable coastal communities of Bangladesh using machine learning techniques. Science of the Total Environment, 742, 140255.

16. Mosavi, A., Sajedi-Hosseini, F., Choubin, B., Taromideh, F., Rahi, G., &Dineva, A. A. (2020). Susceptibility mapping of soil water erosion using machine learning models. Water, 12(7), 1995.

17. Pham, B. T., Le, L. M., Le, T. T., Bui, K. T. T., Le, V. M., Ly, H. B., & Prakash, I. (2020). Development of advanced artificial intelligence models for daily rainfall prediction. Atmospheric Research, 237, 104845.

18. Raza, A. S., Shoaib, M., Faiz, M. A., Baig, F., Khan, M. M., Ullah, M. K., & Zubair, M. (2020). Comparative Assessment of Reference Evapotranspiration Estimation Using Conventional Method and Machine Learning Algorithms in Four Climatic Regions.(CF Alexander Rabinovich, Ed.). Journal of Pure and applied geophysics, 30.

19. Bushra, P., Swapan, T., Susanta, M., Jayanta, M., Pritee, S., Towfiqul, I. A. R. M., &Atiqur, R. (2020). Analyzing trend and forecasting of rainfall changes in India using non-parametrical and machine learning approaches. Scientific Reports (Nature Publisher Group), 10(1).

20. Chen, Y. J., Nicholson, E., & Cheng, S. T. (2020). Using machine learning to understand the implications of meteorological conditions for fish kills. Scientific Reports, 10(1), 1-13.

21. Neelakandan, S., &Paulraj, D. (2020). An automated exploring and learning model for data prediction using balanced CA-SVM. Journal of Ambient Intelligence and Humanized Computing, 1-

22. Li, W., Yin, Y., Quan, X., & Zhang, H. (2019). Gene Expression Value Prediction Based on XGBoost Algorithm. Frontiers in genetics, 10, 1077.

Unsupervised Learning Approach for Clustering Leaf Images

G Chamundeswari

Research Scholar, JNTU Kakinada, Kakinada, A.P., India.

Abstract. With appropriate representation methods, the clustering techniques are found to be efficient with neural networks. The present work aims to propose various feature representation techniques for efficient clustering. The methods used for feature representation in this paper are, a method using random closed set, a method using edge information of input entity, a method that uses Huff transformation and a method that uses boundary moments. A comparative study of these representation methods for clustering the input objects using artificial neural networks, specifically Self-Organizing Map (SOM) is focused.

Keywords – Feature vector, Self Organizing Maps (SOM), Clustering, Neural Networks.

1. Introduction

Clustering is the process of grouping similar objects that have same characteristics. Objects belonging to a cluster have high similarity and are dissimilar to objects of other clusters. Clustering technique [3] needs efficient representation methods. Image segmentation [5] involves the process of dividing the image into segments or regions having similar attributes or characteristics like intensity or texture, colour etc., so as to analyse the acquired information. Using these features the leaf images are clustered. The present work aims to propose appropriate combinations of various representation techniques with Self Organizing Maps (SOM) for future performance. The neural networks work on the model of a human brain. In neural networks the neurons are clustered based on the active neurons by reinforcing the neighbours of the active neurons and suppressing the remaining neurons. The distinct property of SOM is that they can map high dimensional input vectors into spaces with fewer dimensions and preserve the original topology [2]. The next section emphasises on the various representation methods. Among various techniques used, a Boundary Moment based Clustering Technique (BMCT) is used to represent the input image with the moment based feature vector. The second method - An edge based clustering technique is proposed for edge representation. In which, the identified segments of the image are enhanced using an enhancement filter for effective representation. After which, various sub images and edge image are generated. Features are computed for every sub-image and the edge image. In the third method – contact distribution function based clustering technique is proposed for random closed set representation. In this method, the spherical contact distribution function based feature vector, which gives the neighborhood connectivity representation is considered as the input object. The fourth method – line segment based clustering is an approach for evaluating the feature vector which is proposed with the Hough transformation. With the Hough transformation, the points are mapped to the line segment. The line features are considered as the feature vector. In all these methods, the extracted features are further utilized for clustering using Self Organizing Map (SOM). In SOM based clustering technique the evaluation of features is found to be

IOP Conf. Series: Materials Science and Engineering 1074 (2021) 012020 doi:10.1088/1757-899X/1074/1/012020

crucial. The related work is given in second section, methodology in the third and the experimental observations and results are given in the fourth section.

2. Related Work

Grouping of similar objects by splitting up of data is referred to clustering objects that are similar to one another and dissimilar to the objects in other groups, are contained in every group or cluster. Reduction of intra cluster distance [24, 19], maximizing inter cluster distances is the objective. For pattern recognition applications, the data clustering is found to be a crucial stage [13]. Ensemble based clustering method [16] is used to improve the reliability of the clustering method.

It is found that, in computer vision and recognition of patterns, clustering is one of the most interested research areas. To perform the clustering process, various approaches and techniques are available in the literature [23, 20]. Because of the wide variety of applications like segmentation, content based information retrieval and object recognition, it has gained high attention by the researchers. The difficult task is the search of relevant information [17] in unsupervised approach. For solving the combinational optimization problem, a Multi-Cluster Feature Selection (MCFS) approach is used. Using neuro morphological approach [18], the neostirate inter neurons are clustered. The predictor is estimated by it and then analysis of multivariate clusters is performed and then classified into multi labels. An ensemble of clustering tree [21] is used to improve the efficiency of the fuzzy min-max [14] network. An efficient learning model is used by it to improve the performance of the clustering method. Four different types of mean measures are used in the network. To cluster the fuzzy mid-wave infrared image, a multilayer immune network [25] is used. The mechanism of the co-ordination network is used by this. With the immune network, the inter class variance is minimized by it. It is found that the low performance is yielded when compared with other networks by the Euclidean distance based clustering with ART2 network [26]. A steganographic project with the hybrid neural system [15] is used to improve this. The performances of both ART2 and RBF networks are used by it. It is observed that, for analyzing the dynamic patterns, the unsupervised neural network [22] is found to be effective. To analyze the data with the clustering techniques SOM is used.

3. Methodology

The clustering techniques require the efficient representation of the input image. For this, the present work compares the various methods proposed like, a boundary moment based clustering technique (BMCT) [8], an edge based clustering technique (ECT) [9], a contact distribution function based clustering technique (CDFCT) [10], and line segment region representation based clustering technique (LSCT) to give a better choice. The flow chart is as shown in figure.1.

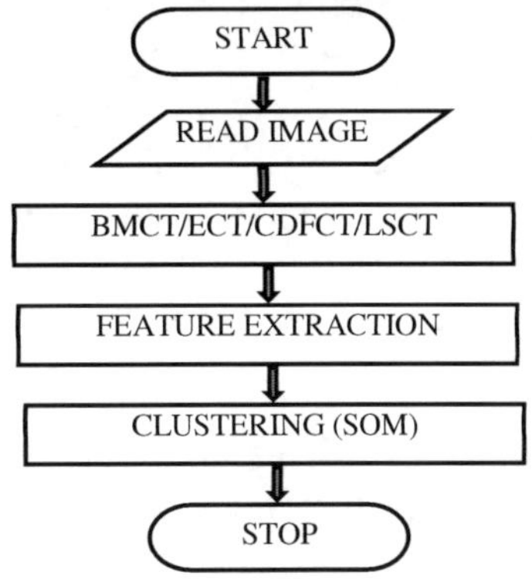

Figure 1. Flowchart for the methodology

A database consisting of 795 leaf image samples from 90 different plant species ranging from 2 to 25 is taken for experimenting.

IOP Conf. Series: Materials Science and Engineering 1074 (2021) 012020 doi:10.1088/1757-899X/1074/1/012020

3.1. Boundary Moment based Clustering Technique with SOM

Hu in 1992 gave the mathematical foundation for 2D moment invariants, which were found to be quick and highly reliable. These moment invariants are invariant with respect to translation, scale and orientation. A boundary moment based clustering technique (BMCT) in which, the boundary of the input image is represented with the moment based feature vector. The boundary of the input image is used for this. Seven invariant features will be evaluated from the boundary of the image, as shown in (1).

$$\mu_{pq} = \sum_x \sum_y \left(x - \bar{x}\right)^p \left(y - \bar{y}\right)^q f(x, y) \tag{1}$$

The invariance is achieved by computing moments that are normalized with respect to the centre of gravity ie., the centre of mass of the distribution is at the origin, which are called the central moments. Seven-moment invariant descriptors are formulated as a feature vector. The SOM is used to perform the process of clustering.

3.2. Edge based Clustering Technique with SOM

To reduce noise and smoothen the image, mean filter is used. Each pixel is replaced with the average of its neighbouring pixels. A local window is considered, of 3 X 3 size as shown in Table 1. The pixels (p1, p2, p4) present to the left top corner of this window are used by this filter as shown in (2).

1	P2	P3
4	P5	P6
7	P8	P9

Figure 2. A Local window for the pixel P5

$$F = (A \times P_1 + P_4) - (B \times P_1) + (C \times P_5 - [D \times (P_2 + P_4)] + [D \times P_1]) \tag{2}$$

Using (3)-(7), the edge image and the sub images are generated

$$S_1 = F_{P_1} - F_{P_8} \tag{3}$$

$$S_2 = F_{P_4} - F_{P_5} \tag{4}$$

$$S_2 = F_{P_4} - F_{P_5} \tag{5}$$

$$S_4 = F_{P_2} - F_{P_7} \tag{6}$$

$$E = \begin{cases} 1 & \left(\left[\sum P_i\right]^2\right) > 1 \\ 0 & otherwise \end{cases} \tag{7}$$

A co-occurrence matrix is generated for each of the edge and sub images. The co-occurrence matrix is a statistical texture analysis method that examines the relationship among the pixels and defines how frequently the combination of pixels are present in the image in a given direction and distance. The feature vector consists of the five features extracted and are then evaluated in the next stage of clustering using SOM.

3.3. Methodology of CDF Based Clustering Technique with SOM

For the texture based object recognition, there are many standard features [6] and random closed sets is one such used in this approach. Random closed set (RCS) can be used as univariate RCS for binary images and multivariate for gray scale images. This representation is described with various features like volume fraction, covariance, cross covariance, Ripley's function and contact distribution. It is found that the contact distribution function based feature vector is efficient for object recognition. The contact distribution function can be either spherical or linear.

Let S be the random closed set and the window is T then the volume fraction (VF) in (8) indicates the area or volume of the window

$$VF = \frac{|S \cap T|}{|T|} = \frac{\nu(S)}{\nu(T)} \quad (8)$$

The covariance (CV) of the S with increasing size t is given in (9).

$$CV(t) = \frac{(|S+t) \cap (T+t)|}{|T+t|} = \frac{\nu(\in_{\{0,t\}}(S))}{\nu(\in_{\{0,t\}}(T))} \quad (9)$$

The Spherical contact distribution function (SCDF) of S with a ball of radius t centered at the origin is given in (10).

$$SCDF(t) = 1 - \frac{\nu(\in_{B(0,t)}(S^C))}{\nu(\in_{B(0,t)}(T))} \frac{\nu(T)}{\nu(S^C)} \quad (10)$$

In [11], these features are used to describe the binary texture efficiently. For clustering, SOM is observed to be efficient.

3.4. Line Segment Region Representation Based Clustering Approach with SOM

The borders between the regions are separated with different grey levels by line segments. The input image must be a thresholded edge image which is calculated using sobel operator. The line segment will be estimated with various angles from 0^0 to 360° using Huff-transformation. With 'n' pixels, the set of pixels that make up straight lines are found.

$$b = -xa + y \quad (11)$$

A single point in X-Y plane gives a line in (a,b) space. With the peaks, the lines are identified and the length of the detected lines are considered as features, which are clustered further using SOM neural network.

4. Results

The experiment has been conducted using different SOM organizations namely grid topology, random topology, triangle topology and hexagon topology. The proposed methods are experimented with various sizes of neighborhoods varying for 3 to 20, whose performance is evaluated for each structure of the topological pattern. In this experiment four measures are used to calculate the distance- The boxdist is the distance given by the position vector of the neuron layer, linkdist is the number of links associated by a neuron, dist is the distance given by the Euclidean measure computed and finally the mandist is the distance given by the Manhattan measure. For each experiment conducted with SOM, associated structure of the topology, the size of the neighborhood, the calculated distance measure and the feature vector is fed.

A confusion table is used to evaluate the performance of a classifier. Wherein the predicted and the actual class labels are compared and the results are measured to evaluate and measure the level of performance of the proposed method, four measures are used, the false positive ratio (FPR), the false negative ratio (FNR), the true positive ratio (TPR) and the true negative ratio (TNR).

For example the following figures 3,4,5 and 6 demonstrates TPR representing all the topology structures, with varying neighborhood sizes and the four mentioned distances that are calculated. By this observation the TPR is at its highest for randtop topology.

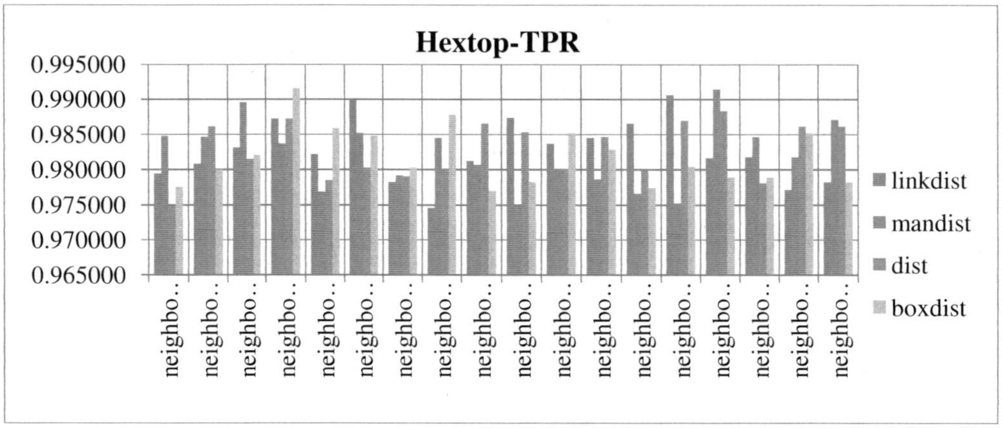

Figure 3. SOM-Hextop-TPR Comparison Result

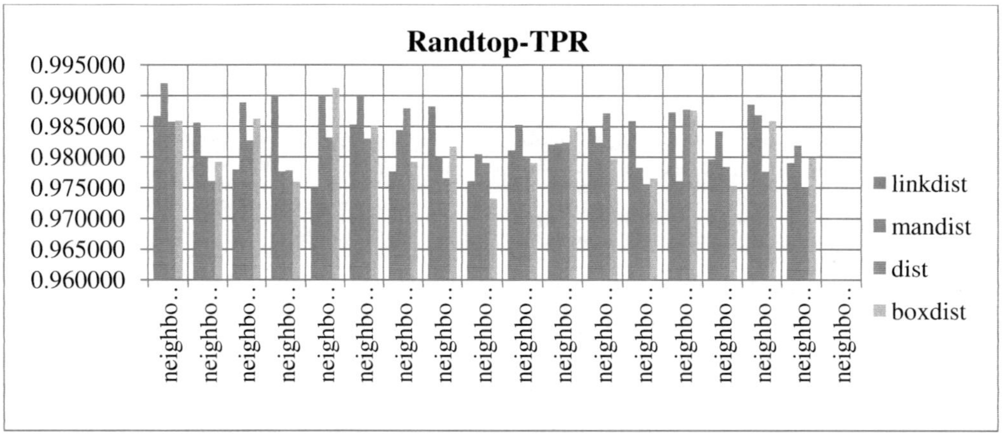

Figure 4. SOM-Randtop-TPR Comparison Result

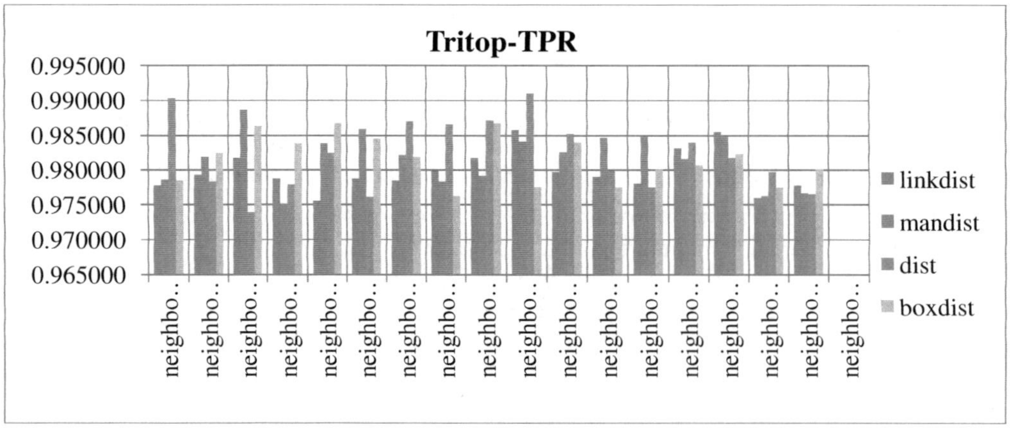

Figure 5. SOM-Tritop-TPR Comparison Result

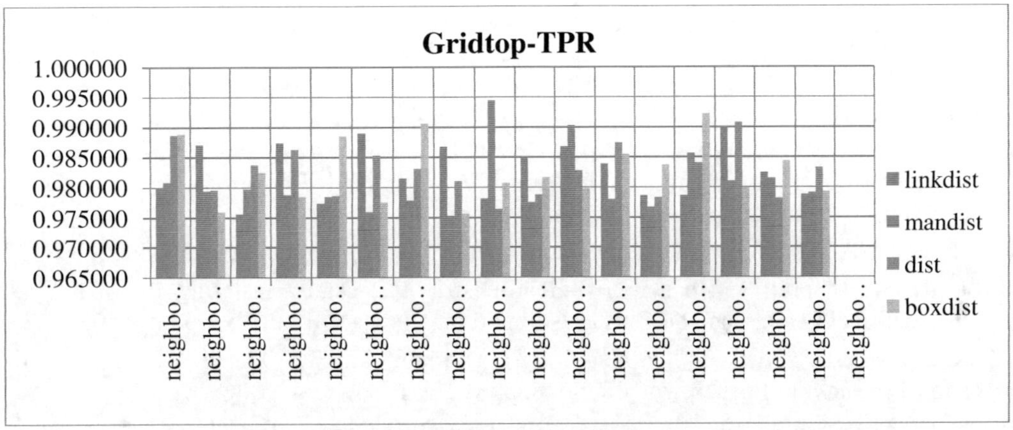

Figure 6. SOM-Gridtop- TPR Comparison Result

In this work the extracted feature vectors is scaled to 1. The result achieved with SOM on various topologies is the normalized one.

The performance measures that are evaluated for Hextop structure of SOM and neighborhood size=3 with Euclidean distance measure of all the methods BMCT, ECT, CDFCT and LSCT are shown in Table 1.

Table 1. Performance evaluation of BMCT

	FNR	FPR	TPR	TNR
BMCT	0.011194	0.003035	0.952521	0.988806
ECT	0.010966	0.025160	0.974840	0.989034
CDFCT	0.011099	0.007640	0.925693	0.988901
LSCT	0.011119	0.017244	0.982756	0.988881

4.1. Comparison of Methods

The various performance metrics used in the comparisons of the proposed methods are Precision and F-Measure. Precision gives the number of correct predictions. The best precision is 1 and the worst is 0. The F measure gives the accuracy when there are uneven class distributions. Similarly the best F measure is 1 and the worst is 0.

Comparing the results of the following four methods with Hexagonal topology, Euclidean distance and neighbourhood size 3, we can conclude that Edge based clustering method (ECT) is more efficient than the remaining three methods as shown in Table 2.

Table 2: Comparison of various methods

METHOD	PRECISION	FMEASURE
BMCT	0.99	0.99
ECT	1	0.98
CDFCT	0.99	0.989
LSCT	0.98	0.985

5. Conclusion

The present work is the effort of quantifying the representation of the input, so that it is more efficient for interpretation using SOM. Of the various methods proposed the BMCT, ECT, CDFCT and LSCT, the ECT is found to be more efficient. The proposed methods can be further extended with the deep neural networks for getting finer clustering results.

References

[1] A. W. P. Fok, H. S. Wong and Y. S. Chen, "Hidden Markov Model Based Characterization of Content Access Patterns in an e-Learning Environment," 2005 IEEE International Conference on Multimedia and Expo, Amsterdam, 2005, pp. 201-204.

[2] Kohenen.T, "The Self-Organizing Maps," Third Ed., Springer, Germany, 2000.

[3] J. Li, D. Li and Y. Zhang, "Efficient Distributed Data Clustering on Spark," 2015 IEEE International Conference on Cluster Computing, Chicago, IL, 2015, pp. 504-505.

[4] R. Nock and F. Nielsen, "On weighting clustering," in IEEE Transactions on Pattern Analysis and Machine Intelligence, vol. 28, no. 8, pp. 1223-1235, Aug. 2006.

[5] S. D. Pandya and P. V. Virparia, "Comparing the application of classification and association rule mining techniques of data mining in an Indian university to uncover hidden patterns," 2013 International Conference on Intelligent Systems and Signal Processing (ISSP), Gujarat, 2013, pp. 361-364.

[6] T. Jabid, M. H. Kabir and O. Chae, "Gender Classification Using Local Directional Pattern (LDP)," 2010 20th International Conference on Pattern Recognition, Istanbul, 2010, pp. 2162-2165.

[7] Yi Tan and Guo-Ji Zhang, "The application of machine learning algorithm in underwriting process," 2005 International Conference on Machine Learning and Cybernetics, Guangzhou, China, 2005, pp. 3523-3527 Vol. 6.

[8] G.Chamundeswari, G.Partha Saradhi Varma and Ch. Satyanarayana, "SOM based Clustering Technique with Boundary Moments," Proceedings of the International Conference on Intelligent Computing and Sustainable System, Coimbatore. DVD Part Number: CFP18O03-DVD; ISBN; 978-1-5386-4344-0.

[9] G. Chamundeswari, G. P. S. Varma, Ch. Satyanarayana, "Contact Distribution Function based Clustering Technique with Self-Organizing Maps", International Journal of Image, Graphics and Signal Processing(IJIGSP), Vol.10, No.3, pp. 9-17, 2018. DOI: 10.5815/ijigsp.2018.03.02

[10] G. Chamundeswari, G. P. S. Varma, Ch. Satyanarayana, "An Edge based Clustering Technique with Self-Organizing Maps", International Journal of Information Technology and Computer Science(IJITCS), Vol.10, No.5, pp.30-39, 2018. DOI: 10.5815/ijitcs.2018.05.03.

[11] Irene Epifanio, Guillermo Ayala, "A Random Set View of Texture Classification," IEEE Transactions on Image Processing, Vol. 11, No. 8, August 2002, pp. 859-867.

[12] Brewster E., Keller J.M., Popescu M., "A new approach for extracting texture features to aid detection of explosive hazards using synthetic aperture acoustic sensing," Proceedings Volume 10182, Detection and Sensing of Mines, Explosive Objects, and Obscured Targets XXII; 101821F (2017).

[13] Abdelkarim Ben Ayed , Mohamed Ben Halima , Adel M. Alimi, "Adaptive fuzzy exponent cluster ensemble system based feature selection and spectral clustering," IEEE International Conference on Fuzzy Systems (FUZZ-IEEE), 1-6, 2017.

[14] B. Goswami, G. Bhandari and S. Goswami, "Fuzzy min-max neural network for satellite infrared image clustering," 2012 Third International Conference on Emerging Applications of Information Technology, Kolkata, 2012, pp. 239-242.

[15] Andrzej Bielecki, Mateusz Wójcik, "Hybrid system of ART and RBF neural networks for online clustering," Applied Soft Computing, Volume 58, September 2017, Pages 1-10.

[16] Dong Huang, Chang-Dong Wang, Jian-Huang Lai,"Locally Weighted Ensemble Clustering," IEEE Transactions on Cybernetics, Volume: 48, Issue: 5, 1460-1473, 2018.

[17] Cai, Deng and Zhang, Chiyuan and He, Xiaofei, "Unsupervised Feature Selection for Multi-cluster Data," Proceedings of the 16th ACM SIGKDD International Conference on Knowledge Discovery and Data Mining, 2010, 333-342.

[18] Ivan Grbatinić, Nebojša Milošević, Bojana Krstonošić, "The neuromorphological caudate–putaminal clustering of neostriate interneurons: Kohonen self–organizing maps and supervised artificial neural networks with multivariate analysis," Journal of Theoretical Biology, In Press, Accepted Manuscript, Available online 21 November 2017.

[19] J. Li, D. Li and Y. Zhang, "Efficient Distributed Data Clustering on Spark," 2015 IEEE International Conference on Cluster Computing, Chicago, IL, 2015, pp. 504-505.

[20] M. Steinbach, G. Karypis, V. Kumar, "A Comparison of Document Clustering Technqiues," In KDD Workshop on Text Mining, 2000.

[21] Manjeevan Seera, Kuldeep Randhawa, Chee Peng Lim, "Improving the Fuzzy Min–Max neural network performance with an ensemble of clustering trees," Neurocomputing, In Press, Corrected Proof, Available online 7 November 2017.

[22] Marina Resta, Michele Sonnessa, Elena Tànfani, Angela Testi, "Unsupervised neural networks for clustering emergent patient flows," Operations Research for Health Care, In Press, Corrected Proof, Available online 26 August 2017.

[23] Pavel Berkhin, "Survey of Clustering Data Mining Techniques," In: Accrue Software, 2003.

[24] R. Nock and F. Nielsen, "On weighting clustering," in IEEE Transactions on Pattern Analysis and Machine Intelligence, vol. 28, no. 8, pp. 1223-1235, Aug. 2006.

[25] Xiao Yu, "Fuzzy infrared image segmentation based on multilayer immune clustering neural network," Optik - International Journal for Light and Electron Optics, Volume 140, July 2017, Pages 959-963.

[26] Y. Peng, S. Liu and Y. Zhang, "Indicator diagram identification based on ART2 neural network and features of moment invariant," 2012 2nd International Conference on Consumer Electronics, Communications and Networks (CECNet), Yichang, 2012, pp. 1075-1078.

An approach for suggestion mining based on deep learning techniques

T Raghunadha Reddy[1], P Vijayapal Reddy[2], T Murali Mohan[3], Raju Dara[4]

[1]Associate Professor, Dept. of CSE, Matrusri Engineering College, Saidabad, Hyderabad, Telangana
[2]Professor, Dept. of CSE, Matrusri Engineering College, Saidabad, Hyderabad, Telangana
[3]Professor of CSE, Principal, Swarnandhra Institute of Engineering and Technology, Narsapuram, AP
[4]Professor, Dept. of CSE, Vignana Bharathi Institute of Technology, Ghatkesar, Telangana

raghu.sas@gmail.com[1], drpvijayapalreddy@gmail.com[2], drtmm512@gmail.com[3], rajurdara@gmail.com[4]

Abstract. An organization often uses forums and social media channels for getting feedback from costumers or users. The ratings of products on rating platforms are a useful feedback to make a product better. The feedback from a customer is in the form of a suggestion which appears in a rating text or is directly asked from the customer. Suggestion mining is a binary classification problem that labels sentences as Suggestion or Non-suggestion. The suggestion mining is similar to sentiment analysis which is associated with common linguistic properties and challenges irrespective of the domain and application. Most of the previous works of suggestion mining proposed rule based methods and a very few developed statistical classifiers by using manually identified features. Recently, several researchers paid attention on deep learning technique based solutions to suggestion mining where features are automatically learned. In this work, various deep learning techniques like RNN, LSTM, Attention based LSTM and GRU are used in the experimentation of suggestion mining. The experiment carried out on the dataset provided in SemEval 2019 suggestion mining competition. The Attention based LSTM achieved best accuracies for suggestion mining when compared with other deep learning techniques.

1. Introduction

Online text is becoming an increasingly popular source to acquire public opinions towards entities like persons, brands, products, services etc. Opinion mining is a research area which investigates the extraction of opinions towards a target entity from the text written by a set of people. The problem of Opinion mining was introduced in 2002, which dealt with the classification of a textual unit not with regard to its topic but with regard to the opinion expressed in the text. Opinion mining is primarily used to refer opinion classification of a text unit into positive, negative, objective or neutral categories. The opinionated text also contains descriptive information like advice, tips, wishes for improvements, suggestions for improvements, warnings and recommendations which are generally overlooked in a sentiment based summary, mostly because it does not contain any sentiment towards the target entity.

Suggestions present in the reviews convey ideas for improvements to the brand owners, or tips and advice to fellow customers about commercial products and services [1]. The suggestion mining is the extraction of suggestions from the source text where suggestions of interest are likely to appear. The task of suggestion mining is similar to sentiment analysis in that the input is the same. However, the output of a sentiment analysis model and a suggestion mining model is different. While sentiment classifiers focus on grouping reviews as positive or negative or neutral, suggestion mining models identify the reviews that have suggestions or non-suggestions to other customers or service providers.

Suggestion mining is used in various applications such as ideas for product improvement, Customer to customer suggestions, Suggestion summarisation, improving sentiment detection and Recommender systems [2]. Suggestions present among the opinions expressed on social media often convey ideas for improvements to a brand owner. These ideas are otherwise sought by the stakeholders through dedicated suggestion boxes or online forms. Reviews, blogs, and other opinionated text about commercial entities also carry tips, warnings, and advice for the future customers. Often future customers seek such advice through dedicated question answering forums, where they have to wait for a considerable amount of people to answer. Suggestion mining can also be employed for summarisation of dedicated forums or surveys used for collecting suggestions around a certain topic or entity. A suggestion mining system can identify the exact sentence where a suggestion is conveyed in a larger piece of text, and thus help in summarising these responses. A significant number of sentences in the benchmark sentiment analysis datasets are labelled as neutral or objective. Often sentiment analysis systems perform poorly in identifying sentiments of neutral and irrelevant sentences. Suggestion mining is useful for determining the sentences which carry suggestions with respect to the target entity. Recommender systems predict user preference by utilizing the content from a user profile and the entity profile to recommend the entities. Suggestion mining systems will be capable of extracting such recommendations, which can then be utilized by recommender systems.

In this work, the experiment carried out with various deep learning techniques such as RNN, LSTM, Attention based LSTM and GRU for suggestion mining. All the techniques used word2vec vectors as input. Accuracy measure is used to evaluate these models. The experiment performed on training dataset of SemEval 2019 suggestion mining challenge.

This work is planned in 6 sections. The existing work in suggestion mining is explained in section 2. The section 3 presents the dataset characteristics and evaluation measures. The section 4 discuss about different types of deep learning techniques which are used in this experimentation. The experimental results are presented in section 5. The section 6 concludes this paper with future scope.

2. Existing Work

Nowadays, a huge number of texts are posted in online reviews or discussion forums. Such media can be a valuable source for obtaining a suggestion about products or services. The obtained suggestion is not only useful for readers but also important information for stakeholders. Indeed, such advice can be used to improving the quality of products or giving helpful recommendations. However, identifying a suggestion from a lot of reviews or comments needs extra effort and time. Moreover, such online texts are mostly in unstructured form. Thus, automatically mining the suggestion from given texts is challenging and significant. Suggestion mining is relatively a new research interest in text classification tasks. Several studies have initiated to mining suggestions from online texts. However, they concluded that it is not easy to identify suggestion texts automatically. In other words, it still has room to improving the classification result in the suggestion mining task.

Masahiro Yamamoto et al., explored [3] the Bidirectional Encoder Representations from Transformers (BERT) to focus on the sub task A of suggestion mining where the training and test datasets from same domain in SemEval-2019 competition. These model parameters were then fine-tuned based on the labeled training dataset. Generally, the BERT training consists of two parts such as pre-training on the general domain corpus and fine-tuning on the target task. In their work, the proposed system consists of three training steps such as pre-training on the general domain corpus, pre-training on the target domain corpus, and fine-tuning on the target task. They further pre-train

IOP Conf. Series: Materials Science and Engineering 1074 (2021) 012021 doi:10.1088/1757-899X/1074/1/012021

BERT model using an unlabelled corpus related to domain of documents from a windows forum and run additional steps of pre-training using these documents. For the classification task, they added one layer to output predictions like suggestion or non-suggestion. They applied the idea of distant supervision for the sub task B of cross-domain suggestion mining where the training and test datasets are from different domains. Distant supervision is a weakly supervised learning framework which tries to automatically generate noisy training examples. Specifically, they used the rule based system which is provided by the task organizer for creating a noisy training dataset and train the model based on them. Proposed system significantly outperforms baseline methods on two subtasks. Their approach achieved the 3rd place in subtask A and the fifth place in Subtask B.

Tirana Noor Fatyanosa et al., explored [4] various machine learning algorithms such as Logistic Regression (LR), Random Forest (RF), Multinomial Naive Bayes (MNB), Sublinear Support Vector Classification (SSVC), Linear Support Vector Classification (LSVC), Variable Length Chromosome Genetic Algorithm-Naive Bayes (VLCGA-NB) and Convolutional Neural Network (CNN). Their system obtains reasonable results of F1-Score 0.47 and 0.37 on the evaluation data in Subtask A and Subtask B, respectively. They identified that the obtained result of all classification methods outperformed that of the baseline for the Non-suggestion class. MNB yielded the best results with 0.95. In terms of F1-Score of the Suggestion class, they found that RF, SSVC, and VLCGA-NB obtained a competitive result outperforming baseline for Suggestion class. The highest F1-Score was obtained by SSVC at 0.47. RF and VLCGA-NB produced F1-Score at 0.29 and 0.31, respectively. Overall, note that MNB and SSVC obtained the best F1-Score for Non-suggestion and Suggestion classes, respectively.

Tobias Cabanski created [5] a model for task A by using the combination of CNN and LSTM networks. This model used input features of BERT word embeddings. For subtask B, the idea is to extend the model from subtask A with a domain discriminator and shared features. Since that adds a lot of parameters to the model, the text classifier has a simpler structure than in subtask A and uses only CNNs for classification. In contrast to subtask A, no labelled data for supervised training is provided, so that additional unlabelled data is taken to apply a cross-domain classification. This is done by using adversarial training of the three model parts label classifier (Model that predicts if an example is a suggestion), domain classifier (Model for the prediction of the domain of an example) and the shared feature representation (Model that applies a transformation on the input features). For subtask A, the best model reached a F1-score of 0.7273 for the test data. This model archived a validation F1-score of 0.875 which is a noticeable difference to the test data score. The model for subtask B reached a final F1-score on the test data of 0.8187.

Aysu Ezen-Can et al., employed [6] a hybrid approach that utilizes a recurrent neural network (RNN) along with rule-based features to build a domain-independent suggestion mining model to identify suggestions within reviews. They used two sets of features such as rule-based and model-generated from word embeddings for suggestion mining. The rule-based features are extracted from the heuristics used in the baseline system for this challenge. The first rule-based feature is using a pattern matching algorithm based on regular expressions. The second rule-based feature utilizes keywords without any patterns such as ‘suggest’, and ‘recommend’. The third rule-based feature relies on part of speech tags. Recurrent neural networks require a mechanism to convert textual input to numerical vectors to be able to perform computations. They used pre-trained word embeddings where each word in the embedding table has a vector of size 100 and Glove embeddings which was trained on Wikipedia 2014 and Gigaword 5 corpora. RNN used a fully connected layer that takes the rule-based features and the review as the inputs. Then two bidirectional LSTM layers follow for modelling the textual input. Before the softmax layer, an LSTM layer takes the advantage of both learned representations form bidirectional layers and the rule based features. With the test dataset, the model obtained 74.49% F1 score where the majority baseline was 57.77%.

Ilia Markov et al., used [7] scikit-learn implementation of the Support Vector Machines (SVM) algorithm which is trained on handcrafted features, function words, sentiment features, digits, and verbs for Subtask A, and handcrafted features for Subtask B. Handcrafted features are manually

IOP Conf. Series: Materials Science and Engineering 1074 (2021) 012021 doi:10.1088/1757-899X/1074/1/012021

selected list of representative keywords and patterns of a suggestion from the training and development data. For Subtask A, we used a set of 57 handcrafted keywords and 77 keywords were used for Subtask B. They used term frequency weighting scheme. They observed that function words, sentiment features, digits, and verbs did not improve the performance of a system for Subtask B and also observed that handcrafted features and function words are the most indicative features in their system. Their system achieved a F1-score of 51.18% on Subtask A and 73.30% for Subtask B.

Junyi Li et al., combined [8] the attention mechanism with the LSTM model. They trained a sequence of neural network models such as long short-term memory(LSTM), TextCNN, convolutional neural network(CNN) and C-LSTM and all these models are combined to get results by soft voting. Among these models, the attention-based LSTM models achieved the best results. The word-embedding used for all models in this task is Word2Vec. Then, the word vectors are fed into the long short-term memory (LSTM) layer. Finally, an attention mechanism (Luong et al., 2015) is added into the neural networks, and the prediction results are output via the softmax activation. The final system achieved an accuracy of 0.6776 and got 14th place in SubTask A. They find that TextCNN achieved a better result than a single convolutional neural network.

3. Dataset Details and Evaluation Measures

In this work, the dataset is taken from SemEval 2019 suggestion mining competition. The dataset for subtask A provides an overall count of 8500 examples, where 6415 examples are labeled as non-suggestion and 2085 as suggestion. Every example contains only one sentence, which could be part of a whole post in the forum where it was extracted. The dataset provided in the Semeval 2019 Suggestion Mining Challenge was highly imbalanced. The Table 1 shows the characteristics of the dataset.

Table 1. The characteristics of Dataset

Dataset	Total No. of Sentences	Total No. of Suggestions	Total No. of Non-Suggestions
Training Data	8500	6415	2085

The researchers of suggestion mining used precision, recall, F1-score and accuracy as evaluation measures to test the efficiency of the proposed approach. In this work, accuracy is used which is the number of test sentences are correctly predicted their suggestion from a set of test sentences.

4. Deep Learning based Approach for Suggestion Mining

The exponential growth of deep learning techniques helped to solve problems across different fields of research. Deep learning techniques extract features from the data automatically without any manual adjustment. Deep learning is a subdomain of machine learning in which algorithms are inspired by the structure of the human brain. Deep learning models are built upon the way the human brain is structured, having several layers and the information passes through each layer to generate an output. Traditional machine learning models are not scalable i.e. the model performance does not improve as we add more data to the training set. On the other hand, deep learning models are scalable. Deep learning model performance increases as you add more data to the training set. The deep learning model can extract features from the data automatically without extensive handcrafted feature engineering.

Deep Neural Networks (DNNs) have shown great ability at performing tasks on a wide range of applications such as image captioning, object detection, and segmentation. A better understanding of cost functions and the amount of training data has helped the neural network model to learn complex structures in images, videos, audios, and texts. The performance of deep learning algorithms is directly proportional to the amount of data available for training. Deep learning algorithms simultaneously learn feature extraction and classification parameters through the process of back-propagation. Vectors are used as an input and output to a deep learning model. These vectors are a bunch of numbers which are learned by train the model to minimize the loss. The data coming from various modalities (image, text, audio, video.) is first converted into its vector representation before passing on to the deep learning models. Therefore, it is important that the vector is good enough to represent the data properly.

There are a lot of deep learning algorithms out there for different problems such as CNNs and RNNs. CNN's are good at extracting features from image data, while RNN's perform well on temporal data. Similarly, variations of recurrent neural networks such as Long-Short Term Memory and Gated Recurrent Unit architectures do a good job extracting useful information from temporal data such as text and time series data. Although, these networks are good at extracting features for a particular modality, learning features across multiple modalities is still a challenging task. In this work, the experiment conducted with RNN, LSTM, Attention based LSTM and GRU for suggestion mining. All deep learning models used the input vectors generated through word2vec model.

4.1. Recurrent Neural Network (RNN)

RNNs are good at extracting features from sequential data such as time series, text, and audio. RNNs consist of an encoder which encodes the data into a vector and a decoder which decodes the vector into a destination data format. RNNs use some information from the previous layer to predict future outcomes. In contrast to feed-forward networks, RNNs have a feedback loop connecting their past decisions, thus ingesting their earlier outputs as input. Due to the inherent sequential nature, RNNs require an extension of back propagation considering the temporal aspect called Back propagation through time (BPTT). In the back-propagation phase of classic RNNs, the gradient interactions between words that are several steps apart in a sequence can get lost/weaken due to vanishing or exploding gradient problem. RNNs such as Long-Short Term Memory [9] and Gated Recurrent Unit [10] consist of a memory cell which helps remember previous layer information. RNN's are widely used in applications such as image captioning [11, 12], and video summarization [13].

The basic RNN takes input from the previous time step and current input to predict the next outcome. $h_0, h_1, h_2, h_3, \dots h_t$ are the inputs to the time steps $t = 0, 1, 2, 3, \dots, t$ which are used along with $x_1, x_2, x_3, \dots, x_t$ to predict the output $y_1, y_2, y_3, \dots, y_t$. The equation (1) is used to compute the current hidden state output by using previous hidden state output and current input.

$$h_t = f\left(W^{hh} h_{t-1} + W^{hx} x_t\right) \tag{1}$$

The Equation (2) passes the hidden state output to tanh function to restrict the output values in the range from -1 to 1.

$$h_t = \tanh\left(h_t\right) \tag{2}$$

The equation (3) computes the final output by multiplying with the weight matrix.

$$y_t = W^{hy} h_t \tag{3}$$

The weights are updated for each time step by back-propagation using the error calculated for that time step.

RNNs do a great job at extracting features from temporal data but they sometimes face the problem of vanishing gradients. The vanishing gradient problem occurs when the value of gradients exponentially decreases as it reaches to the early layers. To tackle this problem, LSTM's and GRU's are used.

4.2. Long Short-Term Memory

Long Short Term Memory Networks (LSTMs) refer to a type of artificial neural network modelled to utilize patterns occurring in data sequences. The generic term for such networks is Recurrent Neural Networks (RNNs). LSTMs make use of sequential information allowing operations over sequences of vectors. The sequences can exist in the input, output or both. Traditional neural networks consider all inputs independent of each other. On the other hand, RNNs do consider a sequential dependency between them. It can also be considered as having a "memory" recording a summary of the previous events in the sequence. This allows the information to persist throughout the network, and making every sequence output to be conditioned on the previous computations as well.

The LSTM model introduces a memory cell, a new structure to handle the past events in the sequence. The Figure 1 shows the internal structure of a LSTM cell. The Neural Network module of LSTM memory cell contains four main elements such as an input gate, a neuron with a self-recurrent connection, a forget gate and an output gate. The recurrent connection in the LSTM NN module uses a weight value of 1.0 to ensure that the state of a memory cell is remain constant moving from one time step to another in the sequence. The forget gate allows the memory cell to remember or forget its earlier state if required. The input gate controls the altering of the memory cell state following the current state incoming signal. On the contrary, the output gate controls the effect of the memory cell state on other neurons.

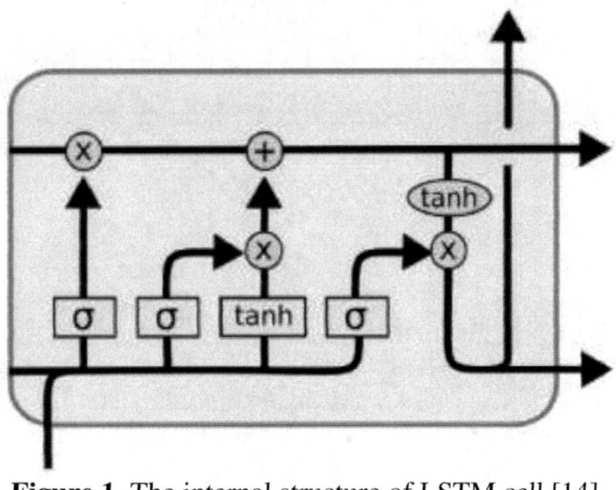

Figure 1. The internal structure of LSTM cell [14].

LSTM's are used for problems containing temporal data. LSTM's solve much of the vanishing gradient problem which is observed in basic RNN's. LSTM's come with a memory cell that helps remember the previous information for longer time steps. They were introduced by Hochreiter and Schmidhuber [9]. The Equation (4) computes the new information that is to be stored in input gate i.

$$i_t = \sigma\left(x_t U^i + h_{t-1} W^i\right) \tag{4}$$

The Equation (5) related to forget gate f which determines the information that is not important for the model to store.

$$f_t = \sigma\left(x_t U^f + h_{t-1} W^f\right) \tag{5}$$

IOP Conf. Series: Materials Science and Engineering 1074 (2021) 012021 doi:10.1088/1757-899X/1074/1/012021

The Equation (6) related to output gate o which provide activation to the final output of the LSTM block.

$$o_t = \sigma\left(x_t U^o + h_{t-1} W^o\right) \tag{6}$$

The Equation (7) is related to internal memory unit which is used to store the previous information.

$$\hat{C}_t = \tanh\left(x_t U^g + h_{t-1} W^g\right) \tag{7}$$

The Equation (8) computes current hidden state output C_t by using the information from previous hidden state and current input.

$$C_t = \sigma\left(f_t * C_{t-1} + i_t * \hat{C}_t\right) \tag{8}$$

The Equation (9) computes final hidden state output by combining C_t and output gate. This hidden state is a vector representation of the data which is then further used for a variety of applications such as classification and captioning.

$$h_t = \tanh\left(C_t\right) * o_t \tag{9}$$

In these equations σ represents the sigmoid function. It outputs values between 0 and 1. The sigmoid function determines the percentage of information to be passed through the gate.

4.3. Attention based LSTM

This model combines the attention mechanism with the LSTM. The attention mechanism is a good solution to the information vanish problem in long sequence input situations. When dealing with machine comprehension problems, the combination of LSTM and the attention mechanism are more effective than they are used individually. Traditional recursive neural networks are ineffective when dealing with very long sentences. The LSTM model is developed to solve the gradient vanishing or exploding problems in the RNN. The LSTM model can alleviate the problem of gradient vanishing, but this problem persists in long range reading comprehension contexts. The attention mechanism breaks the constraint on fix-length vector as the context vector, and enables the model to focus on those more helpful to outputs. After LSTM layer, we used the attention mechanism on the output vectors produced by previous layer. It is proven effective to improve the performance.

In the attention-based LSTM model, all sentences and labels are converted to word vectors by the word embedding layer. These word vectors will be fed to the LSTM layer. Subsequently, the word vector is represented as a hidden vector. Next, the attention mechanism assigns weights to each hidden vector, and the mechanism produces attention weight vectors and weighted hidden representations. Note that the weight vector is mainly obtained by calculating the similarity. An attention weight vector is generated by calculating a sentence vector matrix and a label vector matrix. The attention weight vector is then fed to the softmax layer.

The attention mechanism allows the model to retain some important hidden information when the sentence is long. In our mission, the information of sentences is kept for a relatively long time. Using the standard LSTM may result in the loss of hidden information. To solve this possible problem, we have facilitated the attention based LSTM model.

4.4. *Gated Recurrent Unit*

Fig. 2 displays the internal structure of the GRU network.

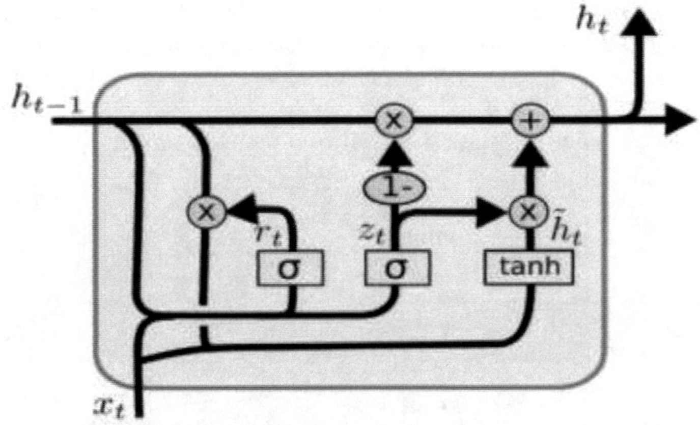

Figure 2. The internal structure of GRU cell [15].

The GRU was first introduced by Chung et al. [10]. GRU is a variant of LSTM. Unlike LSTM, GRU has two gates (reset and update gates). GRU doesn't have a memory unit. It outputs the entire hidden unit without any control. GRU is less complex and computationally more efficient as compared to LSTM. The following Equations explain the mathematical working of GRU. Equation (10) specifies the Update gate z_t which is computed by using the current input and the previous hidden state output.

$$z_t = \sigma\left(x_t U^z + h_{t-1} W^z\right) \tag{10}$$

Equation (11) is related to reset gate which determines how much information about past information to forget. Both, update and reset gate use sigmoid functions to constrain the output between 0 and 1.

$$r_t = \sigma\left(x_t U^r + h_{t-1} W^r\right) \tag{11}$$

In equation (12), $\hat{h}_t$ computes the current memory that is to be passed to final output.

$$\hat{h}_t = \tanh\left(x_t U^h + \left(r_t * h_{t-1}\right) W^h\right) \tag{12}$$

The equation (13) computes the final memory h_t at the time step t.

$$h_t = \left(1 - z_t\right) * h_{t-1} + z_t * \hat{h}_t \tag{13}$$

5. Experimental Results

The hyper-parameters of deep learning models play a vital role in the accuracies of suggestion mining. In the experiments, it was identified that the same model will get different results under different parameter adjustments. Therefore, reasonable adjustment of parameters during the experiment is also a factor in obtaining a good experimental result. In this work, Tensorflow is used to run these models. Tensorflow is a machine learning framework with great documentation for training and configuring neural networks which is developed by Google.

IOP Conf. Series: Materials Science and Engineering 1074 (2021) 012021 doi:10.1088/1757-899X/1074/1/012021

All deep learning models used the following hyper-parameters in the experiment such as number of epoch = 10, Embedding vectors size = 200, Dropout = 0.5, Recurrent Dropout = 0.4, Maximum sequence length = 1200, layers = 3, Attributes = 200 and activation function ="tanh". The table 2 shows the accuracies of suggestion mining when experimented with different deep learning techniques.

Table 2. Accuracies of Suggestion Mining using different deep learning methods

Deep Learning Model	Accuracy
RNN	81.34
LSTM	82.49
GRU	83.82
Attention based LSTM	85.97

Regarding deep learning methods, the Attention based LSTM technique achieved good accuracy of 85.97 for suggestion mining when compared with other deep learning techniques. The LSTM model accuracy is improved when attention layer is added in the network of LSTM.

6. Conclusions and Future Scope

In this work, a solution is proposed for suggestion mining which is used to classify the sentences as suggestions or non-suggestions. The experiment conducted with different types of deep learning models such as RNN, LSTM, Attention based LSTM and GRU. Among these deep learning models, the Attention based LSTM technique achieved good accuracy of 85.97% for suggestion mining.

In the future, we are planned to implement a model by combining multiple deep learning models. We also planned to experiment with machine learning algorithms by identifying a set of stylistic features.

References

[1] Sapna Negi and Paul Buitelaar. 2015. Towards the extraction of customer-to-customer suggestions from reviews. In Proceedings of the 2015 Conference on Empirical Methods in Natural Language Processing, Association for Computational Linguistics, pages 2159–2167, Lisbon, Portugal. ACL.

[2] Sapna Negi, Kartik Asooja, Shubham Mehrotra, and Paul Buitelaar. 2016. A study of suggestions in opinionated texts and their automatic detection. In Proceedings of the Fifth Joint Conference on Lexical and Computational Semantics, pages 170–178, Berlin, Germany. ACL.

[3] Masahiro Yamamoto, Toshiyuki Sekiya, "m_y at SemEval-2019 Task 9: Exploring BERT for Suggestion Mining", Proceedings of the 13th International Workshop on Semantic Evaluation (SemEval-2019), pages 888–892, Minneapolis, Minnesota, USA, June 6–7, 2019.

[4] Tirana Noor Fatyanosa, Al Hafiz Akbar Maulana Siagian, Masayoshi Aritsugi, "DBMS-KU at SemEval-2019 Task9: Exploring Machine Learning Approaches in Classifying Text as Suggestion or Non-Suggestion", Proceedings of the 13th International Workshop on Semantic Evaluation (SemEval-2019), pages 1185–1191, Minneapolis, Minnesota, USA, June 6–7, 2019.

[5] Tobias Cabanski, "DS at SemEval-2019 Task 9: From Suggestion Mining with neural networks to adversarial cross-domain classification", Proceedings of the 13th International Workshop on Semantic Evaluation (SemEval-2019), pages 1192–1198, Minneapolis, Minnesota, USA, June 6–7, 2019.

[6] Aysu Ezen-Can, Ethem F. Can, "Hybrid RNN at SemEval-2019 Task 9: Blending Information Sources for Domain-Independent Suggestion Mining", Proceedings of the 13th International Workshop on Semantic Evaluation (SemEval-2019), pages 1199–1203, Minneapolis, Minnesota, USA, June 6–7, 2019.

[7] Ilia Markov, Eric Villemonte De la Clergerie, "INRIA at SemEval-2019 Task 9: Suggestion Mining Using SVM with Handcrafted Features", Proceedings of the 13th International Workshop on Semantic Evaluation (SemEval-2019), pages 1204–1207, Minneapolis, Minnesota, USA, June 6–7, 2019.

[8] Junyi Li, Haiyan Ding, "Lijunyi at SemEval-2019 Task 9: An attention-based LSTM model and ensemble of different models for suggestion mining from online reviews and forums", Proceedings of the 13th International Workshop on Semantic Evaluation (SemEval-2019), pages 1208–1212 Minneapolis, Minnesota, USA, June 6–7, 2019.

[9] Hochreiter, Sepp, and Jürgen Schmidhuber. "Long short-term memory." Neural computation 9.8 (1997): 1735-1780.

[10] Chung, Junyoung, et al. "Empirical evaluation of gated recurrent neural networks on sequence modeling." arXiv preprint arXiv:1412.3555 (2014).

[11] Jeffrey Donahue et al., "Long-term recurrent convolutional networks for visual recognition and description," in CVPR, 2015, pp. 2625–2634.

[12] Kelvin Xu, Jimmy Ba, Ryan Kiros, Kyunghyun Cho, Aaron Courville, Ruslan Salakhudinov, Rich Zemel, and Yoshua Bengio, "Show, attend and tell: Neural image caption generation with visual attention," in International conference on machine learning, 2015, pp. 2048–2057.

[13] S. Sah, S. Kulhare, A. Gray, S. Venugopalan, E. Prud'Hommeaux, and R. Ptucha, "Semantic Text Summarization of Long Videos," in 2017 IEEE Winter Conference on Applications of Computer Vision (WACV), 2017, pp. 989–997.

[14] https://colah.github.io/posts/2015-08-Understanding-LSTMs/

[15] https://www.data-blogger.com/2017/08/27/gru-implementation-tensorflow/

Prediction and Analysis of Corona Virus Disease (COVID-19) using Cubist and OneR

[1]R.V.S.Lalitha, [1]J.Divya Lalitha, [2]K.Kavitha, [3]RamaReddy.T, [3]Rayudu Srinivas, [4]Challapalli Sujana
[1]Department of C.S.E, Aditya College of Engineering & Technology, Surampalem
[2]Department of C.S.E, Gokaraju Rangaraju Institute of Engineering & Technology(A), Hyderabad
[3]Department of C.S.E, Aditya Engineering College(A), Surampalem
[4]Department of I.T., Pragati Engineering College(A), Surampalem

{rvslalitha@gmail.com,lalitha0517@gmail.com,kavitha.bits@gmail.com, ramatreddy@gmail.com, srinivas.rayudu@aec.edu.in, sujana.challapalli@gmail.com}

Abstract. Inflammatory syndrome usually occurs with the adults who have already affected with corona virus. Multi system syndrome in children may affect different parts of the body like lungs, liver, kidneys and brain. The level of current infection can be known by proceeding with two kinds of tests, namely viral tests and anti body tests. Viral tests determine the level of current infection and antibody tests are about past infection. Though decision on test will be provided by health care units, there will be an uncertainty in evaluation of results. The variation in potential of antibodies will influence the resulting parameters. The effect of inaccurate test results is to be traced, because people may suffer from reinfection in their bodies over a period of time. In this paper the uncertainty analysis is carried out using Cubist and OneR algorithms. In the cubist algorithm leaf nodes can be analyzed using regression models. Accuracy of the predictions are analyzed using contingency table and false positives and true negatives are tracked using confusion matrix. This analysis assists in determining the trans conditional state of the disease with dynamism. With the analysis carried out cubist proves to be the best algorithm.
Keywords: Cubist, OneR, Rule based Learning, Confusion Matrix, Contingency Matrix.

1 Introduction

Corona viruses cause illnesses like cold, respiratory problems such as Severe Acute Respiratory Syndrome (SARS) and Middle East Respiratory Syndrome (MERS). The greatest need of current situation of pandemic diseases is evaluation of test results. In order to make test result vigilant, the analysis of symptoms is to be carried out resourcefully. Even though there are antibodies persisting in the body, there may be a chance of getting reinfected. Since long the attention had been paid towards the people who had developed antibodies for raising immunity levels to the virus. Anti bodies are y shaped proteins that lie in certain regions of the body. Anti bodies are produced by immune cells also known as B cells. They fight against foreign body cells for not getting infected on the specific region that they bind with. There is an imbalance between the virus and the immune cells that fight against virus. Owing to this, the results will vary. In this paper, the dataset is analyzed using two machine learning algorithms one is Cubist and the other OneRule(OneR). The Cubist algorithm is regression based model and target attribute prediction can be done with several predictor attributes. By submitting the dataset to OneR the impact of individual feature attributes can be well analyzed.

2 Related Work

Jian worked on Random Forest and Cubist algorithms for analyzing shear strength rockfill materials. Comparison of the models were done based on Root Mean Square Error(RMSE), Mean Estimation Error(MAE) and sensitivity analysis for prediction analysis[1,15]. Phan compared Multiple linear regression, Cubist Regression and Random Forest algorithms for estimating temperature of air. The dataset was divided into two parts one is calibration and the other is validation using 70% and 30% ratio respectively. Linear and Multiple linear regression models were used for Ta estimation using MODIS Land Surface Time(LST)[2]. Sathishkumar discussed a rule based regression predictive model for predicting bike sharing based on weather data. The model was evaluated using Cubist, Random Forest, classification and regression trees, KNN and conditional interference tree. Global Positioning System(GPS) enabled mobile application was developed for bicycle renting[3]. 36 machine learning algorithms were compared to predict indoor room temperature of a smart building. Finally, they came to conclusion that ExtraTrees regressor is the best for obtaining highest average accuracy[4,10]. The coronavirus was predicted using with regards to accuracy using ARIMA, CUBIST, RIDGE and RF learners. This analysis will assist in decision making systems. Also it was inferred that SVR and stacking ensemble learning are best suitable for forecasting COVID-19 cases[5,11]. Xuan developed highly accurate artificial intelligence model for predicting slope in open pit mines. Author proposed M5Rules-GA model and proved to be the best in comparison with FFA SVR, ANN PSO, ANN ICA, ANN GA and ANN ABC [6,14]. Yaghoub Dabiri proposed a real time left ventricular mechanics simulator for analyzing both passive and active behavior of myocardium during diastole.LV pressures are estimated using XGboost and Cubist. They concluded that Cubist results are good when compared to Gradient Boost algorithm [7]. The transformations of portrait photographs are analyzed using deep learning and image processing techniques. Guanyu Lian worked with online selfie photographs [8,9,12].

3 Methodology

World Health Organization (WHO) emphasized that nobody can give confirmation about the period of protection that antibodies give. Expanding on this point, it is required to track the inaccurate tests by observing the error rates of tests based on the statistical data collected. The uncertainties caused due to inaccurate assumptions can be overcome by analyzing the data with different algorithms. Evidently, the algorithm that yields less error will be considered as solution. The strategic solution derived can

IOP Conf. Series: Materials Science and Engineering 1074 (2021) 012022 doi:10.1088/1757-899X/1074/1/012022

also be reassessed for obtaining better results. The problem is addressed in 2 phases: 1. Applying Cubist algorithm to predict target value with 'n' number of input variables that experiments on rules. 2. Using OneR evaluates accuracy of the prediction of each feature in decreasing order.

3.1. Modeling Using Cubist

The dataset is downloaded from kaggle and used for analysis. This contain the following information:

```
'data.frame':   374 obs. of  9 variables:
 $ id            : int  1 2 3 4 5 6 7 8 9 10 ...
 $ country       : chr  "China" "China" "China" "China"..
 $ gender        : chr  "male" "female" "male" "female"..
 $ age           : int  66 56 46 60 58 44 34 37 39 56 ...
 $ visitingWuhan: int  1 0 0 1 0 0 0 1 1 1 ...
 $ fromWuhan     : int  0 1 1 0 0 1 1 0 0 0 ...
 $ symptom1      : int  0 0 0 0 0 0 0 0 0 0 ...
 $ symptom       : chr  "" "" "" "" ...
 $ death         : int  0 0 0 0 0 0 0 0 0 0 ...
```

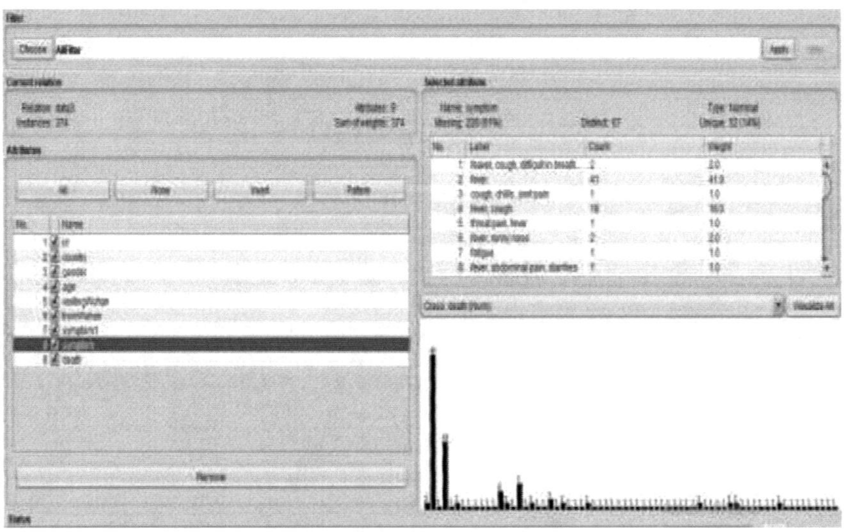

Fig.1.Analysis of symptoms using WEKA.

The frequency count of the symptoms is shown in Fig.1.Cubist algorithm is applied for the dataset chosen, for conditional outcomes. The spread of the disease may occur with people of age>60 and people visiting or coming from Wuhan as per the dataset. Also the deaths may cause with or without symptoms. Considering these factors, the rules are framed based on the current scenario and analyzed the outputs. The impact of disease attack of corona is analyzed with the rules as mentioned below:

```
Rule 1:age > 60 & visitingWuhan == 1 || fromWuhan == 1 &
death == 1[Fig.2]
Rule 2: age > 60 & visitingWuhan == 1 || fromWuhan == 1 &
death == 0[Fig.3]
Rule 3: age > 60 & symptom1 == 0 & death == 1[Fig.4]
Rule 4: age > 60 & symptom1 == 1 & death == 0[Fig.5]
```

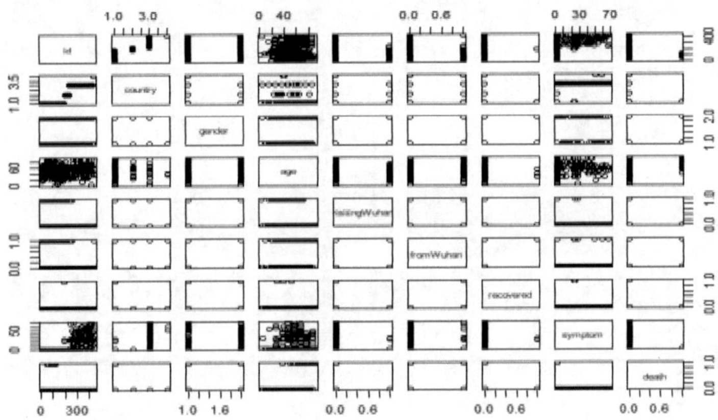

Fig.2. Outcome of Rule 1.

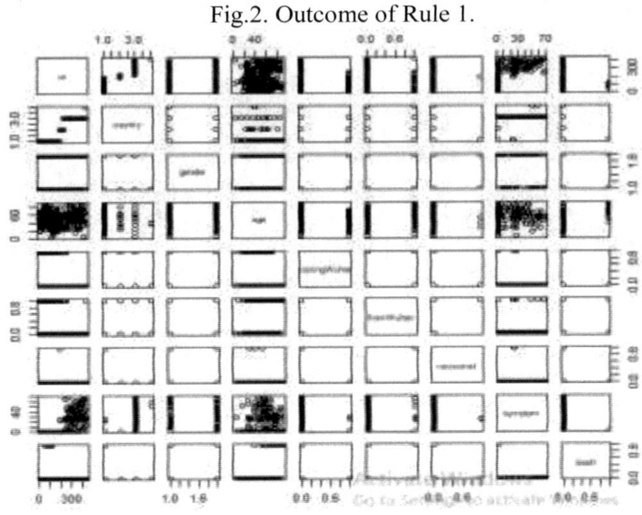

IOP Conf. Series: Materials Science and Engineering 1074 (2021) 012022 doi:10.1088/1757-899X/1074/1/012022

Fig.3 Outcome of Rule 2

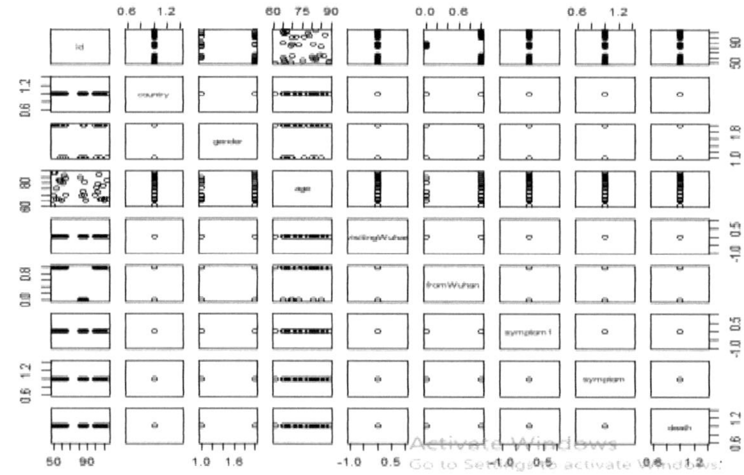

Fig.4. Outcome of Rule 3.

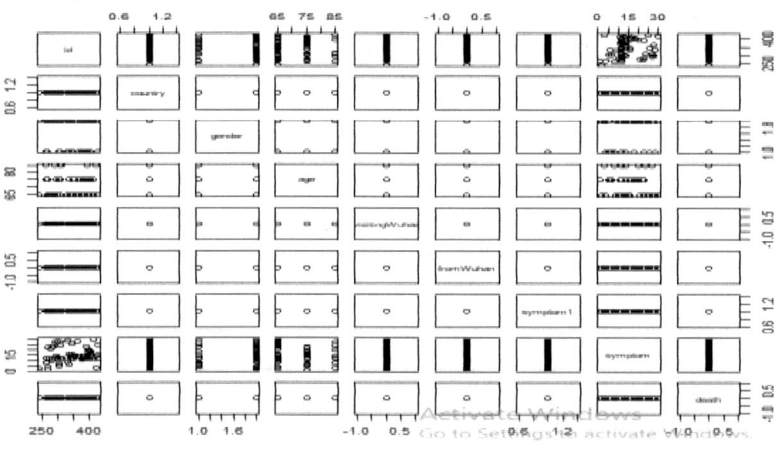

Fig.5. Outcome of Rule 4.

Using cubist method in R, 299 cases are considered (80% training and 20% test data) with mean 0.1 and range 0 to 1 with estimated error 0.0 for applying one rule. Rule 1: [299 cases, mean 0.1, range 0 to 1, estimated err 0.0] Fig.[2-5] represents the rule based analysis based on Cubist algorithm.

3.2. One Rule(OneR)

The implication of disease analysis is computed using OneR for evaluating predictions and is interpreted as below:

```
OneR algorithm for resolving uncertainty:
 For each symptom
  For each value of the symptom
    Count for number of cases becoming positive
    Observe the frequent symptoms for positive cases
 Create rules based on the observations made
 Compute error in each prediction
 Select the predictor for which it is less error prone
```

The following are the observations obtained by implementing OneR algorithm. The OneR evaluation model is depicted in Fig.6. The number of people died off with or without symptoms in Fig.7. and the frequency of death occurred or not in Fig.8.

3.2.1.Contingency table for finding accuracy

Table.1. Contingency Table

	id					
Death	(0,88)	(88,177)	(177,264)	(264,352)	(352,440)	Sum
0	* 63	* 65	* 58	* 72	* 75	333
1	24	15	0	0	0	39
Sum	87	80	58	72	75	372

Contingency table is constructed for evaluation of association rules based on the symptoms. The number of deaths occurred in each interval is listed in Table.1.

IOP Conf. Series: Materials Science and Engineering 1074 (2021) 012022 doi:10.1088/1757-899X/1074/1/012022

3.2.2. Confusion matrix for tracking of false positives and true negatives [Table.2.]

Table 2 Confusion matrix

	Actually negative(0)	Actually positive(1)
Predicted negative(0)	335	35
Predicted positive(1)	2	2

The accuracy based on rules using OneR analysis is 89.52 %(333/372) , error rate is 0.1043(39/374) and Maximum occurrences in each column is represented by '*' using contingency table. Also the accuracy of prediction is 89.57 %(335/374) with confusion matrix. The OneR model is explored as given below:

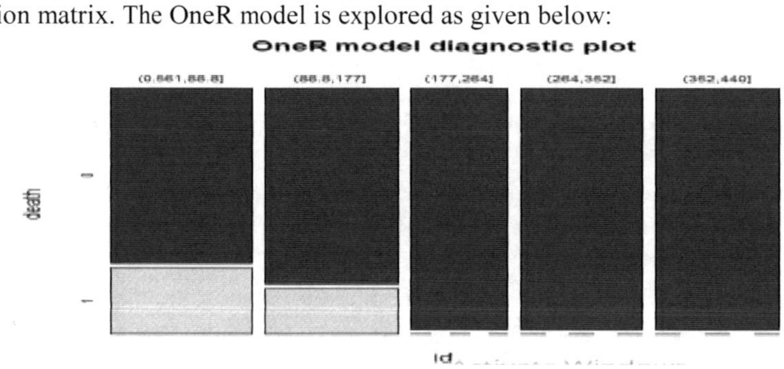

Fig.6. OneR model

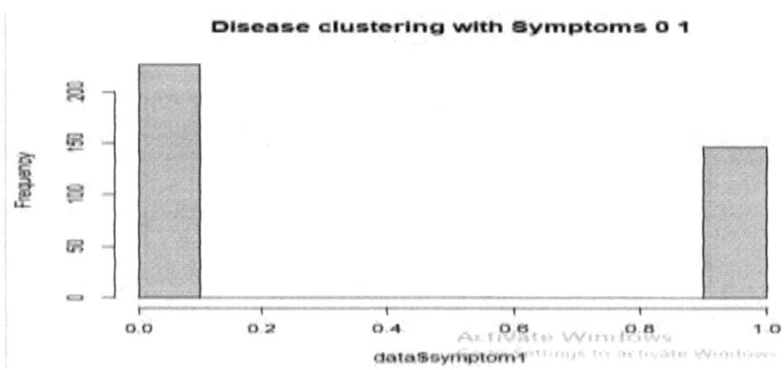

Fig.7. Disease clustering based on symptoms

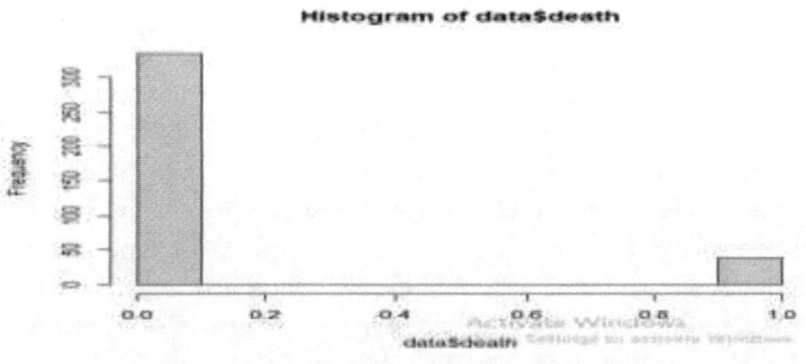

Fig.8. Histogram showing frequency of deaths.

The death rate analysis is calculated by taking bins, which discretizes the numerical data into clusters. In this scenario, number of bins is taken as 2, stating '0' for alive and '1' for death with people having symptoms. Like that we can pose and analyze rules for any dataset with any number of conditions. Data imbalance arises due to uneven distribution of instances for various classes. Data imbalance problem is analyzed in this paper by forming disease clusters based on symptoms (Fig.1 and Fig.7).

4 Results and Discussion

Based on computational analytics explored above the following observations are inferred and are tabulated in Table.3. as listed below.

Table.3. Comparative analysis of Cubist and OneR

Parameter	Cubist	OneR
Purpose	Regression model by pruning subsets of tree based on rules	Baseline for classification using which number of deaths is counted using bins.
Accuracy	Estimated error 0, average and relative errors are 0 and correlation coefficient 1.	Error rate: 0.1043 Accuracy:89.52%
Removed instances	0	Two instances are removed due to missing values
Ordering	Accuracy will be evaluated based on number of rules	Accuracy of the attribute is in decreasing order
False predictions	0	2(372 with Contingency table and 374 with confusion matrix
Evaluation	Evaluation based on training data	Based on total instances
Rules	Only one rule is used. More number of rules can be imposed using 'rules' parameter of cubistcontrol(method.	5 rules are framed based on id attribute.

Conclusion

With the time varying instances both the algorithms competed with each other in the performance point of view. However, since the accuracy takes highest priority, and the corona disease impact will be based on symptoms prediction analysis should be carried out properly. OneR evaluates accuracy based on the chosen attribute in decreasing order. If the individual feature is more of concern, OneR is preferable and Cubist evaluates large datasets in a very less amount of time. Cubist gives better result as the error rate is zero with reference to OneR algorithm. The analysis carried out so far, helps in facilitating baseline performance analysis of statistical data to the authorities involved in decision making.

References

1. Jian Zhou, Enming LI, Haixia Wei, Chuanqi Li, Qiuqiu Qiao and Danial Jashed Armagani, Random Forests and Cubist algorithms for Predicting Shear strengths of Rockfill Materials, Applied sciences,9(8),1621,2019. https://doi.org/10.3390/app9081621
2. Phan Thanh Noi, Jan Degener and Martin Kappas,Comparision of Linear regression, Cubist regression, and Random Forest algorithms to estimate daily air surface temperature from dynamic combinations of MODIS LST data, Remote Sensing,9(5),398,2017. https://doi.org/10.3390/rs9050398
3. Sathishkumar V E, Yongyun Cho, A rule based model for Seoul Bike sharing demand prediction using weather data, European Journal of Remote Sensing, 2020. https://doi.org/10.1080/22797254.2020.1725789
4. Sadi Alawadi, David Mera, Manuel Fernandez-Deglado, Fahed Alkhabbas, Carl Magnus Olsson & Paul Davidsson, A comparison of machine learning algorithms for forecasting indoor temperature in smart buildings, Energy systems, 2020. DOI:https://doi.org/10.1007/s12667-020-00376-x
5. Matheus Herrique Dal Molin Ribeiro, Ramon Gomes da Silva, Viviana Cocco Mariani, Leandro dos Santos Coelho, Short term forecasting COVID-19 cumulative confirmed cases: perspectives for Brazil, Chaos, Solitons and Fractals, 135,2020,109853. DOI:10.1016/j.chaos.2020.109853
6. Xuan Nam Bui, Hoang Nguyen, Yasoon Choi, Trung Nguyen Thoi, Jian Zhou and Jie Dou, Prediction of slope failure in open pit mines using a novel hybrid artificial intelligence model based on decision tree and evolution algorithm, Scientific Reports,10, June 2020, Article number 10,9939. DOI:https://doi.org/10.1038/s41598-020-66904-y
7. Yaghoub Dabiri, Alex der Velden, Kevin L Sack, Jenny S Choy, Ghassan S Kassab and Julius M Guccione, Prediction of Left Ventricular Mechanics using Machine Learning, Frontiers in Physics, September 2019. https://doi.org/10.3389/fphy.2019.00117

8. Guanyu Lian and Kang Zhang, Transformation of portraits to Picasso's cubism style, The Visual computer, 36, 799-807, 2020. DOI:https://doi.org/10.1007/s00371-019-01661-2
9. Qidi Peng, Nan Rao, and Ran Zhao, Covariance based dissimilarity measures applied to clustering wide sense stationary ergodic processes, Machine Learning,108, 2159-2195,2019. DOI:https://doi.org/10.1007/s10994-019-05818-x
10. Pierre Alquier and Benjamin Guedj, Simpler PAC-Bayesian bounds for hostile data, Machine Learning, 107, 887-902, 2018.
11. Qian, Z., Alaa, A.M. & van der Schaar, M. CPAS: the UK's national machine learning-based hospital capacity planning system for COVID-19. Mach Learn (2020). https://doi.org/10.1007/s10994-020-05921-4
12. Heinze-Deml, C., Meinshausen, N. Conditional variance penalties and domain shift robustness. Mach Learn (2020). https://doi.org/10.1007/s10994-020-05924-1
13. Raghavendran, C. V., Satish, G. N., Krishna, V. and Basha, S. M. (2020). Predicting Rise and Spread of COVID-19 Epidemic using Time Series Forecasting Models in Machine Learning. International Journal on Emerging Technologies, 11(4): 56–61.
14. K. Kavitha, S. Laxman Kumar, Pudu Pravalika, Kotte Sruthi, R.V.S. Lalitha, N.V. Krishna Rao, Fashion compatibility using convolutional neural networks, Materials Today: Proceedings, 2020,ISSN 2214-7853, https://doi.org/10.1016/j.matpr.2020.09.365.
15. Dr.R.V.S.Lalitha, Dr.G.Jaya Suma, Analyzing the impact of Corona Symptoms using Correlation, Journal of Critical Reviews, Volume 7, Issue 18, pp 281-287, ISSN 2394-5125, June 2020

IOP Conf. Series: Materials Science and Engineering 1074 (2021) 012023 doi:10.1088/1757-899X/1074/1/012023

Transform Domain Analysis of Multimodal Medical Image Fusion

Ch.Hima Bindu[1],G.Sridevi[2]

[1]Professor, QIS College of Engineering & Technology, Ongole
[2] Professor, Department of ECE, Aditya Engineering College, Surampalem, AP

[1]hb.muvvala@gmail.com, [2]sridevi_gamini@aec.edu.in

Abstract. Fusion of multimodal medical images ensures the quality diagnosis in the field of medical sector. Fused image possess extra substance of data than input individual scanned images. The fusion process can be carried on different clinical scan images wherever more subjective data is required. In this work, we are supposed to discuss the subjective treatment of multiple transform techniques (Discrete Fourier Transform (DFT), Discrete Cosine Transform (DCT), Hadamard Transform (HT) & Discrete Wavelet Transform (DWT)) significance in fusion process. These coefficients are fused with the correlation of the spatial frequency (SF) and visibility (V) factor values. The resultant images are visually and qualitatively verified with standard execution estimates like Peak Signal to Noise Ratio (PSNR), Mean Square Error (MSE) and Root Mean Square Error (RMSE).

1. Introduction

Acute disease recognition raised the need of medical image techniques like Computed Tomography (CT), Magnetic Resonance Imaging (MRI), Positron Emission Tomography (PET), and Single Photon Emission Computed Tomography (SPECT), etc. Medical image fusion divided into single- modal fusion and multimodal-fusion. Multimodal image fusion is the procedure to fuse different medical images in order to improvise better performance, high quality and visual enhancement by reducing redundancy in fused image. Fusion process can be done in spatial domain and transform domain.

Spatial domain is based on the pixels and techniques are averaging, principal Component Analysis (PCA), Brovery method, I-H-S INTENSITY HUE SATURATION. Transform domain fusion is based on frequencies changes, popular transforms are Discrete Fourier Transform (DFT), Discrete Cosine Transform (DCT), Discrete Hadamard Transform (DHT), Discrete Wavelet Transform (DWT), Redundant Wavelet Transform (RWT), Contourlet Transform, Curvelet Transform and Singular Value Decomposition (SVD).

Fusion approaches classified in to 3 categories pixel-two images are fused by transforming into similar, decision –based on region interest- and feature-based on interpreted or labeled data. Muhammad Arif [1] Implemented new method for the medical images using curvelet transform for curve shapes and genetic algorithm for image features and approached detailed Image Enhancement for acute identification. Hadiakramhadi [2] proposed and promoted quality improvement in fused images with multimodal medical fusion technique discrete wavelet transform for (ct computerized tomography and MRI magnetic resonance imaging) by tested and analyzed metrics. Rajalingam [2]

implemented A Novel Neuro –Fuzzy Hybrid Multimodal Image Fusion technique by working with different medical images and produced high quality fusion image by reducing processing time for treatment diagnosis. Sridevi [4] implemented new Algorithm for the reconstruction of images as preprocessing stage, fuzzy sets called Yagers Intuitionistic Fuzzy Compliment Set and worked with several medical modalities (CT, MRI, PET etc) and promoted better contrast for visual recognition. MaruturiHaribabu [5] proposed Novel medical image fusion method with 2D Hartley Transform technique in HSV (Hue, Saturation, Value) color space and produce statistical measurements by Combining PET –MRI. Priyanka Sharma [6] presented the role and fundamentals of fusion in disease location and promoted redundancy in information of fusion in MRI –CT by applying Discrete Wavelet and Curvelet Transform fusion techniques. Sanjay Patsaria [7] presented review on modalities MRI, CT, IR, PET, SET based on organs (BRAIN, LIVER, HEAD, NECK, etc.), fusion in PIXEL, FEATURE, and DECISION levels fusion techniques DWT, DCT etc. Vaishnavi [8] proposed dual tree complex wavelet transform DTCWT method for fusing medical images MRI-PET to reduce redundancy by using complex valued filtering and promoted high quality fused image. RENICAN NIE [9] proposed a fusion frame work based multisource information exchange and coding by using PCNN pulse complex coupled neural network for medical images CT, MRI, PET and approached better perception for the visual disease diagnosis. ANISH VIJAN [10] presented performance analysis on different medical modalities PET, MRI, CT with fusion techniques Discrete Wavelet Transform Dwt, Principle Component Analysis PCT and Laplacian Pyramid Transform.
The rest of the paper coordinated in the method of proposed work and experimental results. At last the paper is finished up with conclusion.

2. Proposedmethod

This part of the paper discusses about the process of fusion with step by step explanation and shown in flow chart 'figure 1'.

1. Consider two multimodal medical scan images of MRI and PET Images are input images of fusion process.
2. Initially RGB color model of PET image is transformed into HIS/HSV color model to extract intensity component (I) of an image by protecting color coordinates.
3. Further the intensity component of PET image and MRI image (Gray Image) are transformed into image coefficient information with the use of DFT, DCT, HT and DWT transform techniques.
4. Especially in DWT transform case the subbands are fused:
   a. The low frequency subband information of two images is fused using average method.
   b. The high frequency subband information of two images are undergoes fusion process with spatial frequency method or visibility factor.
5. For other transform coefficients are (DFT, DCT & HT) fused with spatial frequency method or visibility factor like DWT high frequency subbands.
6. Once the fusion is done apply inverse transform techniques to get the spatial domain information.
7. The resultant image is fused intensity component of two multimodal medical images, later the preserved color components (Saturation and Hue) of PET image is combined with the resultant intensity component.
8. The final resultant fused color image of multimodal images are obtained by converting HIS/HSV color model into RGB color model.
9. The resultant image is having more qualitative and quantitative information than the individual images.
10. The effectiveness of the proposed method is validated with the standard transform domain techniques of (DFT, DCT, HT) with basic performance measures.

2.1. *Fusion rules*

2.1.1. *Spatial Frequency*

The spatial frequency (SF), of an image coordinates are defined as:

$$SF = \sqrt{(RF)^2 + (CF)^2} \quad (1)$$

Where RF and CF are the row frequency and column frequency respectively:

$$CF = \sqrt{\frac{1}{MN}\sum_{i=1}^{N}\sum_{j=2}^{M}\left[I(i,j) - I(i-1,j)\right]^2} \quad (2)$$

$$RF = \sqrt{\frac{1}{MN}\sum_{i=1}^{M}\sum_{j=2}^{N}\left[I(i,j) - I(i,j-1)\right]^2} \quad (3)$$

The neighbouring pixel coordinates are considered for CF and RF computation.

2.1.2. *Visibility*

The clarity of an image can be measured with visibility (V) factor. Here the α =0.6 to 0.7 and μ is the mean value of the image.

$$v_l = \sum_{m=1}^{M}\sum_{n=1}^{N}\left| LL_1(m,n) - \frac{\mu}{\mu^{\alpha+1}} \right| \quad (4)$$

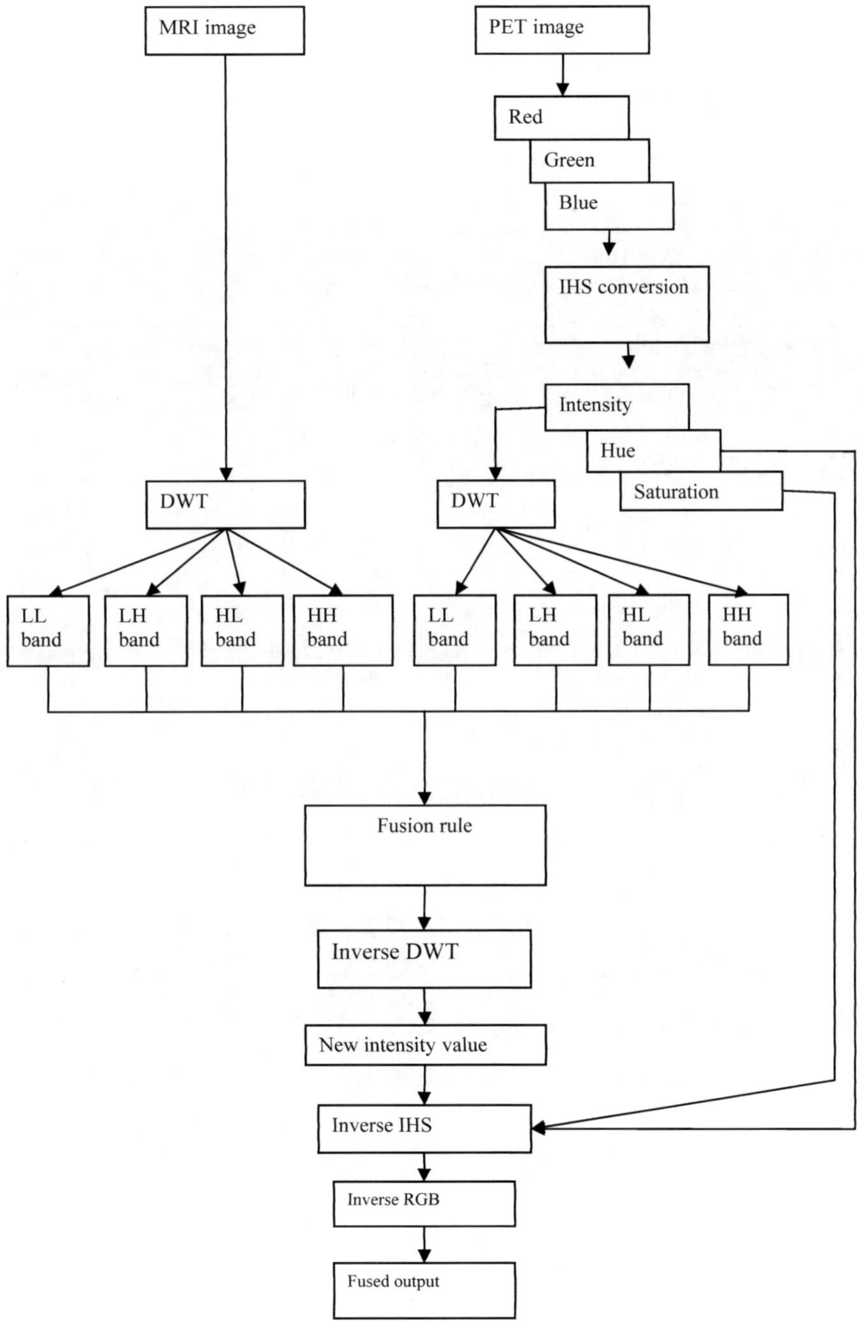

Figure 1. Proposed Work Flow

3. Experimental Results

This experimental session discuss about the qualitative and quantitative analysis of the proposed work. The basic multi modal medical images are collected from [11]. These images are given as input to the proposed process and the results were evaluated with performance measures of PSNR, RMSE, & MSE [12]. The resultant values are tabulated in the Table 1. The following Figure 2-5) show the resultant image starting from basic transform techniques to multi resolutional transform techniques. The results are plotted in graphical way also (Figure 6-7) for more visual analysis.

Images	Methodology	Quality parameters		
		PSNR	RMSE	MSE
	DFT+SF	62.9078	0.1825	0.0333
Fig 1	DCT+SF	62.9078	0.1825	0.0333
	HT+SF	62.9085	0.1817	0.0312
	DWT+SF	62.9241	0.1603	0.0275
	DFT+SF	59.8625	0.2591	0.0671
Fig 2	DCT+SF	59.8629	0.2591	0.0671
	HT+SF	58.8524	0.2753	0.0724
	DWT+SF	60.1873	0.2496	0.0623
	DFT+VIS	60.4103	0.2432	0.0592
Fig 3	DCT+ VIS	62.0785	0.2007	0.0403
	HT+ VIS	58.9445	0.2875	0.0829
	DWT+VIS	62.9635	0.1813	0.0329
	DFT+VIS	57.3661	0.3453	0.1193
Fig 4	DCT+VIS	59.3876	0.2736	0.0749
	HT+VIS	56.4661	0.3830	0.1467
	DWT+VIS	60.3593	0.2447	0.0599

Table 1. Performance analysis of Proposed Work

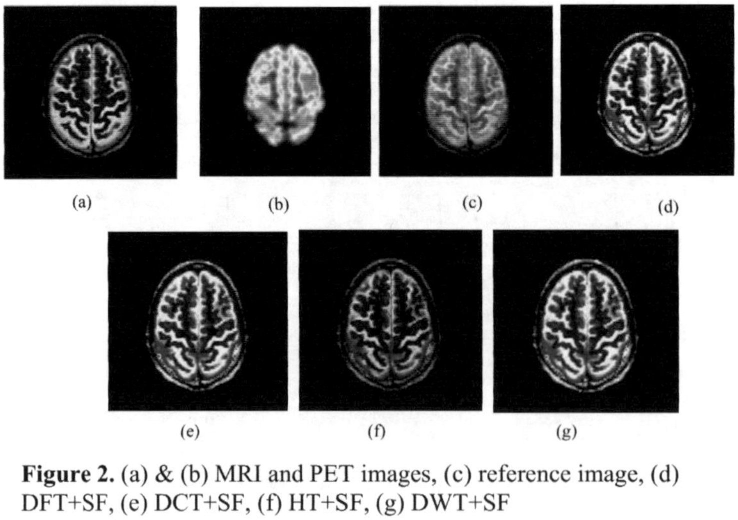

Figure 2. (a) & (b) MRI and PET images, (c) reference image, (d) DFT+SF, (e) DCT+SF, (f) HT+SF, (g) DWT+SF

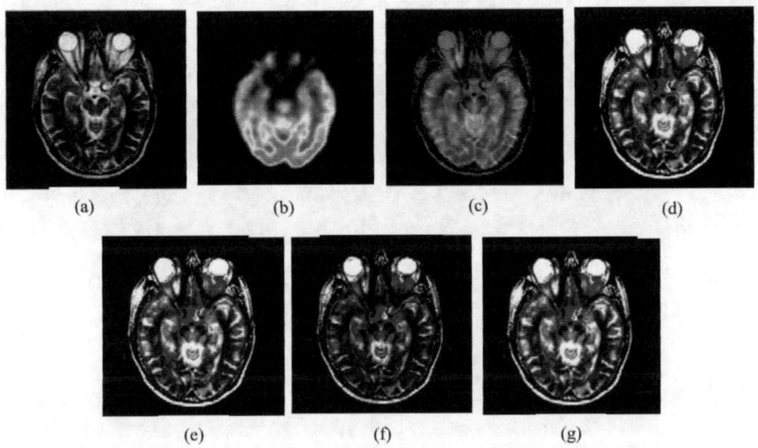

Figure 3. (a) & (b) MRI and PET images, (c) reference image, (d) DFT+SF, (e) DCT+SF, (f) HT+SF, (g) DWT+SF

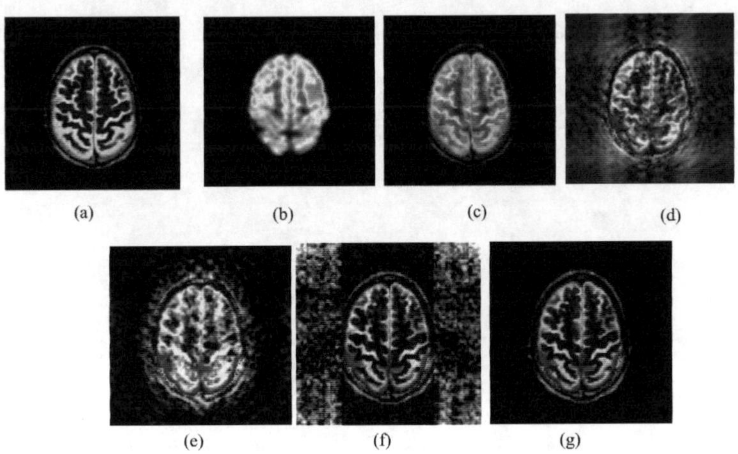

Figure 4. (a) & (b) MRI and PET images, (c) reference image, (d) DFT+SF, (e) DCT+SF, (f) HT+SF, (g) DWT+SF

IOP Conf. Series: Materials Science and Engineering 1074 (2021) 012023 doi:10.1088/1757-899X/1074/1/012023

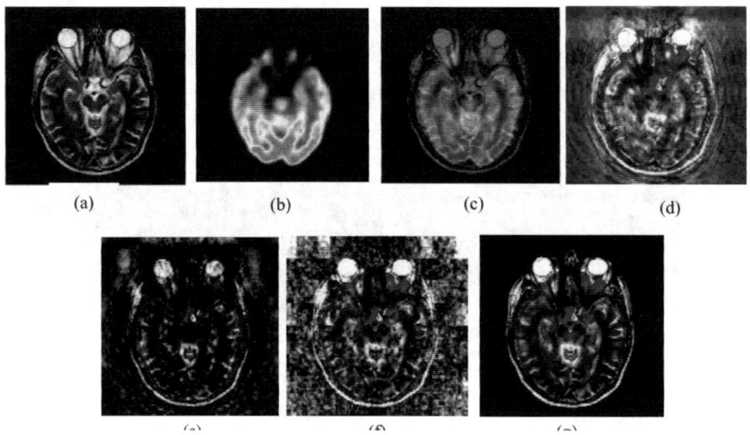

Figure 5. (a) & (b) MRI and PET images, (c) reference image, (d) DFT+VIS, (e) DCT+VIS, (f) HT+VIS, (g) DWT+VIS

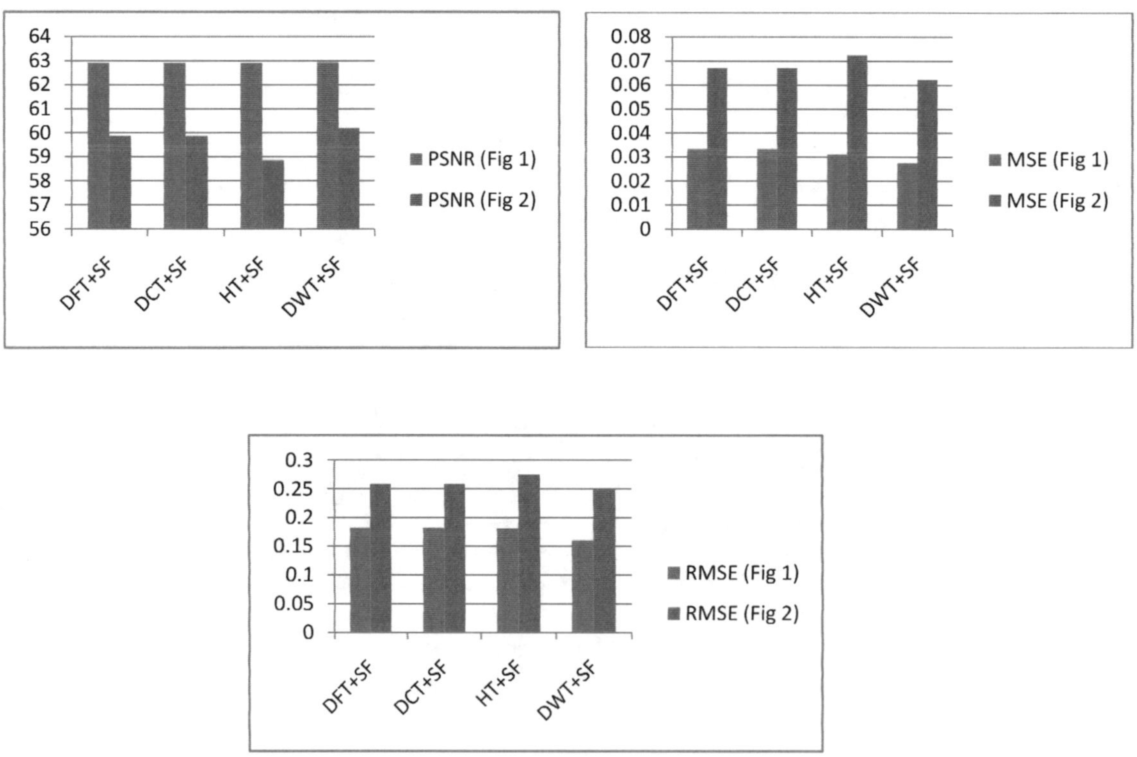

Figure 6. The graphical representation of results with SF factor

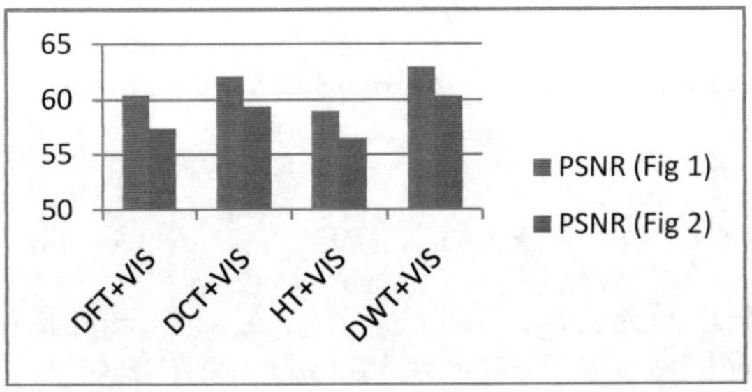

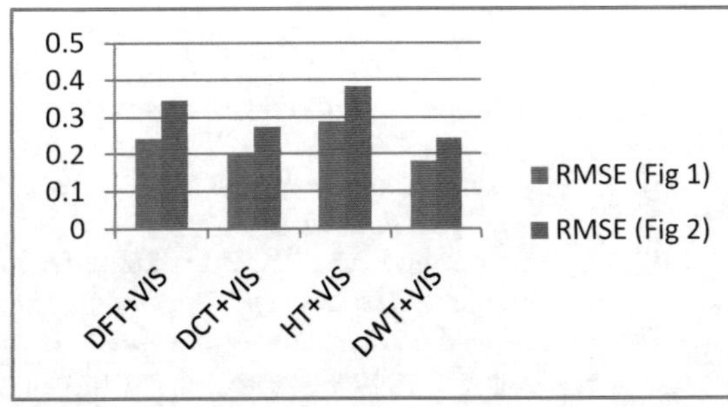

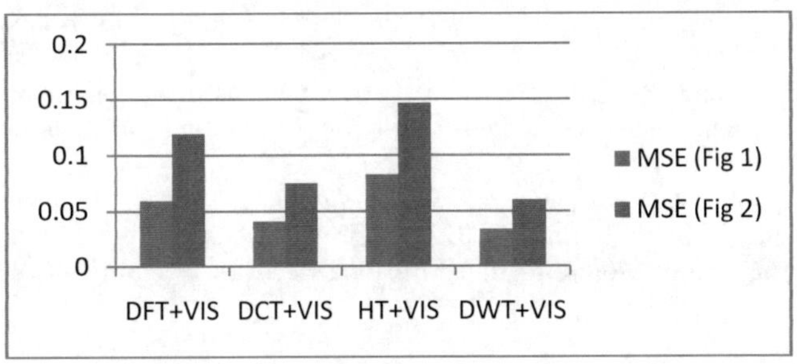

Figure 7. The graphical representation of results with V

4. Conclusion

This work addressed about the fusion of multimodal medical images with multiple transform techniques. The transform techniques are considered with basic level to multi-resolutional level of techniques. The applied fusion methods are SF & V. The outcomes are compared with standard performance measures of PSNR, RMSE & MSE. By observing the experimental results section we can conclude that the multi resolutional transform technique is giving more accurate results in both fusion rules than the other standard transform methods. Further this can be verified with multi-level subband resolution and higher end of multi resolution techniques in future work.

References

[1] Muhammad A and Wang G 2019 Fast curvelet transform through genetic algorithm for multimodal medical image fusion *Springer-Verlag gmbh Germany, part of Springer Nature***24** pp1815-36.

[2] Hadi Akram Hadi and Ziad Saeed Mohameed 2020 Fusion of the multimodal medical images to enhance the quality using discrete wavelet transform *IOP Conference Series Materials Science and Engineering***745** *pp 1-10.*

[3] Rajalingam B and Priya R 2018 Hybrid multimodality medical image fusion technique for feature enhancement in medical diagnosis*One Day National Conference on Internet Of Things - The Current Trend In Connected World***60**.

[4] Sridevi G and Srinivas Kumar S 2019 Image inpainting based on fractional-order Nonlinear

diffusion for image reconstruction *Circuits, Systems, and Signal Processing* **38** pp 3807-17.

[5] Hima Bindu Ch 2017 Visibility based multi modal medical image fusion with dwt *IEEE International Conference on Power, Control, Signals and Instrumentation Engineering* pp1561-1566.

[6] Priyaka Sharma, Neeraj T and Dinesh G 2020 A review: fusion of medical images *D.R S. Research Journal* **10**.

[7] Sanjay P and Manish Dixit 2020 Fusion of medical images in wavelet domain: a hybrid implementation *Article in Computer Modeling in Engineering and Sciences* **122** pp 303-21

[8] Vaishnavi S, Varshini A and Senthil Kumar K 2020 Image fusion based on dtcwt to improve segmentation accuracy in tumour analysis *International Journal of Advanced Development in Science and Technology* **2** pp 1214-19.

[9] Rencan N and Dongming Z 2020 Multi-source information exchange encoding with pcnn for medical image fusion *IEEE Transactions on Circuits and Systems for Video Technology* pp1-4.

[10] Anish V, Parth D and Shruti J 2020 Comparative analysis of various image fusion techniques for brain magnetic resonance images *International Conference on Computational Intelligence and Data Science*.

[11] The whole brain Atlas *http://www.med.harvard.edu/aanlib/*. Accessed 18 Nov 2020.

[12] Ch.Hima Bindu and Satya Prasad K 2012 Performance analysis of multi source fused medical images using multiresolution transforms *International Journal of Advanced Computer Science and Applications* **3** pp 54-62.

Expediting International Student Admission Process Using Data Analytics

Mounicasri Valavala[1], Wasim Alhamdani[2], Hemalatha Indukuri[3]

[1] University of the Cumberlands, Williamsburg KY 40769, USA

mvalavala1456@ucumberlands.edu

[2] University of the Cumberlands, Williamsburg KY 40769, USA

wasim.alhamdani@ucumberlands.edu

[3] S.R.K.R Engineering College, Bhimavaram, Andhra Pradesh 534204, India

indukurihemalatha@gmail.com

Abstract. The student admission process is a critical step from both student and university perspectives. The students will benefit if the universities take admission decisions sooner rather than later, leaving them time to opt for another university in case of rejection. However, with the level of manual work involved in the student admission process, universities often take long durations for the admission decisions. Expediting the admission process using Data analytics starts new research across different tasks in educational institutes. The research presents a model to automate each step of the international student admission process in universities across the USA, using Data analytics. The model's distinguishing capability is that it performs both predictive and descriptive analytics to analyze different student's application sections and adapt to different admissions types with minor changes.

Keywords Student Admissions. Data Analytics. Word Cloud. Machine Learning. Natural Language Processing.

1 Introduction

Student admission is a tedious process for both students and the universities. Nevertheless, the technology helped eliminate the paperwork, making it far better than a decade ago. The admission process depends on the degree, student accomplishments in the selected area, course grades, recommendation letters, Statement of Purpose (SOP), and other factors like nationality. With the increase in globalization, there is tremendous interest to study in prestigious universities across the globe for better opportunities.

The 2020 report on International Educational Exchange discloses that over a million international students admitted to USA universities during 2019/2020, and the United States continue to enjoy international students' top destination position [1]. Hegarty (2014) points out that international student education is a significant industry contributing to the US economy [2]. The admissions continue to grow despite laborious visa processes and stringent immigration rules [2]. Considering the criteria mentioned above, we chose to expedite the international student admission process in the USA using Data analytics, which is a promising solution across different industries.

The student admission process contains a student application that takes a student's personal information, education history, test scores, activities participation, and Statement of Purpose (SOP). Also, most of the universities require recommendation letters that support the applicant's capabilities. Some university website asked the applicants to supply no more than two recommendation letters, as it will take more time for the staff to read and evaluate the recommendation letters [3]. The other documents accepted are transcripts, general or subject test (GRE), English proficiency test (IELTS, TOEFL) score reports.

This research will categorize the application requirements into two categories and provide a holistic analysis of finalizing student admission. The first category is to perform predictive analytics using supervised classification algorithms on education history and score reports. The second category is to apply natural language processing techniques to speed up the analysis of recommendation letters and SOP. We will present the model theoretically and present the implementation results in the upcoming articles.

2 Related Work

The research of using Data Analytics in the education sector is scattered around different academic levels and across different activities within an educational institution. There is extensive research in this area to predict student performance, perform association rule mining from both teacher and student perspectives, and sentiment analysis to decide students' interest levels to admit into a particular university. According to our knowledge, there is limited to no research specific to international student admissions automation using Data Analytics. There is an urgent need for research to implement a data analytics framework in the education domain [4]. The model presented in [4] addressed using data science models that provide information to the instructor about the need for intervention to improve student performance.

A case study presented in [5] presents a predictive model to foresee students' performance based on the given information during Nigeria's admission process. The limitation of this research is that it performs only predictive analytics. Hence, this model is not adaptable to admissions across the globe where the applicant often needs to submit textual information such as recommendation letters and SOP. Robin & Brian (2010) presented research on students' technology usage in undergraduate admissions [6]. However, it covers only the communications means to connect students with the universities.

The authors in [7] provided a review of existing research using predictive analytics to evaluate student performance. According to this survey, most of the research relies mainly on the cumulative grading point average (CGPA) and other internal data assessments [7]. The models presented in [8] predict poor student performance in academics using academic data and socio-economic details. International Data Mining society coined a term called Education Data Mining (EDM) to explore the educational domain's data and draw insights to understand students and the learning environment [8, 9].

The conceptual model discussed in [10] explains how the universities can use data analytics in different stages of student-college association. The authors mentioned that data analytics helps in admission decision making using predictive analytics to predict a student's performance in the future and natural language processing techniques to analyze qualitative content [10]. We consider the author's inputs on data analytics in the admission process to create a data analytics model for the student admission process. The research in [10] extended the conceptual model to implement data analytics during the student and post-student stage.

The prediction model presented in [11] predicts the student enrollment using the admission data, such as the selected academic program, current course plan, intended course plan, age group, residency status etc. The test scores, and school grades, which influence the admission decision, are not considered and the authors mentioned that this should be part of future work [11]. The current research addresses this research gap by including the test scores, school grades, and community activities.

3 Research Contribution

The existing research in this area is confined to predicting student grades within the coursework, technology usage in the education sector, analyzing university and student statistics, and sentiment analysis of social media content. The authors in [10] provided a high-level view of the model for prospective students in universities. We will delve into further details and create a theoretical model for international student admissions in the USA. The proposed model is adaptable to different admission types with minor changes.

The current research will automate the student admission process using predictive and descriptive analytics. The proposed model ensures that it analyzes both the qualitative and quantitative content of a student application and provides the student performance prediction and word clouds to represent the textual information. As a result, the time spent on manual analysis of recommendation letters, essays, and SOPs can be reduced drastically. Also, the predicted performance measure helps the faculty to make an admission decision. We also perform qualitative analysis to get the feedback of the proposed model from the admission departments across different universities.

4 Methodology

The proposed model expedites the student admission process using Machine Learning (ML) and Natural Language Processing (NLP) techniques. We analyze application data using ML and textual content, such as recommendation letters and SOP using NLP. The model contains two main modules, namely the Application data analyzer and qualitative data analyzer, and the model's outcome is input to the admission office to make an admission decision.

The model reduces the extent of manual effort required by the admissions department to make the admission decision. As a result, the turn-around time for an application will be considerably less. The below schematic design gives an overview of the admission model.

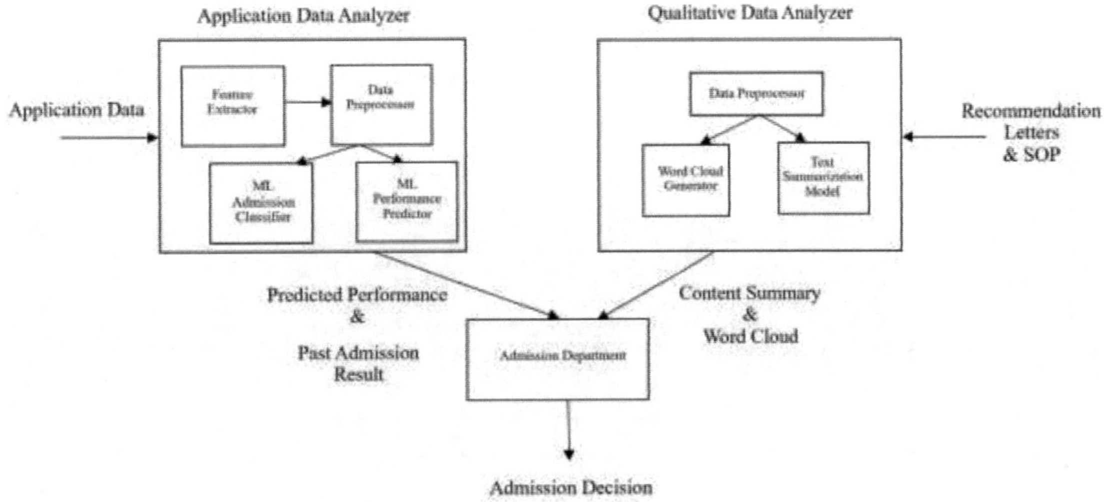

Figure 1: Model Schematic Design

4.1 Application Data Analyzer

The Application Data Analyzer (ADA) contains a feature extractor, data preprocessor, ML admission classifier, and ML performance predictor. ADA takes application data as input and outputs the report with the predicted student future performance and a Boolean value that indicates whether a student with similar academic credentials was admitted to the school. We use regression and classification ML models to generate the ADA outcome. Using ADA eliminates the need for manual review of application data to determine the student eligibility.

4.1.1 **Feature Extractor:** The application contains different sections, such as personal information, family information, education history, test scores, extracurricular activities, and essays. The ML models in the ADA module only needs education history, community activities and test scores. Hence, the feature extractor retrieves the required information from the application data and sends it to the Data preprocessor.

The feature of the Dataset for ADA are Career Interest , Intended Highest Degree, Graduation Class Size, Class Rank, GPA Scale, Rank Weighting, CGPA, Schedule Type, ProgressionCategory, Honors Exists, Honors Title1, Honors Grad Level1, Honors Recognotion1, Honors Title2, Honors Grad Level2, Honors Recognotion2, Honors Title3, Honors Grad Level3, Honors Recognotion3, Honors Title4, Honors Grad Level4, Honors Recognition4, Community Based Participation, and Community Based Participation Count. We create the dataset structure based on the student application in the common app [12]. The Common App provides a common student application for different universities across the USA.

4.1.2 **Data Preprocessor:** Data preprocessing is an essential step before applying the model, as data quality impacts the model accuracy. The Data preprocessor cleans the data in four steps, such as cleaning the missing values, converting the categorical variables, standardizing the data, and handling multicollinearity. We will follow the strategies, such as deleting the record, imputing the missing value, and deleting the column based on the percentage of missing values to clean the missing data. We will use one-hot encoding to handle the categorical variables and standardize the data to fall on a fixed scale to ensure that the impact of values in one column does not dominate the model result. Finally, we will check each column's correlation with other columns in the Dataset using a Heat map and eliminate the highly correlated column to avoid the dual impact of a feature on the model.

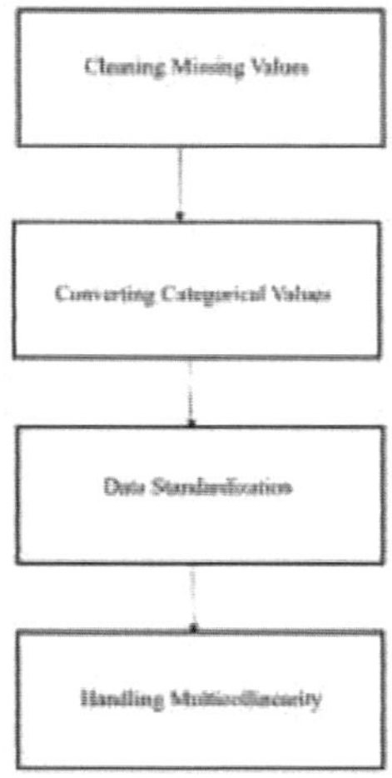

Figure 2: Data Preprocessing for Application Data Analysis

4.1.3 Machine Learning Admission Classifier: The ML classifier follows a supervised learning approach, where the admission history of students in the school constitutes the training dataset. The classifier's outcome is "Yes" or "No," indicating whether a student can be admitted or not. We will use Random Forest to perform the classification, based on the data size and data sparsity.

Random Forest (RF) is an ensemble model that works based on the aggregation of Decision Trees' outcome and is robust enough to handle variance. RF is emerged as a standardized classification algorithm in innovative environments [13]. RF, as a classification model, chooses the label using a voting mechanism, where the output of the highest number of trees is considered.

The advantages of RF are the robustness to overfitting, model effectiveness with variable data sizes and unbalanced datasets, and support to fault tolerance [14]. For our research, we will generate the data to test the model, and the data will be replaced with the real data from the universities when the model is implemented in real-time. RF will ensure that the model fits both scenarios.

4.1.4 Machine Learning Performance Predictor: The ML predictor uses a linear regression model to predict the CGPA of a student based on his/her educational history and test scores. The different variants of linear regression are Bivariate linear regression, where the model can fit two independent variables to predict a dependent variable, and Multivariate linear regression, where the model can handle more than two independent variables to predict a dependent variable. A study to predict the CGPA of engineering students proved that usage of Multivariate linear regression is a promising solution to predict the performance [15-17]. The below equation represents Multivariate linear regression, where ($x_1, x_2 ... xn$) are independent variables, and y is the dependent variable.

$$y = a + b_1x_1 + b_2x_2 + \cdots + b_nx_n$$

The Multivariate regression model's advantages are the support for complex data, personalized nature, and each independent variable's contribution to the result [16]. There is extensive research to predict student performance across different courses, and multivariate regression is proven as an efficient model. The performance predictor uses the Multivariate regression to predict the CGPA of a prospective student in the future. We will perform hyper-parameter tuning to achieve maximum accuracy.

4.2 Qualitative Data Analyzer

The qualitative data analyzer analyzes the textual content, such as Recommendation letters and SOP. The model will drastically bring down the time required to read the content by creating content summaries and word clouds. We use natural language processing techniques to perform these activities. This module's components are a data preprocessor, a word cloud generator, and a text summarization model. The module takes recommendation letters and SOP as input and sends the summary and word cloud generated from this content to the admission office. The below sections will explain the component of this module in detail.

IOP Conf. Series: Materials Science and Engineering 1074 (2021) 012024 doi:10.1088/1757-899X/1074/1/012024

4.2.1 Data Preprocessor: Data preprocessing is a mandatory step for NLP to focus on the required content. The different preprocessing steps are special characters removal, tokenization, stop words removal, and lemmatization. We use all the above-mentioned preprocessing steps for the word cloud generator, whereas we only first three preprocessing techniques for the text summarization model. The reason to skip the lemmatization in the text summarization model is to maintain the word's state in the sentence, which will impact the summary extraction.

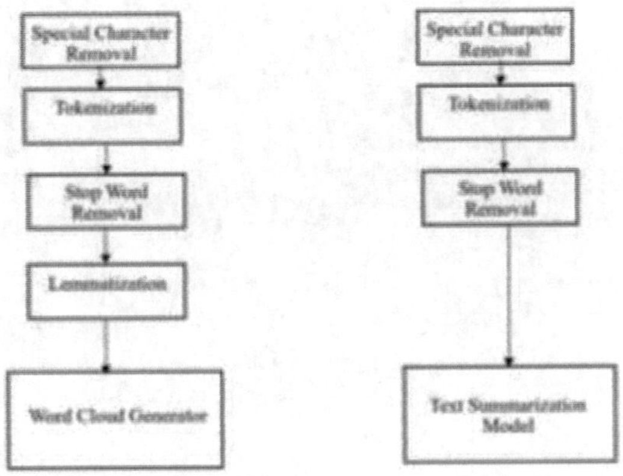

Figure 3: Data Preprocessing Steps for Qualitative Data Analysis

In special characters removal, the preprocessor will remove the symbols, numbers, and any other character that is not an alphabet. In tokenization, the sentence will be broken into words, which will be used as a bag of words for word cloud and text summary generation. Stop word removal eliminates the connecting words in the sentence, such as "is", "a", "the", "an", "to" etc. Finally, lemmatization will take a word to it's base form. For example, the word "studies" is used in a sentence; the lemmatization step converts this word to “study”.

4.2.2 Word Cloud Generator: The word cloud generator generates a word cloud with the key terms in the content. Word Cloud is a graphical representation of words where color, size, and weight represent the features like word frequency. Word clouds are considered as an effective means to draw insight into the content, construct the relationship between different keywords, and quickly retrieve the knowledge from information [18]. There are different word cloud applications, such as student response assessment, content comparison, feedback analysis, etc.

The word clouds can be generated using the reporting tools like Tableau, programming languages like Python, and other open access websites. We use Python to generate the word clouds for quantitative content. We chose to use word cloud over Named Entity Recognition (NER), as NER functionality is restricted to named entities. In the case of student content, identifying named entities alone does not provide the complete picture. We use the WordCloud package of Python to generate the word cloud and Matplotlib to create the word cloud's graphical representation. The below word cloud shows the key-terms in a recommendation letter and gives a high-level view of the student's achievements and skill.

Figure 4: Recommendation letter word cloud

4.2.3 Text Summarization Model: The text summarization model summarizes the content in a recommendation letter or SOP. The two types are summarizers are extraction summarizer and abstraction summarizer. The extraction summarizer extracts the summary based on word frequency count, and the summary is a combination of the sentences with high-frequency words. The abstraction summarizer is a semantic-based summarizer that works based on the content's semantics.

Extractive summarizers often result in better performance than abstract summarizers due to the practical issues with semantic representation and the need for complex analysis techniques. There are no pure abstractive summarizers available today. Extractive summarizer is used before abstractive summarizer [19]. We use an extractive summarizer to summarize the content in recommendation letters and SOP, and the implementation follows the below steps.

- Calculate the word frequency.
- Tokenize the content into sentences.
- Calculate the score for each sentence using the word frequency.
- Identify the sentences with a score greater than the average score of all the sentences in the content.

The model will create a report with the content summary for recommendation letters and SOP. This report will help the admission officer to get a view of the textual content presented by the prospective student. Word Cloud that gives highly used terms and a summary that gives an abstract view of the content will provide a complete overview of the submitted documents, resulting in faster application screening. We will review the outcome of the textual data analysis with the admission officers to get their feedback. Further improvisations will be implemented based on the qualitative analysis of the admission officer's interview outcome.

5 Conclusion

The proposed model speeds up the application process without losing the integrity of the application screening process. It also provides a complete design to cover both qualitative and quantitative content. There is scope to improvise this model based on the feedback from the admissions department. The model helps to manage resources efficiently due to the prediction and text processing reports. We will present the implementation details and results of this model in the next articles of this series.

We can incorporate the design to analyze the extracurricular activities section in future work, contributing to some universities' admission rate. The model can be extended to decide funding to a student and predict student performance in non-academic activities. Though the model is directed towards international student admissions, it can be extended to other admission types.

References

[1] *United States Hosts Over 1 Million International Students for the Fifth Consecutive Year*. (2020, November 16). Institute of International Education. https://www.iie.org/en/Why-IIE/Announcements/2020/11/2020-Open-Doors-Report

[2] Hegarty, N. (2014). Where We Are Now –The Presence and Importance of International Students to Universities in the United States. Journal of International Students, 4(3), 223-235. Retrieved from https://www.ojed.org/index.php/jis/article/view/462

[3] *How to Apply: International*. (2020, October 26). Undergraduate Admissions. https://admissions.unh.edu/apply/international-students

[4] Liu, M.-C., & Huang, Y.-M. (2017). The use of data science for education: The case of social-emotional learning. *Smart Learning Environments*, *4*(1), 112. https://doi.org/10.1186/s40561-016-0040-4

[5] Oladokun, V., Adebanjo, A., & Charles-Owaba, O. (2008). Predicting Students' Academic Performance using Artificial Neural Network: A Case Study of an Engineering Course.

[6] Robin, L., & Brian, F. (2010). Using Technology in Undergraduate Admission: A Student Perspective. *Journal of College Admission*, 10–17. https://eric.ed.gov/?id=EJ893890

[7] Shahiri, A. M., Husain, W., & Rashid, N.'. A. (2015). A Review on Predicting Student's Performance Using Data Mining Techniques. *Procedia Computer Science*, *72*, 414–422. https://doi.org/10.1016/j.procs.2015.12.157

[8] Lopez Guarin, C. E., Guzman, E. L., & Gonzalez, F. A. (2015). A Model to Predict Low Academic Performance at a Specific Enrollment Using Data Mining. *IEEE Revista Iberoamericana de Tecnologias Del Aprendizaje*, *10*(3), 119–125. https://doi.org/10.1109/rita.2015.2452632

[9] *educationaldatamining.org*. (n.d.). Educational Data Mining Society. https://educationaldatamining.org/

[10] Attaran, M., Stark, J., & Stotler, D. (2018). Opportunities and challenges for big data analytics in US higher education: A conceptual model for implementation. Industry and Higher Education, 32(3), 169–182. https://doi.org/10.1177/0950422218770937

[11] Cirelli, J., Konkol, A. M., Aqlan, F., & Nwokeji, J. C. (2018). Predictive analytics models for student admission and enrollment. *Proceedings of the International Conference on Industrial Engineering and Operations Management*, *2018*(SEP), 1395-1403.

[12] *Common App*. (n.d.). Common Application. https://apply.commonapp.org/common/6/33

[13] Couronné, R., Probst, P., & Boulesteix, A.-L. (2018). Random forest versus logistic regression: a large-scale benchmark experiment. BMC Bioinformatics, 19(1), 567. https://doi.org/10.1186/s12859-018-2264-5

[14] Liaw, A. & Wiener, M. Classification, and regression by randomForest (2002). R News 2, 18–22

[15] Rajalaxmi, R. R. Natesan, P. Krishnamoorthy, N. Ponni, S. (2019). Regression Model for Predicting Engineering Students Academic Performance, International Journal of Recent Technology and Engineering, 7(6S3), 71-75.

[16] Elbadrawy, A., Studham, R. S., & Karypis, G. (2015). Collaborative multi-regression models for predicting students' performance in course activities. *Proceedings of the Fifth International Conference on Learning Analytics And Knowledge - LAK '15*, 103–107. https://doi.org/10.1145/2723576.2723590

[17] Khan, W. Z., & Zubaidy, S. A. (2017). Prediction of Student Performance in Academic and Military Learning Environment: Use of Multiple Linear Regression Predictive Model and Hypothesis Testing. *International Journal of Higher Education*, *6*(4), 152. https://doi.org/10.5430/ijhe.v6n4p152

[18] DePaolo, C. A., & Wilkinson, K. (2014). Get Your Head into the Clouds: Using Word Clouds for Analyzing Qualitative Assessment Data. *TechTrends*, *58*(3), 38–44. https://doi.org/10.1007/s11528-014-0750-9

[19] Allahyari, M., Pouriyeh, S.A., Assefi, M., Safaei, S., Trippe, E.D., Gutierrez, J.B., & Kochut, K. (2017). Text Summarization Techniques: A Brief Survey. *ArXiv, abs/1707.02268*.

IOP Conf. Series: Materials Science and Engineering 1074 (2021) 012025 doi:10.1088/1757-899X/1074/1/012025

Epoch interrogation for skin cancer detection using convolutional neural network models

A Poornima[1], M Shyamala Devi [2], M Sumithra[3], Mullaguri Venkata Bharath[4], Swathi[5], S Sathishkumar[6], K Yogesh[7], Shubham Santosh Upadhyay[8], Nitesh Kumar Sah[9]

[1]Assistant Professor, [2] Professor, [3]Assistant Professor Computer Science & Engineering, Vel Tech Rangarajan Dr. Sagunthala R&D Institute of Science and Technology, Chennai.
[4, 5,6,7,8,9]Final Year B.Tech Student, Computer Science & Engineering, Vel Tech Rangarajan Dr. Sagunthala R&D Institute of Science and Technology, Chennai.

a.poornicse@gmail.com

Abstract. Skin cancer is very important notable disease and it is probable to everyone nowadays, it flourishes on the area of body where it exposed to ultraviolet rays. It leads anomalous gain in skin cells. It initiate on various parts of body like face, hand and bottoms of the feet as cautious hole or spot. The initial investigation of anomalous gain is essence to cure the disease at early stage, and it still remains a feasible challenge in the scientific improvements. From the analysis, this paper endeavour to inspect the category of disease with the following improvements. Initially, the skin dataset from ISIC machine archive is utilized for image processing. Secondly, the values of dataset images are normalized by dividing all the RGB values by 255. Thirdly, keras sequential API is used to add one layer at a time, initiating from the input. The CNN can extract the features that are useful for classifying the image, by using the kernel filter matrix. MaxPool reduce the computational cost by down-sampling the image, and the relu activation function is implemented to provide non linearity to the network. The flatten layer is utilized to remodel the final feature maps into 1D vector. CNN model provides accuracy of 94.83% with 3297 images and ResNet 50 model has attained accuracy of 90.78% due to less number of images used for classification. AlexNet model has attained accuracy of 81.8% with 1300 images and GoogleNet V3 inception has attained accuracy of 96% with 3374 images. Finally Vgg16 model has attained accuracy of 97.3% with 5636 samples.

1. Introduction

Convolutional Neural Network is a deep learning approach utilized in all the areas like image recognition, video recognition, image analysis, image classification and natural language processing. The recognition of deep learning approach is where the computer grasp analyse tasks explicitly from images, text or sound. It achieved better efficiency at higher levels than other methodologies that helps consumer to meet the user requirements. The utilization of deep learning improved from automated driving industries to medical devices. In healthcare sector based on the previous history and the disease are predicted to advance the treatment.

2. Literature Review

AI classifiers still have important role in assisting non-dermatologist physicians, since most skin cancer patients will consult them before being transferred to dermatologists. First step is to extract the unwanted features from the skin image like hair and ink markings whenever classification process is initiated to get better accuracy. Data mining technology has secured only 89.5% of accuracy while processing skin image for detecting the cancer [1]. The segmentation neural network model and deep neural network based on google inception has attained high skin lesion segmentation and melanoma diagnosis accuracy [2]. Using machine learning tool it is possible to find the cancer risk levels at early stage based on the score obtained for each factor according to their impacts on skin. Some of the significant factors are Skin Color Turn into pale, formation of bubbles in skin are respectively found top risk factors of cancer [3]. Many of the analyst not specified the train/test images or classification is performed with unequal images which leads to biased classification [4].The ensemble model trained on dermatology datasets classify the images on six classes C1: psoriasis, C2: seborrheic dermatitis, C3: lichen planus, C4: pityriasis rosea, C5: chronic dermatitis, C6: pityriasis rubra and determines the type of disease occurred [5]. The data mining techniques used to predict the disease and to improve the survivability rate regarding death related problems [6]. The initial model involves collection of dataset from ISIC dataset (International Skin Imaging Collaboration). Removal of parameters like glare removal, hair removal and bubbles from the image helps us to identify the parameters like color, size and shape in an efficient way. Then image segmentation and feature extraction is performed to classify the skin images. The extracted images trained with the model like back propagation algorithm and support vector machine to predict the image as malignant or non-malignant melanoma and has attained only 85% of accuracy [7].

This paper acquired annotated dataset statistics by changing the all categorical and Boolean attributes to numerical integers. Then it assemble the data compatible by pre-processing the images that to be fed into the model [8]. Classifiers predict the test image of malignant from the sample images and not benign [9]. The predictions are more effective and reliable when ensemble of different deep models, including deep residual networks and convolutional neural networks (CNNs) is used, in order to identify malignant melanomas, the deadliest type of skin cancer. These kinds of apps may cause harm to the patients by misleading them with incorrect diagnosis [10]. The study illustrates the method of building models and applying them to classify dermal cell images [11]. The skin magnifier with polarized light (SMP) is utilized to analyze the diagnosis of image based on the audio waves and it requires high quality images [12]. This model achieve a very high skin lesion segmentation and melanoma diagnosis accuracy with 2000 dermoscopic images includes 374 melanoma images, 1372 nevus images and 254 seborrhoea keratosis images [13]. The DCNN requires a massive number of images for training and testing to achieve high classification rates. This is a big challenge especially with skin cancer datasets where the number of available labelled images for training and testing is very limited [14]. Several issues remain for the CNN skin tumour classifier to overcome. First, there are no standardized evaluation test datasets to measure the efficacy of CNN classifiers. However, if the test dataset is known in advance, there is a risk of adapting the CNN classifier to the test dataset. Therefore, it might be better to conduct tests by a third-party organization to measure classification efficiency using closed datasets. [15].

3. Proposed Work

The main analysis of this paper is to predict the kind of skin cancer to advance the treatment at initial stage itself. The extracted top features from the convolutional layer are fitted to the classifier to analyze the performance and to differentiate the type of skin cancer. The proposed work of this paper is shown in Figure 1. The working flow of ResNet 50 model to classify the skin cancer image is shown in Figure 2. The Skin Cancer Dataset from the ISCI repository with 1800 images of benign moles and

1497 images of malignant classified moles columns is used for predicting the classification of skin cancer disease with the following contributions.

(i) Firstly, the skin cancer dataset from ISCI repository is subjected with the data processing.
(ii) Secondly, the values of dataset images are normalized by dividing all the RGB values by 255.
(iii) Thirdly, feature of the image is extracted and down-sampled with alternative convolutional layer and max pooling layer.
(iv) Fourth, flatten layer is utilized to remodel the final feature maps into 1D vector.
(v) Fifth, the input image is classified as benign or malignant according to their pixel values.

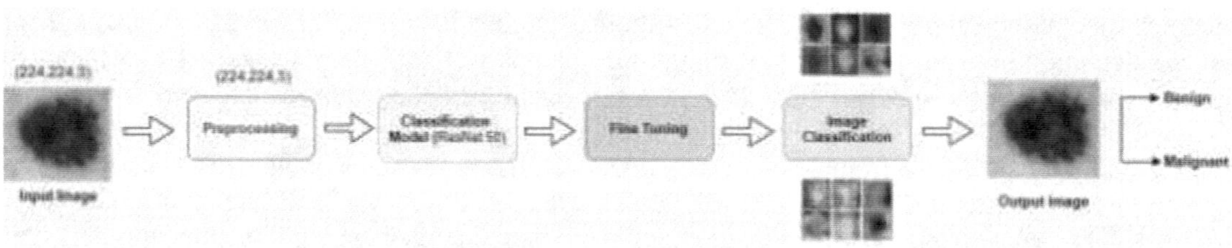

Figure 1. Overall Proposed Workflow

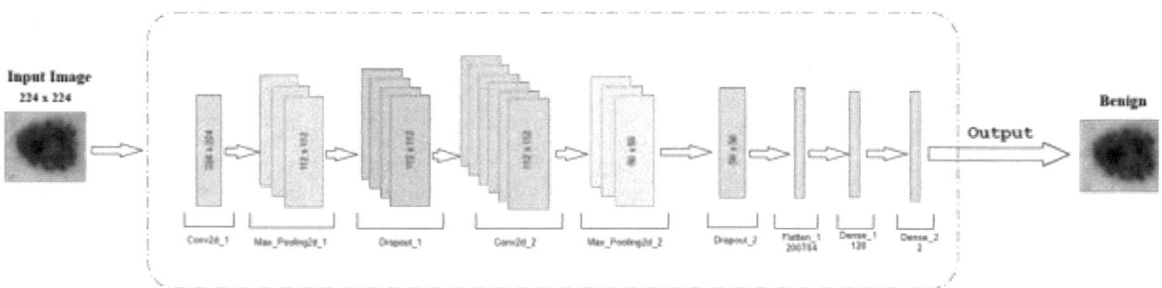

Figure 2. Flow of ResNet Model

4. Implementation Setup

4.1. Dataset Exploratory Analysis

The Skin Cancer Dataset from the ISCI repository with 1800 images of benign moles and 1497 images of malignant classified moles of individual patient records is used for prediction of skin cancer disease using deep learning classification algorithms. The python scripting language is coded in Spyder editor with Anaconda navigator for implementation. The dataset information is shown in Figure 3.

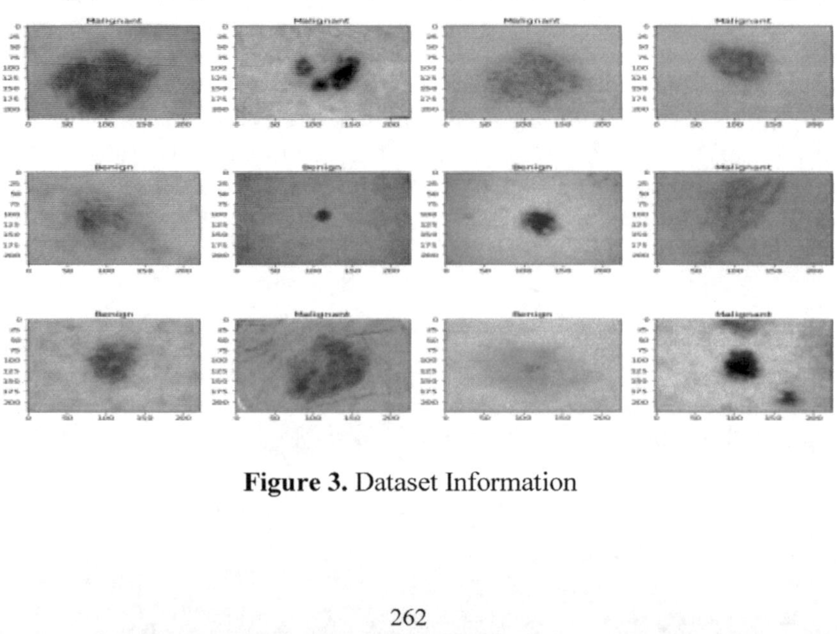

Figure 3. Dataset Information

5. Results and Discussions

The Skin Cancer Dataset from the ISCI repository with 1800 images of benign moles and 1497 images of malignant classified moles of individual patient records is used for prediction of skin cancer disease using deep learning classification algorithms. The model accuracy and model loss of skin cancer is identified with various set of input images and with different epochs. It provides step loss of 0.1494, validation loss of 0.4231, validation accuracy of 0.8201 and the classification accuracy obtained for 50 epochs is 94.83 % is shown in Figure 4 and 5. The output shape of convolutional layer and max-pooling are represented in Figure 6.

Table 1. Analysis of Performance Metric for different Epoch for CNN Model

Epoch No	Step Loss	Accuracy	Validation Loss	Validation Accuracy
1/50	0.9927	0.5974	0.5673	0.7614
2/50	0.5286	0.7634	0.5200	0.6932
3/50	0.4536	0.7757	0.6281	0.7254
4/50	0.4181	0.7933	0.5645	0.7386
5/50	0.4011	0.8075	0.4142	0.7898
6/50	0.3887	0.8075	0.4471	0.7917
7/50	0.3729	0.8132	0.4708	0.7822
8/50	0.3522	0.8288	0.4532	0.7879
9/50	0.3419	0.8369	0.4875	0.7955
10/50	0.3273	0.8530	0.4950	0.8106
11/50	0.3069	0.8563	0.4702	0.7822
12/50	0.3037	0.8549	0.4657	0.7955
13/50	0.2920	0.8729	0.5437	0.7803
14/50	0.2764	0.8777	0.4889	0.7841
15/50	0.2717	0.8757	0.4500	0.7936
16/50	0.2646	0.8867	0.4820	0.8163
17/50	0.2512	0.8938	0.4397	0.7879
18/50	0.2429	0.8957	0.4963	0.7973
19/50	0.2387	0.9009	0.4425	0.8030
20/50	0.2358	0.8943	0.4292	0.8182
21/50	0.2309	0.9047	0.4578	0.8125
22/50	0.2211	0.9094	0.4671	0.8068
23/50	0.2144	0.9137	0.4563	0.8182
24/50	0.2052	0.9142	0.4523	0.8125
25/50	0.2120	0.9090	0.4321	0.8106
26/50	0.1866	0.9256	0.4659	0.8068
27/50	0.1911	0.9251	0.4541	0.8163
28/50	0.1944	0.9227	0.4816	0.8049
29/50	0.1703	0.9322	0.4862	0.8030
30/50	0.1785	0.9317	0.4903	0.8106
31/50	0.1576	0.9393	0.4615	0.8087
32/50	0.1865	0.9251	0.4525	0.8125
33/50	0.1647	0.9360	0.4391	0.8220
34/50	0.1766	0.9317	0.4113	0.8144
35/50	0.1698	0.9312	0.4421	0.8125
36/50	0.1507	0.9436	0.4149	0.8106
37/50	0.1828	0.9279	0.4332	0.8125
38/50	0.1604	0.9431	0.4404	0.8144
39/50	0.1459	0.9426	0.4470	0.8144
40/50	0.1629	0.9374	0.4423	0.8125
41/50	0.1641	0.9360	0.4253	0.8182
42/50	0.1639	0.9374	0.4707	0.8144
43/50	0.1559	0.9445	0.4466	0.8239
44/50	0.1425	0.9436	0.4566	0.8144
45/50	0.1553	0.9440	0.4442	0.8163
46/50	0.1551	0.9426	0.4533	0.8201
47/50	0.1555	0.9388	0.4338	0.8239
48/50	0.1574	0.9407	0.4263	0.8220
49/50	0.1576	0.9403	0.4332	0.8220
50/50	0.1494	0.9483	0.4231	0.8201

IOP Conf. Series: Materials Science and Engineering 1074 (2021) 012025 doi:10.1088/1757-899X/1074/1/012025

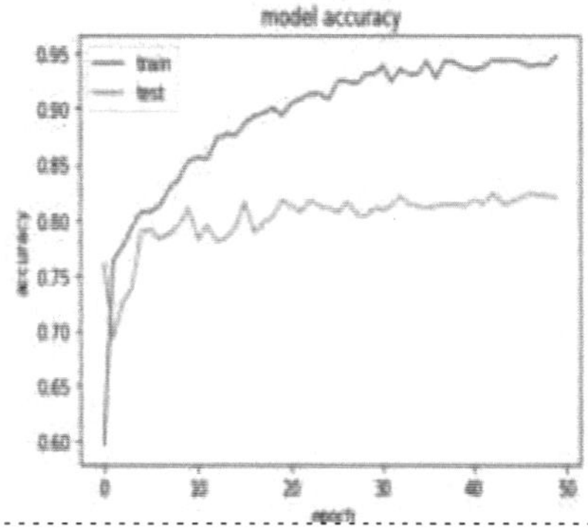

Figure 4. CNN Model Accuracy for Skin Dataset

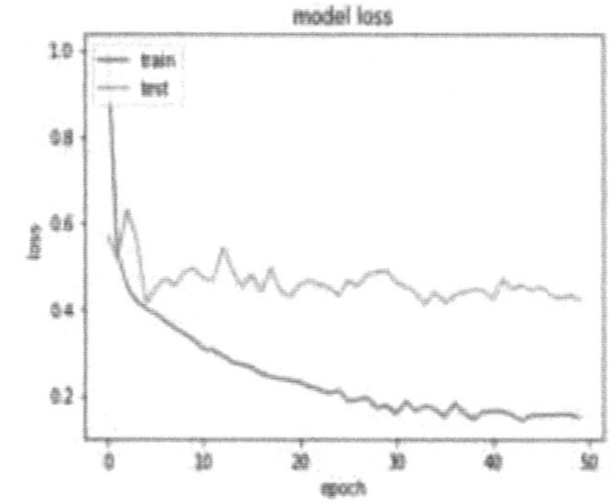

Figure 5. CNN Model Loss for Skin Dataset

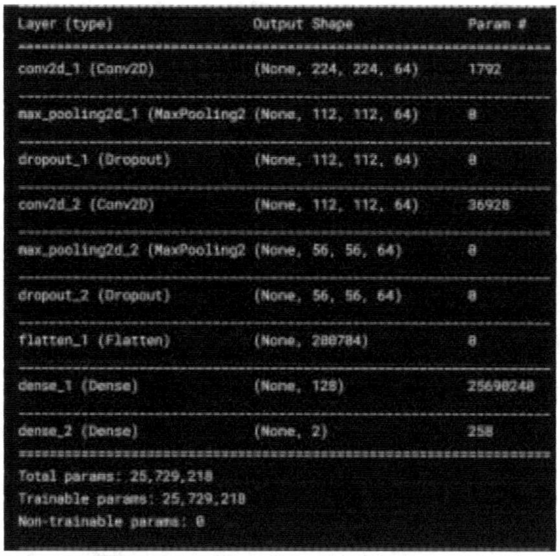

Layer (type)	Output Shape	Param #
conv2d_1 (Conv2D)	(None, 224, 224, 64)	1792
max_pooling2d_1 (MaxPooling2	(None, 112, 112, 64)	0
dropout_1 (Dropout)	(None, 112, 112, 64)	0
conv2d_2 (Conv2D)	(None, 112, 112, 64)	36928
max_pooling2d_2 (MaxPooling2	(None, 56, 56, 64)	0
dropout_2 (Dropout)	(None, 56, 56, 64)	0
flatten_1 (Flatten)	(None, 200704)	0
dense_1 (Dense)	(None, 128)	25690240
dense_2 (Dense)	(None, 2)	258

Total params: 25,729,218
Trainable params: 25,729,218
Non-trainable params: 0

Figure 6. CNN Model Summary

The skin cancer classification is performed by including additional samples by training the model on 1000 samples and validating the model on 26 samples. It provides sample loss of 8.6903, sample accuracy of 0.4430, step loss of 0.2136, validation loss of 8.6741, validation accuracy of 0.4430 and the classification accuracy obtained for 15 epochs is 90.78 % is shown in Figure 7 and 8. The output shape of ResNet 50 model layers are shown in Figure 9.

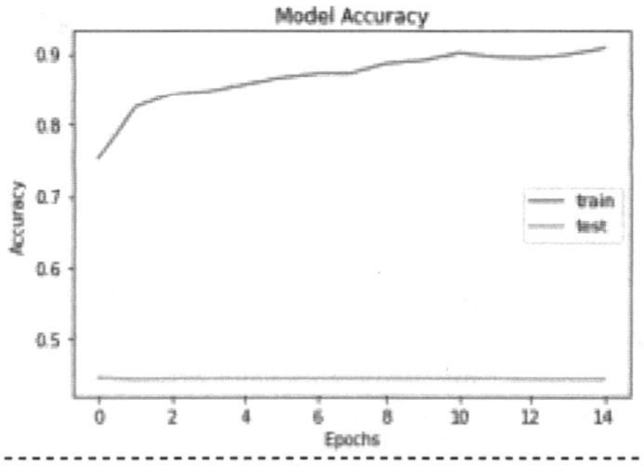

Figure 7. ResNet 50 Model Accuracy

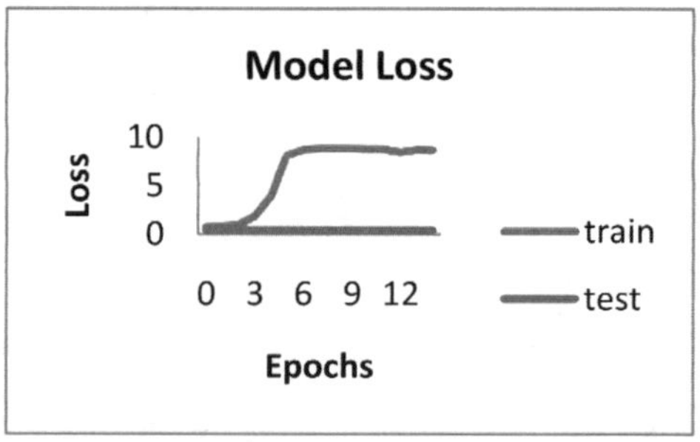

Figure 8. ResNet 50 Model Loss

Table 2. Analysis of Performance Metric for different Epoch for ResNet 50 Model

Epoch No	Sample Loss	Sample Accuracy	Step Loss	Accuracy	Validation Loss	Validation Accuracy
1/15	0.7494	0.4450	0.5355	0.7526	0.7489	0.4450
2/15	0.8283	0.4430	0.4124	0.8241	0.8274	0.4430
3/15	1.0141	0.4440	0.3647	0.8430	1.0123	0.4440
4/15	1.8495	0.4440	0.3308	0.8473	1.8456	0.4440
5/15	3.9576	0.4440	0.3183	0.8571	3.9480	0.4440
6/15	8.1071	0.4440	0.2892	0.8668	8.0870	0.4440
7/15	8.6974	0.4440	0.2871	0.8723	8.6807	0.4440
8/15	8.7863	0.4440	0.2875	0.8735	8.7679	0.4440
9/15	8.8098	0.4440	0.2526	0.8864	8.7908	0.4440
10/15	8.7982	0.4440	0.2561	0.8907	8.7795	0.4440
11/15	8.7727	0.4440	0.2378	0.9010	8.7546	0.4440
12/15	8.7763	0.4440	0.2383	0.8955	8.7582	0.4440
13/15	8.7345	0.4430	0.2447	0.8943	8.4313	0.4430
14/15	8.7145	0.4430	0.2390	0.8986	8.6977	0.4430
15/15	8.6903	0.4430	0.2136	0.9078	8.6741	0.4430

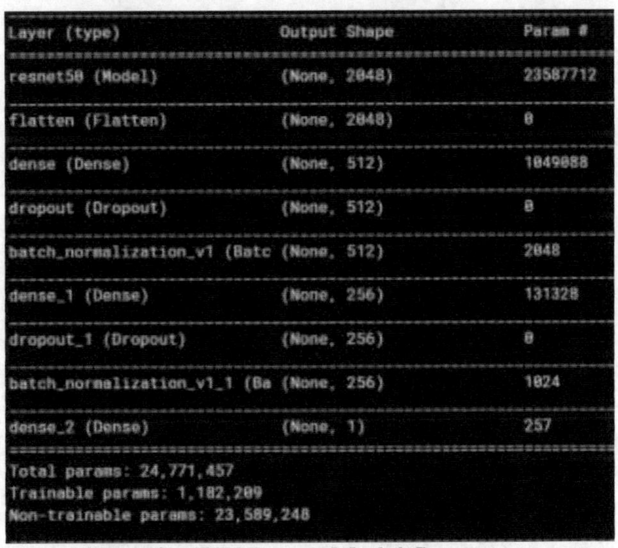

Layer (type)	Output Shape	Param #
resnet50 (Model)	(None, 2048)	23587712
flatten (Flatten)	(None, 2048)	0
dense (Dense)	(None, 512)	1049088
dropout (Dropout)	(None, 512)	0
batch_normalization_v1 (Batc	(None, 512)	2048
dense_1 (Dense)	(None, 256)	131328
dropout_1 (Dropout)	(None, 256)	0
batch_normalization_v1_1 (Ba	(None, 256)	1024
dense_2 (Dense)	(None, 1)	257

Total params: 24,771,457
Trainable params: 1,182,209
Non-trainable params: 23,589,248

Figure 9. CNN Resnet Model Summary

The skin cancer classification is performed by including additional samples by training the model on 5636 samples and validating the model on 149 samples. It provides step loss of 0.0778, validation loss of 0.4804, validation accuracy of 0.8658 and the classification accuracy obtained for 20 epochs is 97.32% is shown in Figure.10 and Figure 11. The output shape of Vgg16 model layers are shown in Figure 12.

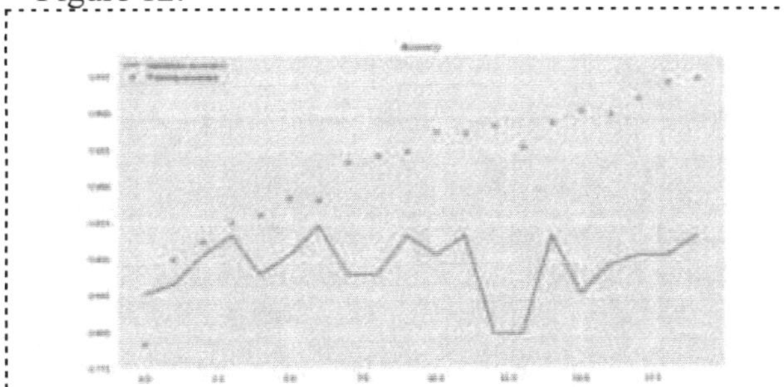

Figure 10. Skin Dataset Vgg 16 Model Accuracy

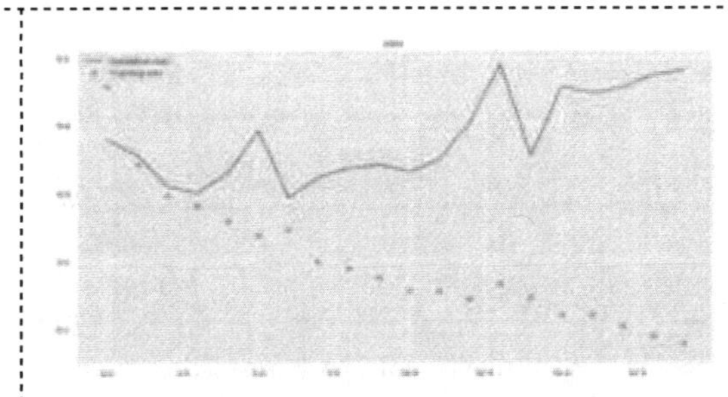

Figure 11. Skin Dataset Vgg 16 Model Loss

IOP Conf. Series: Materials Science and Engineering 1074 (2021) 012025 doi:10.1088/1757-899X/1074/1/012025

Table 3. Analysis of Performance Metric for different Epoch for Vgg 16 Model

Epoch No	Step Loss	Accuracy	Validation Loss	Validation Accuracy
1/20	0.4571	0.7917	0.3790	0.8255
2/20	0.3428	0.8497	0.3552	0.8322
3/20	0.2965	0.8618	0.3096	0.8523
4/20	0.2811	0.8751	0.3003	0.8658
5/20	0.2587	0.8799	0.3302	0.8389
6/20	0.2374	0.8912	0.3913	0.8523
7/20	0.2460	0.8896	0.2930	0.8725
8/20	0.1992	0.9159	0.3226	0.8389
9/20	0.1892	0.9203	0.3363	0.8389
10/20	0.1754	0.9233	0.3412	0.8658
11/20	0.1565	0.9365	0.3310	0.8523
12/20	0.1552	0.9358	0.3487	0.8658
13/20	0.1444	0.9407	0.4019	0.7987
14/20	0.1653	0.9265	0.4892	0.7987
15/20	0.1456	0.9427	0.3551	0.8658
16/20	0.1203	0.9510	0.4559	0.8255
17/20	0.1215	0.9485	0.4468	0.8456
18/20	0.1035	0.9594	0.4547	0.8523
19/20	0.0889	0.9705	0.4744	0.8523
20/20	0.0778	0.9732	0.4804	0.8658

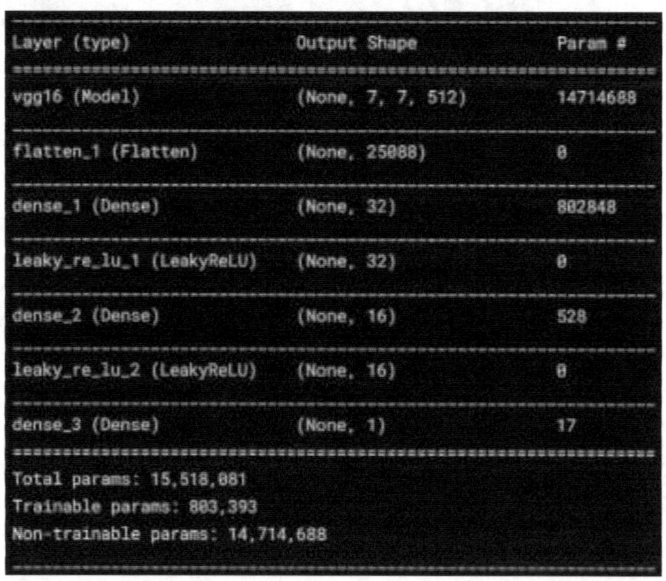

Layer (type)	Output Shape	Param #
vgg16 (Model)	(None, 7, 7, 512)	14714688
flatten_1 (Flatten)	(None, 25088)	0
dense_1 (Dense)	(None, 32)	802848
leaky_re_lu_1 (LeakyReLU)	(None, 32)	0
dense_2 (Dense)	(None, 16)	528
leaky_re_lu_2 (LeakyReLU)	(None, 16)	0
dense_3 (Dense)	(None, 1)	17

Total params: 15,518,081
Trainable params: 803,393
Non-trainable params: 14,714,688

Figure 12. CNN Vgg Model Summary

The skin cancer classification is performed by training the model on 1300 image samples. Feature extraction is done using an AlexNet whose fully connected layer was replaced by convolutional layer and validating the model on 149 samples. It provides step loss of 0.2236, validation loss of 8.7103, validation accuracy of 0.4440 and the classification accuracy obtained for 15 epochs is 81.8% is shown in Figure 13 and Figure 14.

IOP Conf. Series: Materials Science and Engineering 1074 (2021) 012025 doi:10.1088/1757-899X/1074/1/012025

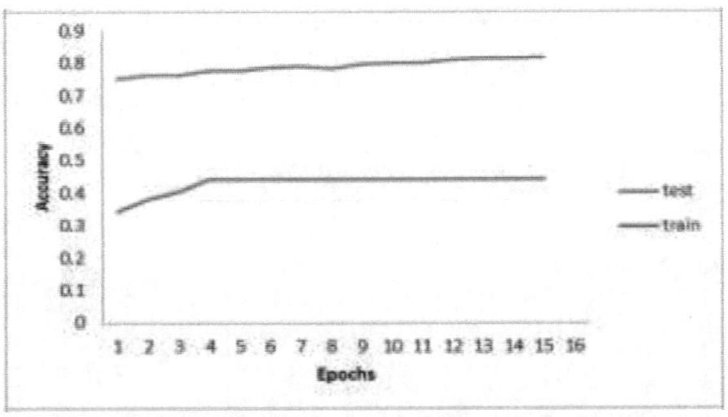

Figure 13. Skin Dataset AlexNet Model Accuracy

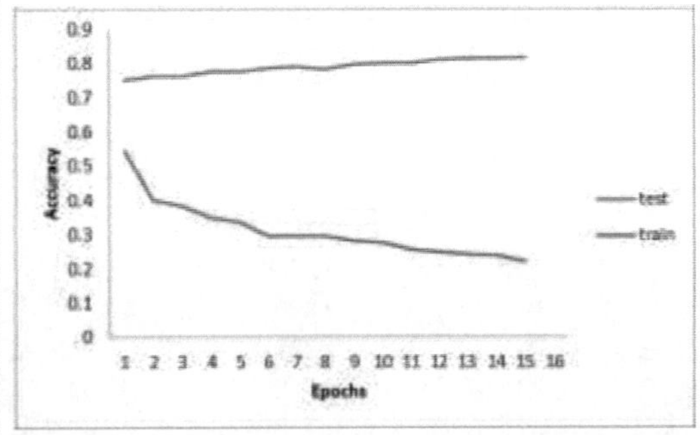

Figure 14. Skin Dataset AlexNet Model Loss

Table 4. Analysis of Performance Metric for different Epoch for AlexNet Model

Epoch No	Sample Loss	Sample Accuracy	Step Loss	Accuracy	Validation Loss	Validation Accuracy
1/15	0.7694	0.3450	0.5455	0.7526	0.7694	0.4450
2/15	0.8083	0.3830	0.4024	0.7641	0.8083	0.4430
3/15	1.0441	0.4050	0.3847	0.7630	1.0441	0.4450
4/15	1.8695	0.4450	0.3508	0.7773	1.8695	0.4450
5/15	3.9876	0.4450	0.3383	0.7771	3.9876	0.4450
6/15	8.1571	0.4450	0.2992	0.7868	8.1571	0.4450
7/15	8.7974	0.4450	0.2971	0.7923	8.7974	0.4450
8/15	8.8163	0.4450	0.2975	0.7835	8.8163	0.4450
9/15	8.8298	0.4450	0.2826	0.7964	8.8298	0.4450
10/15	8.7982	0.4450	0.2761	0.8007	8.7882	0.4450
11/15	8.7727	0.4450	0.2578	0.8010	8.7627	0.4450
12/15	8.7763	0.4450	0.2483	0.8105	8.7663	0.4450
13/15	8.7345	0.4450	0.2447	0.8143	8.7345	0.4450
14/15	8.7545	0.4440	0.2390	0.8166	8.7445	0.4440
15/15	8.7303	0.4440	0.2236	0.8188	8.7103	0.4440

The skin cancer classification is performed by training the model on 3374 image samples. Feature extraction is done using an GoogLeNet Inception v3 model. It provides step loss of 0.0889, validation loss of 0.4744, validation accuracy of 0.8523 and the classification accuracy obtained for 15 epochs is 96% is shown in Figure.15 and Figure 16.

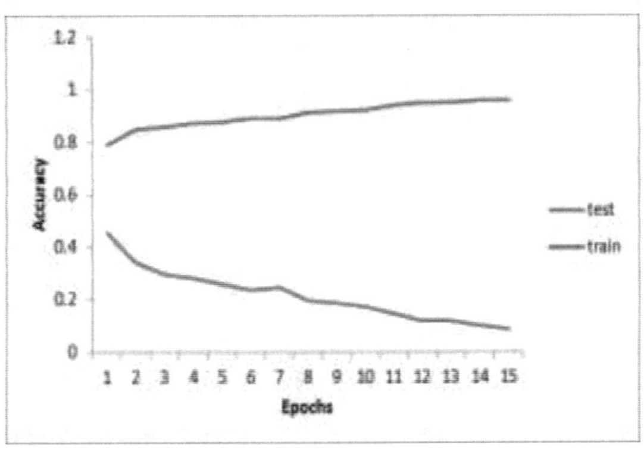

Figure 15. GoogLeNet v3 Model Accuracy

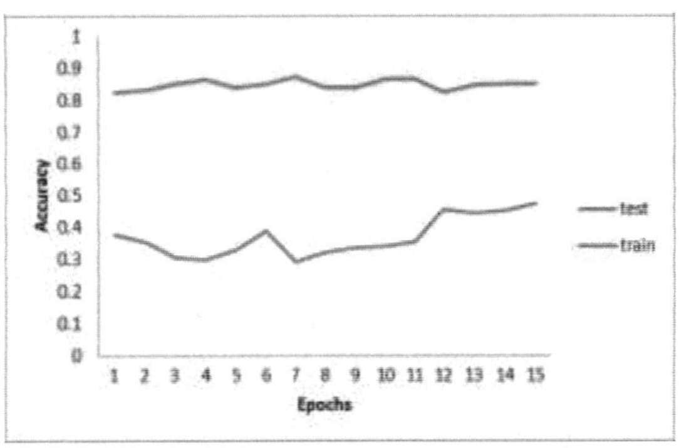

Figure 16. GoogLeNet v3 Model Loss

Table 5. Analysis of Performance Metric for different Epoch for GoogleNet V3 Inception Model

Epoch No	Step Loss	Accuracy	Validation Loss	Validation Accuracy
1/15	0.4571	0.7917	0.3790	0.8255
2/15	0.3428	0.8497	0.3552	0.8322
3/15	0.2965	0.8618	0.3096	0.8523
4/15	0.2811	0.8751	0.3003	0.8658
5/15	0.2587	0.8799	0.3302	0.8389
6/15	0.2374	0.8912	0.3913	0.8523
7/15	0.2460	0.8896	0.2930	0.8725
8/15	0.1992	0.9159	0.3226	0.8389
9/15	0.1892	0.9203	0.3363	0.8389
10/15	0.1754	0.9233	0.3412	0.8658
11/15	0.1456	0.9427	0.3551	0.8658
12/15	0.1203	0.9510	0.4559	0.8255
13/15	0.1215	0.9485	0.4468	0.8456
14/15	0.1035	0.9594	0.4547	0.8523
15/15	0.0889	0.9600	0.4744	0.8523

The final skin cancer classification images for benign and malignant are shown in Figure 17.

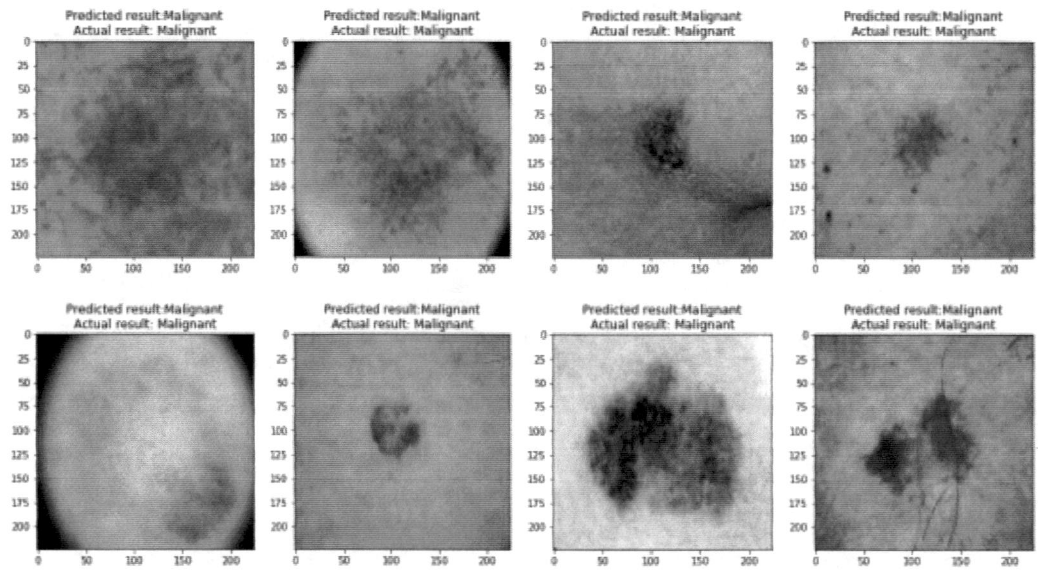

Figure 13. Skin Cancer Predicted Result

6. Conclusion

An attempt is made in this paper to analyze the dataset and the model which provides higher accuracy on skin cancer detection. CNN model provides accuracy of 94.83% with 3297 images and ResNet 50 model has attained accuracy of 90.78% due to less number of images used for classification. AlexNet model has attained accuracy of 81.8% with 1300 images and GoogleNet V3 inception has attained accuracy of 96% with 3374 images. Finally Vgg16 model has attained accuracy of 97.3 with 5636 samples. When the number of samples get increased the deep learning algorithm provides better accuracy at the same time we need to avoid the model leads to over fitting otherwise it will not provide better accuracy on test images.

References

[1] Masood, Dr Syed. (2012). Cancer diagnosis using data mining technology. Life Science Journal. 1. 308-313.

[2] Munir K, Elahi H, Ayub A, Frezza F, Rizzi A. Cancer Diagnosis Using Deep Learning: A Bibliographic Review. Cancers (Basel). 2019 Aug 23;11(9):1235. doi: 10.3390/cancers11091235. PMID: 31450799; PMCID: PMC6770116.

[3] Chan S, Reddy V, Myers B, Thibodeaux Q, Brownstone N, Liao W. Machine Learning in Dermatology: Current Applications, Opportunities, and Limitations. Dermatol Ther (Heidelb). 2020 Jun;10(3):365-386. doi: 10.1007/s13555-020-00372-0. Epub 2020 Apr 6. PMID: 32253623; PMCID: PMC7211783.

[4] Verma AK, Pal S, Kumar S. Classification of Skin Disease using Ensemble Data Mining Techniques. Asian Pac J Cancer Prev. 2019 Jun 1;20(6):1887-1894. doi: 10.31557/APJCP.2019.20.6.1887. PMID: 31244314; PMCID: PMC7021628

[5] Masood, Ammara; Ali Al-Jumaily, Adel (2013). Computer Aided Diagnostic Support System for Skin Cancer: A Review of Techniques and Algorithms. International Journal of Biomedical Imaging, 2013(), 1–22. doi:10.1155/2013/323268

[6] Weli, Z. N. S. (2020). Data Mining in Cancer Diagnosis and Prediction: Review about Latest Ten Years. Current Journal of Applied Science and Technology, 39(6), 11-32. https://doi.org/10.9734/cjast/2020/v39i630555

[7] Rejaul, Md & Royel, Md Rejaul & Jaman, Ajmanur & Masud, Fuyad & Ahmed, Arzo & Muyeed, Abdul. (2020). Machine Learning and Data Mining Methods in Early Detection of Stomach Cancer Risk. Journal of Applied Science Engineering and Technology. 24. 1-8. 10.6180/jase.202102_24(1).0001.

[8] Melbin, K. & Raj, Y.. (2020). Hybrid Machine Learning Approach for Skin Disease Detection Using Optimal Support Vector Machine. 10.1007/978-3-030-34080-3_73.

[9] M, Vijayalakshmi. (2019). Melanoma Skin Cancer Detection using Image Processing and Machine Learning. International Journal of Trend in Scientific Research and Development. Volume-3. 780-784. 10.31142/ijtsrd23936.

[10] Pacheco, A. G. C. and Krohling, R. A., "Recent advances in deep learning applied to skin cancer detection", arXiv e-prints, 2019.

[11] Kadampur, Mohommad. (2020). Skin cancer detection Applying a deep learning based model driven architecture in the cloud for classifying dermal cell images Elsevier Enhanced Reader. Informatics in Medicine Unlocked. 18. 10.1016/j.imu.2019.100282.

[12] Dascalu, Avi & David, Eli. (2019). Skin cancer detection by deep learning and sound analysis algorithms: A prospective clinical study of an elementary dermoscope. EBioMedicine. 43. 10.1016/j.ebiom.2019.04.055.

[13] Chang, Hao. (2017). Skin cancer reorganization and classification with deep neural network.

[14] K. M. Hosny, M. A. Kassem and M. M. Foaud, "Skin Cancer Classification using Deep Learning and Transfer Learning," 2018 9th Cairo International Biomedical Engineering Conference (CIBEC), Cairo, Egypt, 2018, pp. 90-93, doi: 10.1109/CIBEC.2018.8641762.

[15] Fujisawa, Yasuhiro & Inoue, Sae & Nakamura, Yoshiyuki. (2019). The Possibility of Deep Learning-Based, Computer-Aided Skin Tumor Classifiers. Frontiers in Medicine. 6. 191. 10.3389/fmed.2019.00191.

CANONICAL VARIATE FEATURE SELECTION BASED ADAPTIVE ENHANCED WINNOW MAP REDUCE CLASSIFICATION FOR PREDICTIVE ANALYTICS

R. Sivakkolundu[1], V. Kavitha[2]

[1]Hindusthan College of Arts and Science, Coimbatore, Tamilnadu, India
[2]Hindusthan College of Arts and Science, Coimbatore, Tamilnadu, India
sivakkolundu.stud@gmail.com

Abstract. Classification is a key issue to be resolved in data mining. Few research works have been designed for performing predictive analysis through classifying the information on data warehouse. But, classification accuracy (CA) of conventional works was lower when considering a big size of data as input. In order to addresses this drawback, a Canonical Variate Feature Selection based Adaptive Enhanced Winnow Map Reduce Classification (CVFS-AEWMRC) Method is proposed. The CVFS-AEWMRC Method is designed for organizing and classifying the collected and stored data for decision making. Initially, Canonical Variate Feature Selection (CVFS) is carried out in CVFS-AEWMRC Method to select the relevant features for performing the classification. Canonical Variates analysis is a machine learning technique used to find linear combinations of features which have maximum correlation with each other. The features with maximum correlation are selected for performing the classification. Then, Adaptive Enhanced Winnow Map Reduce Classification (AEWMRC) Process is carried out in CVFS-AEWMRC Method to classify the stored data for taking decision. Adaptive Enhanced Winnow technique learns the linear classifier from labeled data samples. Winnow employs the multiplicative scheme for performing the classification process. Winnow learns the hyperplane to classify the data points for decision making. By this way, the data classification is carried out in accurate manner for decision making during the predictive analytics process. Experimental analysis of CVFS-AEWMRC Method is performed on metrics namely feature selection rate (FSR), CA, classification time (CT) and False positive rate (FPR) with number of features and data points.

Keywords: Canonical Variate Feature Selection, Correlation, Decision Making, Features, MapReduce, Prediction Rule, Predictive Analytics

1 INTRODUCTION

In data analysis, historical data storing is significant. A data warehouse is employed in recent decades to perform data analysis and is the basis for analytic reports, data mining and business intelligence which is significant for decision making in organizations, companies, etc. The volume of data from different sources increasing in an exponential pattern complicates data storage and retrieval in data warehouse, results in bottlenecks. In big data, data warehouse management is a prominent issue due to the high rate of data growth and the need for data analysis. Hence a novel method called CVFS-AEWMRC is introduced in this research work.

In [1], Label-Aware Distributed Ensemble Learning (LADEL) was designed to handle big data. However, FSR using LADEL was higher. An Atrak method using Map Reduce-based data warehouse was presented in [2] for big data to enhance query execution speed through employing node independence. But, CA of predictive analytics was not sufficient.

An extension of Technology–Organization–Environment (TOE) framework was presented in [3] for discover barriers and opportunities of Artificial Intelligence (AI). However, the feature selection process was not carried out in efficient manner. To address the data warehouse issues, A new Chabok method was introduced in [4] which utilize two-phased Map-Reduce. Chabok not required the data replication for the join omission.

High-Level MapReduce Query Languages was employed in [5] directly on MR which translated queries to executable native MR jobs. But, MapReduce-based HLQL failed to reduce the computational cost. A semi-automatic design methodology was introduced in [6] for decision making. However, FPR of predictive analytics was higher.

Boolean query generator utilizing common-table expressions was employed in [7] for giving better clinical data warehouse performance. But computational complexity was more. To design big data warehouse, an efficient and real-time power monitoring platform with open source techniques was introduced in [8]. However, time complexity using this algorithm was not reduced.

Hadoop as big data analytic was implemented in [9] with data warehouse to support decision making process. But predictive analytics performance was poor. A novel method was used in [10] to addresses the data warehouses issues and fulfilling requirements of decision-makers. However, FPR using this method was not reduced.

The characteristics of collected and stored data are utilized in this article to construct a novel method called CVFS-AEWMRC for predictive analysis.

- Initially, CVFS method is employed in CVFS-AEWMRC method to choose the relevant features. After that, an AEWMRC is designed in CVFS-AEWMRC for decision making.
- In data warehouse management system, Potential information is predicted using CVFS-AEWMRC method which assists in decision makers for efficient predictive analytics.

2 LITERATURE SURVEY

By considering the distributed storage and parallel processing, a large-scale system was developed in [11] for social media data warehousing with the assist of Hadoop and Storm. A simple low-overhead mechanism was introduced in [12] to evaluate every data portion for efficient big data processing in warehouse-scale computers.

In [13], Game Theory based Materialized View selection (GTMV) was designed to select optimal set of views for materialization and thus increases the data warehouse performance. Specification of the data warehouse was employed in [14] for the decision-making dimension of the Bid Process Information System.

Parallel Processing System was employed in [15] to increase the capability of processing data on warehouse. A novel methodology was developed in [16] for designing data warehousing applications from different sources.

A mixed approach was introduced in [17] for creating XML data warehouses. An integrated optimization method was used in [18] to provide better computational method for distributed data warehouses.

An extract transforms and load strategy were introduced in [19] for unstructured data into data warehouse with map reduce paradigm for big data analytics. With the help of uniform distributed dataset, a novel method was designed in [20] to resolving data warehouse issues in big data.

3 CANONICAL VARIATE FEATURE SELECTION BASED ADAPTIVE ENHANCED WINNOW MAP REDUCE CLASSIFICATION METHOD

The CVFS-AEWMRC Method is designed with aim of performing an efficient data classification for accurate decision making with a minimal time. On the contrary to conventional works, CVFS-AEWMRC Method is proposed by combining the CVFS algorithm and AEWMRC algorithm. The designed CVFS algorithm in CVFS-AEWMRC Method helps to enhance the feature selection performance for effective predictive analytics. Besides to that, AEWMRC algorithm in CVFS-AEWMRC Method supports for classifying the collected and stored data on warehouse by using selected features and thereby enhances CA of predictive analytics.

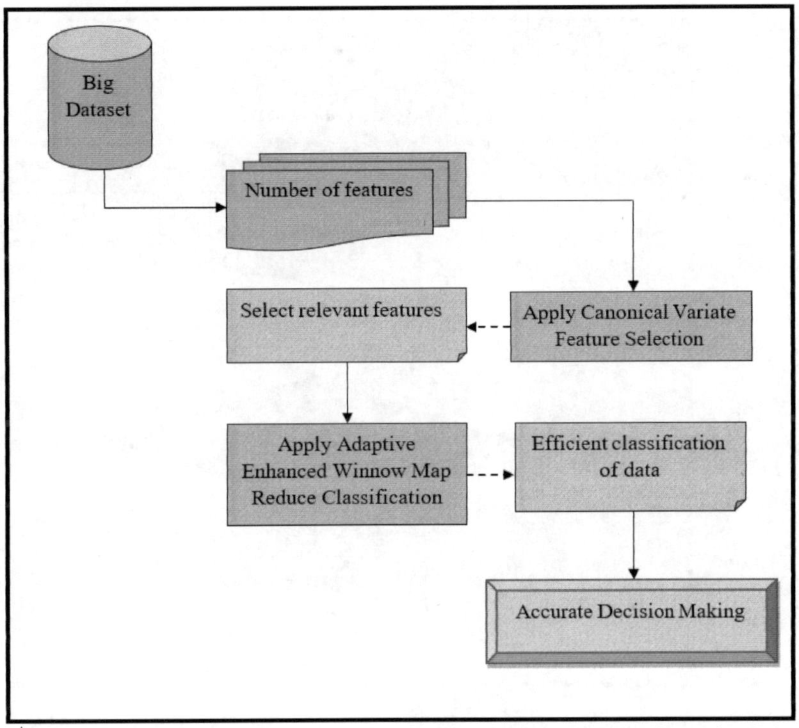

Figure 1 Architecture Diagram of CVFS-AEWMRC Method for Predictive Analytics

Figure 1 presents the flow processes of CVFS-AEWMRC Method. From figure 1, CVFS-AEWMRC Method considers Instacart Market Basket Analysis dataset as input. Followed by, CVFS-AEWMRC Method chooses key features using CVFS algorithm. In CVFS algorithm, Canonical correlation analysis is utilized to discover and compute the associations among two sets of variables for selecting relevant features from big dataset. With the selected relevant features, the collected and stored data are classified for decision making. The detailed processes of CVFS-AEWMRC Method are described in below sections.

3.1 Canonical Variate Feature Selection

CVFS algorithm is designed in CVFS-AEWMRC Method to attain better FSR for predictive analytics. CVFS algorithm is proposed to identify and determine the asso-

IOP Conf. Series: Materials Science and Engineering 1074 (2021) 012026 doi:10.1088/1757-899X/1074/1/012026

ciations among two sets of variables. Besides, CVFS algorithm finds orthogonal linear combinations of the variables within each feature that best explain the variability both within and between features. This helps for CVFS algorithm to extract significant features for accurate predictive analytics.

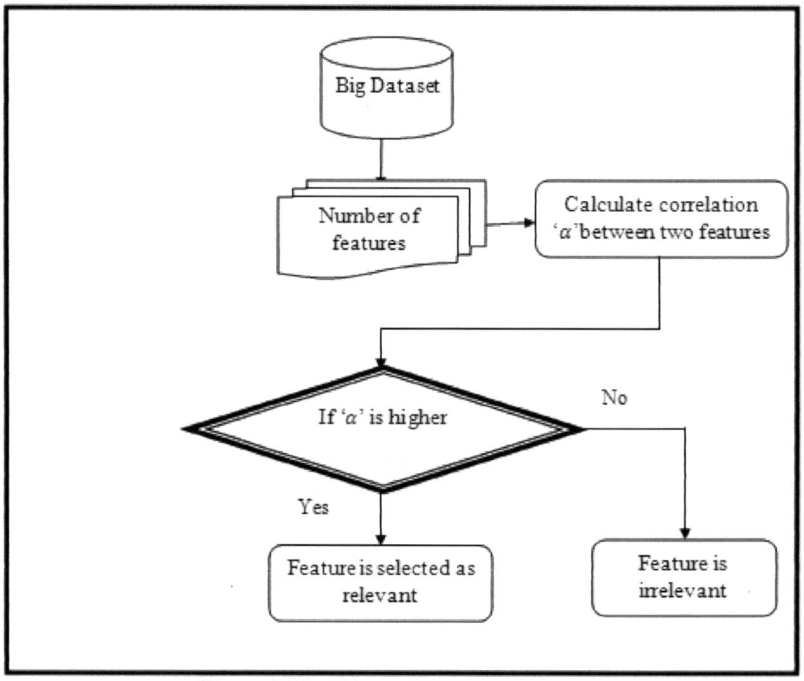

Figure 2 Flow Processes of CVFS algorithm

Figure 2 depicts the flow processes of CVFS algorithm to carry out feature selection process. As demonstrated in figure, the association among two features is computed by Canonical correlation analysis in CVFS algorithm. Let assume two features '$\boldsymbol{a_1}$' and '$\boldsymbol{a_2}$', and there are correlations between the information, and Canonical correlation analysis find out linear combinations of $\boldsymbol{a_1}$ and $\boldsymbol{a_2}$ which have a higher correlation with each other. Accordingly, canonical correlation is mathematically estimated as,

$$\alpha \rightarrow \sum \boldsymbol{a_1}\, \boldsymbol{a_2} = \boldsymbol{Cov(a_1}, a_2) \quad (1)$$

From the above mathematical formula (1), the covariance matrix is evaluated with help of information in features '$\boldsymbol{a_1}$' and '$\boldsymbol{a_2}$'. Followed by, the covariance among two features '$\boldsymbol{a_1}$' and '$\boldsymbol{a_2}$' is mathematically obtained as,

$$\boldsymbol{Cov(a_1, a_2)} = \frac{\sum(a_1 - \overline{a_1})(a_2 - \overline{a_2})}{n} \quad (2)$$

From the above mathematical representation (2), '$\boldsymbol{a_1}$' signifies the values of the data in features '$\boldsymbol{a_1}$' and '$\boldsymbol{a_2}$' designates the values of data in features '$\boldsymbol{a_2}$'. Here, '$\overline{\boldsymbol{a_1}}$'point outs the mean of the features '$\boldsymbol{a_1}$' and '$\overline{\boldsymbol{a_2}}$' indicates mean of the features '$\boldsymbol{a_2}$' and '$\boldsymbol{n}$' indicates a total number of features in big dataset. By using equation (1) and (2), CVFS algorithm calculates the correlation between the two features. Thus, CVFS algorithm extracts the features with higher correlation value as significant for accurately performing data classification process.

```
// Canonical Variate Feature Selection algorithm
Input: Big Dataset, Number of features '{a_1, a_2, ..., a_n}';
Output: Select relevant features for predictive analytics
Step 1: Begin
Step 2:     For each input features 'a_i'
Step 3:         Compute covariance 'Cov(a_1, a_2)' using (2)
Step 4:         Calculate the canonical correlation 'α' using (1)
Step 5:         If (α ishigher), then
Step 6:             feature 'a_i' is selected as relevant
Step 7:         Else
Step 8:             feature 'a_i'is removed as irrelevant
Step 9:         End If
Step 10:    End For
Step 11:End
```

Algorithm 1 Canonical Variate Feature Selection

The process of CVFS is described in algorithm 1. With the help of above algorithmic steps, CVFS-AEWMRC method accurately selects relevant features in a given dataset for accomplishing predictive analytics process with higher accuracy. Thus, CVFS-AEWMRC method improves the FSR of predictive analytics when compared to existing works.

3.2 Adaptive Enhanced Winnow Map Reduce Classification

In CVFS-AEWMRC method, AEWMRC algorithm is designed for performing decision making with enhanced data classification performance with minimal time. AEWMRC algorithm is a machine learning technique. To perform decision making with less time consumption, AEWMRC algorithm is utilized which categorizes the collected and stored data. The AEWMRC is related to the perceptron algorithm. The perceptron algorithm in conventional works utilizes additive weight-update scheme. In contrast to this, AEWMRC applies a multiplicative scheme which helps to perform better data classification process while many dimensions are unrelated (thus it's called winnow). The designed AEWMRC algorithm in CVFS-AEWMRC method is works well while taking a large size of dataset. On the contrary to existing works, AEWMRC algorithm finds a decision hyperplane with help of MapReduce function that used to classify all collected and stored data. AEWMRC algorithm process is shown in below Figure 3.

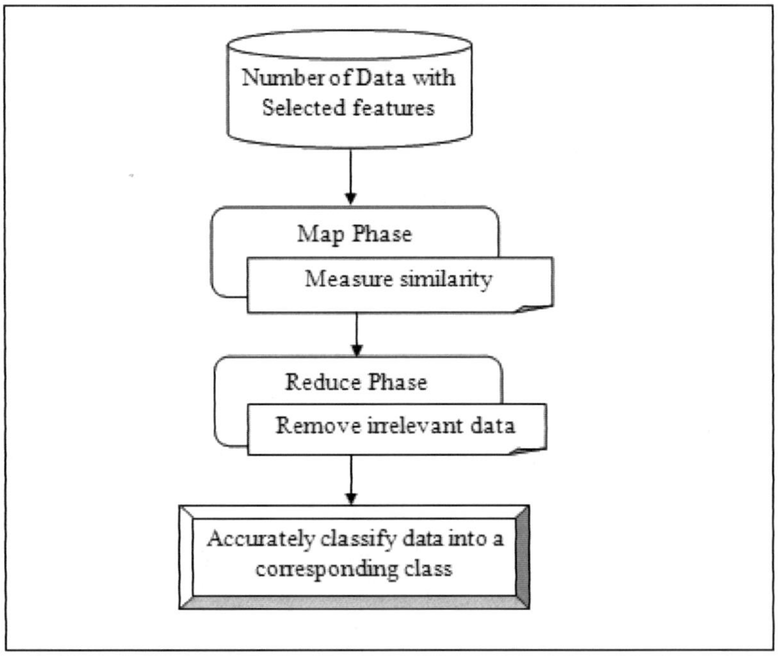

Figure 3 Flow processes of AEWMRC algorithm

Figure 3 shows the overall processes of AEWMRC algorithm. As presented in the above figure, AEWMRC algorithm initially takes a number of collected and stored data '$d_1, d_2, .., d_n$' in warehouse as input where 'n' signifies a total number of

IOP Conf. Series: Materials Science and Engineering 1074 (2021) 012026 doi:10.1088/1757-899X/1074/1/012026

data. The instance space is '$d = \{0,1\}$'X = { 0 , 1 } n {\displaystyle X=\{0,1\}^{n}} i.e. each input data in AEWMRC algorithm is represented as a set of Boolean-valued features. Subsequently AEWMRC give non-negative weights w i {\displaystyle w_{i}} 'w_i' for each data '$i\epsilon\{1,\ldots,n\}$'which are primarily set to '1', one weight for each datai ∈ { 1 , … , n } {\displaystyle i\in \{1,\ldots ,n\}}iiii. Then, AEWMRC algorithm measures the similarity between each data and prediction rule using below,

$$s = \frac{n * \sum d_i * pr_i}{\sum d_i^2 + \sum pr_i^2 - \sum d_1 * pr_i} \quad (3)$$

From equation(3),s denotes the tanimoto similarity coefficient, n denotes the number of data, $\sum d_i * pr_i$ indicates a sum of product of paired scores, $d_i$denote collected data, $\sum d_i^2$represents a sum of squared score of datad_i, $\sum pr_i^2$specifies a sum of squared score of prediction rule 'pr'. Prediction rule includes set of conditions to predict the user's next order. In AEWMRC algorithm, tanimoto similarity value offers output value from 0 to +1. During the mapping process, the AEWMRC algorithm maps all the input data in a big dataset into a corresponding class by using similarity value using below mathematical representation,

$$y^* = \begin{cases} if(w^T s_i > b), then y^* = +1 \\ otherwise y^* = -1 \end{cases} \quad (4)$$

From the above formula (4), 'b' denotes a threshold assigned for similarity value and 'y^*' signifies the classified output of AEWMRC algorithm (predicted output). Together with the weight, the threshold determines a separating hyperplane in the instance space to accurately classify collected and stored data in warehouse. Here, '+1' represents that the product is user's next order whereas '-1'the product is not a user's next order. During reducing phase, AEWMRC algorithm removes the irrelevant data in each class.

For each data classification result, AEWMRC algorithm employs below update rule:

- If an input data is correctly classified, do nothing.
- If an input data is predicted incorrectly and the correct result was '+1', then the AEWMRC algorithm updates each weight multiplied by '2' i.e. '$w_i \leftarrow 2w_i$'only for those data features 'd_i' that are '1'.
- If an input data is predicted incorrectly and the correct result was '-1', then the AEWMRC algorithm updates each weight divided by '2' '$w_i \leftarrow w_i/2$' only for those data features 'd_i' that is '-1'.

With the assist of above processes, the AEWMRC algorithm significantly performs the users in making decision with the collected and stored data.

The algorithmic processes of AEWMRC are presented in below,

// **Adaptive Enhanced Winnow Map Reduce Classification Algorithm**
Input: Number of collected data '$d_i = d_1, d_2, d_3 \dots d_n$'; Actual Output '$y$'; Predicted Output '$y^*$'; weight '$w$';
Output: Achieve higher classification accuracy
Step 1:Begin
Step 2: **For** each collected data 'd_i'
Step 3: Initialize weight $w = 1$
// Map phase
Step 4: Determine similarity between data and prediction rule using (3)
Step 5: maps input data into a corresponding class using (4)
// Reduce phase
Step 6: removes the irrelevant data in each class
Step 7: **End for**
Step 8: **For** each obtained classification result
Step 9: If $(y = +1)$ and $(y^* = -1)$**then**
Step 10: Update each weight $\omega_i \leftarrow 2\omega_i$only for data feature '$d_i$' with '1'
Step 11:Else If $(y = -1)$ and $(y^* = +1)$**then**
Step 12: Update each weight $\omega_i \leftarrow \omega_i/2$only for data feature '$d_i$' with '-1'
Step 13:End If
Step 14: Return accurate data classification result
Step 15: **End For**
Step 16:End

Algorithm 2 Adaptive Enhanced Winnow Map Reduce Classification

Algorithm 2 shows the step by step processes of AEWMRC. By using the above algorithmic steps, CVFS-AEWMRC method accurately classify the data on data warehouse by using selected features with higher accuracy and lower time. Accordingly, CVFS-AEWMRC method enhances the predictive analytics performance to exactly find user's next order than the conventional works.

4 EXPERIMENTAL SETTINGS

To evaluate the performance of CVFS-AEWMRC Method is implemented using MapReduce libraries in Hadoop with assist of Instacart market basket analysis dataset https://www.kaggle.com/c/instacart-market-basket-analysis/data that includes relational set of files describing customers' orders over time. With this data features, CVFS-AEWMRC Method identify the user's next order. This dataset includes 3 mil-

IOP Conf. Series: Materials Science and Engineering 1074 (2021) 012026 doi:10.1088/1757-899X/1074/1/012026

lion grocery samples orders from 200,000 Instacart users. Here, each entity includes of customer, product, order, aisle, etc.

TheCVFS-AEWMRC Method considers number of data features ranges from 5000 to 50000. The performance of CVFS-AEWMRC Method is determined in FSR, CA, CT and FPR. The performance result of CVFS-AEWMRC Method is compared against with two conventional methods [1] and [2]. The experimental evaluation of CVFS-AEWMRC Method is conducted and averagely ten results are depicted in below table and graph.

5 RESULTS

The comparative result of CVFS-AEWMRC Method is discussed and comparison is made with existing [1] and [2]. CVFS-AEWMRC Method is determined with the help of tables and graphs.

5.1 Performance Measure of Feature Selection Rate

In CVFS-AEWMRC Method, 'FSR' determines the amount of time taken for selecting the significant features. The FSR is mathematically obtained using below,

$$FSR = n * time\ (SSF) \quad (5)$$

From equation (5), '$time\ (SSF)$' represents the time consumed for choosing one feature, 'n' denotes a total number of features taken for experimental evaluation. FSR is calculated in milliseconds (ms).

Table 1 demonstrates the results of FSR obtained during the processes of predictive analytics process with three methods namely proposed CVFS-AEWMRC method and conventional Label-Aware Distributed Ensemble Learning (LADEL) [1] and Atrak method [2].

IOP Conf. Series: Materials Science and Engineering 1074 (2021) 012026 doi:10.1088/1757-899X/1074/1/012026

Table 1 Tabulation result of Feature Selection Rate

No. of features	feature selection rate (ms)		
	CVFS-AEWMRC	LADEL	Atrak method
5000	75	95	125
10000	90	100	120
15000	105	117	128
20000	122	136	150
25000	145	173	178
30000	150	177	195
35000	161	186	214
40000	180	200	212
45000	185	203	225
50000	200	225	275

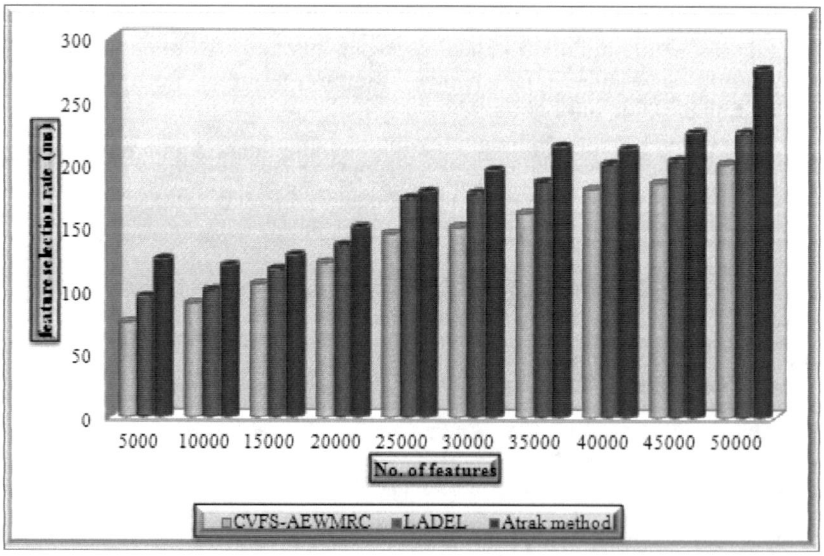

Figure 4 Graphical Result Analysis of Feature Selection Rateversus Number of Features

FSR results of three methods are portrayed in figure 4 with numbers of features ranges from 5000-50000. As depicted in figure 4, CVFS-AEWMRC method achieves minimal FSR in order to efficiently discover the user's next order as compared to existing [1] and [2]. This is due to the utilization of CVFS algorithm in CVFS-AEWMRC method.

During the feature selection process, CVFS-AEWMRC method applies canonical correlation analysis determines the correlation between the two features through estimating covariance matrix. By using this analysis, CVFS-AEWMRC method find outs the features with maximum correlation value as optimal with a minimal amount of time consumption. This supports for CVFS-AEWMRC method to minimize the amount of time required for selecting the significant features from an input dataset when compared to existing Label-Aware Distributed Ensemble Learning (LADEL) [1] and Atrak method [2]. Thus, proposed CVFS-AEWMRC method reduce FSR by 13% and 23% when compared to existing Label-Aware Distributed Ensemble Learning (LADEL) [1] and Atrak method [2] respectively.

5.2 Performance Measure of Classification Accuracy

In CVFS-AEWMRC Method, 'CA' measured as ratio of a number of features which are precisely classified to total number of features. CA is calculated as below,

$$CA = \frac{\delta_{aC}}{n} * 100 \qquad (6)$$

From the above mathematical formula (6), 'δ_{aC}' represents the number of features accurately classified in which 'n' symbolizes a total number of features employed for experimental process. CA of predictive analytics is evaluated in percentage (%).

Table 2 presents the comparative result of CA acquired during the processes of predictive analytics process using three methods namely proposed CVFS-AEWMRC method and traditional Label-Aware Distributed Ensemble Learning (LADEL) [1] and Atrak method [2].

IOP Conf. Series: Materials Science and Engineering 1074 (2021) 012026 doi:10.1088/1757-899X/1074/1/012026

Table 2 Tabulation result of Classification Accuracy

No. of features	Classification Accuracy (%)		
	CVFS-AEWMRC	LADEL	Atrak method
5000	98.5	77.2	74.8
10000	96.75	85.5	84.02
15000	96	82.63	76.13
20000	95.43	86.88	81.01
25000	95.85	88.9	87.96
30000	94.7	90.05	89.153
35000	94.41	90.46	89.923
40000	95.13	88.73	85
45000	94.45	89.05	90.71
50000	94.1	89.52	86.92

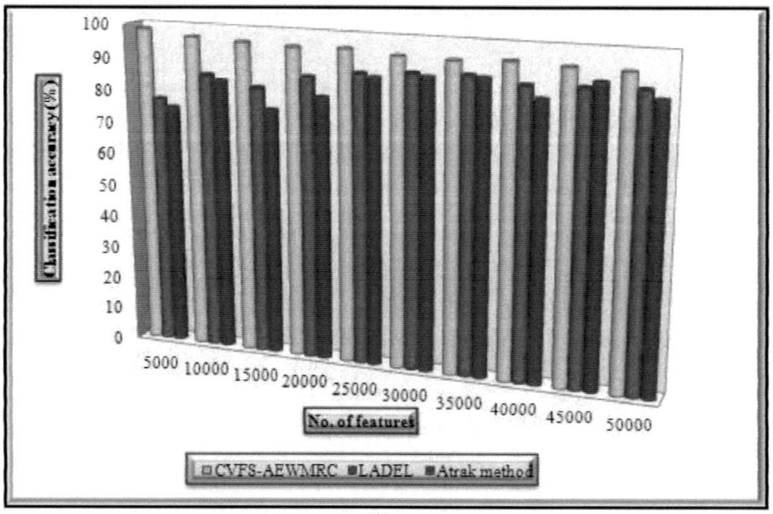

Figure 5 Graphical Result Analysis of Classification accuracyversus Number of Features

CA results of three methods are shown in figure 5 with numbers of features ranges from 5000-50000. As demonstrated in figure 5, CVFS-AEWMRC method gets higher CA in order to predict the user's next order as compared to existing [1] and [2]. This is owing to application of CVFS algorithm and AEWMRC algorithm in proposed CVFS-AEWMRC method. This enhances the ratio of number of features correctly classified as compared to conventional [1] and [2]. Accordingly, proposed CVFS-AEWMRC method increase CA of predictive analytics by 10% and 14% when compared to state-of-the-artLabel-Aware Distributed Ensemble Learning (LADEL) [1] and Atrak method [2] respectively.

5.3 Performance Measure of Classification Time

In CVFS-AEWMRC Method, 'CT' calculates amount of time needed to classify the features for predictive analytics. The CTis mathematically determined using below,

$$CT = n * time\ (CSF) \tag{7}$$

From equation (7), '$time\ (CSF)$' indicates time consumed for classify the single feature, 'n' denotes a total number of features. CT is estimated in milliseconds (ms).

Table 3 depicts the result analysis of CT during the predictive analytics process using three methods namely proposed CVFS-AEWMRC method and state-of-the-artLabel-Aware Distributed Ensemble Learning (LADEL) [1] and Atrak method [2].

Table 3 Tabulation result of classification time

No. of features	Classification Time (ms)		
	CVFS-AEWMRC	LADEL	Atrak method
5000	95	115	150
10000	115	130	160
15000	135	145.5	180
20000	154	196	200
25000	170	197.5	227.5
30000	180	207	255
35000	196	220.5	283.5
40000	220	240	292
45000	229.5	247.5	315
50000	250	275	325

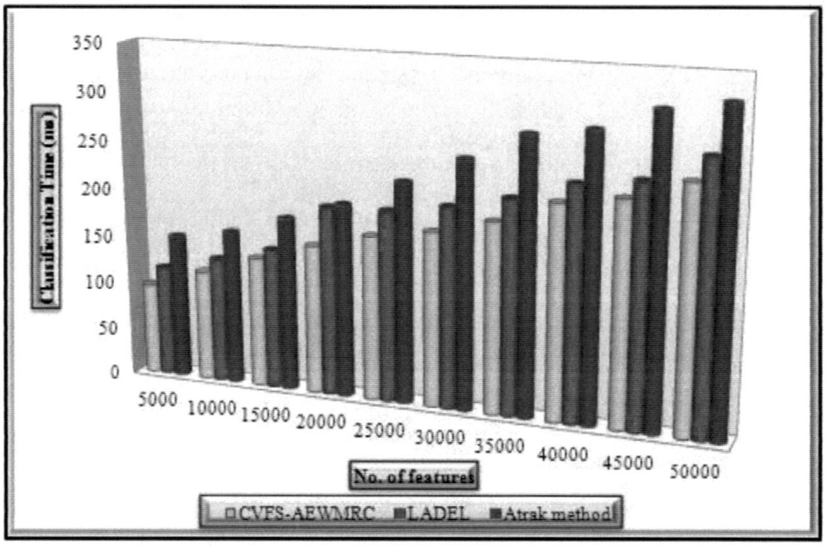

Figure 6 Graphical Result Analysis of Classification timeversus Number of Features

CT results of three methods is illustrated in figure 6 with numbers of features ranges from 5000-50000. As exposed in the above graphical figure, CVFS-AEWMRC method obtains lower CT when compared to existing Label-Aware Distributed Ensemble Learning (LADEL) [1] and Atrak method [2]. This is due to application of CVFS algorithm and AEWMRC algorithm in proposed CVFS-AEWMRC method on the contrary to existing works. This assists for CVFS-AEWMRC method to decrease the amount of time needed to classify the features for predictive analytics when compared to traditional Label-Aware Distributed Ensemble Learning (LADEL) [1] and Atrak method [2]. For that reason, proposed CVFS-AEWMRC method reduces CT of predictive analytics by 12% and 27% as compared to conventional Label-Aware Distributed Ensemble Learning (LADEL) [1] and Atrak method [2] respectively.

5.4 Performance Measure of False Positive Rate

'*FPR*' estimated as the ratio of number of features imperfectly classified to total number of features as input. FPR is measured as follows,

$$FPR = \frac{\delta_{IC}}{n} * 100 \quad (8)$$

From the above mathematical formulation (8), 'δ_{IC}' indicates the number of features wrongly classified in which 'n' signifies a number of features. The FPR of predictive analytics is calculated in percentage (%).

Table 4 illustrates the result of FPR during the processes of predictive analytics using three methods namely proposed CVFS-AEWMRC method and existing Label-Aware Distributed Ensemble Learning (LADEL) [1] and Atrak method [2].

Table 4 Tabulation result of False Positive Rate

No. of features	False Positive Rate (%)		
	CVFS-AEWMRC	LADEL	Atrak method
5000	1.5	22.8	25.2
10000	3.25	14.5	15.98
15000	4	17.37	23.87
20000	4.58	13.12	18.99
25000	4.158	11.1	12.04
30000	5.3	9.95	10.85
35000	5.59	9.54	10.08
40000	4.87	11.27	14.99
45000	5.55	10.95	9.29
50000	5.9	10.48	13.08

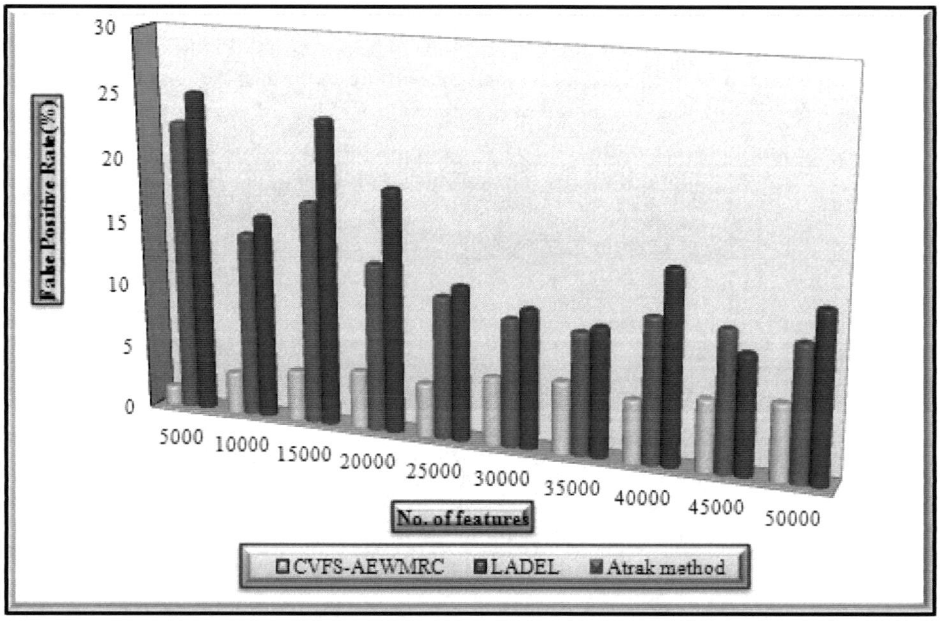

Figure 7 Graphical Result Analysis of false positive rateversus Number of Features

FPR results of three methods is depicted in figure 7 with numbers of features ranges from 5000-50000. As presented in the above graphical illustration, CVFS-AEWMRC method achieves minimal FPR when compared to traditional Label-Aware Distributed Ensemble Learning (LADEL) [1] and Atrak method [2]. This is due to the utilization of CVFS algorithm and AEWMRC algorithm in CVFS-AEWMRC method. This aids for CVFS-AEWMRC method to diminish the ratio of a number of features poorly classified for predictive analytics when compared to traditional Label-Aware Distributed Ensemble Learning (LADEL) [1] and Atrak method [2]. Thus, proposed CVFS-AEWMRC method minimize FPR by 61% and 66% as compared to existing [1] and [2].

6 CONCLUSION

The CVFS-AEWMRC method is proposed with the goal of enhancing the decision-making performance for effective predictive analytics through classifying the collected and stored data on data warehouse. The goal of CVFS-AEWMRC method is attained with the support of CVFS and AEWMRC algorithms. The proposed CVFS-AEWMRC method increases the number of features correctly classified with the assist of AEWMRC algorithm than the conventional works. Also, CVFS-AEWMRC methodminimize the amount of time consumed for choosing the significant features from an input dataset by using the CVFS and AEWMRC algorithmic steps. Besides, CVFS-AEWMRC method lessens the amount of time needed to classify the features for predictive analytics as compared to conventional works. Moreover, CVFS-AEWMRC method lessen the number of features inaccurately classified for predicting the user's next order. The experimental result shows that the proposed CVFS-AEWMRC method presents better predictive analytics performance with higher CA and minimum FSR as compared to conventional works.

References

1. ShadiKhalifa, PatrickMartin, RebeccaYoung, “Label-Aware Distributed Ensemble Learning: A Simplified Distributed Classifier Training Method for Big Data”, Big Data Research, Elsevier, Volume 15, Pages 1-11, March 2019
2. MohammadhosseinBarkhordari, Mahdi Niamanesh, “Atrak: a MapReduce-based data warehouse for big data”, The Journal of Supercomputing, Springer, Volume 73, Issue 10, Pages 4596–4610, October 2017
3. Kamran Mahroof, “A human-centric perspective exploring the readiness towards smart warehousing: The case of a large retail distribution warehouse”, International Journal of Information Management, Elsevier, Volume 45, Pages 176–190, 2019
4. MohammadhosseinBarkhordari and Mahdi Niamanesh, “Chabok: a Map-Reduce based method to solve data warehouse problems”, Journal of Big Data, Springer, Volume 5, Issue 40, Pages 1-25, 2018
5. Marouane Birjali, Abderrahim Beni-Hssane and Mohammed Erritali, “Evaluation of high-level query languages based on MapReduce in Big Data”, Journal of Big Data, Volume 5, Issue 36, Pages 1-21, 2018
6. Lucile Sautot, Sandro Bimonte, LudovicJournaux, “A semi-automatic design methodology for (Big) Data Warehouse transforming facts into dimensions”, IEEE Transactions on Knowledge and Data Engineering, June 2019
7. Daniel R. Harris, Darren W. Henderson, RamakanthKavuluru, Arnold J. Stromberg, Todd R. Johnson, “Using Common Table Expressions to Build a Scalable Boolean Query Generator for Clinical Data Warehouses”, IEEE Journal of Biomedical and Health Informatics, Volume 18, Issue 5, Pages 1607 – 1613, 2014
8. Chih-Hung Chang, Fuu-Cheng Jiang, Chao-Tung Yang, Sheng-Cang Chou, “On construction of a big data warehouse accessing platform for campus power usages”, Journal of Parallel and Distributed Computing, Elsevier, Volume 133, Pages 40-50, November 2019
9. Leo WillyantoSantoso, Yulia, “Data Warehouse with Big Data Technology for Higher Education”, Procedia Computer Science, Elsevier, Volume 124, Pages 93–99, 2017
10. LadjelBellatreche, Sharma Chakravarthy, “A special issue in extending data warehouses to big data analytics”, Distributed and Parallel Databases, Springer, Volume 37, Issue 3, Pages 323–327, September 2019
11. FerdaousJenhani, Mohamed SalahGouider, LamjedBenSaid, “Streaming Social Media Data Analysis for Events Extraction and Warehousing using Hadoop and Storm: Drug Abuse Case Study”, Procedia Computer Science, Elsevier, Volume 159, Pages 1459-1467, 2019

12. Hossein Ahmadvand, MaziarGoudarzi, "Using Data Variety for Efficient Progressive Big Data Processing in Warehouse-Scale Computers", IEEE Computer Architecture Letters, Volume 16, Issue 2, Pages 166 – 169, July 2017
13. Hossein Azgomi, Mohammad KarimSohrabi, "A game theory based framework for materialized view selection in data warehouses", Engineering Applications of Artificial Intelligence, Elsevier, Volume 71, Pages 125-137, May 2018
14. Manel Zekri, SahbiZahaf, SadokBenYahia, "Specification of the data warehouse for the decision-making dimension of the Bid Process Information System", Procedia Computer Science, Elsevier, Volume 159, Pages 756-763, 2019
15. FajarCiputraDaeng Bani, Suharjito, Diana, Abba SugandaGirsang, "Implementation of Database Massively Parallel Processing System to Build Scalability on Process Data Warehouse", Procedia Computer Science, Elsevier, Volume 135, Pages 68-79,2018
16. Khouri Selma,BoukhariIlyès, BellatrecheLadjel, SardetEric,Jean Stéphane, Baron Mihael, "Ontology-based structured web data warehouses for sustainable interoperability: requirement methoding, design methodology and tool", Computers in Industry, Elsevier, Volume 63, Issue 8, Pages 799-812, October 2012
17. ZoubirOuaret, DoulkifliBoukra,OmarBoussaid, Rachid Chalal, "AuMixDw: Towards an automated hybrid approach for building XML data warehouses", Data & Knowledge Engineering, Elsevier, Volume 120, Pages 60-82, March 2019
18. P.Amuthabala, R.Santhosh, "Robust analysis and optimization of a novel efficient quality assurance method in data warehousing", Computers & Electrical Engineering, Elsevier, Volume 74, Pages 233-244, March 2019
19. P.Saravanakumar, M.Athigopal, S.Vetrivel, "Extract Transform and Load Strategy for nstructured Data into Data Warehouse Using Map Reduce Paradigm and Big Data Analytics", International Journal of Innovative Research in Computer and Communication Engineering, Volume 2, Issue 12, Pages 7456-7462, December 2014
20. MohammadhosseinBarkhordari and Mahdi Niamanesh, "Aras: A Method with Uniform Distributed Dataset to Solve Data Warehouse Problems for Big Data", International Journal of Distributed Systems and Technologies (IJDST), Volume 8, Issue 2, Pages 1-14, 2017
21. V.Kavitha, "Clustering Process with Time Series Data Stream", Smart Innovations, Systems and Technologies, 2020.
22. R.Sivakkolundu, V.Kavitha, "Tobit Regressive Based Gaussian Independence Bayes Map Reduce Classifier on Data Warehouse for Predictive Analytics", International Journal of Innovative Technology and Exploring Engineering (IJITEE), ISSN: 2278 – 3075, October 2019, Scopus Indexed.
23. V.Kavitha,"Pre-processed Hierarchical Clustering for Time Series Data Streams", International Journal of Recent Technology and Engineering (IJRTE), Volume-8 Issue-3, September 2019, Pg.251-254, ISSN: 2277-3878

Keyphrase Extraction And Source Code Similarity Detection-A Survey

Nakul Sharma

Department of Computer Science and Engineering

Koneru Lakshmaiah Education Foundation

Vaddeswaram,Guntur-522502, India

nakul777@gmail.com

Prasanth Yalla

Department of Computer Science and Engineering

Koneru Lakshmaiah Education Foundation

Vaddeswaram,Guntur-522502, India

prasanthyalla@kluniversity.in

Abstract. Keyphrase extraction is the starting phase of feature extraction and many other NLP tasks. Hence, keyphrase extraction ensures the separation of essential and relevant information from the rest of the document or corpus. A similarity score is used for detecting similarities between two or more documents. In this paper, a summary of existing research conducted in the field of keyphrase extraction and source code similarity detection is presented. The literature focuses on more concept-based research instead of going towards the domain level. There exist a promising potential for the creation of recommendation systems by employing similarly situated research techniques.

Keywords: Software Engineering, Keyphrase Extraction, Text Mining, Similarity calculation, Source Code Analysis

1. Introduction

Keyphrase Extraction (KE) involves getting the essential and relevant words from a given document or a text [1]. It forms an important component of Text Mining which involves various tasks on the text at hand. The field of text mining also has linkage to Natural Language Processing/Computational Linguistics (CL) which entails the usage of text / speech in completing various tasks. The field of KE finds close application with Information Retrieval, Information Storage, and other such related fields.

Keyphrase or words constitute different types of both print and electronic medium. Whenever there is a presence of text in the form of words, sentences, paragraphs, the role of keyword extraction can come into play. As a result of this, Keyphrase extraction is extensively used in following [22] [30]:-

- Text Classification
- Text Clustering
- Document summarization
- Web mining
- Concept Extraction
- NLP/CL

- Information Retrieval
- Feature Extraction

Source Code Analysis (SCA) forms the basis for program understanding and comprehension. The developers as well the system maintainers are tasked with understanding the source code of the created projects owing to following reasons:

- Reassigned to a new location
- Resource crunch
- New projects assigned to the team.

SCA can also aid in the development of a recommendation system for the developer all its allied roles. These include testers, maintenance engineers in developing a better understanding of the software product. Similarity detection finds its application in source code analysis for detecting duplicate codes or plagiarism. Figure-1 gives different types of existing source code , based on similarity detection techniques [44].

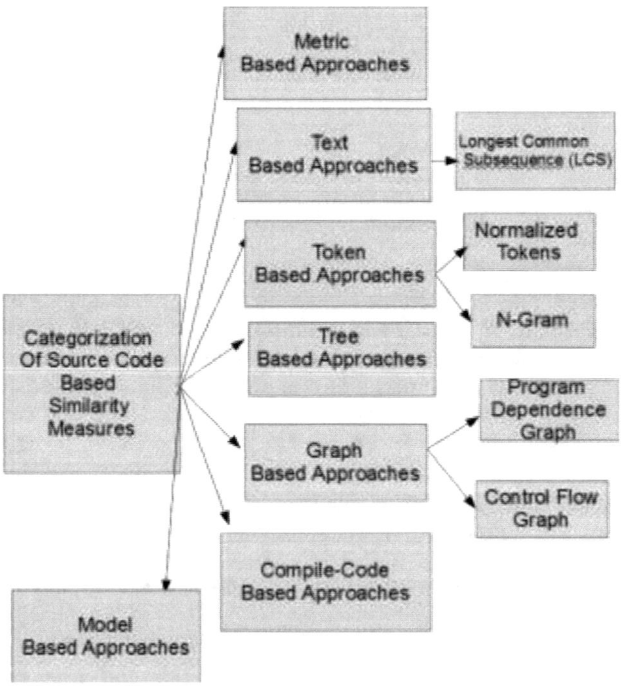

Figure-1 Source Code based Similarity Measures [44].

Token-based approaches are further categorized into single or N-gram according to the size of tokens [8].

Tree-based approaches make use of Abstract Syntax Tree which is part of the model construction phase of SCA. AST helps in detecting location-specific codes within two files. Semantic-based Graph source code similarity measures ignore the programming language constructs. Graph-based approaches can manage minor changes to the source code but have higher complexity as compared to other approaches. Graph-based approaches are further classified as Program-Dependence-Graph and Control-Flow-Graphs. Compiled code-based approaches don't require source code for computing similarity. Model-based approaches detect duplicity in models [8].

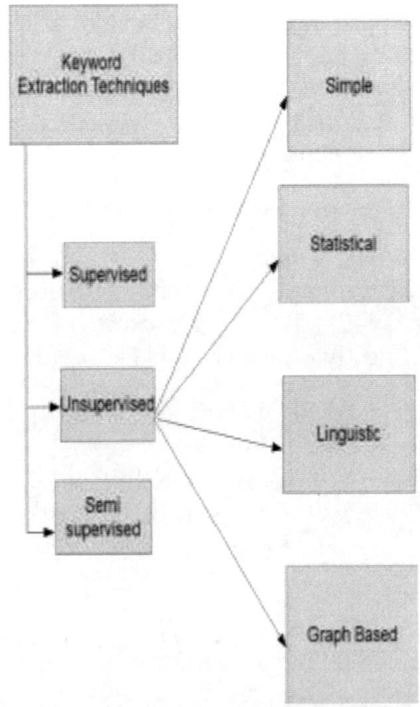

Figure 2: KE Extraction Techniques [2]

Keyphrase extraction consists of supervised, unsupervised and semi-supervised research techniques.

Keyphrase Extraction Models

There are following types of keyphrase extraction models which are present in the literature:

1. TF-IDF
2. Kea
3. TopicRank
4. SingleRank
5. PageRank
6. TextRank
7. ExpandRank [16] [21]

2. Literature Review on Keyphrase Extraction & Similarity-Based Source Code Analysis

2.1. Literature Review of Survey Papers

In the course of undertaking literature survey of the survey papers, following papers came into limelight [1][2][3][4][5][6][17]:-

"An Overview of Graph-Based Keyword Extraction Methods and Approaches" [2]

The authors, in this work, provide an introduction to Keyphrase Extraction basics. The authors provide a classification of KE methods research. This is followed by the categorization of keyphrase extraction techniques. The paper provides a strong mathematical base for graph-based keyword extraction research. The literature review section of the paper provides a paragraph explanation focusing mainly graph on KE techniques [2]. The paper also discusses categories of literature reviews discussed along with the different categories of the KE techniques.

"Automatic keyphrase extraction: a survey and trends" [17]

The paper by Zakariae et. al. provides a major review on the current existing state-of-art research on Automatic Keyphrase Extraction (APKE). The paper also provides different future research directions along with future trends for the prospective researchers [17].

2.2. Literature Review of Keyphrase Extraction papers

Wei et. al. mention employing knowledge graphs for providing semantic relations for keyphrase extracted in the KE task. The authors also provide a framework for combining clustering and graph ranking for the same task of KE. As a result, the author gets better results when compared to traditional approaches [8].

Santhi et. al. propose developing information by using relationships between elements in a graph [9].

Dario et. al. develop an automatic curator system for the purpose of scholarly data access tools. In order to accomplish this, various tools and methods have been employed from the varied fields of CL, cloud, user modeling, and information retrieval [11].

Sujatha et. al. use the following techniques in keyphrase taggers in research papers:-

1. linguistics
2. surface form
3. document-structure

The created tagger is trained with conditional random fields which gives results in consonance with the existing work. In addition, by implementing the feature extraction technique, the system performance increased [12].

Nesi et. al. present a distributed framework for the purpose of running NLP computations and tasks. GATE API is used and build upon the Hadoop platform. The evaluation includes using the real-world corpus of web pages. The framework proposed is distributed as well as parallel environments [13].

Teng-li et al. make use of an unsupervised approach for conducting graph-based ranking and clustering within-collection resources. The graph-based ranking describes significance in-between two words. The topic-based clustering introduces additional linguistic information in words. The clusters assign the right meaning to the appropriate word [18].

Guoxiu He et.al. discuss a keyphrase extraction technique that employs prior knowledge for the extraction of keyphrase from the document. Authors develop a controlled vocabulary using existing document collections in the keyphrase collection phase. Prior probability is used to highlight the need of using TF-IDF and Text Rank jointly. The optimal weights are learned using the machine learning's supervised algorithm Experimental results convincingly show the benefit of employing prior probability for KE [20].

Niraj et. al. propose N-gram filtration method in KE. This is accomplished by using the statistical features and also the co-location strength. Statistical features imply using the centrality scores of words for checking the border nodes. In addition, the N- gram graph is also generated [21].

IOP Conf. Series: Materials Science and Engineering 1074 (2021) 012027 doi:10.1088/1757-899X/1074/1/012027

Dhruva et.al. created a tool named BiLSTM-CRF. Keyphrase extraction is done as a labelling task. This represents input text in form of contextualized embedding. For evaluation contextualized ad fixed word embedding models are used. For comparison, the authors make use of supervised & unsupervised approaches [23].

Santosh et. al. have developed a tool named DAKE. This tool conducts keyphrase extraction at the document level by using BLSTM [24].

Tianlei et. al. propose developing keyphrases by classifying the candidate keyphrases in different categories using graph-based attributes and non-graph-based attributes [25].

Ankit et al address two major issues concerning keyphrase extraction. These include candidate selection for the keyphrase extraction and extraction of necessary features. The authors use a supervised ML algorithm for keyword extraction. For evaluation purpose, SingleRank, TFIDF, Expand Rank is used [26].

Corina et. al. propose a new scoring measure for phrases for calculating the final scores by averaging out the individual words weighted by the frequency of phrases [28].

Keyphrase or keyword extraction or searching also finds its relevance in spatial objects databases and software testing [11] [29]. This is mainly due to the textual nature of the artifacts produced.

SKEM methodology proposed by authors generates graphs based on the semantics of keywords by making use of different centrality measures. The source for keywords extraction is the twitter data. The top ten words are chosen based on ranking and nodes score calculations [31].

Linkal et. al. have devised an unsupervised algorithm for keyphrase-extraction. Authors proposed making use of structural and the semantic information of the document in the process of keyphrase extraction. After creating the graph consisting of potential keywords, the weights are calculated using the relative distance between candidate keyphrases [32].

Mauro et. al. developed Crank approach which uses an unsupervised approach to extract keyphrases from documents. The heuristics and centrality measures are used in assigning weights to the graph [33].

Gollam et. al. proposes TeKET technique which uses lesser statistical knowledge and needs no prior training. The authors also propose a Keyphrase based tree for final candidate selection and an index (CI) for measuring nodes cohesion wrt root [34].

Zheng et. al. develop knowledge graphs through which key phrases can be extracted. The technique finds the latent relationships between two key-terms using knowledge graphs [35].

Swagata et.al. develop a framework for extracting keywords. The complex network developed is used for
extracting features. Node properties are sued to differentiate between keywords ad non-keywords [38].

D. Sowmya et al. use swarm optimization techniques in keyphrase extraction. The process followed includes keyword extraction, clustering of the keyphrases. The process uses intervention at every stage [40].

In the current section, a literature review based on UML diagrams generation, keyphrase extraction, and clustering are discussed. Table-1 gives a comparative analysis of research papers related to UML diagram construction.

Table 1. Literature Review of existing work on Keyphrase Extraction

Reference number	Title of Paper	Advantage of Proposed Work	Disadvantages of Proposed Work
8	A New Multi-lingual Knowledge-base Approach to Keyphrase Extraction for the Italian Language	Authors propose creating a multi-lingual tool for extracting keyphrases of English and Italian language	Authors did not consider keyphrase extraction for different languages. Keyphrase generation can be extended for other languages also.
10	Knowledge-Based Techniques for Scholarly Data Access: Towards Automatic Curation	The authors propose a system that is useful for researchers by using data mining, NLP, clustering, Graph-based approaches.	This is a major project & needs multiple domain level experts for its proper execution
12	A Distributed Framework for NLP-Based Keyword and Keyphrase Extraction From Web Pages and Documents	The authors use keyphrase-extraction in cloud based systems. GATE architecture is employed on cloud. The web crawler is also integrated into the system	The more testing of GATE based applications can be done. Hadoop implemetation can also be extended
13	Building a Construction Project Key-Phrase Network from Unstructured Text Documents	For unstructured text, authors propose document management system, project keyphrase network, support libraries & system for civil engineering projects	The current work can be extended for contextual awareness, interactive interface and controlled vocabulary
14	Identifying Candidate Tasks for Robotic Process Automation in Textual Process Descriptions	Authors identify& classify tasks using BPMN for textual processes	In practice textual descriptions are not the same as provided in the paper.
15	Simple Unsupervised Keyphrase Extraction using Sentence Embeddings	Authors make use of EmbedRank, to extract keyphrases from single document. The unsupervised ML algorithm is used for the process. Sentence embedding is employed in the process.	Unsupervised approach may introduce biasing due to inherit noise in the document.
16	How Document Pre-processing affects Keyphrase Extraction Performance	Authors reassess the performance of five keyphrase extraction models and conclude that document pre-processing affects performance.	More keyphrase extraction models such as ExpandRank, SingleRank can be used for assessment.
18	An Unsupervised Approach for Keyphrase Extraction Using Within-Collection Resources	Authors use graph-based ranking for checking relevance across the words. The topic-based cluster for the purpose of semantic enrichment.	The work is restricted to only scholarly documents
20	Keyphrase Extraction Based on Prior Knowledge	Authors make use of controlled vocabulary, prior-probability & supervised ML for Keyphrase Extraction	None mentioned

21	A graph-based unsupervised N-gram filtration technique for automatic keyphrase extraction	Authors used ngram approach for candidate selection and ranking of candidate keyphrases.	As resources become multilingual, it is possible to extend the authors work.
22	Feature selection, optimization and clustering strategies of text documents	Authors provided a review of clustering and its allied research areas including similarity measures. Authors conclude text clustering research focussed on dynamic & Heterogeneous clustering applications	Community-driven clustering could not be addressed in the given literature.
23	Keyphrase extraction as sequence labeling using contextualized embeddings	Authors make use of BiLSTM-CRF ML technique architecture for keyphrase extraction and including embedding the context.	The authors work can be extended to keyphrase generation
24	DAKE: Document-Level Attention for Keyphrase Extraction	Authors makes use bi-directional LSTM which extracts keyphrases from documents.	The source documents are not multilingual
25	A Graph-Based Keyphrase Extraction Model with Three-Way Decision	Authors propose a three way categorization of candidate phrases before undertaking graph-based clustering	None mentioned

2.3. Literature Review of Research papers Related to Source Code Analysis Based Similarity Index

Recent research in applying Deep Learning to Source Code Analysis deals with using program vector representation. The authors conclude that Deep Learning cannot be applied directly to Back Propagation Networks. Hence, alternatively, code criterion forms the basis of building program representations. The authors also check the studying programming by deep NN. The methodology entails defining granularity at the character, token, AST, statement levels [45].

Plagiarism in source code is a major problem. Hence, its detection provides several research directions. TF-IDF is traditional metrics employed in the field of text mining. Author's hence proposed applying this measure to source-code files. The methodology divides the program into language dependent tokenizes and language-specific indexes [46].

Code similarity calculation finds its application in online architecture and comparison. Some of the source code similarities and general similarities are discussed. The methodology is divided into program collection, candidate selection, plagiarism detection, and in-depth discussion [47].

Binary code similarity identifies similar or identical code segments in binary programs. The authors state that the semantic-based code has varied issues on source code. Hence, the authors propose a hybrid-method to compare binary similarity tools named BinMatch. This tool captures the semantic aspects of binary function both dynamically and statistically. Binmatch's proposed methodology includes test cases, target function, and similarity computation[48].

In academics, similarity calculation has been applied to undertake evaluation and ranking of students. Authors propose the creation of examples using intelligent hypertext. The ontology is employed in the extraction of necessary concepts from examples using similarity calculation. In order to do content enrichment, structural &non-structural approaches are proposed to be used [49].

The lexical or tokens based similarity for been proposed by the authors for short sentences to undertake granularity evaluation. Authors employ on the string of code, Levenstein distance, cosine similarity, hamming distance, ASCII-based hashing, and Round-Karp rolling. Rabin Karp performs better than all others for different repositories size [50].

SCA also has been used to aid the identification of class-errors and to create a recommendation system for giving feedback to users. Similar codes provide assessment criteria to measure common properties. The authors propose code plagiarism detection as the first step & cluster creation as the second step in the methodology [51].

SCA also finds its applicability in clone detection for large scale online repositories. Authors propose a technique called OCD which serves the purpose of code similarity as well as code clone searching [52].

The author presents code-clone detection as an application of SCA. Two different metrics introduced compare program features and control statement. The author's work can be summarized as project code metrics calculation and using these metrics for similarity calculation [53].

SCA has also been in the generation of comments for the existing codes. The tool developed named CloCom uses detection techniques to discover code segments and use their comments in describing codes having a similar theme [54].

Song et. al. compute the similarity between two source codes. Convolution Kernel-function is employed for the ultimate aim of plagiarism detection among a pair of codes [55].

Ivan et. al. use string complexity and compression in source code plagiarism detection. Kolmogorov complexity is employed for computing similarity measures. The authors provide a comparative analysis of the calculated similarity with synthetic benchmarks [56].

Traceability between SDLC artifacts forms a major concern for the IT industry. Hence, Amir et. al. present a language-independent link for connecting source code to requirement documents. The methodology employed entails feature extraction from documents and source code, completion of abbreviation through similarity measure, and using data mining technique to find linkages between documents and source code [57].

In the area of code smell detection, centrality measures are often employed. These include measures such as Euclidean distance being used with Genetic-Algorithm & particle swarm optimization. The code smells are detected by computing the similarity between the systems. The approach detects BLOB, Functional Dependencies, spaghetti code, Data Class, and feature envy. The authors also categorize similarity detection literature as search-based non-search based and model-based code smell detection techniques [58].

Firas et. al. present an approach for detecting similar code snippets based on graph fingerprinting. The methodology proposed first creates graphs from source code. The generated graphs are used for creating unique fingerprinting of the code [59].

Oscar et. al. summarize different source code-based similarity detection techniques research papers. Authors categorize the submitted source code similarity detection technique into structural, non-structural, and hybrid approaches [60].

Zhongehen et. al. develop a structural similarity measure for the representation of UML Class Graph. Similarity calculated is both intra-structural as well as inter-structural. These measures are claimed to be useful for software reuse [61].

Table 2 Advantage and disadvantage of Research Papers on Source Code Analysis Based on Similarity Detection & Code Smell

Reference Number	Title of Paper	Advantage Research of Paper	Disadvantage/Future Scope of Research Paper
45	Building Program Vector Representations for Deep Learning	Author applied vector representations to code analysis.	DL can be applied to complete domain of SCA.
46	TF-IDF Inspired Detection for Cross Language Source Code Plagiarism and Collusion	The author applied vector representations to code analysis. DL can be applied to complete domain of SCA	Identifier renaming disguise to be done based on first occurrence.
47	A Scalable Code Similarity Detection with Online Architecture and Focused Comparison for	Cosine similarity introduced by the authors reduces lead to short running time especially with large token strings.	The techniques proposed can be proposed for some other programming courses as well.

	Maintaining Academic Integrity in Programming		
48	BinMatch: A Semantics-Based Hybrid Approach on Binary Code Clone Analysis	A hybrid method for comparing binary similarity. Evaluation of methodology done on nine real-world projects.	Methodology can be extended for 64-bit s1ystem as well.
49	A study of concept-based similarity approaches for recommending program examples	RS system developed aids in the understanding of the student.	RS system based on other ideas can be developed
50	Assessing lexical similarity between short sentences of source code based on granularity	Differed measures of source code compared and Rabin– Karp rolling hashing is suggested as the best.	Rabin–Karp rolling hashing can be applied to open source software's as well.
51	Automatic programming error class identification with code plagiarism-based clustering	Authors provide a clique of offline analysis error detection and on-the-fly learners feedback.	Different datasets can be used for experimentation can be used such as code-hunt data sets.
52	Code similarity and clone search in large-scale source code data	The thesis discusses the technique for a dual methodology of calculating code similarity and providing clone searching on large code data. OCD and Siamese are tools developed.	Clone direction detection can be automated. New services in cloud-based environments can be given to developer.
53	Code similarity detection through control statement and program features	The proposed methodology is evaluated on C projects of Bellon's benchmark dataset and student lab programs (SLP)	Control Statement can be included in the proposed methodology.
54	CloCom: Mining existing source code for automatic comment generation	Automation in source code comments is achieved.	Yield & Accuracy can be improved by the proposed methodology.
55	Computation of program source code similarity by composition of parse tree and call graph	For Java-based source code plagiarism detection, the proposed method performs better than existing systems. The proposed work can be used for C, C++, Python.	Centrality measures can be included in this work.
56	Detecting Source Code Similarity Using Compression.	Similarity detection is done based on Kolmogorov complexity	Consideration of base code forms the part of the future assignment.

The literature review indicated that for Keyphrase extraction for various natural languages has been conducted. These include English, Arabic, Persian languages [39].

3. Evaluation of System Developed

There are following metrics which are used in keyphrase extraction research:-

1. Preference measure (bpref)
2. Mean Reciprocal Rank (MRR)
3. Precision
4. Recall
5. F-measure [4]

Suitable datasets were also chosen according to the domain of work for evaluation purposes. Keyphrase extraction can also be applied to various phases of SDLC and its artifacts. These include SRS, SDS, Testing artifacts, source code, and maintenance phases [29]. Keyphrase extraction also finds its applicability in clustering approaches and plagiarism detection and similarity calculation [37] [41]. Software Engineering has been applied at the subject level in various domains [43]. Keyphrase Extraction can be applied however at the topic level.

4. Recommendation System

Table 3 Keyphrase-based RS versus Source Code Based RS

Keyphrase-Based RS	Source Code-Based RS
Part of a Larger System including Source Code- based RS	Specific to Source Code
Addressed for larger audience	Addressed for specific audience
Domain-level concepts can be addressed	Domain-level concepts can be addressed

Recommendation Systems (RS) are useful to various stakeholders across SDLC. The systems can aid in quick decision making by providing insights to the changes being made to system at large. Table-IV shows the comparison of keyphrase-Based RS and source code-based RS. The artifacts of SDLC are developed by different stakeholders [62]. RS system can be developed for addressing the common needs of the stakeholders.

5. Conclusion

In the current work, an effort is made to summarize the existing state-of-art work in the field of Keyphrase Extraction and similarity measure based source code analysis. Keyphrase extraction is more related to the pre-processing stage of Source Code Analysis. Similarity measures denote a statistical term of measurement of any file based parameter. The future scope presented for the research papers can be used for conducting future research.

References

[1] A W. Shi, W. Zheng, J. Xu Yu, H. Cheng,L. Zou, "Keyphrase Extraction Using Knowledge Graphs", Data Sci. Eng. (2017) 2:275–288. https://doi.org/10.1007/s41019-017-0055-z.

[2] S. Beliga, A. Meštrovic, S. Martincic-Ipšic, "An Overview of Graph-Based Keyword Extraction Methods and Approaches", Journal of Information and Organizational Sciences, VOL. 39, NO. 1 (2015), PP. 1-20.

[3] F. Bulgarov, C. Caragea, "A Comparison of Supervised Keyphrase Extraction Models", WWW 2015 Companion, May 18–22, 2015, Florence, Italy, ACM 978-1-4503-3473-0/15/05.http://dx.doi.org/10.1145/2740908.2742776.

[4] K. Saidul Hasan, V. Ng, "Automatic Keyphrase Extraction: A Survey of the State of the Art", Proceedings of the 52nd Annual Meeting of the Association for Computational Linguistics, Baltimore, Maryland, USA, June 23-25 2014, pages 1262–1273. http://acl2014.org/acl2014/P14-1/pdf/P14-1119.pdf

[5] T. Gupta, "Keyword Extraction: A Review", International Journal of Engineering Applied Sciences and Technology, 2017. Vol. 2, Issue 4, ISSN No. 2455-2143, Pages 215-220.

[6] S. Siddiqi, Aditi Sharan, "Keyword and Keyphrase Extraction Techniques: A Literature Review", International Journal of Computer Applications, Volume 109 – No. 2, January 2015. ISBN: 0975 – 8887.

[7] Jose Mary, G., & Haritha, D. (2017). A survey on best keyword cover search. Journal of Advanced Research in Dynamical and Control Systems, 9(Special issue 14), 2217-2231. Retrieved from www.scopus.com

[8] D. Innocenti, D. Nart, C. Tasso, "A New Multi-lingual Knowledge-base Approach to Keyphrase

Extraction for the Italian Language", In Proceedings of the International Conference on Knowledge Discovery and Information Retrieval - Volume 1: KDIR, 78-85, 2014, Rome, Italy, ISBN: 978-989-758-048-2.

[9] V. B. T. Santhi, Santhi, M. V. B. T., Sagar Imambi, S., Yamini Devi, J., & Tejaswani, Design and development of information using keyword-element relationship graph - a critical study. International Journal of Engineering & Technology, [S.l.], v. 7, n. 2.7, p. 1020-1024, mar. 2018. ISSN 2227-524X. Available at: <https://www.sciencepubco.com/index.php/ijet/article/view/12211>. doi:http://dx.doi.org/10.14419/ijet.v7i2.7.12211.

[10] D. Nart, C. Tasso, "Knowledge-Based Techniques for Scholarly Data Access: Towards Automatic Curation.", Ph.D. Thesis, Universita degli Studi di Udine, Italy, 2016.

[11] S. Gollapalli, X. Li, P. Yang, "Incorporating Expert Knowledge into Keyphrase Extraction", Proceedings of the Thirty-First AAAI Conference on Artificial Intelligence (AAAI-17), pg.3180-3187,2017.

[12] P. Nesi, G. Pantaleo and G. Sanesi, "A Distributed Framework for NLP-Based Keyword and Keyphrase Extraction From Web Pages and Documents", DOI reference number: 10.18293/DMS2015-024 , URL: http://www.disit.org/axmedis/791/00000-7915eb81-51cc-4285-bf37-5a483323ba91/2/~saved-on-db-7915eb81-51cc-4285-bf37-5a483323ba91.pdf

[13] N. jkovic, Đ., M. Kovacevic, "Building a Construction Project Key-Phrase Network from Unstructured Text Documents.", Journal of Computing in Civil Engineering 31, no. 6 (2017): 04017058.

[14] H. Henrik , H. Aa, Hajo A. Reijers,"Identifying Candidate Tasks for Robotic Process Automation in Textual Process Descriptions." ,In Enterprise, Business-Process and Information Systems Modeling, pp. 67-81, Springer, Cham, 2018.

[15] K. Smires, C Musat, A Hossmann, M Baeriswyl, M Jaggi, "Simple Unsupervised Keyphrase Extraction using Sentence Embeddings", Proceedings of the 22nd Conference on Computational Natural Language Learning (CoNLL 2018), pages 221–229. Brussels, Belgium, October 31 - November 1, 2018. 2018 Association for Computational Linguistics.

[16] B Florian, H Mougard, and D Cram. "How document pre-processing affects keyphrase extraction performance." arXiv preprint arXiv:1610.07809 (2016).

[17] Alami Merrouni, Z., Frikh, B. & Ouhbi, B. Automatic keyphrase extraction: a survey and trends. J Intell Inf Syst 54, 391–424 (2020). https://doi.org/10.1007/s10844-019-00558-9

[18] Li, Teng-Fei, et al. "An Unsupervised Approach for Keyphrase Extraction Using Within-Collection Resources." IEEE Access 7 (2019): 126088-126097

[19] El BazzI, Mohamed Salim, et al. "A graph-based ranking model for automatic keyphrases extraction from Arabic documents." Industrial conference on data mining. Springer, Cham, 2017.

[20] G. He, J. Fang, H. Cui, C. Wu, W. Lu, "Keyphrase Extraction Based on Prior Knowledge", JCDL'18, June 3-7, 2018, Fort Worth, TX, USA, 341, ACM ISBN 978-1-4503-5178-2/18/06. https://doi.org/10.1145/3197026.3203869

[21] N. Kumar, K. Srinathan, V. Varma, "A graph-based unsupervised N-gram filtration technique for automatic keyphrase extraction", Int. J. Data Mining, Modeling and Management, Inderscience Journals Vol. 8, No. 2, 124-143, 2016.

[22] Kousar Nikhath, A., & Subrahmanyam, K. (2019). Feature selection, optimization and clustering strategies of text documents. International Journal of Electrical and Computer Engineering, 9(2), 1313-1320. doi:10.11591/ijece.v9i2.pp.1313-1320

[23] Sahrawat, Dhruva, et al. "Keyphrase extraction as sequence labeling using contextualized embeddings." European Conference on Information Retrieval. Springer, Cham, 2020.

[24] Santosh, Tokala Yaswanth Sri Sai, et al. "DAKE: Document-Level Attention for Keyphrase Extraction." European Conference on Information Retrieval. Springer, Cham, 2020.

[25] Chen, Tianlei, Duoqian Miao, and Yuebing Zhang. "A Graph-Based Keyphrase Extraction

Model with Three-Way Decision." International Joint Conference on Rough Sets. Springer, Cham, 2020.

[26] Guleria, Ankit, Radhika Sood, and Pardeep Singh. "Automatic Keyphrase Extraction Using SVM." Advances in Communication and Computational Technology. Springer, Singapore 945-956.

[27] Florescu C., Caragea C. (2017) A New Scheme for Scoring Phrases in Unsupervised Keyphrase Extraction. In: Jose J. et al. (eds) Advances in Information Retrieval. ECIR 2017. Lecture Notes in Computer Science, vol 10193. Springer, Cham. https://doi.org/10.1007/978-3-319-56608-5_37

[28] Ali C.B., Wang R., Haddad H. (2015) A Two-Level Keyphrase Extraction Approach. In: Gelbukh A. (eds) Computational Linguistics and Intelligent Text Processing. CICLing 2015. Lecture Notes in Computer Science, vol 9042. Springer, Cham. https://doi.org/10.1007/978-3-319-18117-2_29

[29] Chandraprakash, V., Sastry, J. K. R., Sravani, G., Manasa, U. S. L., Khyathi, A., & Harini, A. (2017). Testing software through genetic algorithms – a survey. Journal of Advanced Research in Dynamical and Control Systems, 9(Special Issue 2), 1607-1633. Retrieved from www.scopus.com

[30] CHANGALA, RAVINDRA, And D. RAJESWARA RAO. "A Survey On Development Of Pattern Evolving Model For Discovery Of Patterns In Text Mining Using Data Mining Techniques." Journal of Theoretical & Applied Information Technology 95.16 (2017)

[31] Devika, R., Subramaniyaswamy, V. A semantic graph-based keyword extraction model using ranking method on big social data. Wireless Netw (2019). https://doi.org/10.1007/s11276-019-02128-x

[32] Luo, L., Zhang, L. & Peng, H. An unsupervised keyphrase extraction model by incorporating structural and semantic information. Prog Artif Intell 9, 77–83 (2020). https://doi.org/10.1007/s13748-019-00200-3

[33] Tosi M.D.L., dos Reis J.C. (2019) C-Rank: A Concept Linking Approach to Unsupervised Keyphrase Extraction. In: Garoufallo-u E., Fallucchi F., William De Luca E. (eds) Metadata and Semantic Research. MTSR 2019. Communications in Computer and Information Science, vol 1057. Springer, Cham. https://doi.org/10.1007/978-3-030-36599-8_21

[34] Rabby, G., Azad, S., Mahmud, M. et al. TeKET: a Tree-Based Unsupervised Keyphrase Extraction Technique. Cogn Comput 12, 811–833 (2020). https://doi.org/10.1007/s12559-019-09706-3

[35] Shi, W., Zheng, W., Yu, J.X. et al. Keyphrase Extraction Using Knowledge Graphs. Data Sci. Eng. 2, 275–288 (2017). https://doi.org/10.1007/s41019-017-0055-z

[36] Kolli, S., & Sreedevi, M. (2018). Adaptive clustering approach to handle multi similarity index for uncertain categorical data streams. Journal of Advanced Research in Dynamical and Control Systems, 10(4 Special Issue), 1401-1408.

[37] Bhavani Dasari, Durga; Venu Gopala Rao. K, Dr. Context Similarity Strategy for Text Data Plagiarism Detection. International Journal of Engineering & Technology, [S.l.], v. 7, n. 2.32, p. 14-17, may 2018. ISSN 2227-524X. Available at: <https://www.sciencepubco.com/index.php/ijet/article/view/13517>. doi:http://dx.doi.org/10.14419/ijet.v7i2.32.13517.

[38] Swagata Duari, Vasudha Bhatnagar, Complex Network based Supervised Keyword Extractor, Expert Systems with Applications, Volume 140, 2020, 112876 ISSN 0957-4174, https://doi.org/10.1016/j.eswa.2019.112876.

[39] Wael Etaiwi, Arafat Awajan,Graph-based Arabic NLP Techniques: A Survey, Procedia Computer Science, Volume 142, 2018, Pages 328-333, ISSN 1877-0509,https://doi.org/10.1016/j.procs.2018.10.488.

[40] Wael D. Sowmya, J.I. Sheeba, Keyword Extraction Using Particle Swarm Optimization, Procedia Computer Science, Volume 85, 2016, Pages 183-189,ISSN 1877-0509, https://doi.org/10.1016/j.procs.2016.05.208.

[41] Priya, N., & Sreedevi, M. (2017). A novel similarity based contextual bug localization model for unstructured textual bug reports. Journal of Advanced Research in Dynamical and Control Systems, 9(Special Issue 18), 21-34. Retrieved from www.scopus.com.

[42] Prakash K.B., Dorai Rangaswamy M. (2019) Content Extraction Studies for Multilingual Unstructured Web Documents. In: Chattopadhyay S., Roy T., Sengupta S., Berger-Vachon C. (eds) Modelling and Simulation in Science, Technology and Engineering Mathematics. MS-17 2017. Advances in Intelligent Systems and Computing, vol 749. Springer, Cham. https://doi.org/10.1007/978-3-319-74808-5_58

[43] Pandit D. P. and Pattanik S.R. 2019, "Software Engineering oriented approach for IOT applications: Need of the day", International Journal of Recent Technology and Engineering, vol. 7, no. 6, pp.886-895.

[44] Ragkhitwetsagul, C., Krinke, J. & Clark, D. A comparison of code similarity analysers. Empir Software Eng 23, 2464–2519 (2018). https://doi.org/10.1007/s10664-017-9564-7

[45] Peng H., Mou L., Li G., Liu Y., Zhang L., Jin Z. (2015) Building Program Vector Representations for Deep Learning. In: Zhang S., Wirsing M., Zhang Z. (eds) Knowledge Science, Engineering and Management. KSEM 2015. Lecture Notes in Computer Science, vol 9403. Springer, Cham. https://doi.org/10.1007/978-3-319-25159-2_49

[46] Karnalim, O. (2020). TF-IDF Inspired Detection for Cross-Language Source Code Plagiarism and Collusion. Computer Science, 21(1). doi:https://doi.org/10.7494/csci.2020.21.1.3389

[47] Franclinton, Ricardo, Oscar Karnalim, and Mewati Ayub. "A Scalable Code Similarity Detection with Online Architecture and Focused Comparison for Maintaining Academic Integrity in Programming." (2020): 40-52.

[48] Y. Hu, Y. Zhang, J. Li, H. Wang, B. Li and D. Gu, "BinMatch: A Semantics-Based Hybrid Approach on Binary Code Clone Analysis," 2018 IEEE International Conference on Software Maintenance and Evolution (ICSME), Madrid, 2018, pp. 104-114, doi: 10.1109/ICSME.2018.00019.

[49] Hosseini, Roya, and Peter Brusilovsky. "A study of concept-based similarity approaches for recommending program examples." New Review of Hypermedia and Multimedia 23.3 (2017): 161-188.

[50] Kaur, H., Maini, R. Assessing lexical similarity between short sentences of source code based on granularity. Int. j. inf. tecnol. 11, 599–614 (2019). https://doi.org/10.1007/s41870-018-0213-1

[51] Sébastien Combéfis and Arnaud Schils. 2016. Automatic programming error class identification with code plagiarism-based clustering. In Proceedings of the 2nd International Code Hunt Workshop on Educational Software Engineering (CHESE 2016). Association for Computing Machinery, New York, NY, USA, 1–6. DOI:https://doi.org/10.1145/2993270.2993271

[52] Ragkhitwetsagul, Chaiyong. Code similarity and clone search in large-scale source code data. Diss. UCL (University College London), 2018.

[53] M. Sudhamani, Lalitha Rangarajan, Code similarity detection through control statement and program features, Expert Systems with Applications, Volume 132, 2019, Pages 63-75, ISSN 0957-4174, https://doi.org/10.1016/j.eswa.2019.04.045.

[54] E. Wong, Taiyue Liu and L. Tan, "CloCom: Mining existing source code for automatic comment generation," 2015 IEEE 22nd International Conference on Software Analysis, Evolution, and Reengineering (SANER), Montreal, QC, 2015, pp. 380-389, doi: 10.1109/SANER.2015.7081848.

[55] Song, Hyun-Je, Seong-Bae Park, and Se Young Park. "Computation of program source code

similarity by composition of parse tree and call graph." Mathematical Problems in Engineering 2015 (2015).

[56] Pribela, Ivan, Gordana Rakic, and Zoran Budimac. "Detecting Source Code Similarity Using Compression." SQAMIA. 2019.

[57] Rasekh, Amir Hossein, Seyed Mostafa Fakhrahmad, and Mohammad Hadi Sadreddini. "Mining traces between source code and textual documents." International Journal of Computer Applications in Technology 59.1 (2019): 43-52.

[58] Saranya, G., et al. "Model level code smell detection using egapso based on similarity measures." Alexandria engineering journal 57.3 (2018): 1631-1642.

[59] Alomari, Firas, and Muhammed Harbi. "Scalable Source Code Similarity Detection in Large Code Repositories." arXiv preprint arXiv:1907.11817 (2019).

[60] Karnalim, Oscar, and William Chivers. "Similarity Detection Techniques for Academic Source Code Plagiarism and Collusion: A Review." 2019 IEEE International Conference on Engineering, Technology and Education (TALE). IEEE, 2019.

[61] Yuan, Zhongchen, Li Yan, and Zongmin Ma. "Structural similarity measure between UML class diagrams based on UCG." Requirements Engineering (2019): 1-17.

[62] Nakul Sharma, Prasanth Yalla, “Mapping and Visualization of Source Code: A Survey”, DOI 10.5013/IJSSST.a.21.04.03, 3.1, ISSN: 1473- 804x online, 1473-8031 print, Vol 21, No-3. (https://ijssst.info/Vol-21/No-4/paper3.pdf)

ICT-based Learner-Centric Evolutionary Learning Model: An effective solution to Teaching-Learning process

Smita Vinit Bhoir[a], Sunita Patil[b]

[a]Department of Computer Engineering, K.J. Somaiya College of Engineering, Mumbai University, Ramrao Adik Institute of Technology, Dr. D. Y. Patil Deemed to be University, Mumbai, Maharashtra, India.

[b]Department of Computer Engineering, K.J. Somaiya Institute of Engineering and Information Technology, Mumbai, Maharashtra, India.

[a]smitapatilbe@gmail.com, [b]spatil@somaiya.edu

Abstract. In India around in the second week of March 2020, in order to avoid spread of COVID-19 virus, the Central Governments have declared LOCKDOWN in the country. It results in temporary shutdown of schools and colleges. By looking around, at severity of situation due to spread of the novel corona virus, there is no certainty when schools and colleges will reopen. This is a crucial time for the education field, including board tests, kindergarten admissions, entry tests of various universities and competitive exams. It has greater impact on Education field. It has affected continuous teaching-learning process, student counselling operation, assessment methods of teachers, students' interest towards higher studies in abroad got lower. Information Communication Technology (ICT) may play important role in the lockdown period by providing online learning platforms. Its need in education is discussed in this paper. With the help of power supply, basic digital skills to teachers and students and internet connectivity digital learning can be made possible. Similarly, students those are from low-income groups or presence of disability, etc. for them distance learning program can be included. To mitigate such hazardous situations, ICT-based learning model is essential. In this paper, ICT-based Learner-Centric Blended Education Model is proposed which blends traditional classroom teaching with e-learning where students and teachers both are active, students can interact with teachers; also, teachers can monitor progress of students, etc. After surveying different research papers on Education Models, it has been observed that in learning based on ICT, the Learner-centric environment need to be implemented. In online platform instructors provide course material to students but personalization of these materials is missing. The proposed work presented an ICT-based Learner-Centric Evolutionary Learning Model which suggests appropriate course material to learner, according to learner's profile built by our profile builder module which includes details like learning style, demographic details, recommendations, expectations, likes and dislikes. Our proposed work presented is seamless integration of traditional teaching-learning methods with online teaching-learning methods. The equal importance to teachers is given here to design and recommend study material to different types of learners, their preference is at highest priority in proposed recommendation model. Active engagement of students in learning process increases education graph using technologies and finally it will be helpful in development of our country.

1. Introduction

In India, COVID-19 outbreak affected the structure of education and learning, including teaching and evaluation methods. Only a few private schools and some universities were able to use methods of teaching online. On the other hand, their low-income private and public-school counterparts have shut down completely because their e-learning solutions do not exist. Similarly, there are many students from India who do higher studies abroad. The countries which are affected with such situations are not allowing students to leave these countries. Where the situation persists, the demand for international higher education is expected to decrease over the long term. To manage such crisis, immediate action to ensure continuity of education in schools and universities is vital. Open-source digital learning solutions and software for learning management should be adopted to enable teachers to teach online. Some private schools and colleges started using ICT tools to continue teaching learning process. Still there are issues in Government schools and colleges which are in rural area which could not use these technology-driven tools due to lack of knowledge about technologies, infrastructure and funds. But, with rapid increase in mobile internet users, it will be easy to access and personalize education even in the most remote parts of the country. This can alter the education system and make teaching-learning more effective and give students and teachers a wide range of options. It indicates there is need of Innovative mobile learning models for efficient education delivery. It is also important, through the seamless integration of classroom learning in e-learning modes, to reconsider current delivery and pedagogic methods for higher education and schools to build a unified education system. Different Quality assurance mechanisms and quality standards for online learning need to be established. In this time of crisis, the capacity-building of young minds needs a well-rounded and effective educational practice. It will increase its skilled workforce, productivity, health and well-being in the coming decades and will ensure the progress in India as a whole.

We can’t ignore the fact that technology plays a key role in the education system and only this is the demand for the current situation. Study of importance of Information Communication Technology in education shown that there are various E-learning platforms available in which learner’s are active, but current situation demands both students and teachers must be active, which will give feel of traditional learning in online mode. After analysing various challenges faced in education field by our country an ICT-based Blended Education Model is proposed, in which with basic training of computers students and faculties can be active in teaching-learning process even with mobile internet. Through this model a teacher can monitor student progress, can interact with them through discussion forums, chats, conduct interactive lectures through video streaming, etc. There are number of models present to deliver online education but, in these models,, teacher share same study material to all students without understanding learning style of students. It leads to delivery of content but not quality education which can build quality learners. In order to overcome such issues, in this paper an ICT-based Learner-Centric Evolutionary Learning Model is proposed. Using AI technologies, this model suggests most suitable course materials to Learner by considering Learner's profile which includes learner's learning style, demand, recommendation, reviews; instructor's suggestions, feedback; peer learners reviews and social media users related to domain. Its advantageous feature is increase in active engagement of students in learning process, motivations towards learning new things, also with encourages them to gain and share knowledge across the world. With proposed work it is possible to accomplish Quality Education. It results in development of students, development in Education field ultimately development of India with the help of technology-driven education.

The organization of paper is - section-I discusses ICT, its need in education, various challenges using ICT in education. Section-II discusses existing research work related to ICT in education. Section III elaborates the proposed work. Section IV demonstrates mathematical model of Multi Feature Decision Support System and finally in Section V concluding remarks on proposed work are given.

1.1 Information Communication Technology (ICT)

The Information and Communication Technologies or ICTs is a set of electronic technologies that are versatile, adaptable, enabling and capable of transforming organizations and redefining social relationships when converged into new configurations(Karampa and Paraskeva, 2020). The technological spectrum is increasingly growing and there is a fusion of emerging innovations and traditional media. There is no definitive evidence to prove that learner's performance is superior by using ICTs in the education space, either in developed countries or in developing ones. There is, however, a general opinion among professionals and scholars that introduction of ICTs into education has overall a significant effect on the learning environment. ICTs have the capacity to develop, improve, enrich and expand talents, inspire and educate pupils, help connect school knowledge to job activities, build economic prosperity for tomorrow's staff and enhance teaching learning process(Rico-Bautista, Medina-Cárdenas and Guerrero, 2019).

ICT as a teaching and learning medium refers to the resource itself for training and learning purposes. More than three decades ago, educators were exposed to computers and associated information technology for the purpose of direct teaching and learning. It began with CAL / CBT / CAI, then switched to Digital courseware and finally to Web based technologies and Computer Mediated Communication (CMC) framework(Bhatt, 2008). Using CAI for basic skills drilling and practicing can be highly successful, based on a broad data collection and a long history of use. In classes that use computer-assisted teaching, students typically learn more, and learn faster. This has been shown to be the case in all subject fields, from pre-school to higher education, as well as in normal and special education schools. Efficient teaching includes knowledge delivery, instructor feedback, preparation and student learning assessment. Computer-assisted guidance may be called the use of a computer to have some mixture of these variables. It should be remembered that there is no necessity for all those elements to be supported by the machine. Instead, any variation of these in the learning phase could be sufficient machine interference. The cornerstone of these systems is interactivity, accessibility and learner power. The application of instructional technology to teaching has progressed beyond the use of simple drill and practice applications, and now involves the use of interactive digital devices and sophisticated networking. Students now use technology to interactively study and collaborate on class assignments. They use the Internet for analysis, for project participation, and for correspondence. The new innovations make it possible for students to have more influence of their own learning, to think analytically and objectively and to work together. An growing body of research shows that ICT incorporation of teaching and learning has beneficial results(Twinomurinzi, 2012)(Chukwunonso and Oguike, 2013).

1.2 What is need of ICT in Education?

Information and communication technology are defined as all applications, platforms, content, data, networks and services, digital and those that can be translated into or distributed by digital forms that can be implemented to achieve learning goals, improve access to resources, create capacity, and manage the education system. This will also include systems for digitizing, delivering and maintaining content, designing and implementing capacity building technologies and systems, and providing connectivity. ICT has become an integral part of daily life and all industries, from banking to tourism, now depend heavily on ICT for their transactions. In the year 2005 (On and Practice, 2014) National Curriculum Framework has underscored the value of ICT in education. Many issues were raised that time such as why ICT is needed in education, without computers also learning process was possible then is implementation of ICT in education actually needed? The answer to such questions is the change is important because this is the age of knowledge and technology, an age that requires teachers to promote this knowledge collection and not just teaching('ICT needs and trends in engineering education', 2015).

In India, initially when ICT was introduced in education, it had introduced only use of computers and internet in India. How to use ICT in education was not properly discussed. But only use of computers is not sufficient to improve quality of education unless teachers understand the need for ICT and explain to students. Many teachers refuse to the ICT plunge, they have traditional concept that students can't learn anything without the teacher demonstrating or intervening. But there is need to make realization that students often have their own ideas and experiences this they have learned from daily life; teachers don't acknowledge or use this experience and ideas. This can be done in a major way through ICT. Teachers need to be prepared to promote the learning process, make the experience real, realistic, demanding, and enjoyable and not intimidating. It's incredibly necessary to raising teacher talk and promotes student conversation. There is no need to write anything on the blackboard to be regarded as teachable. The teachers need to realize that knowledge is not difficult to obtain in the 21st century, but it is important skills to coordinate, communicate, and collaborate. So ICT is not just about representing knowledge, it is about communicating, exchanging, and thus learning. ICT offers significant, consuming media that increase the efficiency of teaching-learning. ICTs are a potentially effective resource for expanding educational opportunity, both formal and non-formal, to scattered and remote communities, culturally or socially disadvantaged groups such as ethnic minorities, girls and mothers, people with disabilities and the elderly.

One trait of ICTs is their potential to overcome time and space. Its available anytime anywhere. Online course materials are available to learners 24X7. It removes the requirement for both learners and teachers to be in one physical location. In addition, with Web-based tools like video conferencing the knowledge can be shared with learners that are geo-graphically distributed across the world. ICTs also promote exposure around the globe to qualified users-advisors, consultants, scholars, educators, industry executives, and peers(Jain *et al.*, 2018).

Currently, we are in the era of technology. Thus, one of the most beneficial reasons for using ICTs in the classroom was to better prepare the new generation of students for a workforce where ICTs, particularly computers, the Internet and similar technology, are becoming increasingly omnipresent. Therefore, technical literacy is seen as a strategic advantage in a rapidly globalizing job market. ICT not only develops technical literacy but supports creative thought, higher-order thought and clear reasoning, successful communication(Sinclair and Kalvala, 2015). ICTs' ability to facilitate the learning of these skills is tied to its use as a resource to improve the quality of education, particularly facilitating the move to a learner-centred environment. ICTs can improve the quality of education in a variety of ways: by increasing learner enthusiasm and commitment, by promoting the learning of basic skills, and by improving teacher training (Barrera, Hernández and Luna, 2018). ICTs include audio, videos, multimedia contents which can be used to engage the learner in learning. Rather so than any other form of ICT, networked computers with Internet access will improve learner enthusiasm as it blends the social exposure and interactivity of other ICTs with the ability to communicate with real people and engage in activities in the real ICTs can turn the learning environment into one that is learner-centric.

1.3 Challenges of using ICT in EDUCATION

ICT as a digital tool simplifies and promotes human interaction is not only helpful in many ways but still has many drawbacks. Limitations may be classified as being related to instructors, related to students, and related to technology.

- The mindset of the teacher towards the use of such technologies is important, certain experiences show that some teachers may not have insight as to how helpful technology may be in promoting and optimizing learning due to lack of expertise to deal, while other teachers may have optimistic attitudes to the technology.
- A further weakness could also be teacher reluctance and lack of motivation to use ICT in education.
- Lack of infrastructure and facilities is another problem for Indian education sector back warding.
- If teachers learn some practical skills and ability to play with the pupils, there is a downside of using ICT of education.
- Related learning style would not be the same for all students, so suitable research materials need to be recommended for a range of students

2. Related Work

Various research papers are surveyed related to role of ICT in education and its extensive use till date in same.

In (Bhatt, 2008)(Fung and Yuen, 2012), it is mentioned that due to ICT, education is available anywhere anytime. Authors mentioned that ideal models are required to identify what kind of learner is and accordingly customization of learning materials needs to be done. Many organizations started using Learning Management System. But most of the LMSs don't provide complete learning solutions. In comparison, most LMS don't provide full learning solutions. LMS framework is not sufficient to ensure rational content delivery or personalized the course material to student needs. In (Chukwunonso and Oguike, 2013) authors mentioned that a person got connected to his/her relatives, friends and people across the globe through social networks. This platform can be used in education to develop, discover, evaluate, and manage the information of various domains provided in educational material.

Ease of sharing knowledge across the world is one of the benefits of ICT. But only sharing knowledge is not important but proper data representation and sufficient information management are essential(Satapathy *et al.*, 2015). An ability of identification and representation of relationships between various contents is required for the development of context-aware apps. This knowledge makes it possible to classify important details more accurately. All together creating some sort of system that deals with a person needs capacity to work with imprecision and uncertainty. In order to understand what are expectations of learner, some semantic analysis and fuzzy logic is required. Apparently this dimension is absent in the works under study.

Recently, the use of mobile devices that are having internet connectivity, makes learning possible even outside the school timing. Ubiquitous Learning is closely linked to this, because the learning process can take place almost anywhere and this can be blended into the everyday lives of individuals (Sinclair and Kalvala, 2015). In addition, in recent years, the probability of developing understanding needs of learner's has increased (Tang, Ronghuai and Price, 2016). Specific background data can be explored in mobile learning settings. It is no longer new to provide handheld tools used to promote teaching and learning. The fact that wireless technology is broadly deployed and the proliferation of portable computing devices is rapid has made mobile learning experience possible.

In addition, context-awareness, a term which is useful in enhancing the usability of the learning systems is proposed in this study. The student context may thus include the surroundings of students, location detail, the learning purpose, historical awareness and expectations according to the works in(Ng, Luk and Lam, 2018). Accordingly, in a bid to fulfil this purpose, material adaptation may personalize the learning object. Opinion Mining as Semantic Web definition is also proposed as an idea of machine-processed knowledge in one of the classical works (Szewczyk and Barak, 2019). Its introduction to e-learning provided incentives for developing structures capable of understanding the needs and habits of the students, and choosing learning resources more appropriately. But there is no such system present to provide personalization of course contents to learners who are ultimately at centre point of this ICT-based education.

3. Proposed Work

After observing current LOCKDOWN situation due to COVID-19 and such perilous situations there is need for ICT based education. This paper presented ICT-based Learner-centric Blended Education Model which blends traditional teaching-learning methods with ICT-based teaching-learning methods. For personalization of course contents ICT-based Learner-centric Evolutionary learning model is proposed to which an essential input is learner's profile. Building learner's profile as learning advances is again challenging task. It is overcome by using machine learning techniques.

3.1 ICT-based Learner-centric Blended Education Model

An ICT-based Learner-centric Blended Education Model is given in Figure 1.

Figure 1: ICT-based Learner-centric Blended Education Model

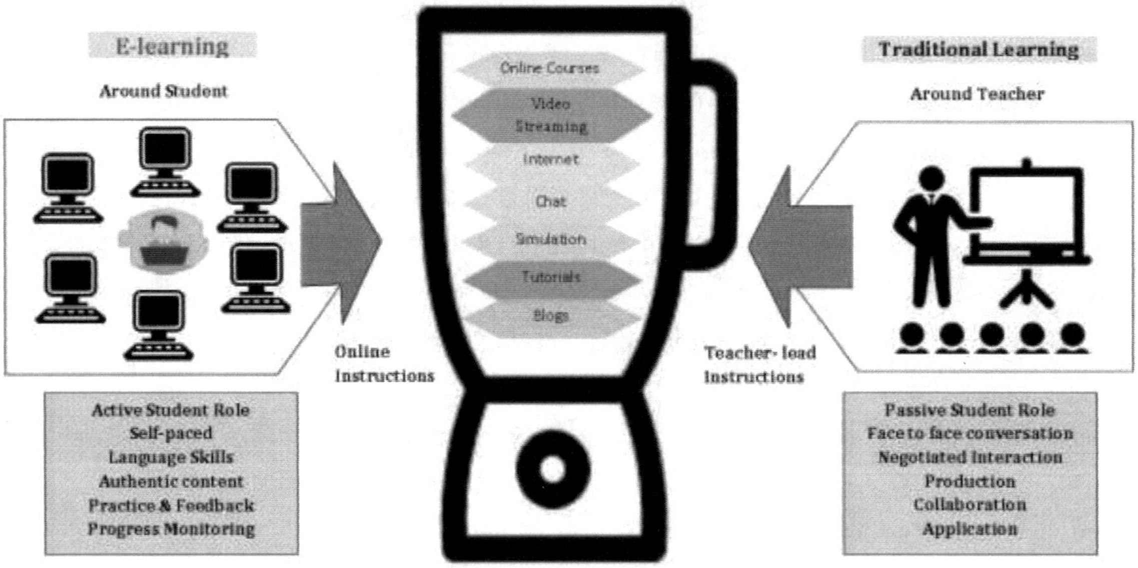

The first thing that springs to mind when there is a contrast between online and conventional classroom education is the computer and the classroom. Many apps and online networking tools are available to make learning online simple and to meet tutors and peers in a moment. This model is designed to help aspirants use different resources to make learning more effective, validated and less time consuming.

The learning style will be different for the different types of the learning mediums. The e-learning is tangentially more autonomous. Learners need to take an active role in going into the information and understanding new things. In achieving the educational objectives, it would be like self-directed approaches. Learner will prepare his / her analysis, rather than relying on the tutors(Ng, Luk and Lam, 2018)(Sinclair and Kalvala, 2015). Though that will be different from the conventional preparation, students should have their mentors to direct and help them to prepare future studies. There will be extra support from peer. They can also be inspired by social pressure to accomplish the impossible targets. Users assume e-learning is better too. But in fact e-learning requires equal commitment and dedication in order to achieve the desired results. There are several e-learning courses accessible under the current conditions of an engaging learning atmosphere involving peer to peer contact and engagement with students and tutors. Traditional learning is primarily about the curriculum in the classroom. It is limited to a specified time limit and a venue. Learners need to attend the lectures, enter the group meetings and engage in other school events intended to support your education. It will be tutor-driven, and the instructors will supervise the learning practices of the learner. They won't have to go through this process in e-learning. The learner would have the freedom to pick all of his / her suitable learning time. The e-learning is easier to use. This is planned to give the users the greatest versatility. But that doesn't mean it would affect the standards of the learning. A learner will get the best education at the convenience of his / her home at any of the suitable times. If learners are involved in online courses in e-learning, then learners may need to develop some basic technology skills. He / she could not be able to obtain an on-line education without any professional expertise. To familiarize with online and e-learning, learners need to grasp search engines, apps, post, text, web page, webinars, chat rooms, video sharing platforms and social networking. In the case of E-learning all these items would be essential. Owing to technological development in recent days, the typical learner would now require certain digital skills for reporting and data collection from the internet. Comprehension of the technology and applications needed to make learning simpler and more efficient. The technological skills would be required in both the conventional and e-learning. Yet the engagement in e-learning should be greater(Morze, Glazunova and Kuzminska, 2018).

After understanding pros and cons of traditional and e-learning, instead of using either traditional learning or e-learning separately, an ICT–based Blended Education Model is proposed. In this model, instructor can instruct or teaches learners through video streaming. The students can raise their issues or clear their doubts through discussion forums, chats, blogs, and reference study material provided by tutor which are accessible at any time anywhere. Instructors can motivate students by giving problem solving based quizzes, discussion forums and students can reply to them hence, learners will be active in learning. For an instructor, due to availability of various ICT tools, evaluation of learner's activities will be also easier so that an instructor can continuously monitor progress of learner. In proposed education model, learner not only takes guidance from instructors also can get experience from peer learners or through social media, various forums and blogs. This will improve problem solving ability of learner and for a given topic learner's thinking will change, he/she will look problem from different perspectives and will try to give best solution.

Currently India is undergoing LOCKDOWN period due to COVID-19, its having greater impact on all the sectors including Education sector. The classroom teaching is not possible and to leave students only on E-learning, it will not be efficient. To continue teaching-learning routine in such situations, the proposed model will be useful where students as well as teachers will be active and they can instruct students through video streaming. They can interact through chats, discussion forums and various ICT-based communication media. Instructors can share their course material prepared like notes, videos, e-books, etc. with students which will be helpful for their studies. An instructor not only can share their specialized area knowledge with their students they can globally share it with Internet technologies. It will definitely useful for growth of our nation as education plays vital role in development of a country.

3.2 ICT-based Learner-Centric Evolutionary learning model

As shown in ICT-based Blended Education Model, the environment should be Learner-centric. An interest needs to be generated in learner, necessary skills to be developed within him/her which will ultimately help to achieve learning objective. But still presenting contents, its delivery, distribution and personalization are still a challenge. Furthermore, vagueness of students while for selection of their preferred courses has not been yet covered. In this paper, an ICT-based Learner-centric Evolutionary Learning Model is proposed, which is architecture of adapting learning materials taking into account learner profiles and their context information. Application of AI technologies helps in effective customization and taking decision for selection of adequate course material for the learner that leads to building truly personalized and evolutionary systems. The proposed learning model is presented taking a full advantage of semantic web, social media reviews, and users' feedback. Ultimately, our research aims to build a comprehensive framework for ICT-based learner-centric models of Evolutionary learning where the use of Multi-Feature Decision Support System is required to understand learner-centric aspects. In addition, the implementation of such a model will serve as a framework for drawing conclusions of immediate practical significance for the design of such applications. It has given equal importance to instructor's efforts in content delivery suitable for learner's personal profile. For example, in fundamental courses the suggestions and constraints imposed by instructors on the content will have a higher priority, influence and precision than the student's preferences, feedback from associate students and associate social media users. With implementation of these strategies, the design of course materials is versatile.

The architecture of the ICT-based Learner-centric Evolutionary Learning Model is presented in Figure 2.

Figure 2: Proposed Architecture of ICT-based Learner-centric Evolutionary Learning Model

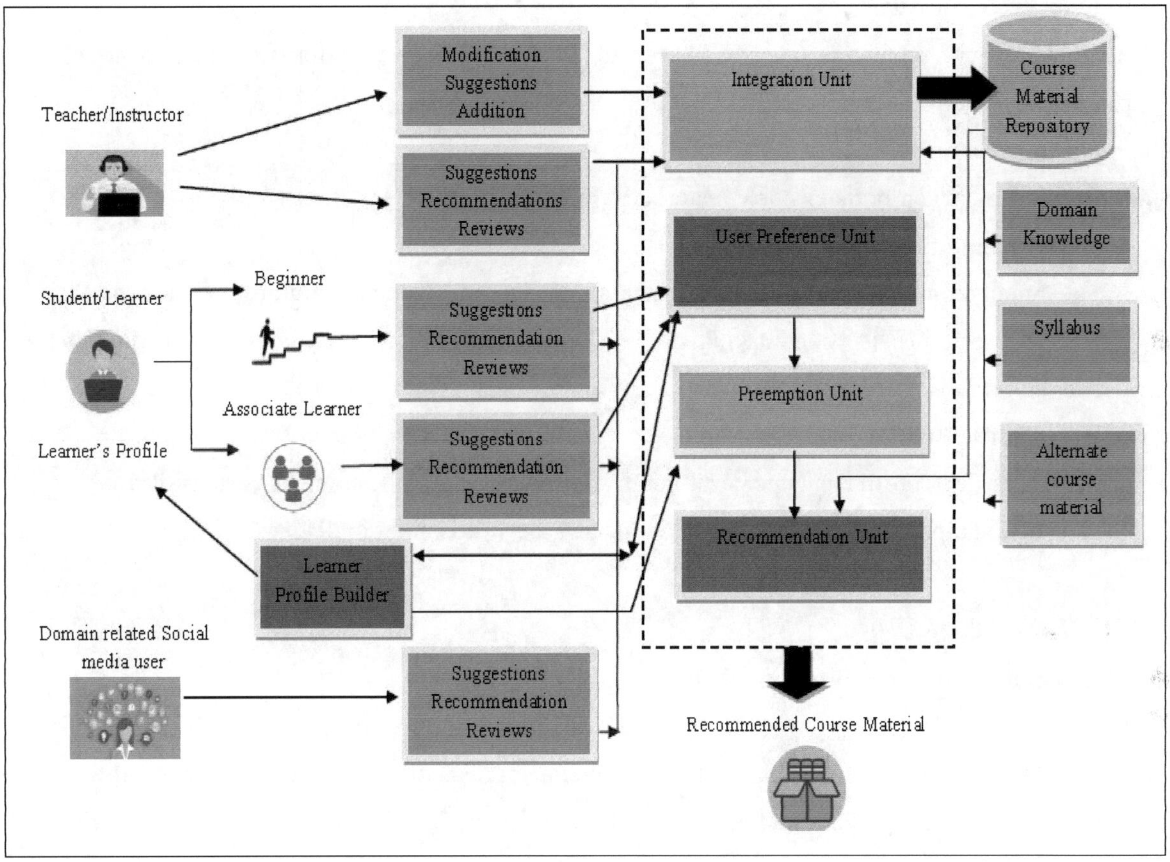

The proposed system understands what the learner wants, his interests, desires. It involves what teachers propose and endorse (suggestions, changes, feedback), and knows what associate learners think about usable content, it also needs domain –related users feedback on social media which can be helpful in recommending alternate course material. Also domain knowledge and syllabus are having equal importance in this system. In this model learner can share his/her review, suggestions, his likes related to topic. Around the same time teachers have the right to track the student and make changes and additions to the content already being used by the student.

The proposed system is divided into 5 major components-

a) Interpretation unit

There is course material repository which keeps on updating. An interpretation unit performs extraction of important options that considers learner's review, recommendations, suggestions; instructors' reviews, suggestions, modifications, addition to material. It is also used to interpret existing material in course repository.

b) User Preference unit

It stores learner's preferences, needs, likes or dislikes which are useful for updating of learner's profile,

c) Preemption unit

It preempts most appropriate course material by analyzing learner's profile.

d) Recommendation unit

It takes input from preemption unit and integrates associated learner' opinions, and instructors' inputs and performs decision making process to suggest most suitable material keeping the learner's profile into account.

e) User Profile Builder

It creates learner's profile and updates it iteratively. A learner profile contains various details such as demographic information, learner's preferences, needs, likes or dislikes

3.3 Methodologies Used

A) Opinion Mining and Semantic Web

Opinion mining in combination with Multi-criteria Decision Support system has shown benefits in dealing of imprecise and ambiguous human judgment decision conditions and models through the use of estimated information(Zhang, Lu and Wu, 2017)(Smita and Bhoir, 2019) . Incorporating opinion mining into the decision-making methods of learners will resolve the issue of unreliability at the point of making decision due to insufficiency in the amount of knowledge students provide. It applies data classification with boundaries that aren't well defined. It is possible to use machine learning approach for analysis and inference, to figure out new essentials and their classification and accuracy to check the class from which they are belonging. These techniques are used to design and build Evolutionary e-Learning models' elements and functions.

B) Creating Learner's Profile

A learner's profile plays a vital role in the proposed system. Figure 3 illustrates the method of building Learner's profile. Initially, demographic data of learners are gathered and their success data are tracked and documented. In web-based platform, web- logs created helps to get learner's progress on given task. Initially while building learner's profile, methods like questionnaires, quizzes should be adopted to know the learner's preliminary information, liking about the course

IOP Conf. Series: Materials Science and Engineering 1074 (2021) 012028 doi:10.1088/1757-899X/1074/1/012028

and his/her learning style. The learner's data is collected from above sources acts as input to the next stage is used as a dataset for further analysis. The data set that has been extracted from the previous stage should be pre-processed using AI-based pre-processing techniques. Now we use Clustering algorithm to identify the learners based on their learning style, behaviors, interests, feedback and comments. (Auddy and Mukhopadhyay, 2015) describes the cluster as a method used to position data items in similar groups without previous knowledge of the meanings of groups. While classification, learning style is important feature of learner which helps for selection of appropriate study material.

Table-I Learning styles and Preferences

Type of Learners	Learning Preferences
Practical Learners	tangible, Practical, Driven by evidence and procedures
Spontaneous Learners	theoretical, inventive, Oriented towards interpretations and hypotheses
Visual Learners	Prefer visual representations contents
Vocal Learners	Prefer written and spoken descriptions
Dynamic Learners	Prefer learning by trying new things, working with others
Deep Learners	Prefer learning through thinking, and working alone
Chronological Learners	Linear, systematic, learn in small incremental steps
Universal Learners	Holistic system thinkers, learn to make great strides

Figure 3: Building Learner’s Profile

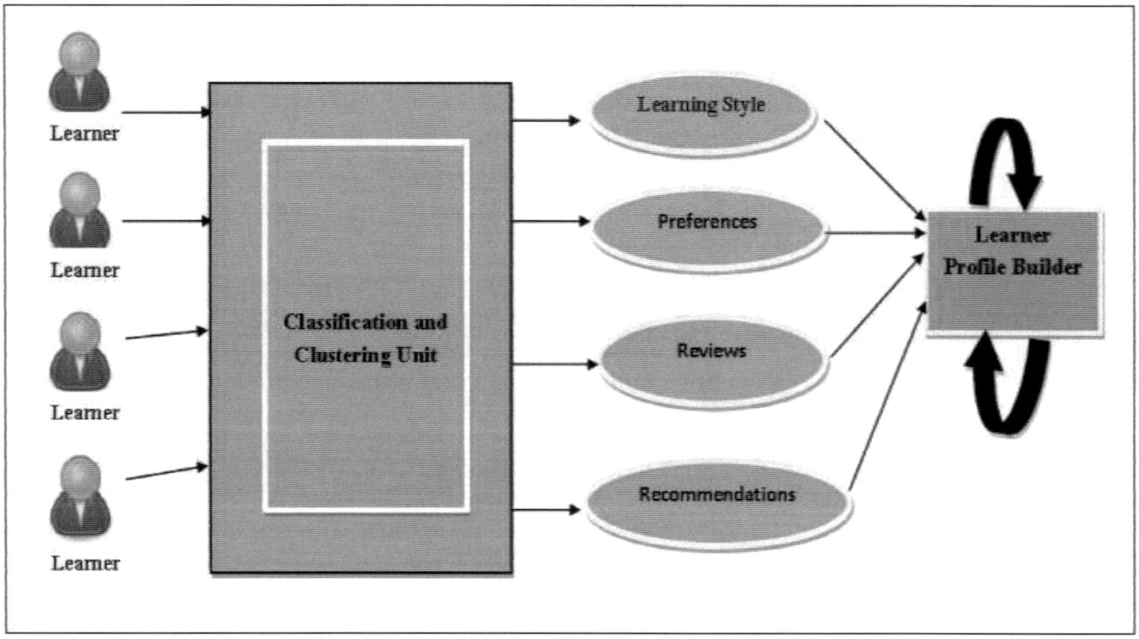

An instructor should produce variety of course materials for different types of learning modes. The suggested system thus includes a variety of ways in which an instructor will track the actions of the students and behave accordingly. Also, the Instructors are able to ask the program for topics pertaining to the content available and learners completing their classes, as well as read reviews received given by students as related to the information they learned. The system also helps instructors to submit answers to student questions, recommend and appropriate alternate materials for classes.

C) Recommendation of Course Materials

The most important aspect of any Learning Model-based system is the method of choosing the most appropriate course materials, i.e. option between several models, various parts, different media etc. For recommendation of appropriate course material various features need to be considered such as- learner's profile, goals, preferences, feedback and review on specific course material. Similarly guidelines, preferences, recommendations, ideas that include reports and experience from the instructors, reviews and recommendations from peer users and social media users related to domain. Overall, the suggested selection processes conduct decision-making by considering instructor's important thinking related to topic, essentials to learner and reviews of associate learner including domain related social media user.

For recommendation of appropriate course material Multi Criteria Decision Support System (MCDSS) is proposed here which uses Hybrid model of Analytical Hierarchy Process and Voting method (which considers user ratings and recommendations). After investigating advantages and

IOP Conf. Series: Materials Science and Engineering 1074 (2021) 012028 doi:10.1088/1757-899X/1074/1/012028

limitations of Analytic Hierarchy Process, Hybrid MCDSS model is proposed which is a Hybrid Model of AHP method and voting system. In the proposed model, after application of AHP, voting from decision makers are involved to finalize decision.

Algorithm for Proposed Hybrid MCDSS-

Step 1: Start

Step 2: Collect reviews, recommendations from various learner's and reviews, recommendations, modifications from instructor

Step 3: Evaluate performance using AHP.

Step 4: Is Consistency Ratio is less than equals to 0.1?, No then Revise comparison judgement and goto step 3 otherwise goto step 5.

Step 5: Create valuation set for available options with respect to criteria

Step 6: Create wide-ranging evaluation matrixStep 7: Perform evaluation and get priority ranking

Step 8: Apply any social voting method on aggregated results.

Step 9: Display resultant ranking of results

4. Mathematical Induction

The working of proposed Hybrid MCDSS model is justified with the help of mathematical model.

1. Describe problem and collect related data. These details are useful for identification of problem, available solutions to problem and other specific parameters related to problem.

2: Apply AHP for quality evaluation. Let Course Content M, for example, be the target of the decision taking, M1, M2, M3, M4 and M5 are the alternatives, and R1: Learner's Recommendation, R2: Instructor's Recommendation, R3:Instructor's Modification, R4: Associate Learner's Modification, and R5: Social Media Consumer Recommendation are the requirements for evaluating those alternatives.

The next move is to use Saaty's Nine Point Scale to create a matrix of comparison in pairs, as example is on Table. II. The priority weights vector W is prepared by identification of beneficiary and non-beneficiary features and then according summing each matrix column and then dividing each matrix variable with the sum of its column. The Consistency Ratio (CR) is then calculated as

$$CR = \text{Consistency Index (CI) /Random Index (RI)}$$

IOP Conf. Series: Materials Science and Engineering 1074 (2021) 012028 doi:10.1088/1757-899X/1074/1/012028

If the CR is less than 0.01, then decision is accepted else need to re-adjust the decision matrix till the condition become acceptable. The value of CI is 0.03039 dependent on the matrix of the pairwise relation in Table-II.

Table-II Pairwise Comaprison Matrix

	R1	R2	R3	R4	R5
R1	1	1/5	1/4	1/4	2
R2	5	1	2	2	5
R3	4	1/2	1	1	3
R4	4	1/2	1	1	3
R5	1/2	1/5	1/3	1/3	1

Table-III RI values

N	1	2	3	4	5	6	7	..	15
RI	0.00	0.00	0.58	0.90	1.12	1.24	1.32	..	1.59

4. Apply suitable optimization technique for decision making such as Simple Additive Method, Weighted Product method and get ranking of alternatives.

5. Establish overall priority ranking

After performing AHP ranking of alternatives are obtained then number of voting methods can be applied. Here for explanation plurality voting method is applied(in which alternative with maximum 1 is having highest priority), we got M5 is the winner, and the order of results is: M5 > M3 > M1 > M2 > M4.

Table-IV Voting Table

	M1	M2	M3	M4	M5

R1	3	4	1	5	1
R2	2	1	4	3	5
R3	1	2	1	5	1
R4	1	3	5	2	1
R5	5	2	1	4	1

6. Get final rankings of results

In this step the alternatives are ranked from highest preference to lowest and decision is finalized.

5. Conclusion

The LOCKDOWN declared in India as preventive measure against spread of COVID-19 virus resulted in temporary shutdown of schools and universities. After analyzing severity of such hazardous conditions there is need to blend digital technology with traditional teaching. The proposed ICT-based Learner-Centric Blended Education Model helps to face such situations and maintain continuity in teaching–learning process. The work proposed has given importance of ICT in education, challenges to use ICT in education and solutions to overcome same. An ICT-based Learner-Centric Blended Education Model ensures active engagement of students and teachers, helps in maintaining Quality Education. Also, an ICT-based Learner-Centric Evolutionary Learning Model proposed here applied AI technologies for suggestion of appropriate course material to learner. A profile builder module is heart of ICT-based Learner-Centric Evolutionary Learning Model, which builds learner's profile that includes learning style, demographic details, recommendations, expectations, likes and dislikes. For recommendation of appropriate course material, hybrid model of Multi Feature Decision Support System is given here, which is integration of AHP and voting method. One of the main challenges of educators worldwide, and especially in India, is to make the present generation of learners ready for the future. In 2020, the average age of youth is expected to be 29 years. However, if basic education is not accomplished across India, India's large youth population would not be of benefit to the economy. A country's development relies largely on the education system. Another important measure of economic growth is literacy. In the case of India, it is still developing nation because its educational system is missing behind many other developed nations, it is crucial that we recognize the need for education and its position in Indian youth carving future. The work proposed here is beneficial in satisfying such requirements.

References

[1] Auddy, A. and Mukhopadhyay, S. (2015) 'Data mining on ICT usage in an academic campus: A case study', *Lecture Notes in Computer Science (including subseries Lecture Notes in Artificial Intelligence and Lecture Notes in Bioinformatics)*, 8956, pp. 443–447. doi: 10.1007/978-3-319-14977-6_48.

[2] Barrera, Á. P. M., Hernández, A. A. R. and Luna, J. E. O. (2018) 'ICT skills in university teachers, the knowledge, use and pedagogical appropriation of these technologies', *Communications in Computer and Information Science*, 870, pp. 249–259. doi: 10.1007/978-3-319-95522-3_21.

[3] Bhatt, R. M. (2008) 'ICT for success of education from an Indian perspective', *IFIP International Federation for Information Processing*, 269, pp. 147–155. doi: 10.1007/978-0-387-09657-5_9.

[4] Chukwunonso, F. and Oguike, M. C. (2013) 'An Evaluation Framework for New ICTs Adoption in Architectural Education', *International Journal of Informatics and Communication Technology (IJ-ICT)*, 2(3), pp. 183–189. doi: 10.11591/ij-ict.v2i3.5285.

[5] Fung, H. and Yuen, A. (2012) 'Factors affecting students' and teachers' use of LMS - Towards a holistic framework', *Lecture Notes in Computer Science (including subseries Lecture Notes in Artificial Intelligence and Lecture Notes in Bioinformatics)*, 7411 LNCS, pp. 306–316. doi: 10.1007/978-3-642-32018-7_29.

[6] 'ICT needs and trends in engineering education' (2015) *Proceedings of 2015 International Conference on Interactive Collaborative Learning, ICL 2015*, (September), pp. 146–149. doi: 10.1109/ICL.2015.7318016.

[7] Jain, S. *et al.* (2018) 'Smart university-student information management system', *Proceedings of the 2017 International Conference On Smart Technology for Smart Nation, SmartTechCon 2017*, pp. 1183–1188. doi: 10.1109/SmartTechCon.2017.8358555.

[8] Karampa, V. and Paraskeva, F. (2020) *Smart learning environments: A blend of ICT achievements and smart pedagogy for the world sustainable development*, *Advances in Intelligent Systems and Computing*. Springer International Publishing. doi: 10.1007/978-3-030-25629-6_75.

[9] Morze, N., Glazunova, O. and Kuzminska, O. (2018) *Training of E-learning managers at universities*, *Communications in Computer and Information Science*. Springer International Publishing. doi: 10.1007/978-3-319-76168-8_5.

[10] Ng, K., Luk, C. and Lam, W. (2018) *Blended Learning. Enhancing Learning Success*. Springer International Publishing. doi: 10.1007/978-3-319-94505-7.

[11] On, I. C. T. and Practice, A. (2014) 'A survey of the impacts of information communication technology (ict) on architectural practice in nigeria', 6(2), pp. 137–156.

[12] Rico-Bautista, D., Medina-Cárdenas, Y. and Guerrero, C. D. (2019) 'Smart University: A Review from the Educational and Technological View of Internet of Things', *Advances in Intelligent Systems and Computing*, 918, pp. 427–440. doi: 10.1007/978-3-030-11890-7_42.

[13] Satapathy, S. C. *et al.* (2015) 'Emerging ICT for Bridging the Future - Proceedings of the 49th

Annual Convention of the Computer Society of India (CSI) Volume 1', *Advances in Intelligent Systems and Computing*, 337, pp. 541–547. doi: 10.1007/978-3-319-13728-5.

[14] Sinclair, J. and Kalvala, S. (2015) *Engagement measures in massive open online courses, Communications in Computer and Information Science*. doi: 10.1007/978-3-319-22629-3_1.

[15] Smita, M. and Bhoir, V. (2019) 'An Efficient FAKE NEWS DETECTOR'.

[16] Szewczyk, R. and Barak, M. (2019) *Advances in Intelligent Systems and Computing 421 Embedded Engineering Education*. doi: 10.1007/978-3-030-40274-7.

[17] Tang, Q., Ronghuai, H. and Price, J. K. (2016) *ICT in Innovational K-12 Education in Global Context*.

[18] Twinomurinzi, H. (2012) 'The role of ICT in sustainable and responsible development: E-skilling', *IFIP Advances in Information and Communication Technology*, 386 AICT, pp. 90–99. doi: 10.1007/978-3-642-33332-3_9.

[19] Zhang, X., Lu, C. and Wu, D. (2017) 'A Model Based on the Factor Analysis for Assessing the ICT Development in Basic Education and Regional Comparison', *Proceedings - 5th International Conference on Educational Innovation through Technology, EITT 2016*, pp. 227–231. doi: 10.1109/EITT.2016.52.

[20] Ingole, P., Bhoir, S. and Vidhate, A.V., 2018, March. Hybrid model for text classification. In *2018 Second International Conference on Electronics, Communication and Aerospace Technology (ICECA)* (pp. 7-15). IEEE.

IOP Conf. Series: Materials Science and Engineering 1074 (2021) 012029 doi:10.1088/1757-899X/1074/1/012029

Design of implantable cardioverter defibrillator using low power subthreshold digital circuit

T Vasudeva Reddy [1], B Nareshkumar [2], R Anirudh Reddy[3], P Kavitha Reddy[4]

[1]Associate Professor, ECE Department, B V Raju Institute of Technology, Narsapur, Telangana, India
[2,3,4] Asst. Professor ECE Department, B V Raju Institute of Technology, Narsapur, Telangana, India

[1]vasu.tatiparthi@bvrit.ac.in,[2]nareshkumar.b@bvrit.ac.in,[3]anirudhreddy.r@bvrit.ac.in,[4]kavithareddy.p@bvrit.ac.in

Abstract. Due to tremendous Technological advancements researchers and scientists have proposed designs operating under subthreshold and near threshold regions. With recently, growing the need for low power consumption of devices, the low power circuits have become more eye-catching to the design engineers. But the act of subthreshold design has become more profound to the PVT variations due to exponential relationship of current in Subthreshold region depends on supply voltage(Vgs) and threshold voltage(Vth). The motivation for low power electronics has been derived from the subthreshold operation & its driving forces. In this paper the subthreshold design of source coupled logic circuit is described with its functionality under subthreshold region and performance in static and dynamic mode of operation and used as Implantable cardioverter defibrillator under 28nm technology.

1. Introduction

Now a day's human life span is reduced by many factors mainly due food, social life and other factors that leads to unhealthy. Most common problem of heart attack and failure. So to improve the life span of person and protect from heart attack, pacemakers are introduced to control heart rhythms. The functionality of pacemaker is used to speed up slow heart rates and sometimes they used to slowdown fast heart rates. Pacemakers standardize functionality of bouncing, de-bouncing also to regulate heart rates. When pacemakers are inserted in body it cannot be replaced every time, operating normally life span of 5 year's maximum. Here in this paper we discuss the circuit that takes minim power there by extend the lifetime of battery for portable pacemaker electronic circuit. This is the most encouraging motivation to design a memory with low power functioning in subthreshold operation and power consumption and dissipation [1-3]. Functionality is at the cost of performance depends on technology. Circuits that are modelled under subthreshold withdraw the limitation by modelling devices with sizing and alternative techniques. In this paper, we have described the SRAM implementation using GDI based technique takes the feedback signal boosted and compared by internal circuit that leads to low power consumptions of the design and thereby increasing the life span of battery that is used for functioning life time [4].

1.1. Design & Simulation of Traditional Subthreshold SRAM

Implementation of traditional 8T SRAM using CMOS design to improve the SNM (static noise margin). The SNM and stability can be improved direct path between bit cell and the storage node cell to improve the leakage, dynamic and static power, this can be obtained by charging and discharging of the single bit line represented in Figure 1

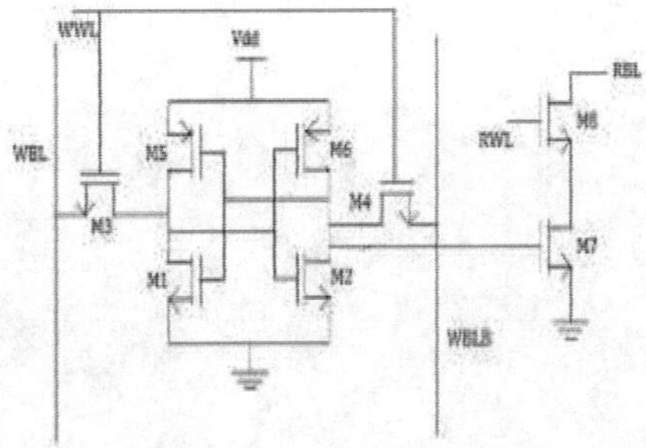

Figure 1. 8T SRAM used as memory in Defibrillator

In order to address functionality read and write operations are separated by stacking of transistor that are applied to the 6T SRAM. A memory circuit with the Cmos technology transistor under 32nm is showed y shown in Figure. 2.

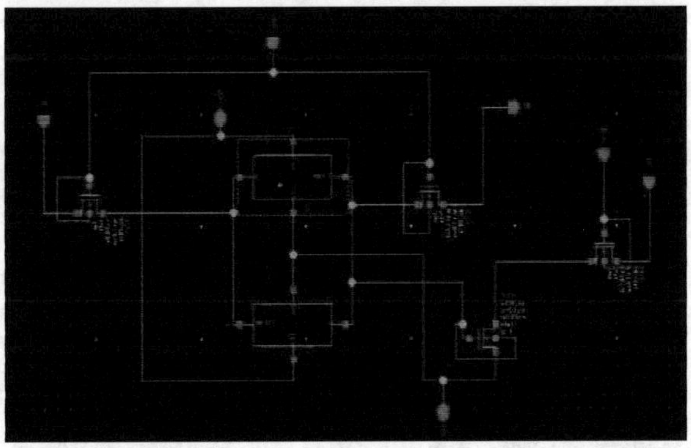

Figure 2. Improved 8T SRAM schematic diagram

1.1.1. Write state functionality

The performance and stability of SRAM is improved by providing isolation between read, write bit line. An 8T model is useful to write operation and for reading operation. Set of stack transistors are used to control read and write operation by Write Word Line (WWL) & Read Word Line(RWL). During a read operation, read bit line (RBL) is pre-charged and WWL maintas as high logical state and that depends on value stored in cross coupled inverters and the operation is charges & discharging [5].

1.1.2. Read state functionality

The read state functionality defines the RBL discharges, treated as bit as s '1' else it is treated as 0. In this case storage nodes are completely isolated from write bit lines. Therefore, the stability of SRAM is enhanced. But during a write operation, the WBL & WBLB lines are precharged to predetermined value. Then asserting write word line and nodes attain the corresponding values. During write operation, WBL and WBLB lines are precharged to predetermined voltages. Then, declaring the write

IOP Conf. Series: Materials Science and Engineering 1074 (2021) 012029 doi:10.1088/1757-899X/1074/1/012029

word line WWL and nodes inserts the values from the bit lines. Therefore, the stability is increased and the metal density, dynamic and leakage powers are greatly reduced [6].

1.1.3. Transient model analysis
The transients' analysis is represented as below from Figure 3, with the supply voltage as 0.4v. either of the transistors will be on, as the CMOS is operating under subthreshold region.

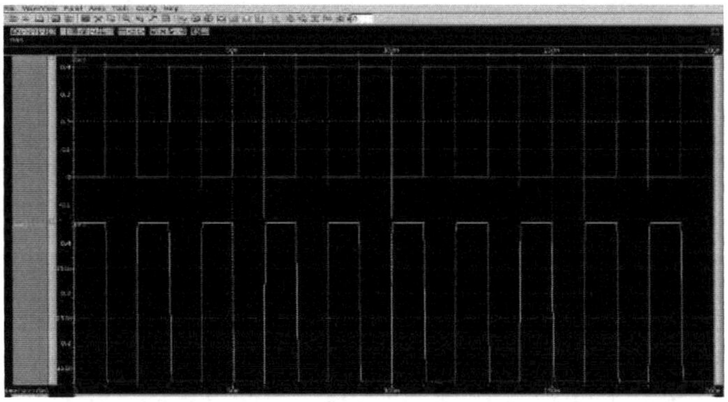

Figure 3. 8T SRAM Transient analysis

1.1.4. Dc Model Analysis
The operating point of the dc analysis is analyzed from DC characteristics, when a DC load line crosses the middle of dc characteristics and is represented by the below Figure 4.

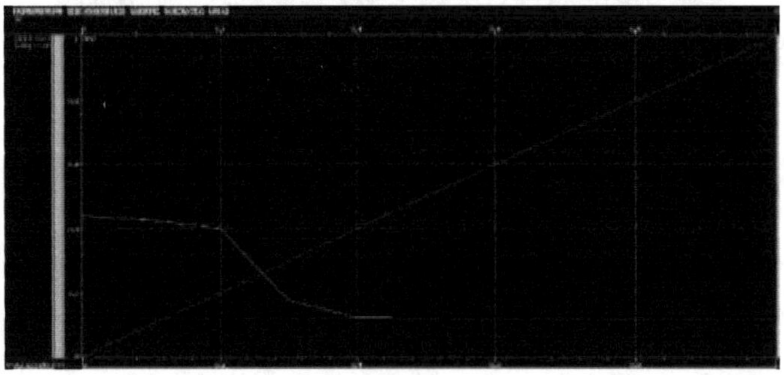

Figure 4. 8T SRAM DC analysis

Dynamic power is marked in read and write operation and non-functionality of the memory is also treated static characteristics. Dynamic power is major, when the circuit is operating in the threshold region of operation, where the technology is degrading and reduction in the supply voltage leads to static power or leakage power, which will dominate the dynamic power at when the technology is enhancing [7].

1.2. SRAM power analysis under Subthreshold
In generally, the power consumptions are idle. But in stand by mode, the static power and the leakage power is dominating. Therefore we need Subthreshold design SRAM power and its and the performance is estimated in 32nm.

IOP Conf. Series: Materials Science and Engineering 1074 (2021) 012029 doi:10.1088/1757-899X/1074/1/012029

1.2.1 ANALYSIS OF STATIC POWER DISSIPATION

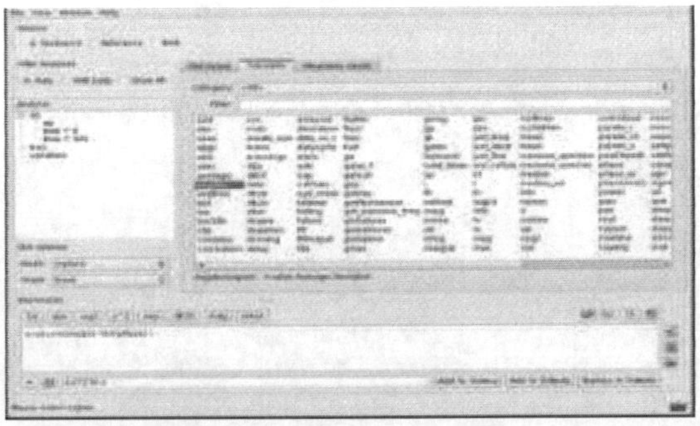

Figure 5. Static Power Dissipation 4.6727nw

In digital design static power dissipation is important parameter. As the complexity of the circuit is keep on increasing nowadays, a SRAM is used in most of the cache memory design indicated by in Figure 5.

1.2.2 Total Power (Tpd) Dissipation

From Figure.6 the total power of the circuit is estimated by considering the static, leakage and dynamic power at a different mode of operation like cut off, active and saturation. Normally the dynamic power is maximum, when the supply voltage of above threshold (Vgs>Vth)[8]. When in sub-threshold region, leakage power dominates as the supply voltage reduced than the threshold (Vgs<Vth) voltage and leakage power are dominating. Therefore, there will maximum power dissipation across the circuit.

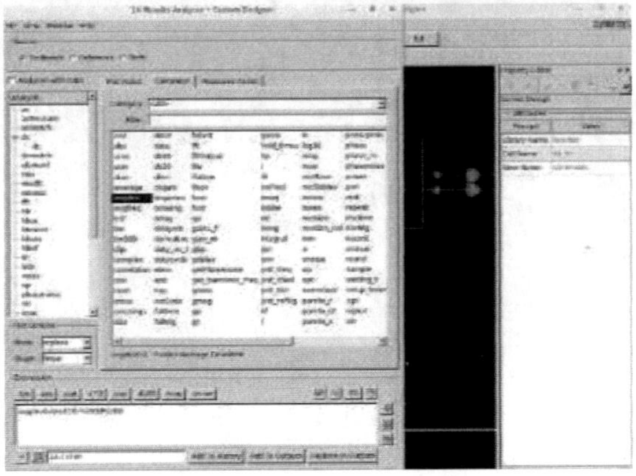

Figure 6. Total Power of 1.377 µw

1.3. SRAM delay analysis

In Subthreshold design SRAM delay is estimated in 32nm and the performance of delay is showed below

1.3.1 Rise time analysis

The rising time of the SRAM is estimated by considering the 2 % to 90 % of the output response, where the 2% is the minimum value and 90 % of the maximum value of the output. The difference in the VoH to the VoL, Where the VoH is the maximum voltage in the total output and the VoL is the lowest voltage in the output voltage and vice versa in fall time denoted in the Figure 7 & Figure 8.

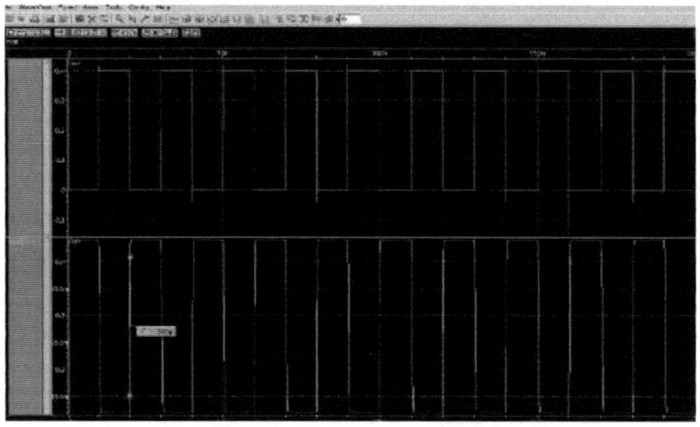

Figure 7. Rise state of 205 Ps

1.3.2 Fall time

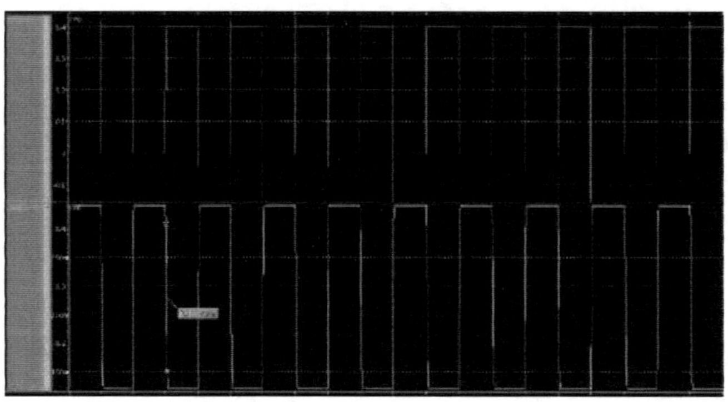

Figure 8. fall time of 42 Ps

2. Subthreshold GDI Inverter Design

GDI Operation
To understand the analogy of the basic GDI cell, for the low swing issues based on the F1, and that will be improved and advanced for another GDI functions.
A below table 1 represents the full set of logical functions of GDI, when the values of A=0, B=0, the value of the output F1=Vtp instead of 0v. Because of reduced characteristics of high to where the low swing of operation of PMOS Transistors.

When A=0 and B=1, the demand that transition occurs only about 50%, The GDI cell is acts as a inverter when A=1 and B=0, which is largely used for the logic level restoration and finally when A=1 and B=1 GDI Cell is acts as CMOS inverter by producing 0 as an output voltage, [88,90].In some of the cases when the value VDD=1, without swing drop, GDI cell is acting like the inverter and recovering the voltage wings, and also it allows self-restoration in many cases. SRAM is design with only two transistors' [11] and low power dissipation & less delay by reducing the complexity of the circuit and the static noise margin can be greatly reduced.

IOP Conf. Series: Materials Science and Engineering 1074 (2021) 012029 doi:10.1088/1757-899X/1074/1/012029

2.1. DESIGN OF GDI MODEL

The Basic GDI function shows the simple GDI Based cell to realize the basic logic functions with a simple Boolean equation that is represented by circuit diagram Figure 9 and Figure 10 is denoted by schematic model.

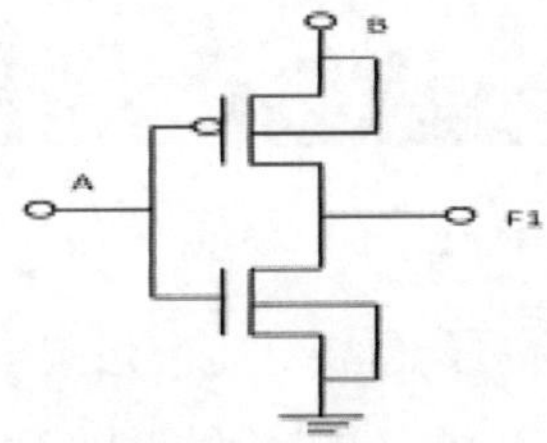

Figure 9. GDI design

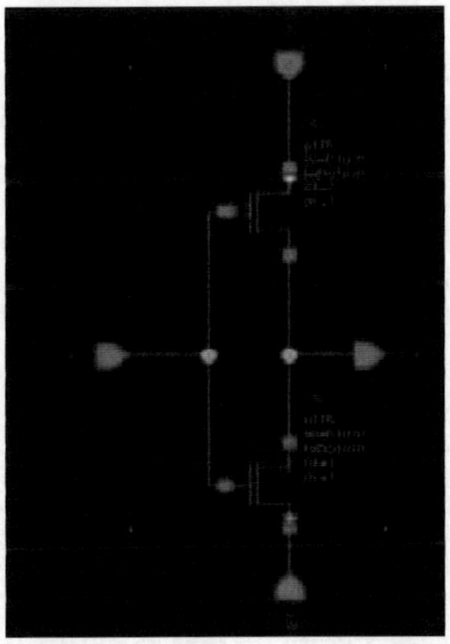

Figure 10. Schematic diagram of GDI Cell inverter

Implementation of all the logic functions is not possible with traditional P-well CMOS, but it can be implemented using SOI Technology. Table 1 CMOS logic will take only two transistors per function. GDI cell has a variation in of low swing voltage, low static power dissipation, less in complexity, less no of transistors [7-9].

Table 1 Truthtable of GDI Inverter

G	P	N	OUT	Output
1	0	0	pMOS Transmission gate	F0
1	0	1	CMOS Inverter	1
1	1	0	nMOS Transmission gate	0
1	1	1	CMOS Inverter	0

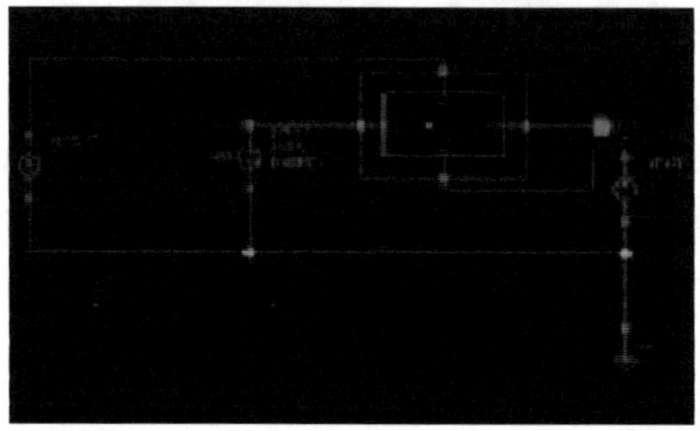

Figure 11. GDI Test bench circuit

A Cell view representation is indicated below; it is used to estimate the performance characteristics of transient & DC analysis of GDI. The below Figure 11 indicates the schematic view of the inverter in GDI. Here in GDI Technology, the input pin is given as VPLUSE of voltage and ground given as VDC values as inputs and test bench model [8-9].

2.2. Operation
Operation of SRAM defines by considering read and write operation by applying the supply under subthreshold operation.

2.2.1 Write state operation
Write operation 0 into the SRAM, the word line should make to 1 and Bit line has to apply '0' VDD is applied to bitline Bar (BL) & Write Word Line (WWL) is inserted to make M1, M4 off. Hence value of Bit Line stores value of '0', stores the value at 'Q'as showed in Figure 12.

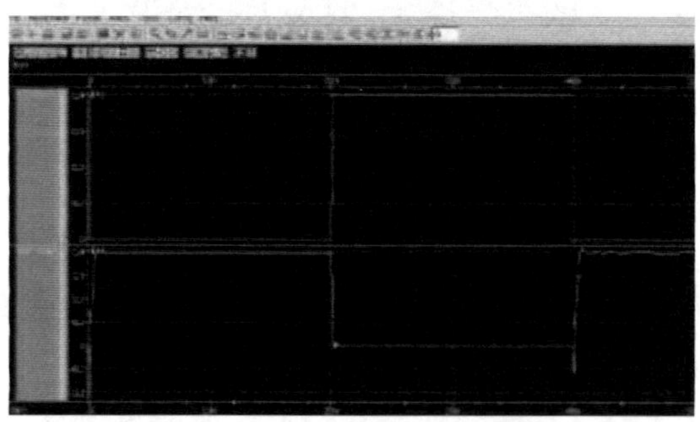

Figure 12. write operation

2.2.2 Read state operation
Read Word Line (RWL) drives the access transistor M2 makes on. When the value stored at Q is equal to 0, then transistor M1 is off, as RBL is directly connect to ground and discharges through M5, M6. Finally, value stored in Q is 0 and QBAR is 1 indicated in Figure 13.

Figure 13. Read operation

2.3. Modelling of SRAM using GDI technique

The GDI based SRAM indicates the Figure 14 indicate the schematic diagram of SRAM GDI. Both VDD and Vss are treated as input pins and reduces the reduce the dissipation of power. The SRAM cell using GDI techniques, is designed by considering four transistors are cross coupled to form inverters and another two transistors acts as access transistors [9]. To separate read and write operation.

2.3.1 GDI based SRAM design.

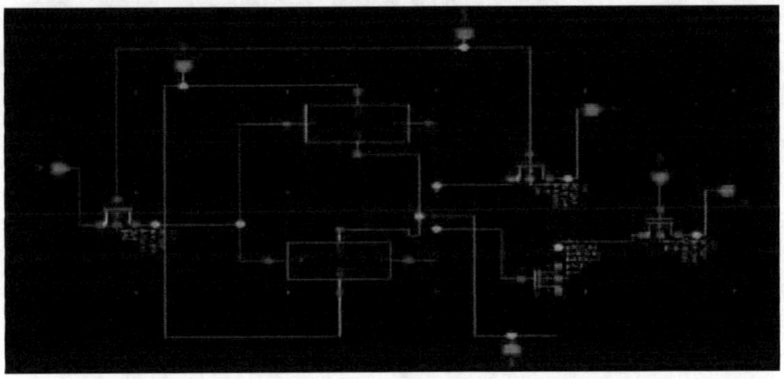

Figure 14. Subthreshold GDI Based SRAM

8T SRAM Cell view of Subthreshold GDI Based SRAM Library Cell using 32nm under subthreshold SRAM Cell diagram represented in Figure 14.

2.3.2 Write state Operation

The write state defines the value stored in cell earlier When GDI logic acts as Inverter & buffer circuit, GDI CMOS takes the value and restores in it. Also treats other two inputs are also 1 and 0. At this point of inputs 0,0[8-9]. A small amount of low voltage swing occurs with BL & WWL values are 0,0which are effected by low swing voltages represented by Figure 15.

Figure 15. GDI Read Operation

2.3.3 Read Functionality

Figure 16. GDI write operation

During the read, BL & WWL is 0,1 & 1,1 respectively. when values WWL and RBL are 1,1 GDI circuit is acts as an inverter. Considering BL and VDD are low, a high vtp is values is on state ie 1v instead of off state ie 0. is represented instead of 0v. The reason being here is poor value of 1 to 0 swing represented by Figure 16.

2.4. GDI characteristics

GDI based subthreshold SRAM is designed & analyzed the functionality with transient and dc analysis in 32nm and described below.

2.4.1 Transient Analysis

The analysis of transient characteristics is obtained by considering the pulse as input and the amount of voltage s are treated as 0.4 for a duration & some other duration will be treated as 0, with a subthreshold region of operation. considering the voltage drop at the input & current across that input node will parameters to estimate the functionality of transient analysis described in Figure 17.

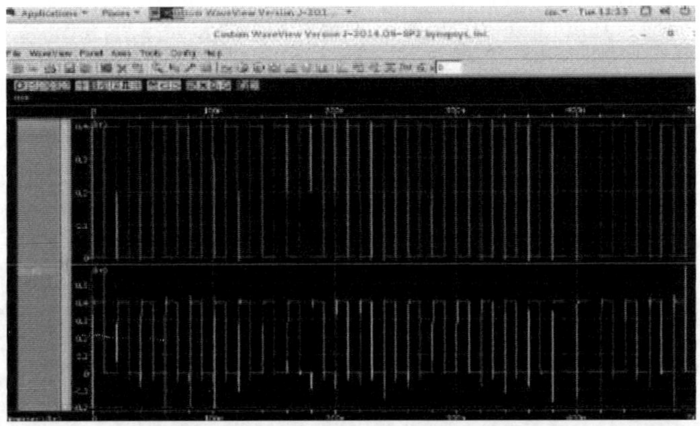

Figure 17. GDI Transient Analysis

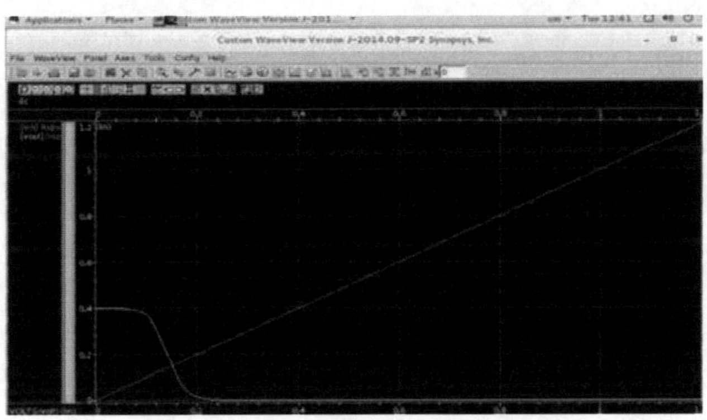

Figure 18. DC Analysis

Figure 18 denotes the DC analysis of SRAM cell using GDI. The above diagram indicating the dc analysis of GDI SRAM, where the dc load line is the middle point of the characteristics. This results can be obtained by considering the input values of dc with minimum current across it. Another variation is the technology based on the width and length of the transistor. A PMOS Transistors are designed in such a way that width is changed to 32nm, as it has less mobility of electrons and majority of holes. Therefore, the width dimensions are reduced by maintaining the length value are constant. As NMOS Transistors has the higher mobility of charge carriers of electron those are maintained at width value is 32nm.

2.5. GDI Power Analysis

GDI power analysis is estimated by Static, dynamic, total power and that has to need to consider under subthreshold operation of GDI SRAM.

2.5.1 Static Power Dissipation

Static power dissipation is the power due direct path between supply voltage that directly connects to ground. In GDI when the transistor is under subthreshold operation, the diode under reverse biased current that leads to parasitic capacitance that leads to leakage current. The leakage current forms between drain and source diffusion and also in the bulk of MOS regions. The of leakage current occurs when Supply voltage between gate and substrate is less than the threshold value Vgs <Vth. ie

transistor operating under subthreshold region of operation. Figure 19 showed below represents static power dissipation

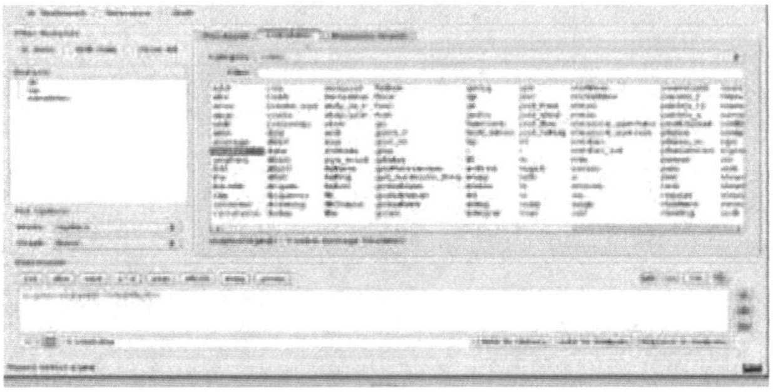

Figure 19. Power Dissipation in Static mode 0.6686μw

2.5.2 Total Power (TPd) Dissipation

Switching activity of the transistor resulting to dynamic power. Normally the power across the transistor during switching mode is high. But in the subthreshold design of operation, leakage power is dominating the dynamic power. The amount of current across the transistor of load capacitance is changing the state from High to low & low to high. The power in total (TPd) of the circuit is given by Total Power is evaluated as sum of the static power and dynamic Power. Figure 20. power dissipation in Total is 0.46692 μA.

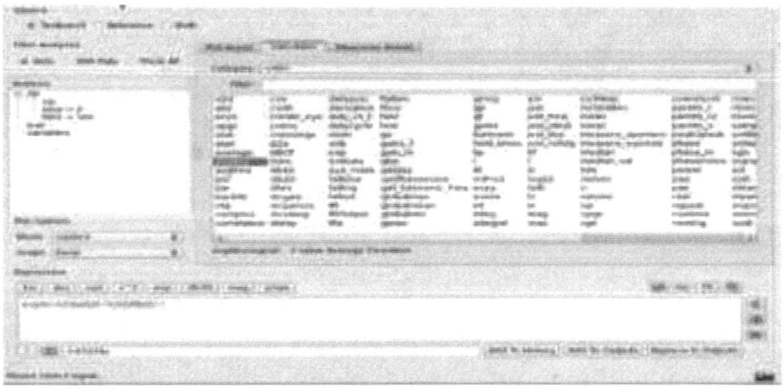

Figure 20. Total power (Tpd) dissipation of 0.46692 μw

From Figure 20 clearly states that there is more dynamic power & the amount of static is reduced drastically in the total power of SRAM cell. Because of power reduction techniques are applied, in GDI technique.

2.6. Delay Analysis

Dynamic response is the time taken change the state from 2% to 90% treated as rising Time (tr), and Fall Time (tf) is time taken to fall from 90% to 2%, average rate of rising and fall time (trf) (tr + tf)/2 is called edge Rate. Delay is the delay is an average of (tpHL + tpLH) /2, Contamination Delay (tcd) is the delay of Min time of the input crossing 50%.

2.6.1 Rise Time (Tr)

Time taken to raise the output signal from 2% to 90% of the max value is treated as rising time period and represented by Tr denoted by Figure 21

IOP Conf. Series: Materials Science and Engineering 1074 (2021) 012029 doi:10.1088/1757-899X/1074/1/012029

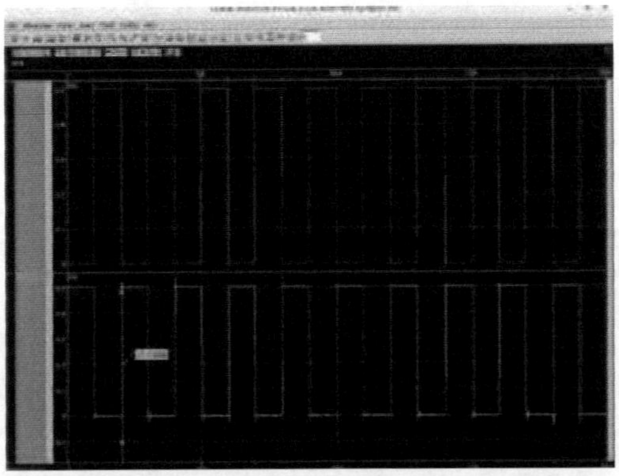

Figure 21. Rise Time delay of 27 ps

2.6.2 Fall Time(Tf) Analysis

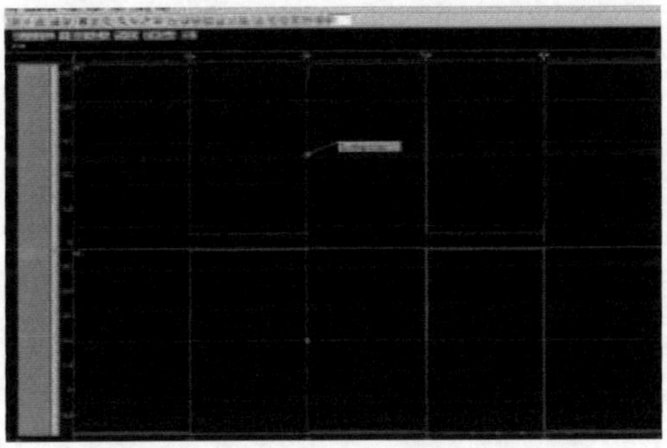

Figure 22. Fall time of 0.015ns

The times to fall the output change the state from 90% to 2% of the max value that represented as fall time and indicated by tf. denoted by Figure 22

3. Results

Table 2. Comparison of SCL and GDI

Analysis	SCL	GDI
Total power nw	13.77	4.6606
Delay µs	0.615	0.122
Power delay product	8.4685	0.5685

Table 2 represents the analysis and comparison of Source coupled logic design vs Gate diffused Input SRAM design with their performance parameters of Total power in nano watts, Delay micro watts and Power delay product.

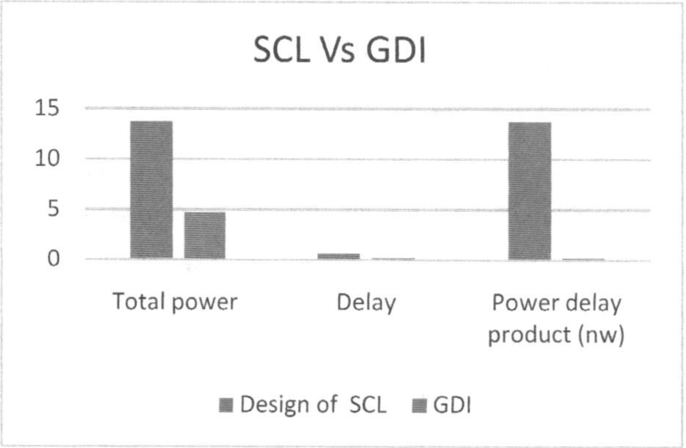

Figure 23. Comparative analysis of Power Delay and power delay product

Comparison of Power, delay and power delay product is analyzed leakage power is gradually reduced by a 34%, Delay is decreased by 20% and finally power delay product reduced by 7% indicated in Figure 23.

4. Conclusion

Subthreshold GDI SRAM of power(Pd), delay(Td), power delay product(PDP) is analyzed in 32nm and described. Design & analysis of single bit SRAM and GDI SRAM under subthreshold operation with same length of the transistors at 32nm technology. The characteristics, the threshold voltage of the transistor. considering the threshold(0.49v), applying Vss<Vth (0.4). Hence this leads to operates the SRAM under Subthreshold Operation & also changing width of the transistor to 32nm technology. The performance and functionalities are obtained on the basis of static, dynamic power and also rise,fall time. From the parameters mentioned, power & delays can be analyzed. These power and delay values are greatly reduced in GDI SRAM comparing with SRAM. So low power SRAM based GDI model can be used in cardioverter defibrillator of pacemakers as the power hungry of the circuit is less, GDI SRAM can be used as internal circuit in pacemaker.

References

[1] T Vasudeva Reddy, Designing Of Schmitt Trigger-Based Architecture 8T SRAM Of 256 bit cells under 45nm technology for low power applications, 2016 International Conference on Engineering and Technology (ICET), IEEE

[2] T Vasudeva Reddy, Design strategy & analysis of Subthreshold SRAM in power & delay for wearable applications, 2017 2nd International Conference on Communication and Electronics Systems (ICCES),IEEE

[3] T. Vasudeva Reddy, Performance & functionality of novel Subthreshold SRAM's using low power techniques for SoC designs, 2018 3rd International Conference on Communication and Electronics Systems (ICCES), IEEE

[4] T. Vasudeva Reddy, Design and estimation of power and delay using high Vth nmos under subthreshold logic operation of SRAM onebit model, **DOI:** 10.1109/SCOPES.2016.7955750

[5] T. Vasudeva Reddy, ,Analysis & design of robust ultra-low power subthreshold SRAM models, IEEE, 2016 International Conference on Control, Instrumentation, Communication and Computational Technologies (ICCICCT), IEEE,**DOI:** 10.1109/ICCICCT.2016.7987945

[6] T Vasudeva Reddy, Sub-threshold SRAM bit cell topologies for ultra-low power applications, International Journal of Scientific & Engineering Research, Volume 5, Issue 8, August-2014 957 ISSN 2229-5518.

[7] K. Amalraj et al presented a paper on " Low Voltage Low Power SRAM design based on

Schmitt Trigger technique Nano Scaled Low Power Leakage 10. ST-based SRAM " in 2012 International Conference on Emerging Trends in Electrical Engineering and Energy Management (ICETEEEM), 978-1-4673-4634-4/12, IEEE
[8] Single-ended Schmitt-trigger-based robust low-power SRAM cell by, Sayeed Ahmad, et al., Journal,IEEE Transactions on Very Large Scale Integration (VLSI) Systems,Volume24,Issue 8,Pages 2634-2642,Publisher,IEEE.
[9] Design of Low Power 6T-SRAM Cell and Analysis for High-Speed Application by G. Shivaprakash1, Vol 9 (46), December 2016.

IOP Conf. Series: Materials Science and Engineering 1074 (2021) 012030 doi:10.1088/1757-899X/1074/1/012030

Activation Layers Implication of CNN Sequential Models for Facial Expression Recognition

M. Shyamala Devi[1], Ankita Sagar[2], Karan Thapa[3], Mreenmoy Hazarika[4], P. Swathi[5]

[1]Professor, Computer Science & Engineering, Vel Tech Rangarajan Dr. Sagunthala R&D Institute of Science and Technology, Chennai.
[2, 3, 4, 5]Final Year B.Tech Student, Computer Science & Engineering, Vel Tech Rangarajan Dr. Sagunthala R&D Institute of Science and Technology, Chennai.
shyamalapmr@gmail.com

Abstract. Facial Expression Recognition is the critical part in the human emotional detection in the field of image processing. The application tends to soft or hard real time application based on the power of expression detection of idle image or videos. The Social communication with any object can be done by verbal or non-verbal format. Expression and Emotion detection completely rely on the non-verbal communication and facial expression. Machine Learning play an important role in the recognition of facial expression. An attempt is made in this paper to analyse the performance of Convolutional neural network models with diverse activation layer for the performance evaluation.Firstly, the facial expression dataset is extracted from the website http://www.consortium.ri.cmu.edu/ckagree/, http://app.visgraf.impa.br/database/faces/ is subjected with the data processing. Secondly, the data analysis is done for the distribution of expression image in the training and testing dataset. Thirdly, the facial expression images are detected with HAAR cascade and then the images are cropped with (350, 350). Fourth, the facial expression images is applied with normalized and the bottleneck features are created for training and testing data. Fifth, the training dataset is fitted with convolutional sequential neural network models with various activation layers like Sigmoid, Elu, Relu, Selu, Tanh, Softsign and Softplus. Sixth, the performance analysis is done with loss and accuracy for all the epoch of all CNN models for all the activation layers. Experimental results show that CNN sequential model with Relu activation layer is found to have the accuracy of 100%.

1. Introduction

Machine learning technology is used in all the areas of image processing in the applications like feature extraction of parts from the body, angular motion detection of body parts, identification of features from the image, counting of objects from the photo image and video, analysis of type of objects in the image, identification of outcome in gaming application and emotion identification of the object. Facial Recognition is the challenging task in the field of virtual reality where the people are working in various diverse applications like emotion identification, feature extraction, person identification, biometric identification, and fingerprint identification and classification and expression detection.

2. Literature Review

This paper deal with the comparative study of all the available dataset for facial expression images that are captured through idle photo image and video image. It also explores all the research insight that are carried in the field of facial expression recognition and evaluation [1]. Human computer interaction is the major trending in the field of virtual reality and multimedia. This paper deals with emotional identification of the people interacting the system by the extraction of feature algorithm [2].]. Machine learning models identify and predict potential high-cost patients and explore the key variables of the forecasting model, by comparing differences in the predictive performance of different variable sets Interpretable regression method based on Dempster-Shafer theory, using the Evidence Regression model to predict the cost for insurance [3]. This paper explores the application of facial expression recognition in the field of driving. The hazardous damage occurs due to the unsafe driving by the drivers of the vehicle. It occurs due to the bad habits of the drivers by the alcohol intake. This may also be due to the frustration and depression of the drivers due to problems in their family. Such unaware feelings of the drivers can somehow identify by their facial expression reaction while they drive. This paper provides a solution for identifying the facial expression of the vehicle drivers to avoid the accident [4].

This work is dedicated in finding the action points in the face features to arrive the solution in finding the facial expression recognition. The expression of the face completely depends on the motion of the facial parts [5]. This paper develops the software for the automatic identification of the person face and their expression at real time [6]. It explores the complete facial expression recognition methods and techniques in the real time applications. For example in the crime cases, the police may get the video of the event happened. In such case, we need a face recognition system to find the target person [7]. This paper deals with normalization and the identification of human features in the face recognition objects by using the discrete cosine transform in logarithm domain [8]. This paper analyzes the use of evolutionary algorithm for the features that are extracted from the face. The inner differences between each facial feature are synthesized by using the genetic algorithm [9]. The facial expression recognition can be done and the efficiency can be improved by using machine learning classification algorithms. There is a possibility of combining the facial features for the facial expression recognition during real time and it is carried out by the support vector machine classification algorithm [10].

3. Our Contributions

The main analysis of this paper is to analyze the performance of Convolutional neural network models with diverse activation layer for the performance evaluation. The overall architecture of this paper is shown in Figure.1.

(i) Firstly, the facial expression dataset is extracted from the http://www.consortium.ri.cmu.edu/ckagree/, http://app.visgraf.impa.br/database/faces/ is subjected with the data processing.
(ii) Secondly, the data analysis is done for the distribution of expression image in the training and testing dataset.
(iii) Thirdly, the facial expression images are detected with HAAR cascade and then the images are cropped with (350, 350).
(iv) Fourth, the facial expression images is applied with normalized and the bottleneck features are created for training and testing data.
(v) Fifth, the training dataset is fitted with convolutional sequential neural network models with various activation layers like Sigmoid, Elu, Relu, Selu, Tanh, Softsign and Softplus.
(vi) Sixth, the performance analysis is done with loss and accuracy for all the epoch of all CNN models for all the activation layers.

IOP Conf. Series: Materials Science and Engineering 1074 (2021) 012030 doi:10.1088/1757-899X/1074/1/012030

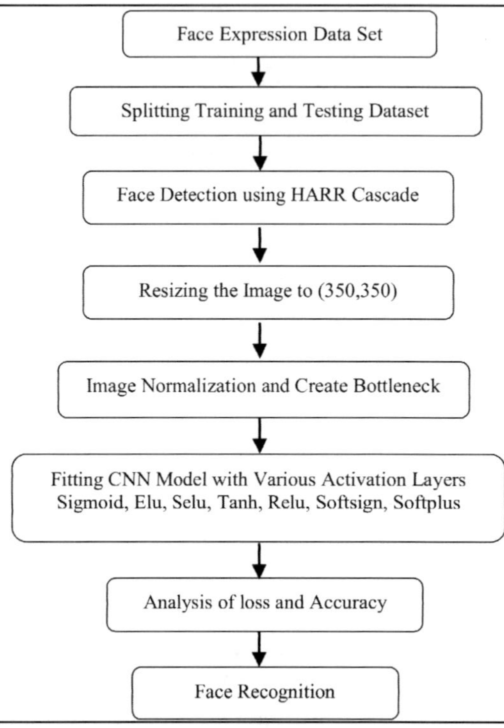

Figure 1. Overall Proposed Workflow

4. Implementation Setup

4.1. Dataset Exploratory Analysis

The facial expression dataset is extracted from the http://www.consortium.ri.cmu.edu/ckagree/, http://app.visgraf.impa.br/database/faces/ is subjected with the data processing. The python scripting language is coded in Spyder editor with Anaconda navigator for implementation. Data analysis is done for distribution of expression image in training and testing dataset and is shown in Figure. 2 – Figure.4.

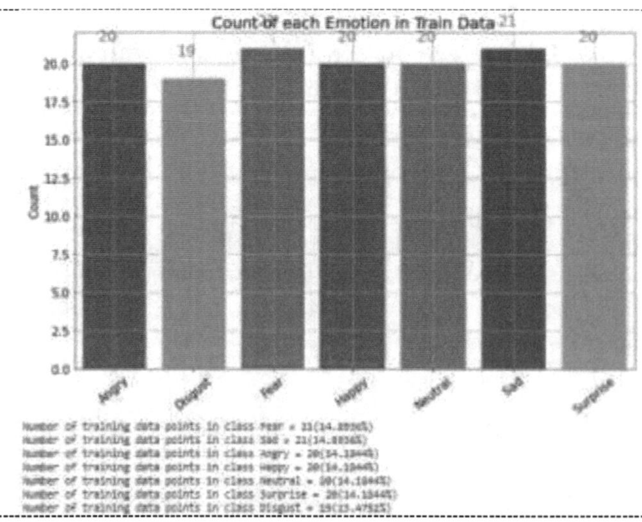

Figure 2. Distribution of Emotion of Training Data

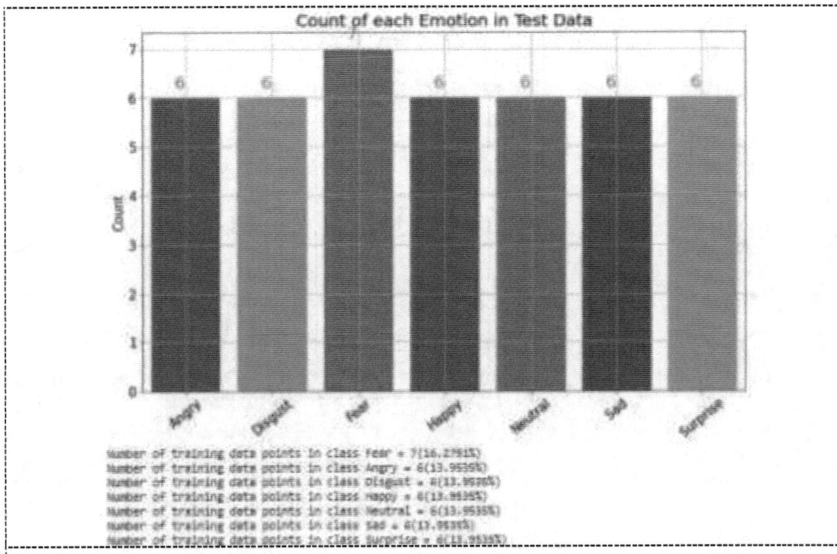

Figure 3. Distribution of Emotion of Testing Data

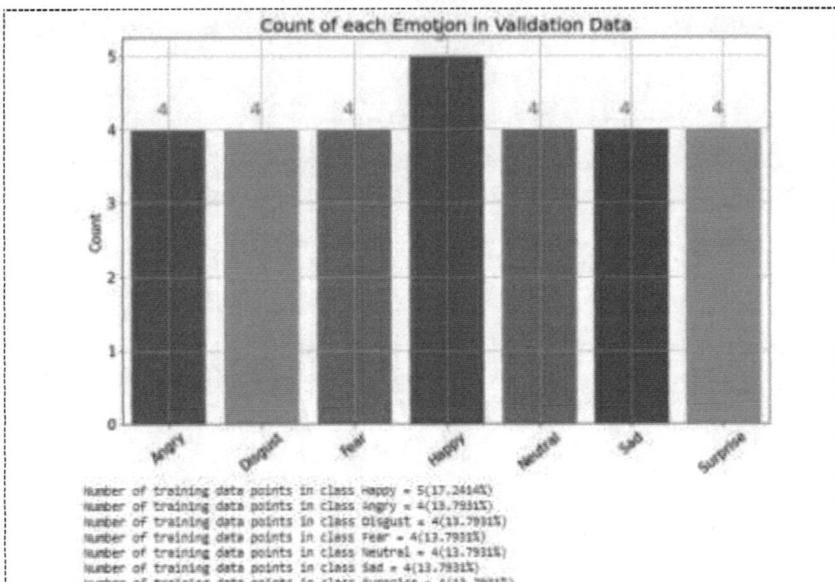

Figure 4. Distribution of Emotion of Validation Data

5. Results and Discussions

The facial expression images are detected with HAAR cascade and then the images are cropped with the size of (350, 350). The facial expression images is applied with normalized and the bottleneck features are created for training and testing data. The training dataset is fitted with convolutional sequential neural network models with Sigmoid activation layer. Table 1 shows the performance metric for the different Epoch with Sigmoid activation layer. The performance analysis of Epoch VS Accuracy and Epoch VS Loss is shown in Figure. 5-Figure. 6.

IOP Conf. Series: Materials Science and Engineering 1074 (2021) 012030 doi:10.1088/1757-899X/1074/1/012030

Table 1. Analysis of Performance Metric for different Epoch of SIGMOID

Epoch	Comb_Train_Loss	Comb_Train_Accuracy	CVHuman_Loss	CVHuman_Accuracy
1	2.055966	0.142857	2.15062	0.114286
2	1.694149	0.3	2.02234	0.0
3	1.627381	0.342857	1.936379	0.0
4	1.510428	0.4	1.883323	0.242857
5	1.523984	0.428571	1.870232	0.3
6	1.431663	0.457143	1.86721	0.3
7	1.481572	0.442857	1.873386	0.3
8	1.492955	0.414286	1.869783	0.314286
9	1.497246	0.342857	1.903426	0.314286
10	1.444708	0.428571	1.907336	0.271429
11	1.507477	0.385714	1.893588	0.257143
12	1.66884	0.3	1.922621	0.285714
13	1.430753	0.428571	1.827565	0.371429
14	1.83239	0.357143	1.878896	0.142857
15	1.554315	0.3	1.788014	0.314286
16	1.281836	0.542857	1.755762	0.485714
17	1.373914	0.631429	1.763743	0.457143
18	1.270418	0.697143	1.941378	0.214286
19	1.461964	0.704286	1.833551	0.214286
20	1.233822	0.791429	1.914558	0.271422

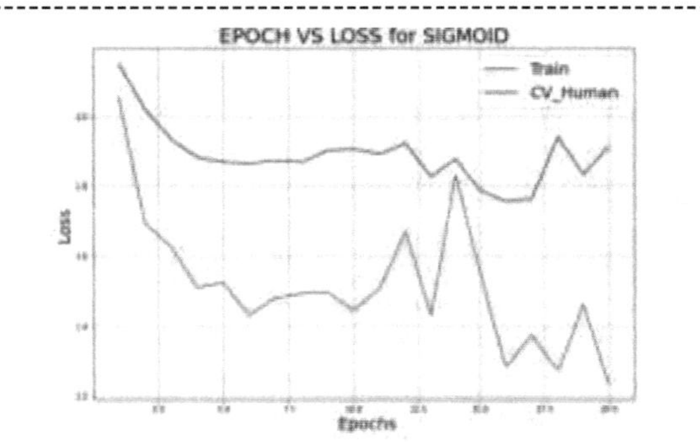

Figure 5.EPOCH VS LOSS for SIGMOID

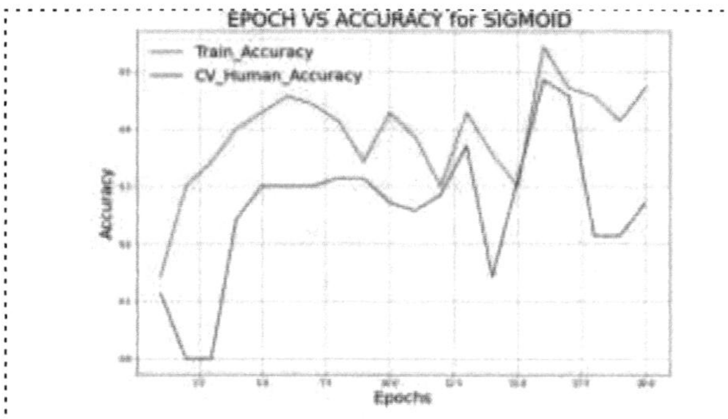

Figure 6. EPOCH VS ACCURACY for SIGMOID

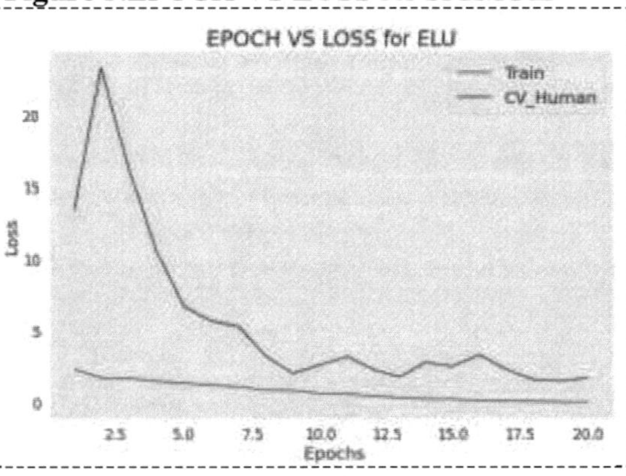

Figure 7.EPOCH VS LOSS for ELU

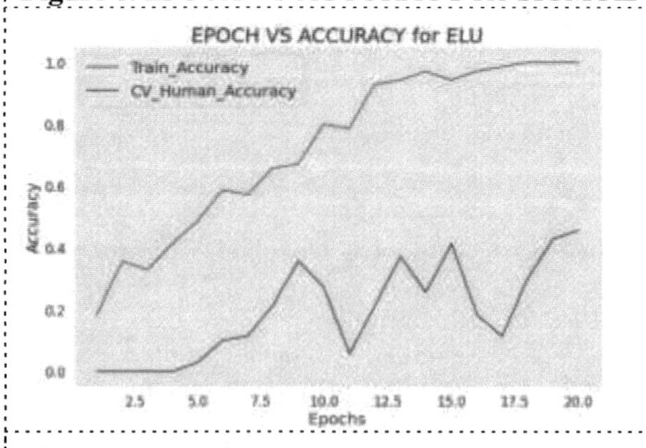

Figure 8. EPOCH VS ACCURACY for ELU

The training dataset is fitted with convolutional sequential neural network models with Elu activation layer. Table 2 shows the performance metric for the different Epoch with Elu activation layer. Performance analysis of Epoch VS Accuracy, Epoch VS Loss is shown in Figure. 7-Figure. 8.

Table 2. Analysis of Performance Metric for different Epoch of ELU

Epoch	Comb_Train_Loss	Comb_Train_Accuracy	CVHuman_Loss	CVHuman_Accuracy
1	2.303932	0.185714	13.380573	0
2	1.719868	0.357143	23.328525	0
3	1.673203	0.328571	16.39051	0
4	1.487079	0.414286	10.606923	0
5	1.375292	0.485714	6.738069	0.028571
6	1.204241	0.585714	5.717521	0.1
7	1.106813	0.571429	5.330394	0.114286
8	0.866124	0.657143	3.219572	0.214286
9	0.839699	0.671429	2.029013	0.357143
10	0.624861	0.8	2.621637	0.271429
11	0.611234	0.785714	3.189922	0.057143
12	0.443013	0.928571	2.274216	0.214286
13	0.333843	0.942857	1.771692	0.371429
14	0.229817	0.971429	2.786089	0.257143
15	0.242833	0.942857	2.510457	0.414286
16	0.142365	0.971429	3.314649	0.185714
17	0.134086	0.985714	2.281677	0.114286
18	0.099402	0.98	1.601833	0.3
19	0.058849	0.98	1.482906	0.428571
20	0.044989	0.99	1.698972	0.457143

The training dataset is fitted with convolutional sequential neural network models with Elu activation layer. Table 3 shows the performance metric for the different Epoch with Tanh activation layer. The performance analysis of Epoch VS Accuracy and Epoch VS Loss is shown in Figure. 9-Figure. 10.

Table 3. Analysis of Performance Metric for different Epoch of TANH

Epoch	Comb_Train_Loss	Comb_Train_Accuracy	CVHuman_Loss	CVHuman_Accuracy
1	2.162163	0.171429	2.389536	0.2
2	2.008764	0.185714	2.345485	0.157143
3	2.31146	0.114286	2.060338	0.185714
4	1.903016	0.257143	2.075945	0.2
5	2.027521	0.185714	2.123116	0.271429
6	2.192716	0.128571	2.336197	0
7	2.044704	0.171429	2.610682	0
8	2.081102	0.114286	2.711213	0
9	2.021846	0.214286	2.415396	0
10	2.02971	0.348571	2.312553	0
11	1.997615	0.392857	2.248497	0
12	1.975284	0.457143	2.115869	0.042857
13	1.891716	0.49	2.026604	0
14	1.959064	0.537143	2.248666	0.042857
15	1.95536	0.6	2.343453	0.2
16	1.972222	0.652857	2.317549	0.157143
17	2.005251	0.7	2.326868	0.114286
18	4.950364	0.757143	2.418729	0
19	5.970067	0.794286	2.418652	0
20	6.916914	0.817143	2.845998	0

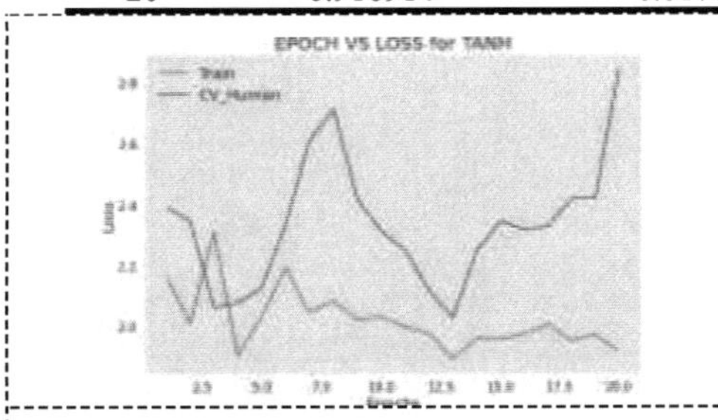

Figure 9.EPOCH VS LOSS for TANH

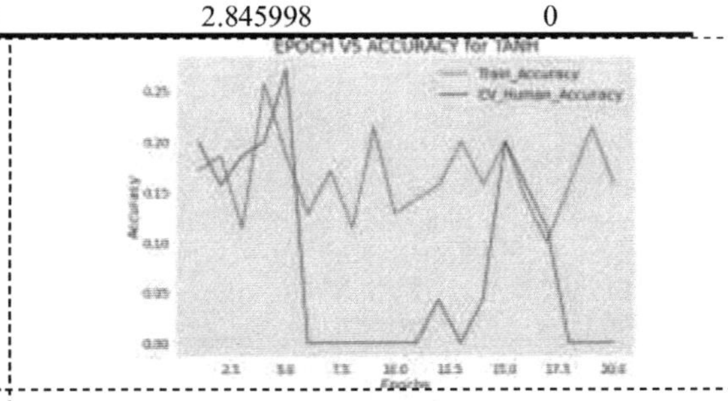

Figure 10. EPOCH VS ACCURACY for

IOP Conf. Series: Materials Science and Engineering 1074 (2021) 012030 doi:10.1088/1757-899X/1074/1/012030

TANH

The training dataset is fitted with convolutional sequential neural network models with Elu activation layer. Table 4 shows the performance metric for the different Epoch with Selu activation layer. The performance analysis of Epoch VS Accuracy and Epoch VS Loss is shown in Figure. 11-Figure. 12.

Table 4. Analysis of Performance Metric for different Epoch of SELU

Epoch	Comb_Train_Loss	Comb_Train_Accuracy	CVHuman_Loss	CVHuman_Accuracy
1	2.559849	0.185714	4.90646	0.185714
2	1.983575	0.242857	9.61076	0.3
3	1.475739	0.428571	13.452377	0.3
4	1.262215	0.514286	8.895803	0.3
5	1.322643	0.514286	4.371711	0.285714
6	1.102371	0.571429	3.892981	0.185714
7	0.957996	0.657143	3.54792	0.2
8	0.796078	0.771429	3.509347	0.2
9	0.631038	0.771429	2.292966	0.5
10	0.689604	0.742857	2.271856	0.457143
11	0.544641	0.814286	2.006077	0.385714
12	0.397772	0.871429	1.38054	0.442857
13	0.264411	0.942857	1.206447	0.457143
14	0.296198	0.928571	1.489429	0.4
15	0.218547	0.914286	1.690097	0.414286
16	0.163724	0.971429	2.512727	0.128571
17	0.143616	0.985714	3.048205	0.185714
18	0.103006	0.985714	2.525985	0.314286
19	0.111789	0.957143	2.655915	0.314286
20	0.080792	0.96	4.199882	0.114286

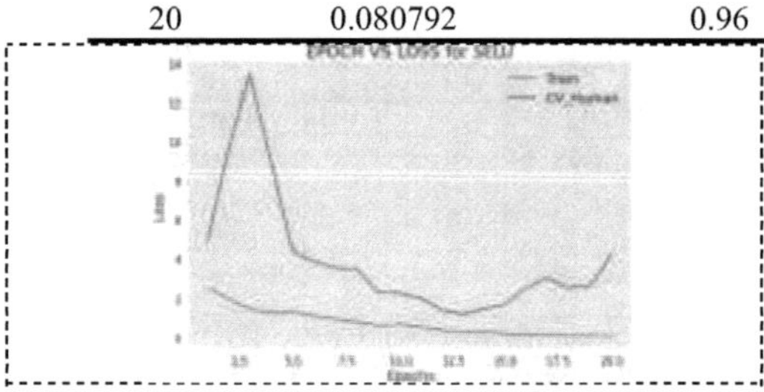

Figure 11.EPOCH VS LOSS for SELU

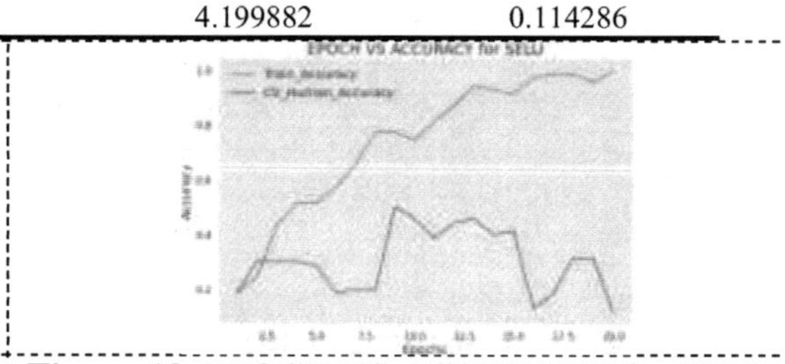

Figure 12. EPOCH VS ACCURACY for SELU

The training dataset is fitted with convolutional sequential neural network models with Elu activation layer. Table 5 shows the performance metric for the different Epoch with softsign activation layer. The performance analysis of Epoch VS Accuracy and Epoch VS Loss is shown in Figure. 13-Figure. 14.

Table 5. Analysis of Performance Metric for different Epoch of SOFTSIGN

Epoch	Comb_Train_Loss	Comb_Train_Accuracy	CVHuman_Loss	CVHuman_Accuracy
1	1.935812	0.271429	1.949712	0.2
2	1.892569	0.228571	1.897579	0.3
3	1.783687	0.257143	1.909734	0.3
4	1.716713	0.4	2.066091	0.3
5	1.700327	0.342857	1.959024	0.3
6	1.758143	0.271429	1.944506	0.3
7	1.637998	0.428571	2.053937	0.271429
8	1.644241	0.49	2.230492	0.142857
9	1.843184	0.542857	2.055639	0.3
10	1.857275	0.657143	1.935686	0.3
11	1.639422	0.657143	2.150152	0.285714
12	1.686175	0.7	2.481322	0.028571
13	1.565544	0.728571	2.778958	0.2
14	1.579589	0.742857	2.918343	0.22
15	1.576361	0.742857	2.956654	0.24
16	1.640967	0.8	2.716831	0.24
17	1.539839	0.81	2.12868	0.23
18	1.589249	0.824286	2.097143	0.25

19	1.531202	0.82	2.240794	0.26
20	1.532373	0.844286	2.356979	0.27

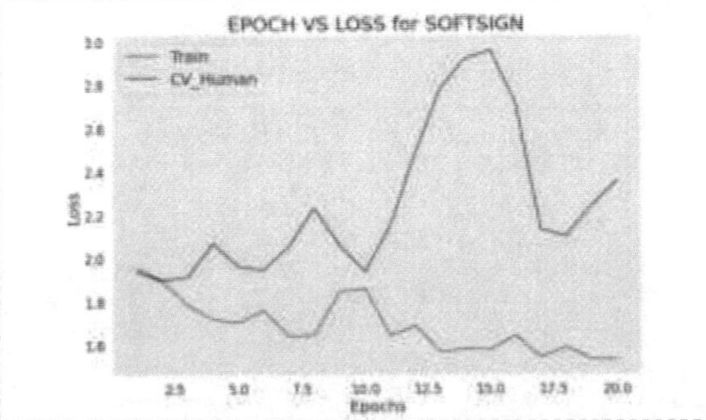

Figure 13.EPOCH VS LOSS for SOFTSIGN

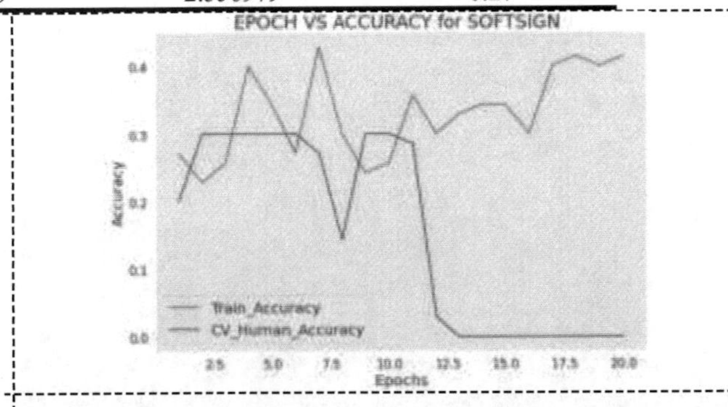

Figure 14. EPOCH VS ACCURACY for SOFTSIGN

Vv

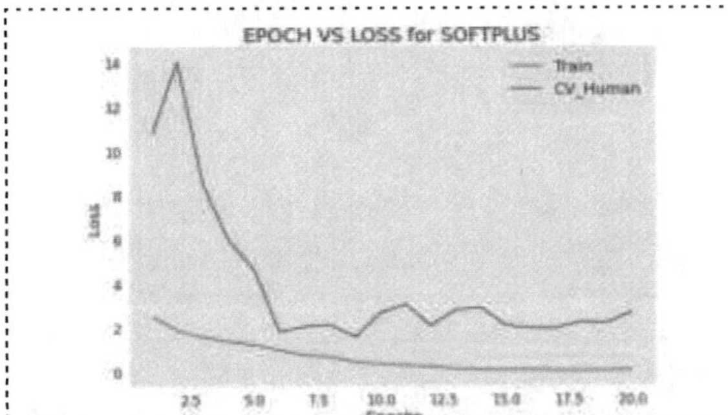

Figure 15 .EPOCH VS LOSS for SOFTPLUS

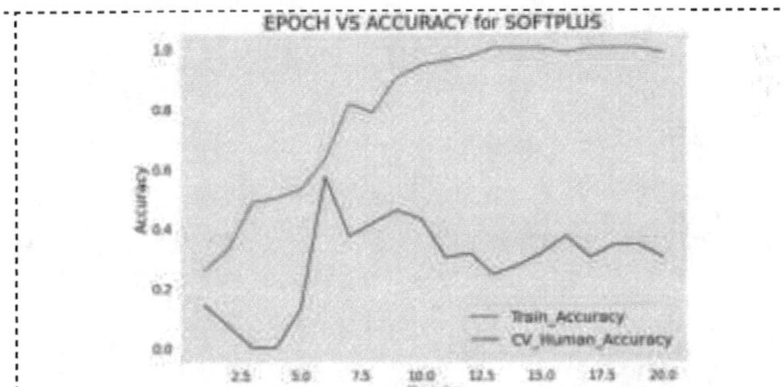

Figure 16. EPOCH VS ACCURACY for SOFTPLUS

The training dataset is fitted with convolutional sequential neural network models with Elu activation layer. Table 6 shows the performance metric for the different Epoch with Softplus activation layer. Performance analysis of Epoch VS Accuracy and Epoch VS Loss is shown in Figure. 15-Figure. 16.

Table 6. Analysis of Performance Metric for different Epoch of SOFTPLUS

Epoch	Comb_Train_Loss	Comb_Train_Accuracy	CVHuman_Loss	CVHuman_Accuracy
1	2.521988	0.257143	10.769857	0.142857
2	1.858173	0.328571	14.013214	0.071429
3	1.586269	0.485714	8.450514	0
4	1.371293	0.5	5.912697	0
5	1.22164	0.528571	4.612834	0.128571
6	0.994899	0.628571	1.815635	0.571429
7	0.759172	0.814286	2.047843	0.371429
8	0.679914	0.785714	2.116582	0.414286
9	0.456746	0.9	1.590425	0.457143
10	0.361784	0.942857	2.683431	0.428571
11	0.291873	0.957143	3.042206	0.3
12	0.229183	0.971429	2.075521	0.314286
13	0.141551	1	2.790561	0.242857
14	0.114004	1	2.889609	0.271429
15	0.072381	1	2.089473	0.314286
16	0.084483	0.985714	1.952255	0.371429
17	0.057751	1	1.957056	0.3
18	0.054487	1	2.250003	0.342857
19	0.059941	1	2.20055	0.342857
20	0.089052	0.985714	2.690978	0.3

IOP Conf. Series: Materials Science and Engineering 1074 (2021) 012030 doi:10.1088/1757-899X/1074/1/012030

The training dataset is fitted with convolutional sequential neural network models with Elu activation layer. Table 7 shows the performance metric for the different Epoch with Relu activation layer. The performance analysis of Epoch VS Accuracy and Epoch VS Loss is shown in Figure. 17-Figure. 18.

Table 7. Analysis of Performance Metric for different Epoch of RELU

Epoch	Comb_Train_Loss	Comb_Train_Accuracy	CVHuman_Loss	CVHuman_Accuracy
1	2.303362	0.157143	10.111909	0.271429
2	1.999738	0.257143	9.292259	0.3
3	1.843891	0.357143	5.913297	0.3
4	1.657243	0.314286	7.447931	0.3
5	1.639881	0.342857	5.799656	0.257143
6	1.515211	0.314286	5.677063	0.2
7	1.355517	0.471429	3.825484	0.142857
8	1.279170	0.471429	2.265329	0.085714
9	1.358319	0.457143	6.458704	0
10	1.229157	0.542857	7.298427	0
11	0.982870	0.600000	4.836370	0.128571
12	0.911965	0.657143	3.400242	0.257143
13	0.771945	0.742857	2.544070	0.114286
14	0.663263	0.828571	1.747203	0.242857
15	0.562241	0.857143	1.785336	0.2
16	0.441646	0.928571	2.436385	0.2
17	0.347169	0.971429	2.715715	0.2
18	0.285406	0.971429	2.387037	0.314286
19	0.262842	0.957143	2.294834	0.328571
20	0.197343	1.000000	3.707844	0.2

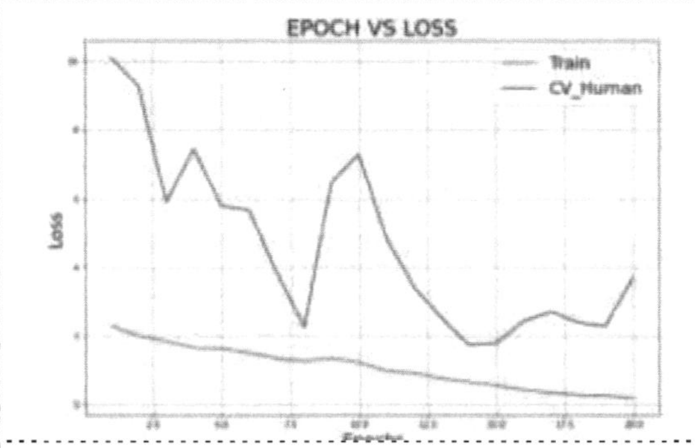

Figure 17. EPOCH VS LOSS for RELU

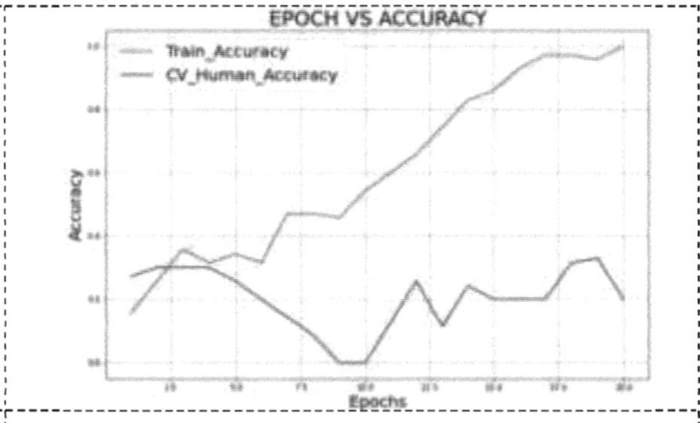

Figure 18. EPOCH VS ACCURACY for RELU

Table 7 shows the Accuracy performance metric for the different Epoch with all activation layers of CNN modes and the analysis is shown in Figure. 19.

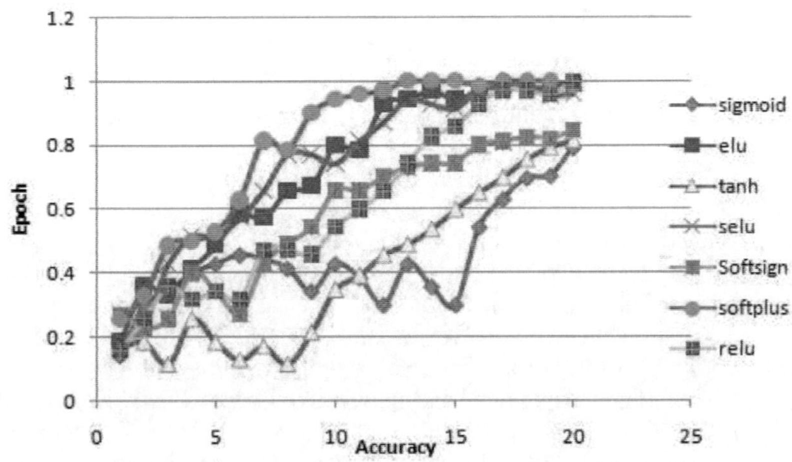

Figure 19. Accuracy Analysis of Activation Layers

Table 8. Accuracy Analysis of Activation Layers of CNN Models

Epoch	sigmoid	elu	tanh	selu	Softsign	softplus	relu
1	0.142857	0.185714	0.171429	0.185714	0.271429	0.257143	0.157143
2	0.3	0.357143	0.185714	0.242857	0.228571	0.328571	0.257143
3	0.342857	0.328571	0.114286	0.428571	0.257143	0.485714	0.357143
4	0.4	0.414286	0.257143	0.514286	0.4	0.5	0.314286
5	0.428571	0.485714	0.185714	0.514286	0.342857	0.528571	0.342857
6	0.457143	0.585714	0.128571	0.571429	0.271429	0.628571	0.314286
7	0.442857	0.571429	0.171429	0.657143	0.428571	0.814286	0.471429
8	0.414286	0.657143	0.114286	0.771429	0.49	0.785714	0.471429
9	0.342857	0.671429	0.214286	0.771429	0.542857	0.9	0.457143
10	0.428571	0.8	0.348571	0.742857	0.657143	0.942857	0.542857
11	0.385714	0.785714	0.392857	0.814286	0.657143	0.957143	0.6
12	0.3	0.928571	0.457143	0.871429	0.7	0.971429	0.657143
13	0.428571	0.942857	0.49	0.942857	0.728571	1	0.742857
14	0.357143	0.971429	0.537143	0.928571	0.742857	1	0.828571
15	0.3	0.942857	0.6	0.914286	0.742857	1	0.857143
16	0.542857	0.971429	0.652857	0.971429	0.8	0.985714	0.928571
17	0.631429	0.985714	0.7	0.985714	0.81	1	0.971429
18	0.697143	0.98	0.757143	0.985714	0.824286	1	0.971429
19	0.704286	0.98	0.794286	0.957143	0.82	1	0.957143
20	0.791429	0.99	0.817143	0.96	0.844286	0.985714	1

6. Conclusion

An attempt is made in this paper to analyze the performance of Convolutional neural network models with diverse activation layer for the performance evaluation. The facial expression images is applied with normalized and the bottleneck features are created for training and testing data. The training dataset is fitted with convolutional sequential neural network models with various activation layers like Sigmoid, Elu, Relu, Selu, Tanh, Softsign and Softplus. Performance analysis is done with loss and accuracy for all the epoch of all CNN models for all the activation layers. Experimental results shows that CNN sequential model with Relu activation layer is found to have the accuracy of 100%.

References

[1] Huang Y Chen F Lv S and Wang X 2019 *Facial Expression Recognition: A Survey.* Symmetry vol 11 1189.

[2] Cowie R Douglas Cowie E Tsapatsoulis N Votsis G Kollias S Fellenz W and Taylor J G 2001 *Emotion recognition in human-computer interaction* IEEE Signal Process Mag vol 18 p 32–80

[3] Jabon M Bailenson J Pontikakis E Takayama L and Nass C 2011 *Facial expression analysis for predicting unsafe driving behavior* IEEE Transactions on Pervasive Computing vol 10 p 84–95

[4] Tian Y I Kanade T and Cohn J F 2001 *Recognizing action units for facial expression analysis*

IEEE Trrans. Pattern Anal. Mach. Intell. 2001, 23, 97–115.

[5] Samal A and Iyengar P A 1992 *Automatic recognition and analysis of human faces and facial expressions: A survey* Pattern Recognition journal vol 25 p 65–77

[6] Deshmukh S Patwardhan M and Mahajan 2016 *A Survey on real-time facial expression recognition techniques*. IET Biometric Journal vol 5 p 155–163

[7] Chen W Er M J and Wu S 2006 *Illumination compensation and normalization for robust face recognition using discrete cosine transform in logarithm domain* IEEE Transaction System. Management Cybernatics vol 36 p 458–466

[8] Yu J and Bhanu B 2006 *Evolutionary feature synthesis for facial expression recognition* Pattern Recognition Letter vol 27 p 1289–1298

[9] Tsai H H and Chang Y C 2018 *Facial expression recognition using a combination of multiple facial features and support vector machine* Soft Computing vol 22 p 4389–4405

[10] Donato G Bartlett M Hager J Ekman P and Sejnowski T 1999 *Classifying facial actions*. IEEE Transaction on Pattern Analysis and Machine Intelligence vol 21 no 10 p 974–989

[11] Pantic M and Rothkrantz L 2000 *Expert system for automatic analysis of facial expression* Image Vis. Comput vol 18 no 11 p 881–905

Predicting User's Web Navigation behaviour using AMD and HMM Approaches

Rita Roy[1], G Appa Rao[2]
Research Scholar[1], Professor[2]
Department of CSE, GITAM Deemed to be university
Ritaroy89@gmail.com[1], apparao.giduturi@gitam.edu[2]

Abstract. The academics were introducing various techniques in web usage mining to reduce user latency time to improve their web performance. The server-side web log data helps to identify the most appropriate pages, based on the user's request. Analyzing web log data creates difficulties since it comprises exhaustive web page data. This paper proposes a new technique in which the web log data we can preprocess in order to extract sequence and navigation patterns useful to predict. In this paper, we cluster users into communities labelled by the websites most frequently visited, to discover their preferences and to describe website reorganisation. The URL sequence navigated by a user during a session. Session denotes a user's browsing pattern. Multiple users' sessions are clustered using the hierarchical clustering technique to analyse the sequence in their navigation patterns. The two model predictions are Adaptive Mahanabolis Distance (AMD) and Hidden Markov model (HMM), providing a list of web pages of interest to the user. We validate the proposed framework on NASA, Clarknet and serc datasets web log files.

Keywords: Prediction model, clustering, User Sessions, Adaptive Mahanabolis Distance

1. Introduction

As a practice, Internet surfing logs on internet services have been a crucial source of information to evaluate hidden preferences of users and to conclude their authentic lives. With a better knowledge of the mobile usage pattern, network service companies can offer more customised offerings and enhance the efficiency of the service. The browsing preferences of users in fields like community design, smartphone ads, transportation, schooling etc. are often beneficial [1–3]. Web use mining relates to the identification of interesting web access trends from the background of web data. Typically, an organization's Database servers monitor its users' web behaviour in a file named log file. Three forms of log files are available, which can be used for web mining. On the web, network and proxy clients, log files are processed. It allows mining quite complicated by providing over one place to store information concerning navigation patterns of users. Only when one requires data from all three, log they will achieve file forms. This is because the server-side may not include archives of the webpage accesses stored on the proxy site or on the client's side. Besides the application log sheet, which includes detailed information on the proxy server. Therefore, the client-side page queries are absent. However, it is challenging to gather all customer information. Most algorithms therefore only work on the server side.

While the user accessing the website, the user transactions are monitored and stored as log. The log contains unstructured format, the conversion of unstructured to structured format by preprocessing technique. The weblog comprises various entries like IP address of user, date/time, categories of product and status code. The preprocessing is a three-step process. The data cleaning cleans the irrelevant data in a log. The user identification is identified by IP address of the specified system. It classifies the session identification based on session management. Identifying user needs and user interest is a laborious task. With the help of log, the user needs and interest can identify by user navigation pattern. They can analyse the user navigation pattern based on the previous user navigation that was stored on the weblog. Using pattern discovery, the processed log is to convert it into sequence and sub sequence of similar pattern. By using forward and backward technique, it can generate subsequence. They group the sub sequence pattern to form a cluster that will helpful to identify the user needs. Some commonly used data mining algorithms for Web usage mining are association rule mining, sequence mining and clustering.

Clustering studies how domain expertise can enhance the efficiency of clusters. It presents domain information as a collection of constraints to apply to the clusters. Orthodox methods to semi-supervised (or restricted) clustering use one of three restrictions. Next, it may update an established clustering algorithm to take into consideration. This method has been adopted by COPKMeans [4], among the first clustering algorithms to solve pairs of restrictions. Second, a distance metric focused on the limitations [5] may be taught, during which it uses the metric in a conventional cluster analysis. Third, the above two techniques may be merged and the so-called hybrid methods developed [6].

2. Related works

In this section, we briefly review some of the methods that are proposed for clustering for webpage prediction techniques with relevant information:

In [7], the authors implemented an intuitive difference test to quantify the issues of site users with expanded web user sessions. It bases the suggested assessment of the disparity of usage between two web user sessions on the page importance, the syntactic page URL structure, and the hierarchical website structure. This suggested they have used logical dissimilarity test for research with the K-Medoids clustering algorithm and contrasts the findings to other independent dissimilarity measurements. Two unchecked cluster validity indexes assessed the significance of the created clusters. The findings from the experiment show that intuitively enhanced session dissimilarity metrics are more practical and superior to the other separate cluster validity index dissimilarity measurements. In [8], authors mentioned how site log data on the server side allows users to find the most suitable pages. Web log data review presents difficulties since it contains enough details about a web page. They developed a novel strategy for pre-processing Web log data to derive a predictable series of events and navigation patterns. They scan every URL in the server log data into web-based tokens. Tokens for URL identification are individually defined. The URL sequence, which a user navigates for 30 minutes, is known as a session. Session reflects a user's browsing routine. It groups multiple user sessions using the hierarchical clustering method to evaluate the sequence occurrence of the browsing trends.

Each cluster recognises the session as the representative since it includes as many pages as possible in the sequence, whereas the representative session is a subset for other sessions in the cluster. Session model navigation trends are helpful to predict the user request pages most suitable. In [9], the authors found that site user session clustering is highly necessary to recognise their internet browsing behaviour. Users with common browsing activities and study by domain specialists in discovered user classes may yield valuable and constructive feedback. A conglomerate clustering technique is proposed to classify site users' clusters based on their browsing activity from the Web server access logs. The algorithm provided incorporates interesting concepts from the subtractive and associated c-mean clustering approach. This algorithm operated in the first step and automatically identified the number of clusters dependent on each relational data and its respective centres' successive subtractive potential density value (centroid). In the second step, it allocates fuzzy membership values to fuzzy clusters from some kind of relational matrix. The proposed method is an improvement in session dissimilarity matrix from the NASA web server log data that is freely available.

In [10], the authors presented an alternative method to creating sessions, the first big phase in network use mining. The suggested approach will get all potential Site users' full navigation sequences. Finding suggest that when our current method is used, previous methods in web-based mining applications are outperformed, for example next-page prediction.

In [11], Relational Medoid-based fuzzy relational clustering (FRC) algorithms are common to the analysts and work better than oriented FRC. In medoid FRC, however, medoid collection is random and often contributes to conflicting outcomes. This paper suggests a subtractive medoid sorting approach focused on the SMS-FRC relational clustering. Geometry and density of peer-

IOP Conf. Series: Materials Science and Engineering 1074 (2021) 012031 doi:10.1088/1757-899X/1074/1/012031

referential dissimilarity values in the SMS-FRC algorithm are favoured over arbitrary initial values of medoids. The SMS-FRC is used to classify clusters of user sessions based upon the browsing behaviour of server log info. The definition of enhanced sessions is used to extract the intuitive matrix of expanded dissimilarity dependent on the page relevance.

In [12], the authors suggest a method to decrease the user's search time and meet the user's predicted intent website (request) through an increased prediction accuracy through the combination of the clustering fuzzy c-means and the model recommendation vector order Markov. At first, methods that allow are performed by applying a web log preprocessing method accompanied by a c-means clustering method to classify patterns of similarity. After that, it renders the recommendation of the web page utilising a model of variable orders to predict access to the next web page by reducing search time and improved predictive accuracy. In [13] the authors recommend a contribution to solving prediction space and to increase precision by integrating CPT+ and PageRank algorithms. In [14] the authors present a global eye monitoring index descriptive named heat map entropy and distinguish its webpage aesthetic predictive benefit. All the above works are inadequate for non-convex data, comparatively vulnerable to surfaces, quickly drawn to optimum local data, the sum of clusters required to be preset and the product of a cluster sensitive to the number of clusters. Our framework that can cope with them and results in optimum efficiency, thus is implemented by implementing the Adaptive Mahanabolis Distance algorithm and the K-means clustering algorithm that can be addressed later.

3. Material and Methods

A system has been proposed, in which session identification is utilized and implemented based on the user who spent time on each page, and also, we analyze the number of webpages accessed by an individual user in a particular session. Potential and non-potential web users were identified from the web server log file. Adaptive Mahanabolis Distance clustering algorithm and K-means clustering algorithm (for comparison purpose) were used to identify the frequently accessed pages. We focus on these models on weight variables, such as time spent on the webpage and the maximum amount of page accesses. The analysis also reveals that performing the AMD cluster algorithm is better than the clustering algorithm of K-means for web log results. The analysis proves the AMD has a substantial advantage in terms of the execution period over the K-means clustering algorithm. The aim of this paper is to effectively predict and prefect a website in order to minimize latency. The overview of the pre-fetch and prediction unit is shown in Figure 1 with necessary details and records for the phase after pre-processing. We have established unique users based on the unique IP address. After we have established the particular user, web users have been classified by web session. A web session is a period the web user takes on a single website. The web session begins from the user's login period to the site's leaving time. We may develop session Ids to store those variables as web visitors are going through the website sites. Before presenting the log file analysis technique, this portion of work is a minor part in considered objectives, which minimize the log file size and cluster the frequently accessed webpages by considering time spent on the page as a weight factor.

Algorithm 1: Data Cleaning
Input: Raw Web log Data
Output: Pre-processed Web log Data
1. ***begin***
2. Read Web log data from log file
3. Parse web log data into tokens
4. ***If*** (web log data! =null) *then*
5. Assign group based on the following Categories
6. ***If*** (Web log data.url= "*.jpg,*.gif, *.png*") *then*
7. Assign as Unwanted Images Files
8. **else if** (Web log data.url= "*.flv,*. avi, *. mpg, *.mpeg")
9. Assign as Unwanted Video Files

10. **endif**
11. **endif**
12. Save Records
13. Repeat until the last record
14. ***end***

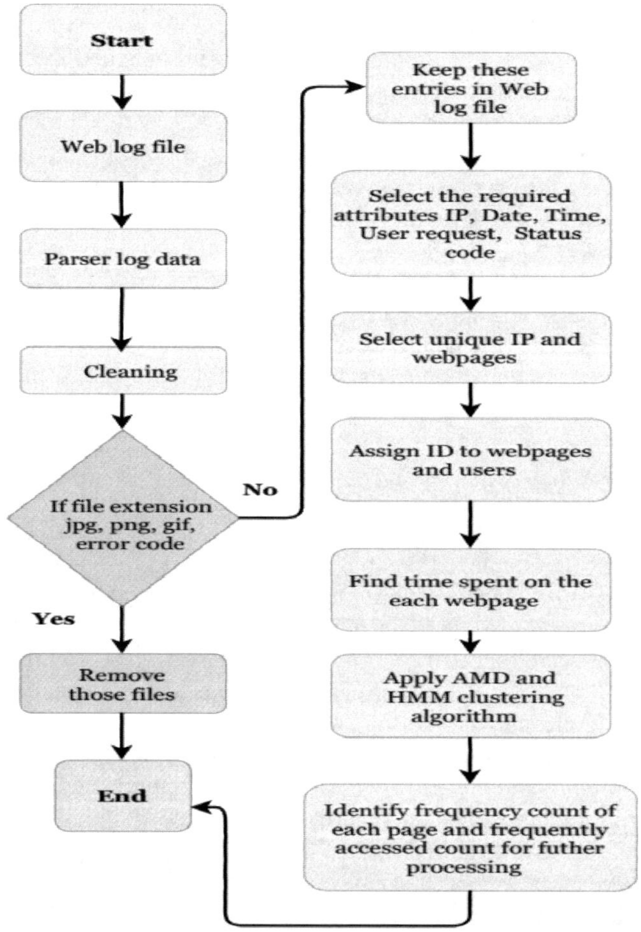

Figure 1: Proposed Model of the System

Algorithm 2 explains user identification algorithm. We can identify Web users from the web server log file constructed on their IP address. If the same IP address exists, we categorize them as same users. If the IP address is different, then the user is different and is identified from the user identification algorithm.

Algorithm 2: User Identification
Input: Pre-Processed Web log data
Output: Frequently accessed webpages
1. ***begin***
2. Read the Pre-Processed Web log data
3. Compare the IP address of two consecutive records
4. ***If*** (IP address = same) then
5. Assign as the Same User
6. ***else***
7. Distinct User

IOP Conf. Series: Materials Science and Engineering 1074 (2021) 012031 doi:10.1088/1757-899X/1074/1/012031

8. ***endif***
9. ***end***

Each clustering algorithm seems to have an objective function that directs the search for potential solutions. This function assesses candidate solutions and shows how good (bad) each would be. The method then seeks to optimize (reduce) the function value. If we provide the algorithm with a certain number of constraints, we request that the algorithm only takes into account solutions which completely fulfil the constraints. Clustering is one of the data processing methods used for grouping related data. The algorithm Adaptive Mahanabolis Distance (AMD) depends on iteratively recalculating the distance between Mahalanobis and a neighborhood center.

Algorithm 3: **Clustering using Adaptive Mahanabolis Distance(AMD)**
Input: Pre-Processed Web log data
Output: Frequently accessed webpages

1. ***begin***
2. Assume an isotropic Mahalanobis distance ($\Sigma = I$)
3. Find the closest K neighbors by computing the Mahalanobis distance
4. Using K neighbors, compute the covariance
5. ***If*** covariance = Mahalanobis distance go to step 6
6. ***else*** go to step 2 and re-compute the Mahalanobis distance
6. Record the points as neighbors of dataset
7. ***en***d

Algorithm 4 depicts the proposed algorithm for session identification. It identifies the potential and non-potential users based on time stamp and number of pages accessed by the user. Webpages in which the user spent time more than the threshold value (i.e. 60 seconds) is considered. The web users who accessed more the three pages were considered as potential users and the users who access less than three pages were considered as non-potential users.

Algorithm 4: Session Identification
Input: User Identified Web log data
Output: Maximum time spent on the webpage and the potential web user
1. ***begin***
2. Let $P \leftarrow$ Potential User;
3. Let $NP \leftarrow$ Non-Potential User;
4. Let $T \leftarrow$ Threshold Value;
5. Let $Pi \leftarrow$ first entry of log file in a particular session;
6. Let $Pj \leftarrow$ second entry of log file in the same session;
7. Compare two consecutive entries within the session;
8. *if* timestamp (Pi) is < timestamp (Pj) then
9. The difference between the timestamp Pi&Pj should be greater than T;
10. ***endif***
11. ***if*** each user accessed more than three pages within the session then
12. Assign P;
13. ***else***
14. Assign NP;
15. ***endif***
16. ***end***

After identifying the potential users from web server log file using Algorithm 4 and then applied clustering algorithms to find frequently accessed webpages.

IOP Conf. Series: Materials Science and Engineering 1074 (2021) 012031 doi:10.1088/1757-899X/1074/1/012031

3.1 Framework of Prediction and Prefetching:

As Figure 2 shows, the suggested model is applied relying on the log file of the web server. We have collected it on the NASA website [15]. The raw web log data gathered includes irrelevant data. It must then be pre-processed and grouped we know such that commonly visited webpages. We tracked each web user interaction and its actions on a specific website. Once their actions and interest have been established, the process of prediction begins by examining the previous web user behavior. Table 1 displays the user navigation sample list. The accessed user 1 series was recorded from the table as P1→P3→P5→P7→P9 etc.

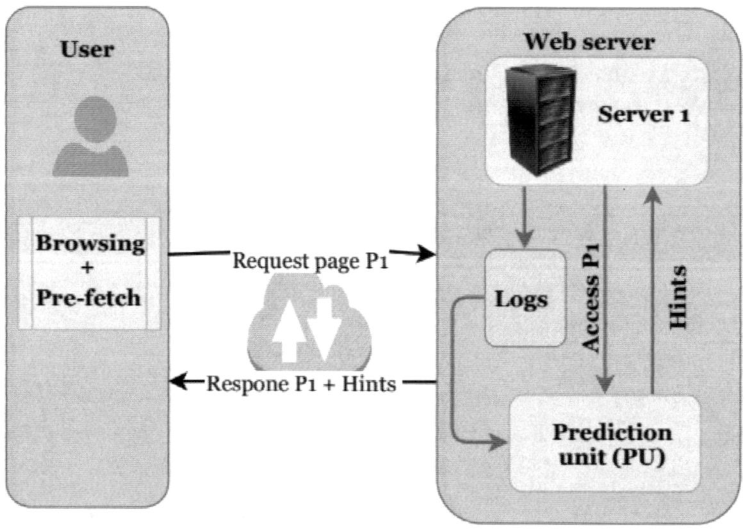

Figure 2: Prediction and Prefetching Framework

Table 1: User Navigation

User 1	User 2	User N
P1	P1	P1
P3	P2	P6
P5	P3	P7
P7	P4	P8
P9	P5	P9

Let us consider a website comprising *w* number of webpages, namely up_1, up_2, ... up_n and number of clusters C_1, C_2,....C_n. Let Web Server Log File be referred to as *WSLF*. The pre-processed log file contains the user id, name of the page and time spent. The pre-processed log file is $WSLF_1$, $WSLF_2$..... $WSLF_n$. Calculate the overall access rate for each cluster, determined by taking into consideration the amount of times the page requires for a cluster. We arrange the web pages according to the method suggested. After analyzing the use of web pages for the time needed (every day, weekly), we generate the clusters based on web pages regularly visited. Calculate the average web page access level for each cluster. We propose an algorithmic architecture for the reorganization of the built environment. We propose an algorithm for reorganizing web pages. After the webpages are frequently accessed, threshold frequency pages are expected, and the number of users viewed by the same Hidden Markov Model (HMM) page is stated to be ideal for discovering several sites to be loaded and it can be submitted to a browser as a hint for pre-fetching.

$$Thr_f = \text{Frequency count (FC)} > \text{(threshold) } \lambda \qquad (1)$$

Web pages frequently visited were detected by examining a user's previous behavior. Prediction is not dependent on a previous user's regularly visited websites. Predictions focused solely on frequent count contribute to mis-prediction, and there would be no satisfactory prediction outcomes from this method.

3.2 Hidden Markov Model to Predict User's Browsing Behavior:

After processing the data unit, we need to build an effective model that contains the parameter structure of the model. The consistency of the model depends on the state transition diagrams and the related parameters. At first, we assign a value between 0 and 1 to arbitrarily assign model parameters (transition likelihood A), satisfying $a_{11} + a_{12} = 1$ and $a_{21} + a_{22} = 1$ (see Figure 3). And the original distribution of odds, $\pi 1$ and $\pi 2$ ($\pi 1 + \pi 2 = 1$). We use the Baum-Welch algorithm to reach an acceptable parameter. We use parameters $\lambda = (A, B, \pi)$ to simultaneously reveal secret states when browsing the website through the Viterbi algorithm.

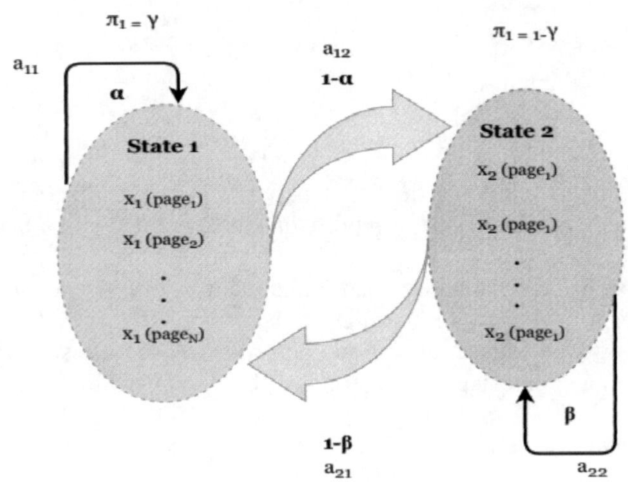

Figure 3: The training phase of HMM

Here N separate web pages and two hidden states are present in our method, called S_1 and S_2. The $x_1(P_1)$ notation shows the possibility that page P_1 is S_1. The Baum-Welch algorithm describes the formula as:

$$\xi_t(S_1, S_2) = \frac{\alpha_t(S_1) a_{12} x_2(o_{t+1}) \beta_{t+1}(S_2)}{P(O \mid \lambda)} \quad (2)$$

Furthermore $\sum_{t=1}^{T-1} \xi_t(S_1, S_2)$ is the expectation probability of all state transit from S_1 to S_2. Thus, the probability of state S_1 at time t, $\gamma_t(S_1)$, can be represented as: $\sum_{j=S_1}^{S_2} \xi_t(S_1, j)$ .We can estimate the new model parameter by the following equations:

$$\pi_1^* = \gamma_1(S_1) = \sum_{j=S_1}^{S_2} \xi_t(S_1, j) = \xi_1(S_1, S_1) + \xi_1(S_1, S_2) \quad (3)$$

$$\pi_2^* = \gamma_1(s_2) = \sum_{j=S_1}^{S_2} \xi_t(s_2, j) = \xi_1(s_2, s_1) + \xi_1(s_2, s_2) \quad (4)$$

The estimated transition probability $a^*_{S_1,S_2}$ thus is ratio of expected numbers of transition from S_1 to S_2expected numbers of transition from S_1 $= \dfrac{\sum_{t=1}^{T-1} \xi_t(s_1, s_2)}{\sum_{t=1}^{T-1} \gamma_t(s_1)}$. And $x_1\left(page_N\right)^*$ ratio of expected number of times in states with page$_N$ to expected number of times in states S_1

$$= \frac{\sum_{t=1, o_t = page_N}^{T-1} \gamma_t(s_1)}{\sum_{t=1}^{T-1} \gamma_t(s_1)}$$

When a user joins the web (log server records o1). Our approach can predict the user's purpose automatically. Similarly, as he selects on the first tab, our process simultaneously guesses the goal and so on. We generate tests to validate our approach in this article. We are using NASA details as a benchmark (https:/www.kaggle.com/souhagaa/nasa-access-log-dataset-1995). This information is a log file for the Kennedy Space Center Logs that is recorded in first service access (1 July to 31 July 1995) and second server log (August 1st to August 31st 1995). This is because it is the forecast of a platform offering links to aerospace quotations, full-text online files, and pictures and videos. Conference articles, journals, patents, study findings, photographs, films, and technical videos, STIs produced or sponsored by NASA to display user purpose, we understand it is difficult to carry every intention behind users while searching the Web. We are therefore concerned with the final discovery emblem (page) deciding if our system might discover its secret condition. The pages beyond the threshold frequency shall signify the most commonly visited page and the cumulative number of users visiting the same page.

4. Results and Discussion

Weblog file provides details on user behaviour of the website. Any of the various log types found [16,17] are SEC.gov, NASA and ClarkNet.

```
199.72.81.55 - - [01/Jul/1995:00:00:01 -0400] "GET
/history/apollo/ HTTP/1.0" 200 6245
unicomp6.unicomp.net - - [01/Jul/1995:00:00:06 -0400] "GET
/shuttle/countdown/ HTTP/1.0" 200 3985
199.120.110.21 - - [01/Jul/1995:00:00:09 -0400] "GET
/shuttle/missions/sts-73/mission-sts-73.html HTTP/1.0" 200 4085
burger.letters.com - - [01/Jul/1995:00:00:11 -0400] "GET
/shuttle/countdown/liftoff.html HTTP/1.0" 304 0
199.120.110.21 - - [01/Jul/1995:00:00:11 -0400] "GET
/shuttle/missions/sts-73/sts-73-patch-small.gif HTTP/1.0" 200
4179
burger.letters.com - - [01/Jul/1995:00:00:12 -0400] "GET
/images/NASA-logosmall.gif HTTP/1.0" 304 0
burger.letters.com - - [01/Jul/1995:00:00:12 -0400] "GET
/shuttle/countdown/video/livevideo.gif HTTP/1.0" 200 0
205.212.115.106 - - [01/Jul/1995:00:00:12 -0400] "GET
/shuttle/countdown/countdown.html HTTP/1.0" 200 3985
d104.aa.net - - [01/Jul/1995:00:00:13 -0400] "GET
/shuttle/countdown/ HTTP/1.0" 200 3985
129.94.144.152 - - [01/Jul/1995:00:00:13 -0400] "GET / HTTP/1.0"
200 7074
unicomp6.unicomp.net - - [01/Jul/1995:00:00:14 -0400] "GET
/shuttle/countdown/count.gif HTTP/1.0" 200 40310
unicomp6.unicomp.net - - [01/Jul/1995:00:00:14 -0400] "GET
/images/NASA-logosmall.gif HTTP/1.0" 200 786
unicomp6.unicomp.net - - [01/Jul/1995:00:00:14 -0400] "GET
/images/KSC-logosmall.gif HTTP/1.0" 200 1204
d104.aa.net - - [01/Jul/1995:00:00:15 -0400] "GET
/shuttle/countdown/count.gif HTTP/1.0" 200 40310
d104.aa.net - - [01/Jul/1995:00:00:15 -0400] "GET /images/NASA-
logosmall.gif HTTP/1.0" 200 786
```

Figure 4: NASA web server Log File

Host	Timestamp	Timezone	Request	Status Code	Bytes
in24.inetnebr.com	[01/Aug/1995:00:00:01	-0400]	GET /shuttle/missions/	200	1839
ix-esc-ca2-07.ix.netcom.com	[01/Aug/1995:00:00:09	-0400]	GET /images/launch-lo	200	1713
slppp6.intermind.net	[01/Aug/1995:00:00:10	-0400]	GET /history/skylab/sky	200	1687
piweba4y.prodigy.com	[01/Aug/1995:00:00:10	-0400]	GET /images/launchme	200	11853
slppp6.intermind.net	[01/Aug/1995:00:00:11	-0400]	GET /history/skylab/sky	200	9202
slppp6.intermind.net	[01/Aug/1995:00:00:12	-0400]	GET /images/ksclogosm	200	3635
ix-esc-ca2-07.ix.netcom.com	[01/Aug/1995:00:00:12	-0400]	GET /history/apollo/im	200	1173
slppp6.intermind.net	[01/Aug/1995:00:00:13	-0400]	GET /history/apollo/im	200	3047
133.43.96.45	[01/Aug/1995:00:00:16	-0400]	GET /shuttle/missions/	200	10566
kgtyk4.kj.yamagata-u.ac.jp	[01/Aug/1995:00:00:17	-0400]	GET / HTTP/1.0	200	7280

Figure 5: Pre-processed web server Log File Log File

A web server log file is a text file published by the web server as operation. Log files gather a broad range of details about the site server knowledge queries. Examples of captured and saved data include date, time, client IP address, referrer, user agent, name of operation, server name, Network key, etc. Figure 4 displays the NASA log file example for a site server. Table 2 describes the data as before by deleting both suffixes, such as *.jpg, *.css, *.gif, and so forth. The file does not include these suffixes. After cleaning the files, it reduces the file size. It often omits incomplete URLs from first picture archives, multimedia files.

Table 2: Summary of Pre-Processed Data

S. No	**Records (NASA 1995)**	**Count**
1	Records in Original log file	3,461,613
2	Records in Cleaned log file	4,77,830
3	Attributes in Original log file	15
4	Attributes in a Cleaned log file	8

Running K-means clustering algorithm and AMD clustering algorithm for the same dataset reveal that AMD managed to outperform the k-means in any data size of the transaction Furthermore, AMD clustering algorithm shows an incredible difference in time performance when the size of the dataset grows larger. Table 3 shows the enhanced performance of the AMD clustering algorithm when compared with K-means.

Table 3: Execution time of Clustering Algorithms

Datasets	**Time Taken to Cluster**	
	K-means Clustering Algorithm (Seconds)	AMD Clustering Algorithm (Seconds)
Dataset 1 (NASA)	0.23	0.06
Dataset 2 (SEC.gov)	0.41	0.18
Dataset 3 (ClarkNet)	0.66	0.21

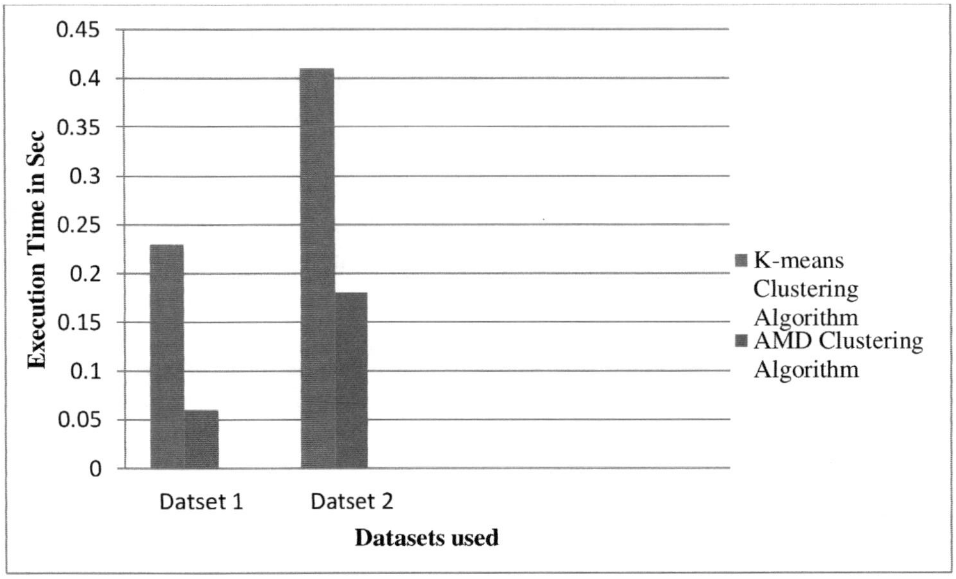

Figure 6: Execution time of K-means Vs AMD Clustering Algorithm

Figure 6 shows the results revealed the enhanced performance. The enhancement is in the sense of execution time taken to complete the entire process from reading the content of the log file to the finding of the most frequently accessed webpages. Here, "apollo.html" has the full AMD value that is prefixed. Figure 7 displays the NASA Kennedy Space Centre AMD sample search data.

Table 4: NASA dataset webpages and its AMD value

Pages	Frequency Count	Sequential Hit	AMD
missions.html	1881	112	17.27
mission-sts-69.html	1313	124	11.07
apollo.html	1110	52	21.53
history.html	889	52	18.91
images.html	845	87	11.19
mission-sts-70.html	828	46	19.75
apollo-13.html	783	74	11.32
movies.html	698	73	9.36
winvn.html	584	76	7.22
stsref-toc.html	577	42	13.58

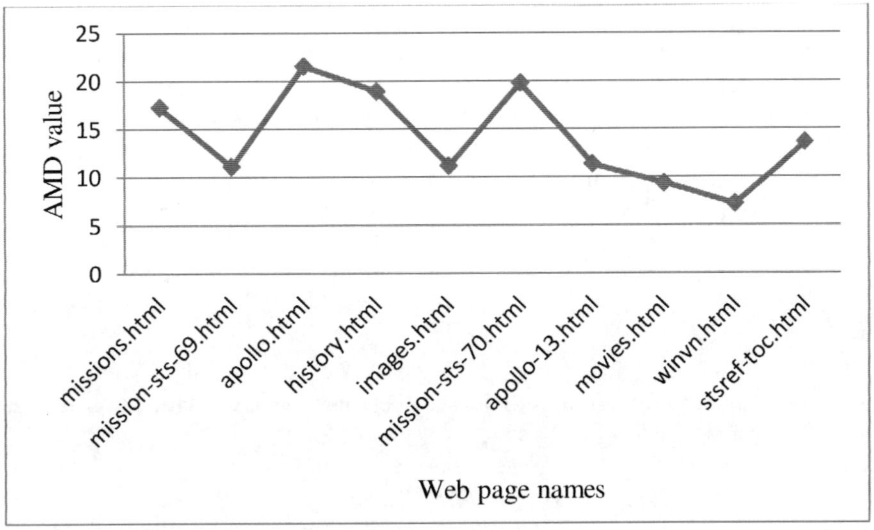

Figure 7: AMD Search of adjacent pages

Table 5: Different datasets accuracy values

Dataset	Number of Records	Number of Records (After Pre-process)	Accuracy
DS1 – NASA	34,61,612	9,33,212	81%
DS2 – SEC.gov	13,56,862	3,52,893	79%
DS3 – ClarkNet	33,28,587	7,63,234	70%

Figure 8 shows the HMM Prediction Algorithm's graphical representation of precision. The HMM prediction Algorithm is implemented and tested in different datasets. The experimental findings are more reliable. It offers the site users optimally predicted pages that reduce user latency.

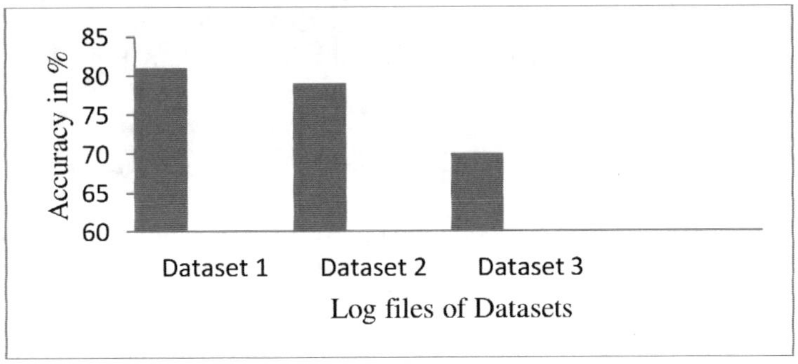

Figure 8: Accuracy of HMM Prediction Algorithm

5. Conclusion

In this article, we apply several strategies for cluster sessions depend on the navigation patterns which generate clusters containing session participants and their sub-set patterns of navigation. The research path began with the initial extraction stage for the pattern analysis of the web log file. The derived log file was pre-processed user control by utilising the number of sites the user visited and time spent on the page attributes. The efficient development of the clustering algorithm AMD shows that the clustering algorithm K-means explodes. For pattern study, the clustered data is used to address the first considered problem of the reorganisation of the website. Through introducing the AMD clustering algorithm, we established commonly visited web pages and recent access to web pages. The study provides a new direction for recognising web pages often accessed from web server log file, suggesting another new strategy, namely HMM algorithms, for the identification of commonly accessed Web log pages with an improved time complexity.We hope to find a successful way in the future selection of parameters. We expect to find evidence of the convergence of the algorithm introduced.

References

[1] N. Lin, G. Chen, K. Zheng, and Y. Tang, "Clusoaf: A cluster-based semantic oriented analyzing framework for user behaviors in mobile learning environment," in Human Centered Computing. Springer, 2015, pp. 340–351.

[2] M. C. Gonzalez, C. A. Hidalgo, and A.-L. Barabasi, "Understanding individual human mobility patterns," Nature, vol. 453, no. 7196, pp. 779–782, 2008.

[3] N. Bicocchi and M. Mamei, "Investigating ride sharing opportunities through mobility data analysis," Pervasive and Mobile Computing, vol. 14, pp. 83–94, 2014.

[4] Wagstaff, K., Cardie, C.: Clustering with instance-level constraints. In: 17th Int'l Conf. on Machine Learning. (2000) 1103–1110.

[5] Kiri Wagstaff, Claire Cardie, Seth Rogers, and Stefan Schroedl. Constrained K-means Clustering with Background Knowledge. In Proc. Of the Eighteenth International Conference on Machine Learning, pages 577–584, 2001

[6] Eric P. Xing, Andrew Y. Ng, Michael I. Jordan, and Stuart Russell. Distance metric learning, with application to clustering with side information. In Advances in Neural Information Processing Systems 15, pages 505–512, 2003.

[7] Sisodia, Dilip & Verma, Shrish & Vyas, O. (2016). Augmented intuitive dissimilarity metric for clustering of Web user sessions. Journal of Information Science. 43. 10.1177/0165551516648259.

[8] Anupama, D.S. & Gowda, Sahana. (2015). Clustering of Web User Sessions to Maintain Occurrence of Sequence in Navigation Pattern. Procedia Computer Science. 58. 558-564. 10.1016/j.procs.2015.08.073.

[9] Sisodia, Dilip & Verma, Shrish & Vyas, O. (2016). A conglomerate relational fuzzy approach for discovering web user session clusters from web server logs. 8. 1433-1443.

[10] Bayir, Murat & Toroslu, Ismail. (2016). Finding All Maximal Paths In Web User Sessions. 15-16. 10.1145/2872518.2889351.

[11] Sisodia, Dilip. (2017). A subtractive medoids selection based fuzzy relational clustering of augmented web user sessions. Journal of Intelligent & Fuzzy Systems. 33. 2259-2268. 10.3233/JIFS-17122.

[12] Thirumoorthy, Gopalakrishnan & Sengottuvelan, P. & Bharathi, A. & Ramasamy, Lokeshkumar. (2018). An approach to webpage prediction method using variable order markov model in recommendation systems. Journal of Internet Technology.

[13] Nguyen, Da & Tan, Hanh & Pham, Duy. (2019). Improving Webpage Access Predictions Based on Sequence Prediction and PageRank Algorithm. Interdisciplinary Journal of Information, Knowledge, and Management. 14. 027-044. 10.28945/4176.

[14] Gu, Zhenyu & Jin, Chenhao & Chang, Danny & Zhang, Liqun. (2020). Predicting webpage aesthetics with heatmap entropy. Behaviour & Information Technology. 1-15. 10.1080/0144929X.2020.1717626.

[15] https://ti.arc.nasa.gov/tech/dash/groups/pcoe/prognostic-data-repository/

[16] https://www.sec.gov/dera/data/edgar-log-file-data-set.html

[17] https://www.Clarknet/data/edgar-log-file-data-set.html

Reducing Complexity in PTS Scheme using Optimization Techniques to reduce PAPR in OFDM Systems

R.Sarankumar[1], K. Sumathi[1], G. Sai Sravanthi[1], B. Venkateswarlu[1], P. Rajasekar
Department of Electronics and Communication Engineering,
[1]QIS Institute of Technology, Ongole, Andhra Pradesh
[2]Narayana Engineering College, Gudur, Andhra Pradesh
rmsarankumar@gmail.com

Abstract: OFDM is a technique used to combat Inter Symbol Interference (ISI) in wireless communication systems. OFDM offers high reliable data rate over frequency selective channels. But the drawback is high Peak to Average Power Ratio (PAPR) which will affect the power amplifier in the RF front end. Partial Transmit Sequence (PTS) has proven to be the efficient technique to minimize the PAPR in OFDM systems and it does not affect the Bit Error Rate (BER) much. But PTS in turn suffers from high search complexity. The optimization algorithms can offer a tradeoff between the search complexity and PAPR reduction. In this work the phase factor optimization in PTS scheme is converted into a one dimensional optimization problem. Genetic Algorithm (GA), Simulated Annealing (SA) and Particle Swarm Optimization (PSO) are applied to the problem defined. They are also compared in terms of PAPR reduction and complexity. The fitness functions are chosen according to the optimization technique.

Key Words: Genetic Algorithm (GA), Low complexity PTS, OFDM, PAPR reduction, Partial Transmit Sequences (PTS), PSO, SA

1. INTRODUCTION

OFDM is a powerful modulation technique used to reduce Inter Symbol interference (ISI) in wireless communication systems. It gives better performance in frequency selective fading environments. But the major problem of OFDM systems is high PAPR. Numerous methods have been discussed in literature to minimize the PAPR. But most of the techniques either increase the complexity of the system or deteriorates its bit error rate (BER) performance.

Designing an algorithm to reduce PAPR without affecting the BER performance and increasing the complexity is a great challenge. Among various techniques PTS technique has proved to be the best scheme in terms of PAPR reduction and PTS scheme does not deteriorate the BER performance much. But PTS increases the complexity of the system. Hence various techniques have been proposed to reduce complexity in PTS scheme. The optimization algorithms such as genetic algorithms (GA), particle swarm optimization (PSO), simulated annealing (SA) can be employed to reduce the so called search complexity in PTS technique. All these algorithms give only sub-optimal solution for the phase factor optimization problem in PTS and hence reduce the PAPR value. But the complexity of PTS is decreased when they are employed for phase factor optimization.

In [1] the overview of PAPR reduction schemes is given. The complexity and description of PTS are studied from [1]. In [2] a reduced complexity PTS algorithm with only two weighting factors $\{1, -1\}$ is proposed. In [3] the signals at the middle stages of an N-point radix FFT using decimation in time (DIT) or decimation in frequency (DIF) are considered for sub-block partitioning. The computational complexity is further decreased by a method known as decomposition PTS (D-PTS) in [3]. In [4], it is discussed that generation of cost function Q_n is based on addition the power of samples in time-domain at time n in each sub-block. As associated with conventional PTS pattern, the technique discussed in [4] attains nearly the similar PAPR reduction performance with reduced computational complexity.

In [11] genetic algorithm based PTS schemes are proposed. It uses the reciprocal of PAPR as fitness function. In [13] Simulated Annealing based PTS is given where the fitness function is the difference between the PAPR and is desired value. In [14] Particle Swarm Optimization (PSO) based PTS is proposed where each phase factors are updated using the velocity equations. In [15], compressive sensing approach is discussed to reduce PAPR and BER is preserved. Joint Iterative filtering and companding is presented in [16] to mitigate PAPR with reduced computational complexity. In [17], PPAR better performance is achieved by addition of mapping signal to the OFDM signal.

2. SYSTEM MODEL AND PROBLEM DEFINITION

2.1 Definition of PAPR

PAPR of an analog signal $x(t)$ is the ratio of its maximum power to its average. It is given by (1).

$$PAPR\{x(t)\} = \frac{\max |x(t)|^2}{E\{|x(t)|^2\}} \qquad (1)$$

Suppose $s(t)$ is a continuous OFDM signal obtained after the A/D converter. The PAPR of $s(t)$ can be estimated using the samples before the A/D converter. It is shown in [8] that the PAPR of the discrete OFDM sequence is a better estimate of that of the continuous signal after Analog to Digital converter when it is up-sampled by a factor of $L \geq 4$. The discrete OFDM signal PAPR is denoted as (2)

$$PAPR\{\mathbf{x}(n)\} = \frac{\max |\mathbf{x}(n)|^2}{E\{|\mathbf{x}(n)|^2\}} \qquad (2)$$

Where $\mathbf{x}(n) = [x[0], x[1], \dots, x[N-1]]$ and N is the number of sub-carriers i.e., number of OFDM symbols in one block. For simulation purposes the (2) is applied to $\mathbf{x}(n)$ after up-sampling it.

2.2 Complementary Cumulative Distribution function

PAPR reduction schemes do not give the same percentage of PAPR reduction for different inputs. So a useful quantity called CCDF is well-defined. It is the complement of the cumulative distribution of the PAPR i.e., probability that the PAPR will be greater than a given value $PAPR_0$. If a scheme is better than any other scheme in PAPR reduction sense then its CCDF curve will be to the left of the other. In the simulation results the CCDF of the PTS and GAPTS will be compared. It is given by (3)

$$CCDF(PAPR_0) = P(PAPR > PAPR_0) \qquad (3)$$

2.3 Partial Transmit Sequences (PTS)

PTS is a probabilistic method which is widely found in many research works [2], [3], [4], etc. The PTS method is depicted in Fig 1. The PTS method partitions N symbol input data to V number of disjoint sub blocks as shown:

$$\mathbf{X} = [\mathbf{X^0}, \mathbf{X^1}, \mathbf{X^2}, \dots, \mathbf{X^{V-1}}]^T \qquad (4)$$

Here, $\mathbf{X^i}$ are known as sub blocks which are of same size and are successively situated. Then each portioned sub block can be multiplication with the complex phase factor $b^v = e^{j\phi v}$, $= 1,2,\dots,V$, after applying the IFFT to provide

$$\mathbf{x} = \mathrm{IFFT}\left\{\sum_{\mathrm{v}=1}^{\mathrm{V}} b^{\mathrm{v}} \mathbf{X}^{\mathrm{v}}\right\} = \sum_{\mathrm{v}=1}^{\mathrm{V}} b^{\mathrm{v}}.\mathrm{IFFT}\{\mathbf{X}^{\mathrm{v}}\}$$

IOP Conf. Series: Materials Science and Engineering 1074 (2021) 012032 doi:10.1088/1757-899X/1074/1/012032

$$\mathbf{x} = \sum_{v=1}^{V} b^{v}\mathbf{x}^{v} \quad (5)$$

here $\{\mathbf{x}^{v}\}$ is known as PTS. The selection of phase vector in such a way that PAPR is minimized and it is denoted as

$$[b^1, \ldots, b^V] = \underset{[b^1,\ldots,b^V]}{\operatorname{argmin}} \left(\max_{n=0,1,\ldots,N-1} \left| \sum_{v=1}^{V} b^{v} x^{v}[n] \right| \right) \quad (6)$$

The PAPR vector is in time domain is denoted as

$$\tilde{\mathbf{x}} = \sum_{v=1}^{V} \tilde{b}^{v}\mathbf{x}^{v} \quad (7)$$

In common, the choice on the phase factors $\{b^v\}_{v=1}^{V}$ is restricted to a set of values to diminish complexity of searching. The allowable phase factors can be denoted as

$$\mathbf{b} = \left\{ e^{\frac{j2\pi i}{W}} \middle| i = 0,1, \ldots, W-1 \right\} \quad (8)$$

The computational complexity of searching rises in exponential manner with rise in amount of sub blocks. The PTS method needs V IFFT processes to every data set and Roof ($\log_2 W^V$) bits of side data. The PAPR value of PTS technique is modified by both number of sub-blocks and sub-block partitioning. The three different methods for sub block dividing are adjacent, interleaved, and pseudo-random. Out of these, the best result is obtained through pseudo-random method.

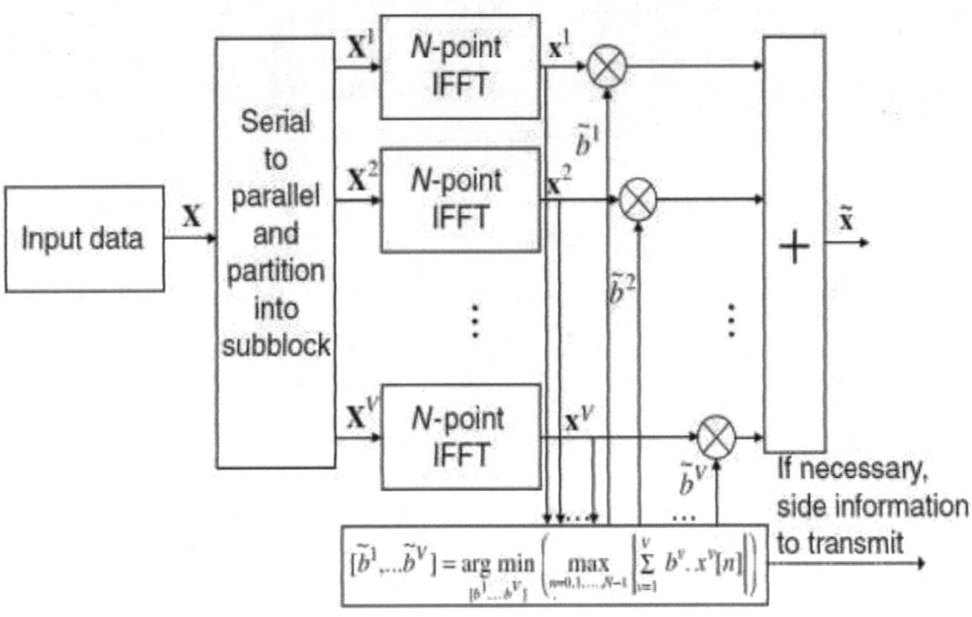

Fig. 1. Block diagram of Partial Transmit Sequence technique.

3. PROBLEM FORMULATION

The objective of this work is to reduce the complexity while minimizing the PAPR by choosing appropriate $b^1, b^2, \ldots, b^V$ given by

$$PAPR\{\mathbf{x}(b^1, b^2, \ldots, b^V)\} = \frac{\max |\mathbf{x}(b^1,b^2,\ldots,b^V)|^2}{E\{|\mathbf{x}(b^1,b^2,\ldots,b^V)|^2\}} \quad (9)$$

IOP Conf. Series: Materials Science and Engineering 1074 (2021) 012032 doi:10.1088/1757-899X/1074/1/012032

Ordinary PTS gives the optimal solution for the above problem but has high search complexity. The optimization algorithms discussed in this paper aim at finding the optimal solution for (9) with less complexity. They provide a trade-off between the complexity and PAPR reduction.

Equation (9) is a V dimensional optimization problem. Since the phase factors can take only discrete values, a whole number B in $[0, W^{V-1}]$ can be assigned to a particular phase vector $b = [e^{-2\pi\phi_1}, e^{-2\pi\phi_2}, \ldots, e^{-2\pi\phi_V}]$ such that $\phi_1\phi_2 \ldots \phi_V$ is the base V representation of the number B i.e., $(B)_{10} = (\phi_1\phi_2 \ldots \phi_V)_V$. Where $\phi_1, \phi_2, \ldots, \phi_V$ are the digits of the baseV number. Hence the V dimensional optimization problem is now converted into a one dimensional problem.

4. GENETIC ALGORITHM BASED PTS (GAPTS)

In this technique the phase optimization is done using genetic algorithm. It is a population based method. Though it gives sub-optimal phase factors its search complexity is less than the traditional PTS. The transmission of side information is required in GAPTS.

4.1 Fitness function

Normally when Genetic Algorithm is employed for PTS the fitness function **F** for a fixed **X** is chosen as the reciprocal of PAPR of the sequence **x** combined using the particular possible set of phase factors $b^1, b^2, \ldots, b^V$.

$$\mathbf{F}(b^1, b^2, \ldots, b^V) = \frac{1}{\log_{10} \mathbf{PAPR}[\mathbf{x}(b^1, b^2, \ldots, b^V)]} \qquad (10)$$

Which can be converted into one dimensional function as $b^1, b^2, \ldots, b^V \rightarrow B$ and hence

$$\mathbf{F}(\mathrm{B}) = \frac{1}{\log_{10} \mathbf{PAPR}[\mathbf{x}(\mathrm{B})]} \qquad (11)$$

Where $\mathrm{B} \in [0, W^{V-1}]$. The reciprocal of PAPR is taken to convert the minimization problem to maximization problem. Now objective is to minimize the objective function. The search space is the set of all possible W^{V-1} phase factors. In this work the fitness function is modified as $\mathbf{F_1}$which is given by (11)

$$\mathbf{F_1} = \mathbf{F} + \boldsymbol{\lambda}(\mathbf{F} - \boldsymbol{\mu}) \qquad (12)$$

Where $\boldsymbol{\mu}$ and $\boldsymbol{\lambda}$ are the parameters. The maxima and minima of **F** and $\mathbf{F_1}$occur for the same values of B because

$$\frac{d\mathbf{F_1}}{dB} = (1 + \lambda)\frac{d\mathbf{F}}{dB}$$

4.2 simple three step genetic algorithm:

A simple three step genetic algorithm is given below

1) Initialization.
2) Reproduction.
3) Cross-over.
4) Mutation.
5) Count=Count+1.
6) If Count<G. Go to step 2)
7) Stop.

4.3 Initialization and coding:

Suppose if W=4 and V=4. Then there are four phase factors given by $b = [1, j, -1, -j]$. If we fix $b^1 = 1$. Then there are $W^{V-1} = 64$ possibilities of choosing four phase factors among the four available phase factors. So 7-bit binary code is assigned to all the possible phase factor sets. Initial population of some size P is chosen. For example if P=4, four random distinct binary strings are chosen and their fitness values are calculated using (11). This set of strings in the initial population is subjected to reproduction, cross-over and mutation. After these three steps we will get the new population which is again subjected to the three steps. This process is repeated till the number of generation is reached.

4.4 Reproduction:

Mean of the fitness values of the population are calculated and each fitness value is divided by the total value. Based on these values a roulette algorithm is used to obtain the reproduced population. There are many ways to implement roulette algorithm in software.

One is to obtain the Expected count of all the strings and round off them as

$$\text{count(k)} = \text{Round(fitness(k)/mean\_fitness)}.$$

where 'k' is the population index. The new population contains the kth string count(k) times after reproduction. It is possible that the sum of all the count values is not equal to the population size. In that case if the sum is less than population size, some random candidates are chosen to bring the sum of count values equal to the population size. If the sum is greater than population size then those number of minimum valued candidates are dropped.

4.5 Crossover and mutation:

The strings are paired randomly. A position is chosen randomly and the bits after that positions are exchanged between the chosen pair. Random numbers can be generated in software (For example using function randi in MATLAB). In hardware pseudo random numbers can be generated using linear feedback shift registers. Mutation is the random flipping of bits. There are few problems in using GA with PTS. The PAPR objective function is too complicated having too many peaks. GAPTS gives suboptimal solution. Hence its PAPR performance is less compared to conventional PTS.

5. Simulated Annealing Based PTS (SAPTS)

In SAPTS the phase factor optimization is done using simulated annealing algorithm. It is a nature inspired meta-heuristic optimization algorithm. Simulated Annealing algorithm uses a steady reduction in the probability of accepting poorer solutions as it discovers the solution space. Accepting poorer solutions is a basic behavior of Meta-heuristics as it permits for a additional widespread search for the best solution [13].

5.1 Simulated annealing algorithm:

It is an iterative procedure which stops after the stopping criterion is met. The temperature is initialized to a value and decremented in every iteration using a cooling schedule which is a function of iteration variable. It the temperature becomes less than the T_{stop} the algorithm is stopped. In every iteration, a random new solution is chosen and its fitness value is calculated. If the new fitness value is greater than that of the previous solution the new solution is accepted. Else if the acceptance probability is larger than a random number in (0,1) the new solution will be accepted. The acceptance probability is also a function of the iteration variable

Initialize T_0 , T_{stop} , C , D, h, α and β
$k = 1$
Choose the initial solution as $\hat{b} = [1\ 1\ 1\ 1\ \ldots]$

Calculate $\hat{f} = f(\hat{b})$
While $(T(k) > T_{stop})$
    $k = k + 1$
    $T = T(k)$ // Update T according to the cooling schedule
    Perturbation is used to obtain a new solution $\tilde{b}$
    Calculate $\tilde{f} = f(\tilde{b})$
    $\Delta E = \tilde{f} - \hat{f}$
    If $(\Delta E < 0)$
        $\hat{b} = \tilde{b}$
        $\hat{f} = \tilde{f}$
    Else
    Generate an uniform number r in $(0,1)$
    Calculate the transition acceptance probability $Prob$
    If $(Prob > r)$
        $\hat{b} = \tilde{b}$
        $\hat{f} = \tilde{f}$
    End if
End while

5.2 Fitness function and search space:

The fitness function **F** for a fixed **X** is the PAPR of the sequence **x** combined using the particular possible set of phase factors $b^1, b^2, \ldots, b^V$.

$$\mathbf{x}(b^1, b^2, \ldots, b^V) = \sum_{i=1}^{V} b^i \mathbf{IFFT}(\mathbf{X}^i) \quad (13)$$

$$\mathbf{F}(b^1, b^2, \ldots, b^V) = \log_{10} \mathbf{PAPR}\,[\mathbf{x}(b^1, b^2, \ldots, b^V)] \quad (14)$$

$$PAPR\{\mathbf{x}(b^1, b^2, \ldots, b^V)\} = \frac{\max|\mathbf{x}(b^1,b^2,\ldots,b^V)|^2}{E\{|\mathbf{x}(b^1,b^2,\ldots,b^V)|^2\}} \quad (15)$$

Now objective is to minimize the objective function. The search space is the set of all possible W^{V-1} phase factors.

5.3 Cooling schedule:

SA is an iterative algorithm which runs using a cooling schedule. Cooling schedule is essentially a function of the iteration variable. The temperature is initialized to a value and it is slowly decremented using the cooling schedule. When the temperature reaches a predefined value called T_{stop} the algorithm stops. The cooling schedule used in this work is in equation (4).

$$T(k) = T_0 exp(-C\sqrt[D]{k}) \quad (16)$$

Where T_0 is the initial temperature. C and D are parameter. C and D are chosen as 0.98 and 2 respectively [13]. For a given number of iterations 'k' we can determine the value of T_{stop} for a given T_0 using (4).

5.4 Perturbation:

This is the technique used to obtain the new candidate solution. Perturbation essentially generates the new solution using some distribution. Since the phase factors can only take discrete values a whole number B in $[0, W^{V-1}]$ is assigned to each possible phase factors $b = [e^{j2\pi\phi_1}, e^{j2\pi\phi_2}, \ldots, e^{j2\pi\phi_V}]$ such that $\phi_1\phi_2 \ldots \phi_V$ is the base V representation of the decimal number i.e., $(B)_{10} = (\phi_1\phi_2 \dots \phi_V)_{10}$. In this work perturbation is done using discrete uniform distribution. A random number in $[0, W^{V-1}]$ is generated and converted into equivalent phase vector.

5.5 Acceptance probability:

The acceptance probability is decreasing function of the iteration variable k. Hence the probability that the bad solution will be accepted in the successive iterations decreases. The acceptance probability used in this work is given by (5)

$$Prob(k) = \left[1 - \frac{h.\Delta E}{T(k)}\right]^{\frac{1}{h}} \qquad (17)$$

6. PARTICLE SWARM OPTIMIZATION BASED PTS (PSOPTS)

PSO is a computational method which provides the best solution for a given problem by iteratively trying to provide candidate solution with respect to quality. The general particle swarm optimization is given below [14].

Let $f: \mathbb{R} \rightarrow \mathbb{R}$ be the cost function that is to be reduced. The PSO algorithm is as follows:

- For every particle $i = 1, \dots, S$ do:
  1) Set the location of particle $\boldsymbol{x_i} \sim \boldsymbol{U(b_{lo}, b_{up})}$, here $\boldsymbol{b_{lo}}$ and $\boldsymbol{b_{up}}$ represent the boundaries of the search space.
  2) Reset the particle's best recognized position to its original location: $\boldsymbol{p_i} \leftarrow \boldsymbol{x_i}$
  3) If $\boldsymbol{f(p_i)} < f(g)$, set the group's best location: $\boldsymbol{g} \leftarrow \boldsymbol{x_i}$
  4) Reset the speed: $\boldsymbol{v_i} \sim \boldsymbol{U}(-|\boldsymbol{b_{up}} - \boldsymbol{b_{lo}}|, |\boldsymbol{b_{up}} - \boldsymbol{b_{lo}}|)$

  Until a termination criterion is met repeat:

  For each particle $i = 1, \dots, S$ do:
  1) Pick random numbers: $r_p, r_g \sim U(0,1)$
  2) For each dimension $d = 1, \dots, n$ do:
     Update the particle's velocity using
     $\boldsymbol{v_{i,d} \leftarrow \omega v_{i,d} + \varphi_p r_p (p_{i,d} - x_{i,d}) + \varphi_p r_g (g_d - x_{i,d})}$
  3) Set speed of particle: $\boldsymbol{x_i} \leftarrow \boldsymbol{x_i} + \boldsymbol{v_i}$
  4) If $\boldsymbol{f(x_i)} < f(\boldsymbol{p_i})$ do:
     set best recognized location of the particle: $\boldsymbol{p_i} \leftarrow \boldsymbol{x_i}$

     If $\boldsymbol{f(p_i)} < f(\boldsymbol{g})$ update the swarm's best known position: $\boldsymbol{g} \leftarrow \boldsymbol{x_i}$
- At this moment, $\boldsymbol{g}$ shows the best identified result.

6.1 Problem formulation for PSO PTS:

The cost function used is the PAPR itself. The discrete phase factors are assigned with a whole number as given by (14). Hence the dimension is one and the search space is discrete. The updated velocities are rounded off before adding them to the previous solution.

7. SIMULATION RESULTS

Simulations are done in MATLAB. The complementary cumulative distribution function (CCDF) of PTS, GAPTS, SAPTS and PSOPTS schemes are obtained which is shown in figs (2-5). The details of the simulations are as follows. Number of sub carriers = 64. Number of iterations = 3000. (3000 independent data sets are generated randomly). Modulation used: 16QAM.Number of sub-blocks V = 4. Number of possible phase factors W=4. Population size $Np = 4,6,8$ and Number of generations $Ng = 4,5,8$ for GA and PSO. For SA the parameters C and D are chosen as 0.98 and 2 respectively. For PSO the acceleration constants c1 and c2 are chosen as 2.3 and 1.6 respectively and

$\omega = 0.4$ The CCDF graphs for various schemes and various complexities are given in the following figures. Fig 2 shows the CCDF graphs of GAPTS for different complexities. These graphs are compared with the normal PTS and OFDM system which does not employ any PAPR reduction technique. From Fig. 2 it is evident that GA provides a tradeoff between complexity and PAPR reduction. Increasing the complexity further may improve the PAPR performance but the limit of complexity is W^{V-1}. If we increase the complexity further then we won't get the advantage of using an Optimization Algorithm. The CCDF graph for this limiting case ($Ng \times Np = W^{V-1}$) is plotted for reference. It corresponds to $Ng = 8, Np = 8$. Similarly the CCDF graphs are of SAPTS and PSOPTS for various complexities are given in Fig.3 and Fig.4 respectively.

All the three schemes provide a trade-off between PAPR reduction and complexity. Fig. 5 shows the CCDF graphs of GAPTS for the two different fitness functions fit1 and fit2 given by (11) and (12) respectively. It is clear that using the proposed fitness function the PAPR performance is increased. Finally all the three schemes for same complexity using the corresponding fitness functions are compared in Fig. 6 which shows that SAPTS performs well for the given set of parameters. The fitness function chosen for GAPTS for this comparison is (12).

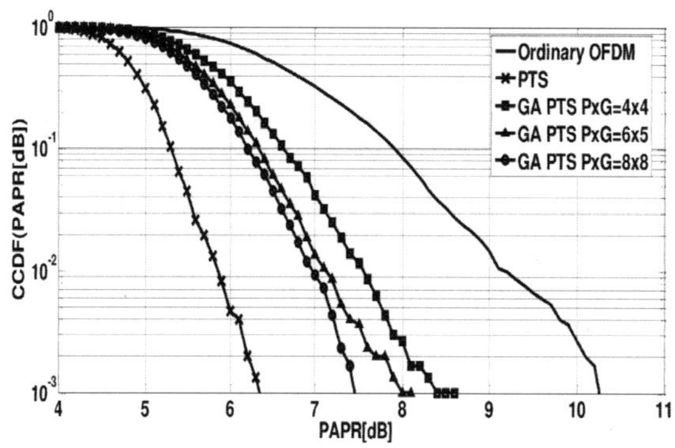

Fig. 2. Comparison of various complexities GA.

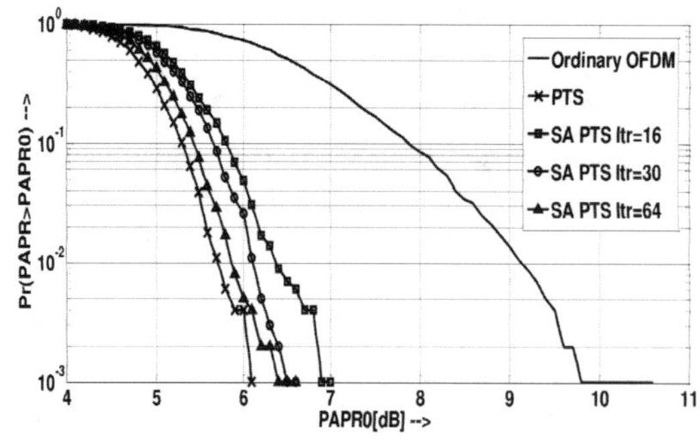

Fig. 3. Comparison of various complexities SA.

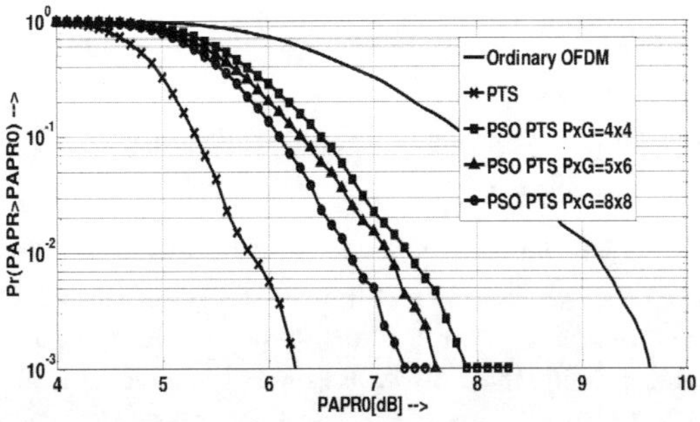

Fig. 4. Comparison of various complexities PSO.

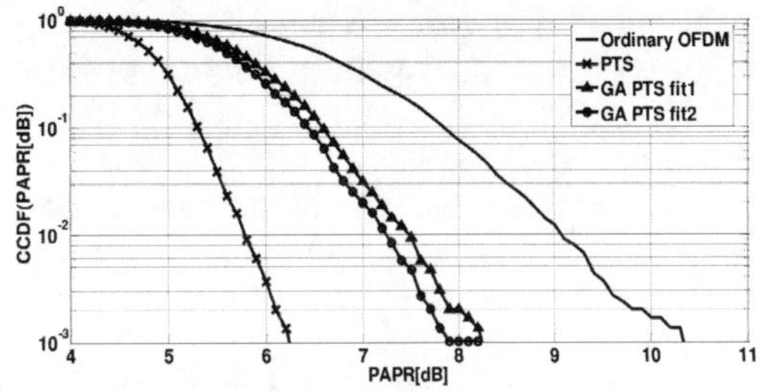

Fig. 5. Comparison of various fitness functions in GA.

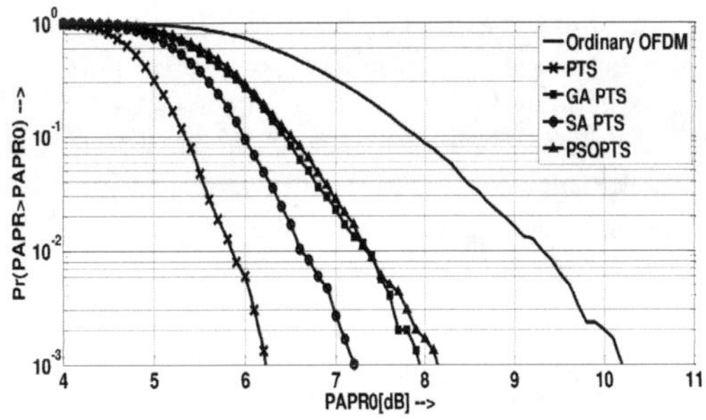

Fig. 6. Comparison GA, SA and PSO for the same complexity.

CONCLUSION

SAPTS has high PAPR performance compared to the other schemes for same complexity. Normal PTS has high PAPR performance but also high computational complexity. There are many other techniques to reduce complexity in PTS. Most of them have less PAPR performance. A few modifications can be made in these algorithms to improve its PAPR performance.

REFERENCES

1. Han, S.H., and Lee, J.H., 2005. An overview of peak-to-average power ratio reduction techniques for multicarrier transmission. IEEE Wireless Communication., 12(2), 56–65
2. L. Yang, R. S. Chen, Y. M. Siu, and K. K. Soo, MARCH 2006. PAPR Reduction of an OFDM Signal by Use of PTS With Low Computational Complexity, IEEE TRANSACTIONS ON BROADCASTING, VOL. 52, NO. 1
3. Abolfazl Ghassemi, T. Aaron Gulliver, MARCH 2008. A low-complexity PTS-based radix FFT method for PAPR reduction in OFDM systems. IEEE TRANSACTIONS ON SIGNAL PROCESSING. VOL.56, NO 3

IOP Conf. Series: Materials Science and Engineering 1074 (2021) 012032 doi:10.1088/1757-899X/1074/1/012032

4. Sheng-Ju Ku, Chin-Liang Wang, Chiuan-Hsu Chen, AUGUST 2010. A reduced complexity PTS-based PAPR reduction scheme for OFDM systems, IEEE TRANSACTIONS ON WIRELESS COMMUNICATIONS, VOL 9, NO 8
5. Pooria Varahram, Wisam F. Al-Azzo, Borhanuddin Mohd Ali, A Low Complexity Partial Transmit Sequence Scheme by Use of Dummy Signals for PAPR Reduction in OFDM Systems, IEEE Transactions on Consumer Electronics, Vol. 56, No. 4, November 2010.
6. L.Yang, K.K.Soo, S.Q.Li and Y.M.Siu, PAPR reduction using low complexity PTS to construct of OFDM signals without side information. IEEE TRANSACTIONS ON BROADCASTING,VOL. 57,NO. 2,JUNE 2011.
7. Tao Jiang, Weidong Xiang, Paul C. Richardson, SEPTEMBER 2007. Jinhua Guo, and Guangxi Zhu, " PAPR Reduction of OFDM Signals Using Partial Transmit Sequences With Low Computational Complexity", IEEE TRANSACTIONS ON BROADCASTING, VOL. 53, NO. 3
8. Clerc, M. (2012), Standard Particle Swarm Optimization" *HAL open access archive*.
9. David E.Goldberg., 2006, Genetic Algorithm in search, Optimization and Machine Learning, Pearson Education.
10. Chang Eon Shin, Kyung Soo Rimand Youngok Kim, MARCH 2013. A weighted OFDM signal scheme for PAPR reduction in OFDM signals, IEEE TRANSACTIONS ON VEHICULAR TECHNOLOGY,VOL.63, No 8
11. Liang H, Chen Y, Cheng C, 2009. A modified genetic algorithm PTS technique for PAPR reduction in OFDM systems. In the Proceedings of 2009 15th Asia-Pacific Conference on Communications, Shanghai, 182–185
12. CHEN Zhi, ZHANG Shuai, YANG Lin, JIA YingYing & LI ShaoQian, June 2014. Optimal phase searching of PTS using modified genetic algorithm for PAPR reduction in OFDM systems, Science china.
13. Tao Jiang, Weidong Xiang, Paul C. Richardson, Jinhua Guo, and Guangxi Zhu, SEPTEMBER 2007. PAPR Reduction of OFDM Signals Using Partial Transmit Sequences With Low Computational Complexity, IEEE TRANSACTIONS ON BROADCASTING, VOL. 53, NO. 3
14. Ho-Lung, Hung, Aug. 2010. Using evolutionary computation technique for trade-off between PAPR performance and computational complexity in OFDM systems, Elsevier.
15. Ghanbar Azarnia, Abbas Ali Sharifi, Hojjat Emami, Dec 2020, Compressive sensing based PAPR reduction in OFDM systems: Modified orthogonal matching pursuit approach, ict express, vol.6, no. 4
16. V.Sandeep Kumar, Sept 2020, Joint Iterative filtering and companding parameter optimization for PAPR reduction of OFDM/OQAM signal, International Journal of Electronics and Communications, Vol. 124.
17. J.Karthiga, G.Thenmozhi, M. Rajkumar, Oct. 2020, PAPR reduction of MIMO-OFDM system with reduced computational complexity SLM scheme, International Journal of Electronics and Communications

Multi Facial Expression Recognition (MFER) for Identifying Customer Satisfaction on Products using Deep CNN and Haar Cascade Classifier

DNVSLS Indira[1], L Sumalatha[2], Babu Rao Markapudi[3]

[1]Associate Professor, [2]Professor and Director of Evaluation, [3]Professor and Head

[1]Dept. of Information Technology, [3]Dept. of CSE

[2]Dept. of CSE, Jawaharlal Nehru Technological University Kakinada (JNTUK), Kakinada.

[1,3]Gudlavalleru Engineering College, Gudlavalleru – 521356

[1,2,3]E-mail: indiragamini@gmail.com, lsumalatha@jntucek.ac.in, baburaompd@gmail.com

Abstract. Face Expression is one of the most normal, remarkable and a general sign for individuals to convey on their enthusiastic states and it is not restricted to national borders, linguistics and gender. This article presents the modeling of a framework that plans to foresee the fulfillment of a customer through his facial feelings. The cutting edge innovation of Facial Expression Recognition framework is the consumer satisfaction estimation. MFER, a Novel procedure is proposed in this paper for identifying consumer satisfaction levels. This sound methodology of client satisfaction estimation is an alternative option of the ordinary method of gathering clients' reaction. This model must anticipate client's behavior in the dynamic cycle. To expect consumer trustworthiness, we have characterized mathematical highlights of the face by utilizing Deep CNN and Haar Cascade Classifier. The kinds of consumer fulfillment are classified as satisfied, not-satisfied and neutral. Our framework shows a decent exhibition, testing it on the FER2013 dataset. Our MFER –Multi Facial Expression Recognition procedure identifies multiple objects in the same image which consists of same and different expressions.

Keywords: MFER, SVM, Feature Extraction, Deep CNN, Haar Cascade Classifier, Recommendation Systems

1. Introduction

Facial expression investigation includes estimation and acknowledgment of articulation. Computerization of Facial expression prediction comprises of three ideas initial one Face Detection (FD), second one Facial Feature Extraction (FFE) lastly the Facial Expression Recognition (FER)[2]. Computerized Facial Recognition Systems have the ability to distinguish people from their facial highlights. Facial feelings are significant elements in human correspondence that assist us with understanding the goals of others. As a rule, individuals derive the passionate conditions of others, for example some expressions like anger, joy and sadness, utilizing facial expressions and vocal tone[4][5]. In this way, it is regular that exploration of facial feeling has been picking up part of consideration over past a long time with applications in the perceptual and psychological sciences, but also in affective computing and computer animations.

Facial Expression Recognition (FER) [13] is the way toward identifying human feelings from outward appearances. Individuals can perceive the articles by observing them with less exertion. Given a picture,

IOP Conf. Series: Materials Science and Engineering 1074 (2021) 012033 doi:10.1088/1757-899X/1074/1/012033

making the machines to perceive and comprehend the item is a difficult assignment. These image recognition application models can be used to detect the expression or emotion in the face. One of the applications of identifying the emotion of a person is opinion mining to measure the customer satisfaction about a product[6]. These customer satisfactions can be used for further recommendation on products, movies and for any business intelligence applications when users give their opinions or reviews as a facial expression image file. This article presents literature survey in the next section. In the sub sequent sections it covers the importance of Haar cascade classifier in the FER in phase 3 and then proposed MFER procedure and finally results and references.

2. Literature Survey

Face is one of the main techniques for human correspondence. It accepts a central occupation in each and every social affiliation. Facial enunciations are non-verbal insinuations to emotions. To start with, face securing is a preparing stage to naturally find the face area in the information pictures. The subsequent stage is to extricate and speak to facial changes brought about by outward appearances. At long last, the arrangement task permits to derive the outward appearances. The extraction strategies are utilized to portray face appearances and changes on a face [1].

The creators in [2][3] characterized facial segments from which they extricated HOG include descriptors. The goal of this work is to perceive consequently the six fundamental feelings just as the nonpartisan state by applying appearance-highlight strategies like LBP and its variations to new explicit face-areas. Different authors discussed and worked on different methods. The following table gives brief information about the algorithms mentioned in the reference articles.

Table 1: Survey on different articles which discussed various methods for FER

Reference Nos.	Year	Methods Discussed
[1]	2014	LTP, MTP, CLBP.
[2]	2015	HOG , PCA , LDA and Wavelet
[3]	2015	MDA and SVM
[4]	2016	EEG signal
[5]	2016	Event Related Potentials (ERP's)
[6]	2016	Harris Algorithm
[7]	2017	CNN
[8]	2017	FERS
[9]	2018	LBP and SIFT into CNN
[10]	2018	Ryerson Audio-Visual Database of Emotional Speech and Song (RAVDESS)
[11]	2019	Deep CNN
[12]	2019	DLP-CNN
[13]	2019	SVM

IOP Conf. Series: Materials Science and Engineering 1074 (2021) 012033 doi:10.1088/1757-899X/1074/1/012033

2.1. Existing Linear Support Vector Machine for Recognizing Face

All most all authors till now they used SVM for recognizing face in the given image. SVM or Support Vector Machine is a straight model for arrangement and relapse issues. It can take care of direct and non-straight issues and function admirably for some handy issues[13]. The possibility of SVM is straightforward. SVM identifies the ROI (regions of interest) in the face, which consists of eyes (left, right), eyebrows(left, right), mouth, nose and distance between eyebrows These features are fed to SVM to recognize the face in the image. Our proposed method identifies multiple faces in the same image with different expressions using haar_cascde classifier.

3. PROPOSED METHODOLOGY

When arranging a modified system, three things are thought of; they are face recognition, facial features extraction, and characterization of articulations or expressions. At first, face recognition is a phase to normally discover the face district in the given pictures. The accompanying stage is to discover facial changes achieved by articulations. Finally, the arrangement determines the outward appearances that are the expressions. Here are three significant strides for building this model. One is face recognition which identifies the person who is giving the response and the later one is feature extraction by identifying geometric features and the final one is emotion recognition which indicates the type of response from that person.

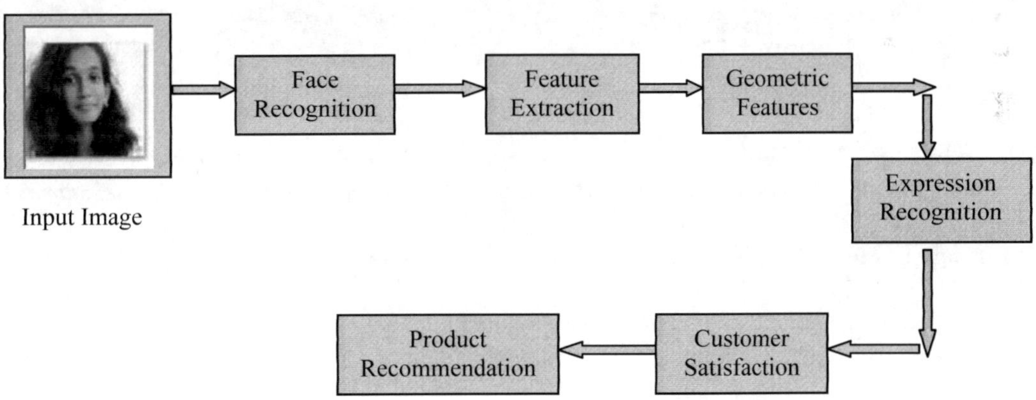

Figure 1 : Flow diagram of proposed work

The proposed framework actualizes a Deep CNN model and a classifier for recognizing expressions by building system. The reason for MFER is used to identify the customer satisfaction on a purchased product when he/she gives opinion as an image file like an expression for further recommendations.

3.1. Face Recognition using Haar_Cascade Classifier

Haar_Cascade Classifier is used to detect objects in an image in a set of machine learning algorithms. It was proposed by Paul Viola and Michel for recognizing faces in the given image. Haar features are different from primary features of images those are color, text and shape. Generally these Haar features are Edge, Line and Rectangle like in convolution networks. It is just like kernel in CNN. Every feature obtained from this is a single value. This can be achieved by subtracting sum of pixels in white area by sum of pixels in black area in one edge.

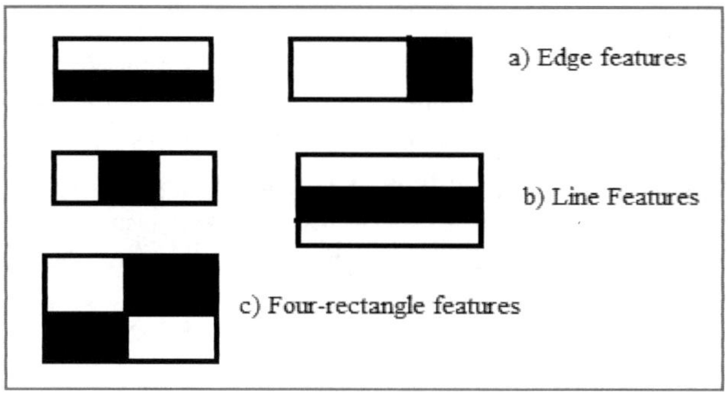

Figure 2: Haar Features

This algorithm passes into four steps. 1. Selection of Haar Features 2. Integral Image Creation 3. Training of input images dataset with ADAboost algorithm. 4. Finally Applying Cascading Classifier for detecting face image and non-face image.

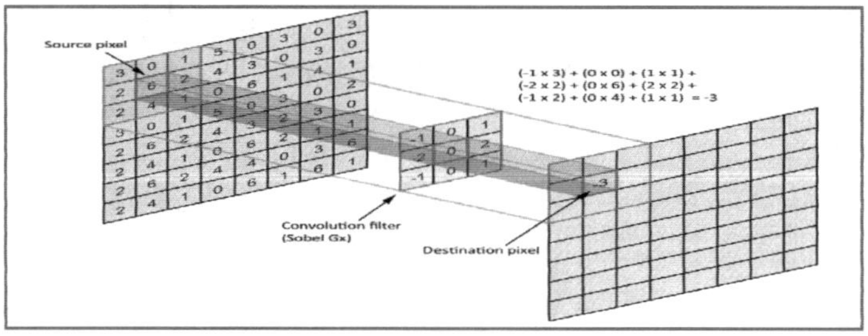

Figure 3: Calculating output one pixel value using four rectangular edge feature in the input image

3.2. Proposed MFER Steps in Python

1. Collect the data from Kaggle, the data set is FER2013.
2. Load the data set.
3. Perform data visualization on the data set.
4. Perform data preprocessing on the data set to remove the missing values and outliers.
5. Split the data set into Train and Validation set.
6. Define the model
   A. Sequential => linear stack of NN layers -> feed forward cnn
   B. layers -> for almost any nn
   C.CNN layers -> convolution2D , maxpooling2D
7. Perform the model summary using Model.Summary().
8. Data Augmentation is done by taking a batch and applying some series of random transformation to increase generalizability of model.
9. Neural networks learn more complex features from input image.
   A. input layer => feeding input

B. first hidden layer => only learn local edge patterns

C. each subsequent layer => filters => learns more complex representations

D. classify

0. Execute the Deep CNN Using Relu function and ***Haarcascade*** classifier,

Then

Relu: $y=\max(0,x) \rightarrow \partial f/\partial x=1$ if $x>=0$

else 0

11. Test on some sample images.
12. Save the model.
13. Store the expressions in the Excel file.
14. Calculate the average value for the product preview obtained from the input image expressions.
15. Recommend the Product based on the satisfaction level.

3.3. The System Architecture of MFER

It consists of the following steps:

1. Data Preparation
2. Feature Extraction using Deep CNN
3. Visualization
4. Training and Testing of Data
5. Identifying Expressions
6. Result with the Polarity of the Person like- Satisfactory, Not Satisfactory and Neutral

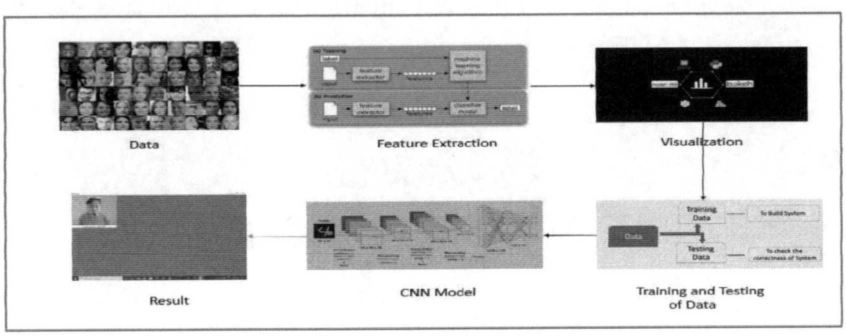

Figure 4: System Architecture for Facial Emotion Recognition

3.3.1 Data Preparation

Initially, the data set to train the emotions is resourced from FER2013 which contains seven categories of emotions such as happy, fear, sad, anger, disgust, surprise and neutral. The required data is only any of the three categories which define the ratings of satisfactory, not satisfactory and neutral.

Figure 5: Sample Data from FER2013

3.3.2 Feature Extraction using Deep CNN

The CNN utilizes various algorithms which calculate the measures of features of face. This is called feature extraction. Each dimension of eye, nose, brows, mouth, jaw line and so on. Different fuzzy logic techniques are used to find the distances like Euclidian distance.[11]

The customized pictures that are being perceived are spared in a folder in a similar catalog. These are the enlisted understudies pictures gathered prior to getting to the criticism framework. The advantage of this methodology is a lot more noteworthy authentic productivity as it accomplishes best in class face acknowledgment execution. The method relies upon learning an Euclidean embedding per picture using a significant convolution organization. The system is set up with the ultimate objective that the squared L2 eliminates in the embedding space clearly identify with go up against comparability. FaceNet[10] utilizes a profound convolution organization. Using FaceNet algorithm we extracted geometric features by highlighting ROI on faces. This method calculate 8 different features in the image which has eyes, nose, mouth, eyebrows etc.

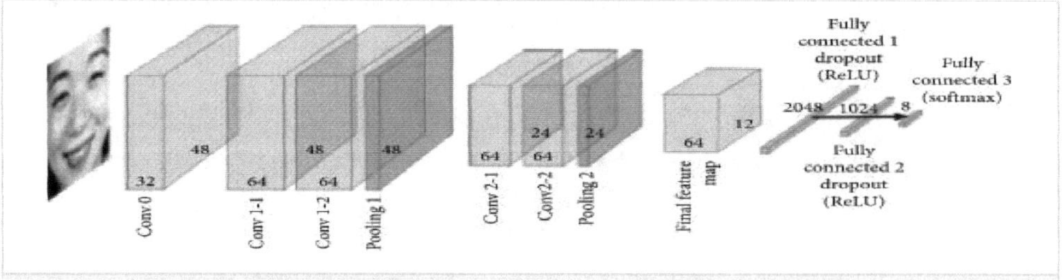

Figure 6: Feature Extraction using Deep CNN from an input Image.

3.3.3. Visualization

Data portrayal is the control of endeavouring to understand statistics by placing it in a visual setting so models, examples and connections that may for no situation be recognized can be revealed. Python offers different exceptional graphing libraries that come squeezed with heaps of different features. Despite if you have to make savvy, live or significantly changed plots python has a surprising library for us.

3.3.4. Training and Test Data

Almost all Data mining applications use supervised data classification. For this the entire data is classified into training and test data sets. Here in this article the data set that is used is FER2013. It consists of nearly 3500 records. We used 3000 records as training data and 500 records for test data. Apart from that we collected some images manually from different websites that expresses opinion on products and own data prepared by taking photos of our friends by giving opinion on product by facial expression. This model works beautifully on test data which gives about 89% accuracy for some expressions.

3.3.5. Identifying Expressions

Outward appearance recognition as a rule brings about classes as per either the Facial Actions Coding System (FACS) or the six fundamental outward appearances. The information layer gets six feelings and the yield layer speaks to 6 potential facial appearances.

3.3.6. Result with The Polarity of the Person

The result of the model is a box on the face of a person with emotion on the top left of the image.
These six emotions are classified in into three customer satisfaction levels. Happy and Surprise are classified as Satisfactory, Sad, Angry and Disgust are Completely Not Satisfactory and Neutral are Neutral class.

4. Results and Discussion:

Figure 7: Output from the proposed MFER on test data. Which shows the multiple expressions on each face in the given input image.

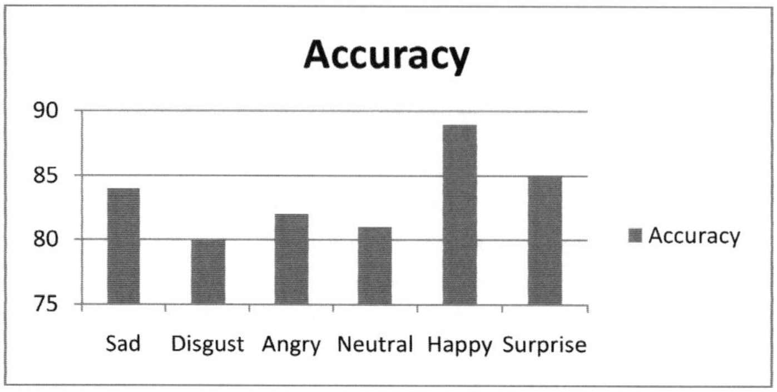

8: All six expressions accuracy on test data. Identified happy expression is nearly 89% accurate using Deep CNN algorithm.

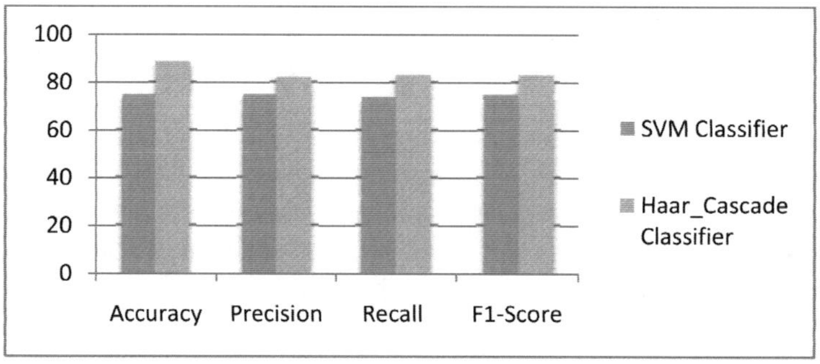

Figure 9: Comparison of Proposed Haar_Cascade Classifier with SVM for Detecting Faces in Images.

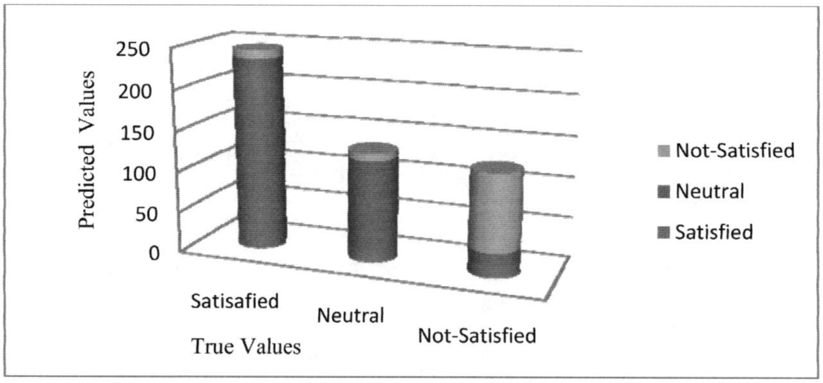

Figure 10: Kinds of Customer Satisfaction Levels.

Facial Emotion Recognition is crucial in many ways providing details about the emotional state. Many researches and studies about Emotion Recognition are used for recognizing the emotions are performed by learning about six basic emotions. Emotions in real life are very complex, so the future scope for this work is to build larger databases and create a powerful model that recognizes all basic and secondary emotions. This paper uses Deep CNN on Kaggle FER2013 dataset which contains 3500 images for emotion recognition. Accuracy rate of about 78% has been achieved. The article identifies expressions of multiple faces in a single image. This type of identification is not possible for the system which are in use.

IOP Conf. Series: Materials Science and Engineering 1074 (2021) 012033 doi:10.1088/1757-899X/1074/1/012033

5. Conclusion

As facial expression recognition technology improves in accuracy, the range of its application will grow, both in the innovation improves in precision, the scope of its application will develop, both in mechanical turn of events and past. The model is intended to take input or audit from a client devouring an item. The survey about the item helps in improving the item qualities. A Facial Expression Recognition framework is intended to take certifiable reaction from the purchaser or client who uses an item. The real reaction is taken as outward appearance, which indicates what is running in the psyche of a client. The report made so ever from the reaction of the client improves the profitability in a business. This framework at times clarifies that what sorts of individuals are utilizing the item and how they are expecting the conveyance or how the administration could be stretched out to arrive at clients desires.

This work can be applied in various divisions or units such as education, business and software. In software field it works between a manager and his subordinates while analyzing their skills. Proposed MFER outputs are compared with existing SVM classifier output. Our model gives best performance in initial stage of Face Recognition (FE) then the next steps automatically improves accuracy for finding customer satisfaction on products when the test data is sample of consumer interest on goods.

6. Future Scope

Machine learning models are biased in accordance to their training data. In my specific application our trained model that adopts Haar_cascade classifier along with Deep CNN is used to fine the emotions. This work can be integrated with speech recognition project and can be extended to find emotions or expressions from video.

References

[1] F. Ahmed, H. Bari and E. Hossain, "Person-independent facial expression recognition based on compound local binary pattern (clbp)", Int. Arab J. Inf. Technol., vol. 11, no. 2, pp. 195-203, 2014.

[2] P. Carcagni, M. Coco, M. Leo and C. Distante, “Facial expression recognition and histograms of oriented gradients: a comprehensive study”, SpringerPlus, vol. 4, no. 1, pp. 1, 2015.

[3] S. Kumar and M. K. Bhuyan, “Neutral expression modeling in feature domain for facial expression recognition”, IEEE Recent Advances in Intelligent Computational Systems (RAICS), 10-12 Dec. 2015.

[4] A. M. Bhatti, M. Majid, S. M. Anwar and B. Khan , “Human emotion recognition and analysis in response to audio music using brain signals”, Computers in Human Behavior 65:267-275, Dec 2016

[5] P. Rajhans, S. Jessen, M. Missana and T. Grossmann , “Putting the face in context: Body expressions impact facial emotion processing in human infants”, in Developmental Cognitive Neuroscience, Vol 19, 115-121, 2016.

[6] Vikrant Chaugule, Abhishek D, Aadheeshwar Vijayakumar, Pravin Bhaskar Ramteke, and Shashidhar G. Koolagudi, “Product Review Based on Optimized Facial Expression Detection”, 978-1-5090-3251-8, IEEE, 2016.

[7] B Hasani and M H. Mahoor, "Spatio-Temporal Facial Expression Recognition Using Convolutional Neural Networks and Conditional Random Fields[C]", IEEE International Conference on Automatic Face & Gesture Recognition, pp. 790-795, 2017.

[8] U. Ayvaz, H. Gürüler and M. O. Devrim, "USE OF FACIAL EMOTION RECOGNITION IN E-LEARNING SYSTEMS", Information Technologies and Learning Tools, vol. 60, no. 4, pp. 95, sept. 2017.

[9] B. Yang, J. Cao, R. Ni and Y. Zhang, "Facial Expression Recognition Using Weighted Mixture Deep Neural Network Based on Double-Channel Facial Images", IEEE Access, vol. 6, pp. 4630-4640, 2018.
[10] S R Livingstone and FA Russo, "The Ryerson Audio-Visual Database of Emotional Speech and Song (RAVDESS): A dynamic multimodal set of facial and vocal expressions in North American English", PLoS ONE, vol. 13, no. 5, pp. e0196391, 2018.
[11] N. Veeranjaneyulu, Jyostna Devi Bodapati, " Facial Emotion Recognition Using Deep Cnn Based Features ISSN: 2278-3075, vol. 8, no. 7, May 2019.
[12] W. Deng and S. Li, "Reliable crowdsourcing and deep locality-preserving learning for unconstrained facial expression recognition", IEEE Transactions on Image Processing, vol. 28, no. 1, pp. 356-370, 2019.
[13] Moulay Smail Bouzakraoui , Abdelalim Sadiq, Abdessamad Youssfi Alaoui, "Appreciation of Customer Satisfaction Through Analysis Facial Expressions and Emotions Recognition", 978-1-7281-1232-9, IEEE, 2019.

Extracting Facial Features in Filter Modified Images using Neural Networks

M Sumithra[1], M Shyamala Devi [2], A Poornima[3], Shatadru Sarkar[4], Harish Patel[5], Preethi Mahankali[6], Boggula Navya[7]

[1] Assistant Professor, Computer Science & Engineering, Vel Tech Rangarajan Dr. Sagunthala R&D Institute of Science and Technology, Chennai.
[2] Professor, Computer Science & Engineering, Vel Tech Rangarajan Dr. Sagunthala R&D Institute of Science and Technology, Chennai.
[3] Assistant Professor, Computer Science & Engineering, Vel Tech Rangarajan Dr. Sagunthala R&D Institute of Science and Technology, Chennai.
[4, 5,6]Final Year B.Tech Student, Computer Science & Engineering, Vel Tech Rangarajan Dr. Sagunthala R&D Institute of Science and Technology, Chennai.
sumithra@veltech.edu.in

sumithra@veltech.edu.in

Abstract. Facial identification plays a vital role in this digital era. Recent advancements in technology provide various filtering effects while capturing the images. Extracting the facial features from the filter modified image becomes challenging. To identify the important facial attribute values from those images, special features extraction techniques needs to be applied. Facial points are the key factors to recognize the face. When these factors are altered, facial detection becomes crucial. Initially apply the image pre-processing techniques to reduce the noise level in the images then perform feature selection. To overcome the curse of dimensionality, Occam's razor approach is followed to limit the selection of attributes. In different stages the features are going to be filtered by replacing the weaker attributes. Later Pearson Correlation, a filter based correlation technique is applied to evaluate the deviated numerical value from the target. The deviated features can be treated with Lasso effect to suppress the poor weighted features. Finally through recursive feature elimination, highest influencing attribute factor can be identified. In this approach some of the facial feature which is used in training the neural network can be discarded based on the deviation rates. In terms of accuracy the proposed system can recognize the dataset with 70% of accuracy with filter effected images.

1. Introduction

Recent technical advancement clearly presents the necessity for extracting the facial feature in images to facilitate the security related applications. Images may be enhanced with filters using the end device special features like makeup recommendations, scar removal, tattoos, disguising features etc., these enhancements beautify the images of the person which in turn changes the originality of the exact facial features.

IOP Conf. Series: Materials Science and Engineering 1074 (2021) 012034 doi:10.1088/1757-899X/1074/1/012034

Different techniques available for image processing concentrates on the major facial angle points using which the identity of the person can be found. Due to change in these features, accuracy varies while retrieving the identity of the image. So images can be treated with correlation techniques to find the predominant factors representing the originality of the image identity.In order to observe the deviation measures, two categories of features are identified as primary attributes and secondary attributes. An original image is treated formerly with primary set of attributes and then with the secondary attributes. Images are further can be scaled and rescaled to achieve further accuracy in measuring the deviation factors to identify the most dominating attributes revealing the identity of the person in the filter applied images.

2. Literature Review

In [1] the facial changes were spotted using the generation scheme with multidimensional feature vector comprises of nearly 26 dimensions each representing the facial feature. But the drawback is the curse of dimensionality where the selection of the feature and its multiplicity happens in exponential factor for training the system. To overcome it, an Occam's razor approach can delimit the selection of the feature based on their weight deviations.

Face attributes: The feature extraction system uses multitask learning technique based on neural network to combine and evaluate the different facial attributes into a single face image proposed in [2]. Whereas the available methods concentrated on single factor based expression existence [11], using this it provides the robustness in differentiating and resembling the images. It is difficult to capture spatial-temporal information of expressions with slight motion. The images can't be neglected based on their variation in their spatial or temporal data provided with the image datasets [4].

In the future, a more powerful structure to model the motion of facial critical areas will emerge. It utilizes specific advanced methods as it is essential in security source and in applications where the image of a person is disguised to avoid the identity [3]. Head pose classification is used in face recognition where the input image to be a front face. But during this pandemic the situation controversy the case to wear the face mask where the face identity becomes the crucial thing. Face classification techniques adopts colour texture analysis of images and line portrait. It considers the H channel in HSV (hue, saturation and value) Colour space to extract the parameters [4]

[5] States that the facial features extraction requires less or no images pre-processing techniques. As the image quality is not a major constraint in recognizing the feature variations pre-processing steps can be diluted. Most of the eight important facial points are identified including eyebrows, eyes, jaw etc., and trained with dataset using landmark detector. It helps in identifying the weak attributes [11].

Face recognition algorithm shows variation depends on the frequency of images identified with similar Patterns [12]. Sampling methods may vary depends on the intensity of the images and the area considered for bounding box to abstract the features [15].

Related works on facial recognition shows the images considered for applying machine algorithms can be unlikely to be processed with pre-processing methods. Most of the important factors to be considered in facial recognition are the eyebrows, lips, eyes, jaw, nose and eyes. Some of the features may contribute more than the other features. The selection of features depends on the parameters considered to differentiate the images among the common dataset of similar images. Based on these observations few more feature reduction techniques also applied to improve the available pattern of image classification in facial recognition.

IOP Conf. Series: Materials Science and Engineering 1074 (2021) 012034 doi:10.1088/1757-899X/1074/1/012034

3. Our Contributions

Facial feature extraction can be performed can be done using machine learning techniques. In this approach a raw image is processed through several stages to extract the facial features. These extracted features are in turn utilised to identify the guised image. Usually an image can be altered using the filtering effects like embedding special symbol like characters, emojis, colour scale variations etc.,

1. A raw image can be enhanced noise removal techniques using Weiner filter and CLAHE (Contrast-limited adaptive histogram equalization).
2. HOG (histogram Equalization techniques) and SVM are employed for face detection among the classes of variety of objects in the image.
3. Apply instance and semantic segmentation feature extraction techniques to observe the facial features.
4. A pixel-level classification of an object can be spotted with landmarks using angle tool.
5. Apply Co-relation techniques to compare the deviated futuristic value.
6. Diminish the less weighted factors to identify the dominant feature by applying lasso effect and Recursive feature elimination techniques.
7. The identified features can be utilised to test the filtered effected image and calculated the accuracy level variation.

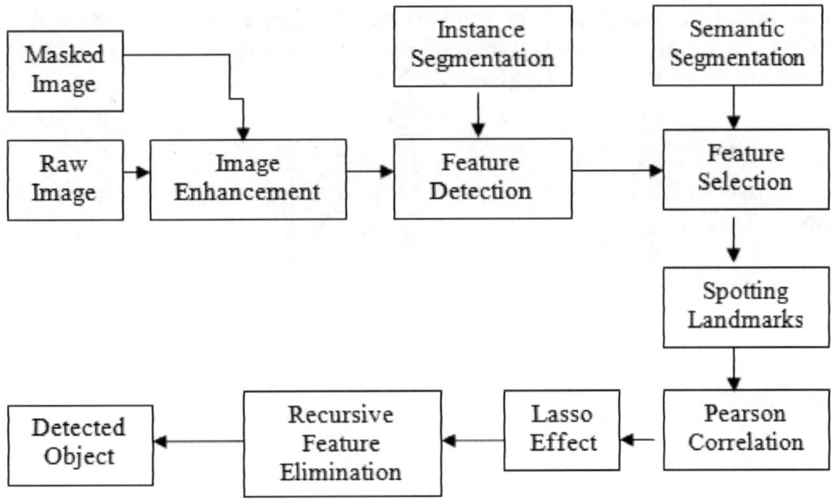

Figure 1. Architectural View.

4. Implementation Setup

4.1 Face detection

HOG extractor performs the task of face detection and SVM can be applied to improve the accuracy of detection. The rectangular portion of the face is margined and the angular points represents the facial landmarks can be spotted. A real time image can serve better in this sample training of vector machines for image recognitions.

IOP Conf. Series: Materials Science and Engineering 1074 (2021) 012034 doi:10.1088/1757-899X/1074/1/012034

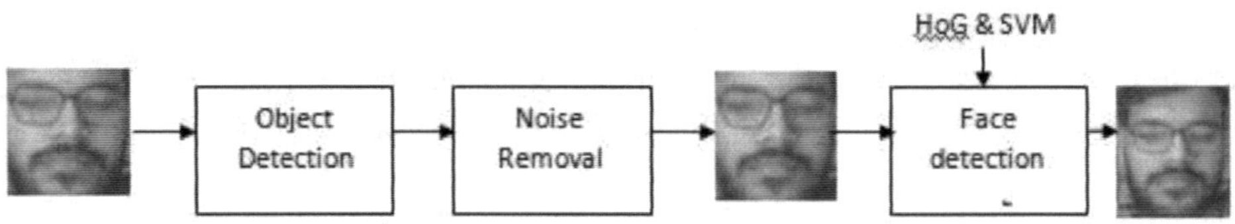

Figure 2. Face detection using SVM classifier.

Initially the presence of face in an image is identified by object detection approaches [2] and then the sample image is treated with SVM in this case to detect the category of the image among the trained samples. The HOG descriptor gives the gradient orientation of the image according to the frequency count in the localized range.

4.2 Feature Selection

To extract the features first confirm the image for the presence of facial portion. Instance segmentation leaves the instances of all classes individually about the variety of objects appears in the image. After identifying the instances for facial objects apply semantic segmentation to perform the pixel-level classification. Deep learning concepts are widely available in abundance regarding image classification to perform these segmentations.

Figure 3. Feature selection.

4.3 Facial Landmarks

The landmark point can be marked near eyebrow corners both inner and outer and in eyes, eye corners, and eye inner and outer rings, upper lips from centre point, corner tips and similarly from lower lip. A centre point in nose can be traced for finding the forehead distance from the middle of the face area. Apart from that cheeks, chin, temple points can also be spotted to find the facial features.

Figure 4. Locating facial landmarks.

4.4 Filter Effects images

The neural network is equipped with the images before applying filters. Apply the filter effects like masking, disguising the lower face like showing different angles of opened mouth, apply tattoos etc., Identify the landmarks from the effects applied images.

After identifying the landmarks, using Pearson co-relation find the deviated numerical distance of the original image. Initially the attributes consider for finding the distance measures can be the entire landmarks spotted on the original image. Compare the deviation range and identify the factors playing vital role in the masked image.

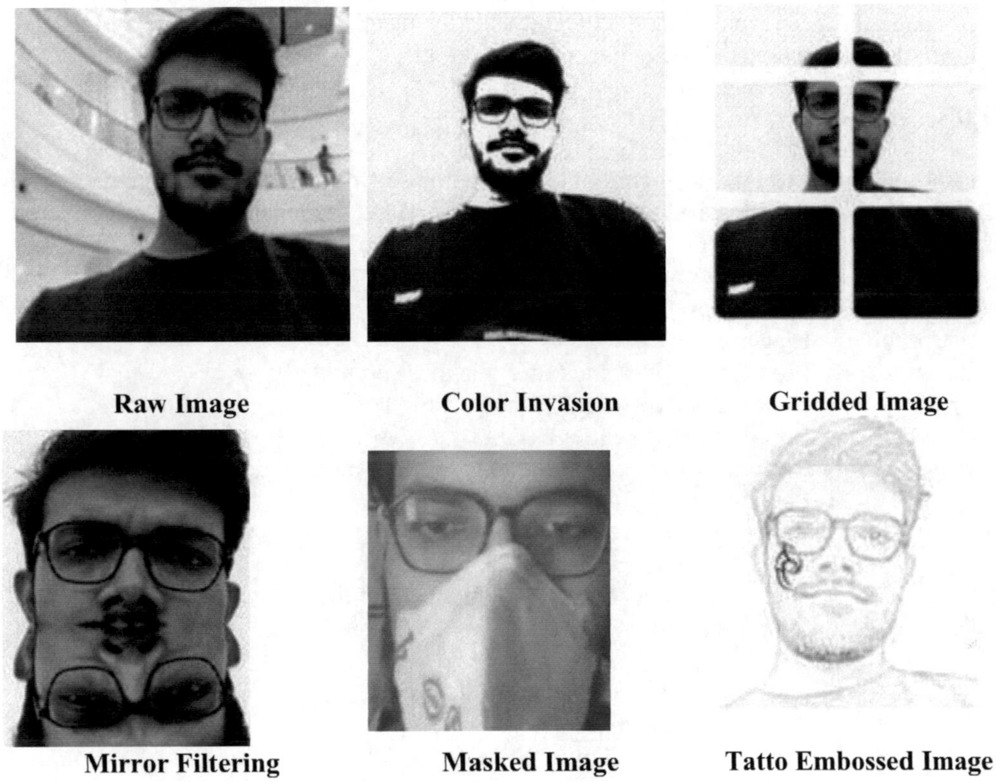

Figure 5. Filter applied images

Using lasso effect filters the features which are of less importance. Apply over the features recursively to remove the unweighted features from the attribute list. Recursive feature elimination technique can be implemented to remove the features and to end up with obtaining higher influencing features.

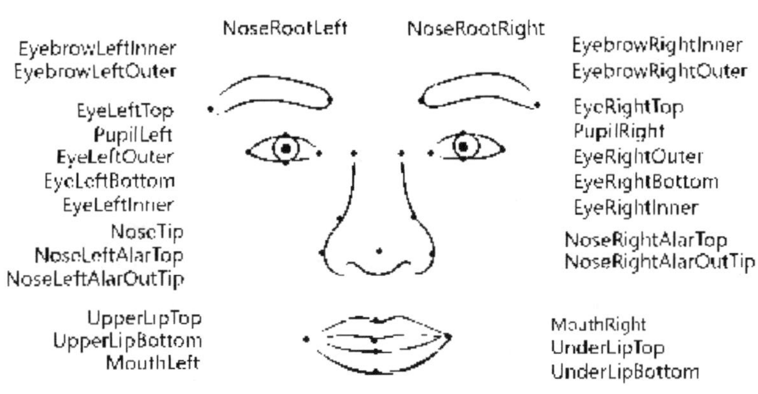

Figure 6. Facial Attributes.

The images can be altered using various filtering effects but the feature spots located are adjusted with appropriate weights to identify the similar images. The attributes may vary from image to image depends on the types of filter applied. It many any other object inclusion like in Figure 4. Tattoo image or some flipping effects as in Fig.3 Mirrored images.

Initially consider all the factors forming the facial features. As the neural network will be trained with the enhanced image for face recognition, the dataset used for identifying the face can be tuned using the lasso technique and recursive feature elimination method.

5. Results and Discussions

The system is analysed using the sample image of a person for self demonstration. Here the facial features considered as primary features are given in the table 1 and the secondary features considered are mentioned in table2.

Table 1. Primary Features.

Attributes	**Categories**
Eyes Corner	Distance
Eyebrows	Location of inner and outer thickness points
Eye ball	Distance between centred eye position
Nose Length	Elevation level near end points
Cheeks	Closeness between nose end and cheeks angular points
Chin	Length of the chin
Lip	Coherence between the lip layers

As the filter effects are not only meant to change the colour scale of the image, deviating attributes are likely considered as secondary parameters.

Table 2. Secondary Features.

Facial Feature	**Categories**
Skin tone	Light, Medium, Dark
Face structural shape	Square, Rectangle, Triangle, Oval, Broaden
Eye colour	Brown, black, blue
Eye Size	Small, Normal, Large eyeball
Lips shape	Short, Widen, Compact
Nose point	Sharp, Blunt, Bend

The deviation factor of the image featured vector values can be calculated using Pearson's correlation where the linear correlation between each and every two variables preferably the primary attributes values can be measured. The range of the deviation lies in the range of +1 to 1 showing the positive and negative deviation measures. For sampling the images, positive deviation factors are considered as given in the table 3.

Most of primary features are considered from the landmark points to train the system amongst which some data are considered for representing the variation for reference purpose. Here it is clearly visible the masked portions are showing more variations. Those attributes can be considered as less dominant attributes by the neural network to consider in recognizing the filtered images.

IOP Conf. Series: Materials Science and Engineering 1074 (2021) 012034 doi:10.1088/1757-899X/1074/1/012034

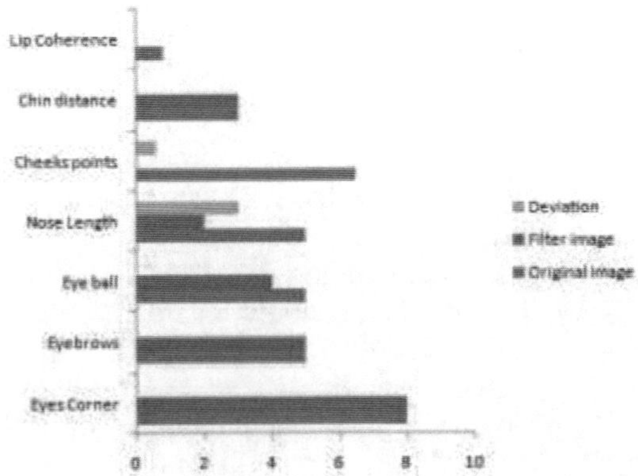

Figure 7. Deviation measure using Primary features.

Table 3. Deviation through Pearson Correlation.

	Original Image	Filter image	Deviation
Eyes Corner	8	8	0
Eyebrows	5	5	0
Eye ball	5	4	0
Nose Length	5	2	0.3
Cheeks points	6.5	0	0.6
Chin distance	3	3	0
Lip Coherence	0.8	0	0

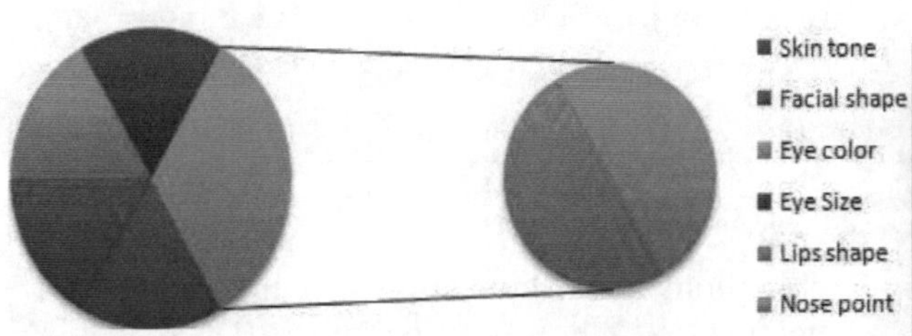

Figure 8. Extracting the features causing deviation.

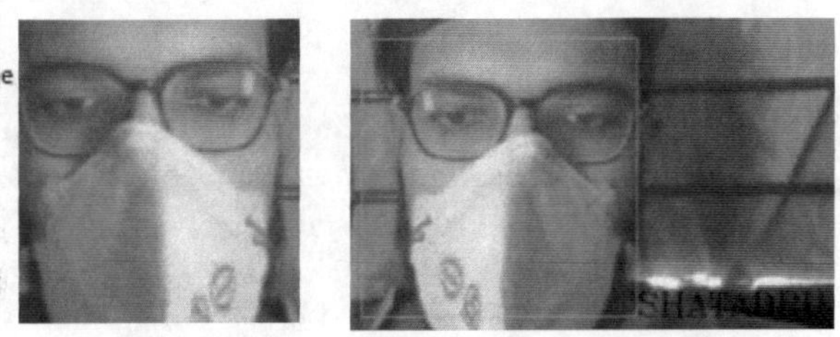

Figure 9. Filtered image extraction.

The attributes causing deviation between the original image and the filtered applied images can then be separated using the secondary features for further accuracy. The top features are extracted from the HOG treated image and then the same image can be rescaled to measure the deviation factors. Zooming or minimizing the image can be happened while applying the filters hence the scaling also considered as a source for differentiating the images.

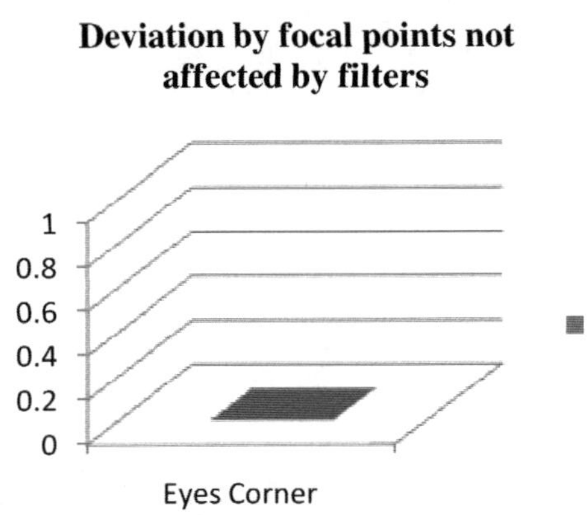

Figure 10. Individual attribute deviation.

Table 4. Deviation measure between original and masked image.

	Original Image	Filter image	Deviation
Eyes Corner	8	7	1
Eyebrows	5	5	0
Eye ball	5	3	2
Chin distance	3	2	1

Similar observations can be found for individual focal point when that landmark is not disturbed by the effects applied to the image.

Conclusion

Though machine learning algorithms are improved day-by-day in image recognition, a new approach is put forth to recognize the guised images by the application of various filtering effects. A neural network is trained to compare the filtered images with the masked images to identify the dominant among the facial features identifying attributes and consider those attributes in recognizing the filtered image. In this method a sample image trained with different effects applied and the accuracy level of the recognition varies between 70 percent than recognizing the Original image.

References

[1] Anima Majumder, Laxmidhar Behera, Senior Member, IEEE, and Venkatesh K. Subramanian *Automatic Facial Expression Recognition System Using Deep Network-Based Data Fusion* IEEE transactions on cybernetics 2016

[2] Luigi celeno, Seimol Balcana 2016 *Fine-grained face annotation using deep Multi-Task CNN* IEEE transactions on Image processing

[3] Kaihao Zhang, Yongzhen Huang, Member, IEEE, Yong Du, Student Member, IEEE, and Liang Wang, Senior Member, IEEE *Facial Expression Recognition Based on Deep Evolutional Spatial-Temporal Networks*

[4] Shuang Li, Xin Ning, Lina Yu, Liping Zhang, Xiaoli Dong *Multi-angle Head Pose Classification when Wearing the Mask for Face Recognition under the COVID19 Coronavirus Epidemic*

[5] S. Singh and F. Nasoz, "Facial Expression Recognition with Convolutional Neural Networks," 2020 10th Annual Computing and Communication Workshop and Conference (CCWC), Las Vegas, NV, USA, 2020, pp. 0324-0328, doi: 10.1109/CCWC47524.2020.9031283.

[6] Facial Feature Extraction Using Frequency Map Series in PCNN

[7] G.-C. Luh and C.-Y. Lin, "*PCA based immune networks for human face recognition*," *Applied Soft Computing*, vol. 11, no. 2, pp. 1743–1752, 2011.

[8] Z. Sun, J. Li, and C. Sun, "*Kernel inverse Fisher discriminant analysis for face recognitio*n," *Neurocomputing*, vol. 134, pp. 46–52, 2014.

[9] I. Marques and M. Graña, "Face recognition with lattice independent component analysis and extreme learning machines," *Soft Computing*, vol. 16, no. 9, pp. 1525–1537, 2012.

[10] "Application: Facial Expression Recognition" in Machine Learning in Computer Vision. Computational Imaging and Vision, Dordrecht:Springer, vol. 29, 2005.

[11] C. Sagonas, E. Antonakos, G. Tzimiropoulos, S. Zafeiriou and M. Pantic, "300 faces In-the-wild challenge: Database and results. Image and Vision Computing (IMAVIS)", *Special Issue on Facial Landmark Localisation "In-The-Wild"*, 2016.

[12] C. Sagonas, G. Tzimiropoulos, S. Zafeiriou and M. Pantic, "A semiautomatic methodology for facial landmark annotation. Proceedings of IEEE Int'l Conf. Computer Vision and Pattern Recognition (CVPR-W)", *5th Workshop on Analysis and Modeling of Faces and Gestures (AMFG 2013)*, June 2013.

[13] L. Best-Rowden, H. Han, C. Otto, B. F. Klare and A. K. Jain, "Unconstrained face recognition: Identifying a person of interest from a media collection", *IEEE Trans. Inf. Forensics Security*, vol. 9, pp. 2144-2157, Dec. 2014.

[14] Y. Huang, Z. Wu, L. Wang and T. Tan, "Feature coding in image classification: A comprehensive study", *IEEE Trans. Pattern Anal. Mach. Intell.*, vol. 36, no. 3, pp. 493-506, Mar. 2014.

[15] Z. Lei, D. Yi and S. Z. Li, "Local gradient order pattern for face representation and recognition", *Proc. 22nd Int. Conf. Pattern Recognit. (ICPR)*, pp. 387-392, Aug. 2014.

IOP Conf. Series: Materials Science and Engineering 1074 (2021) 012035 doi:10.1088/1757-899X/1074/1/012035

A New Feature Based Online Rating System for E-Commerce Websites

D. R. Kumar Raja[1], **M. Ramu**[2] and **AKULA CHANDRA SEKHAR**[3]

[1] Associate Professor, Sree Vidyanikethan Engineering College, Tirupathi, AP, India.
[2] Assistant Professor, Sree Vidyanikethan Engineering College, Tirupathi, AP, India.
[3] Professor, Avanthi Institute of Engineering and Technology, Cherukupally(V), Vizianagaram District, India

E-mail: kumarraja.raj@gmail.com

Abstract. Online reviews and ratings are considered extremely important in any business. Any customer who wants to purchase a particular product online will have a look on the ratings before placing an order, any person who wants to book a hotel or restaurant in a particular location will have a glance on online ratings, any person who wants to choose a best hospital will check ratings. The question which arises here is how far the ratings can be reliable? The ratings cannot be trusted completely as they are not feature specific. The overall rating leaves the people in chaos and they cannot conclude that on which basis the ratings have been awarded. To overcome this problem feature based online rating is introduced. Where ratings for different features have been awarded separately. It helps people to take decisions easily and can purchase a product of their requirement. It helps in the growth of the business as customer satisfaction is equally important and it increases the quality of service.

1. Introduction

Online ratings are one of the most trusted sources of consumer confidence in e-commerce decisions. Due to the ease and availability of internet and due to the hectic schedule customer checks the durability, sustainability and availability while sitting at home. When it comes to online ratings, our herd instincts combine with our susceptibility to positive "Social influence". When we see that the other people have appreciated a particular book, enjoyed a hotel or restaurant or liked a particular doctor and rewarded them with a high online rating, this can cause us to feel the same positive feelings about the book, hotel, restaurant or doctor and to likewise provide a similarly high online rating. Every customer will check reviews and ratings before purchasing the product [3,5,7]. For Example, before purchasing a product customer will do a survey. This is usually done by knowing from a person who is using it, by the respective person's online review. If person is satisfied with one feature and still it increases the overall rating of the product.

But the actual problem which arises here is if the person is taking a decision based on online rating, the respective person might be looking for other features rather than on which the rating is given [4]. Hence it gives a mere dissatisfaction to the customer who purchased it and leaves customers in a dilemma that the products should be purchased or not, based on online rating.

2. Related Work

With the fast development of World Wide Web availability, our lives nowadays bank more and more on the Internet. Online systems permit an oversized variety of users to act with one another and provide thousands of films, lots of films and books, billions of web content for them to decide on [2]. Although plenty of helpful on-line objects are out there, accurately ranking their quality is quite difficult. Therefore, several online websites (such as Ebay, Amazon, and Netflix) introduce the supposed rating system, within which users can evaluate objects by giving distinct ratings [6]. To more or less decide the quality of a particular object, a user can refer to the historical ratings the object received [1]. Let us take an example of an e-commerce website such as Amazon. Suppose if someone is looking for a mobile phone with a good camera quality and fortunately if the purchased mobile phone excels in it. The respective person will reward a good rate in online review system.

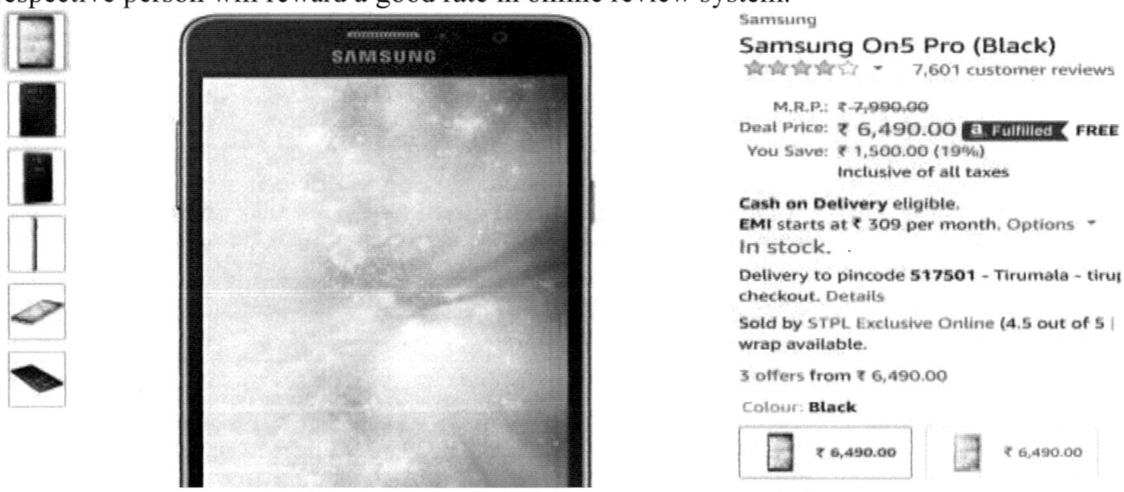

Fig 1: Product details in e-commerce website

Figure 1 shows that the customers who purchased the product have rewarded with ratings based on their experience. The overall rating of product is 4 stars out of 5 stars. The rating can be depended on any feature such as Camera quality, internal memory, battery life etc.

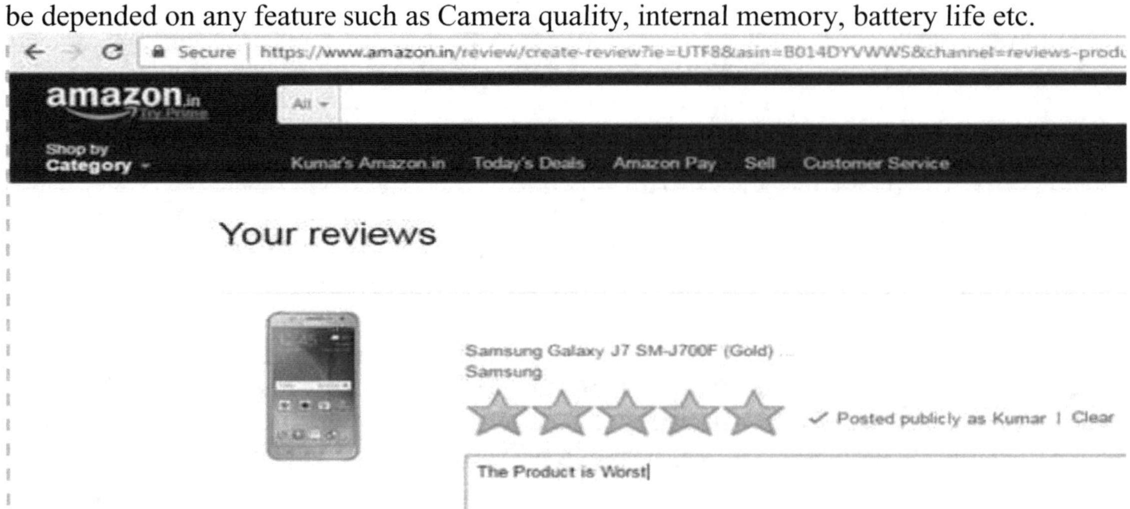

Fig 2: Customer rating system

Figure 2 shows that the customer is satisfied with the product and thus have rewarded 5 star rating.

Let us take other example of hospital online rating system. Figure 3 shows that the patients who have visited this hospital have rewarded with stars based on hospitality, hygiene, workers, equipment's, doctors etc. It is clearly shown in the figure that the hospital overall rating is 4 and a half out of 5 stars.

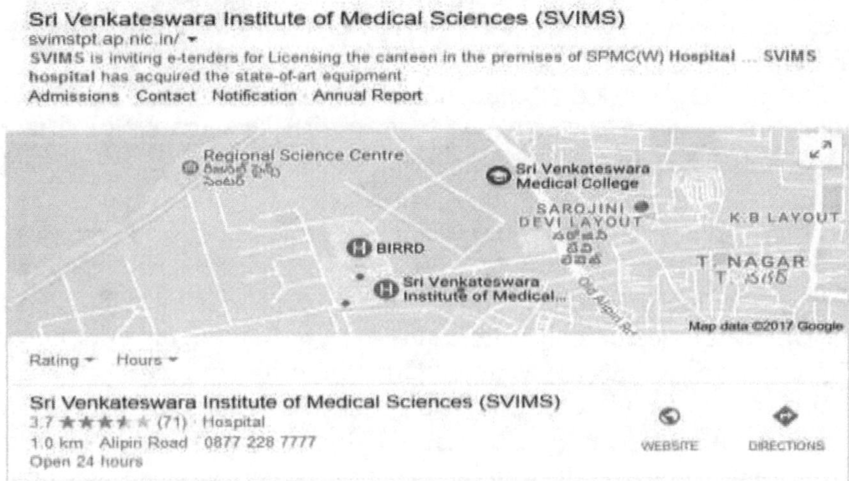

Fig 3 : Star rating given by the customer

Figure 4 below shows the hotel booking website. The overall ratings below are 5 stars for resorts and can be based on features such as resort location, hospitality, food quality, cleanliness etc.

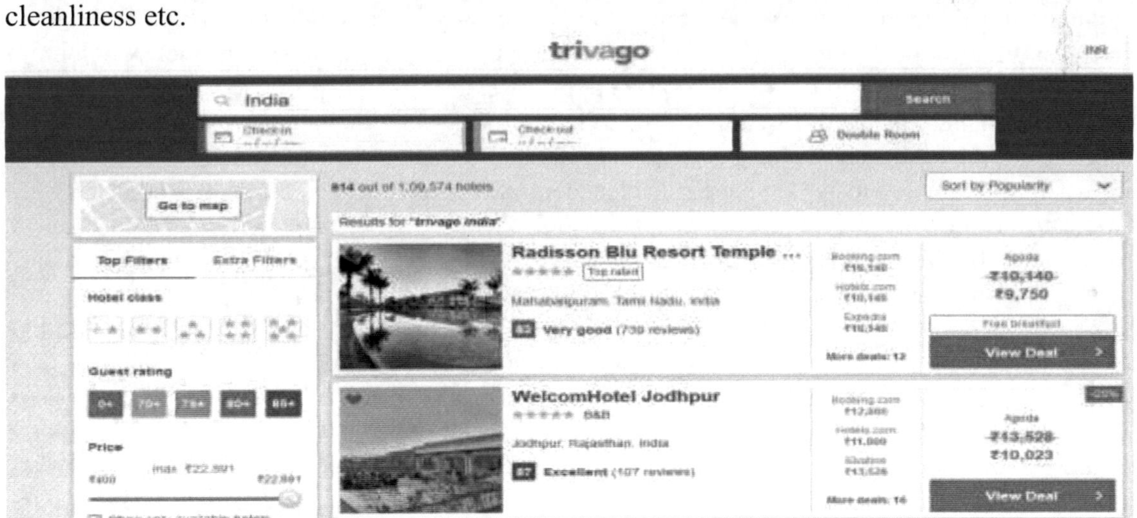

Fig 4: Hotel booking website

Above are the few examples for online rating systems in websites where online ratings are not feature specific. Before purchasing any product online in an e-commerce, the customer will check online ratings to know the general opinion. Before booking any restaurant or resort the person will have a glance on the resorts with a good overall rating. Before going to hospital for treatment a person will have a sneak peek on the online rating of the hospital. Nowadays online ratings is gaining popularity everywhere.

IOP Conf. Series: Materials Science and Engineering 1074 (2021) 012035 doi:10.1088/1757-899X/1074/1/012035

3. Problem Definition

Online rating is nowadays considered one of the most important sources of information for purchasing a product online, for choosing a best hospital for treatment, booking affordable resort. During the analysis of existing work, few problems are encountered. Suppose a person wants to travel and is looking for hotels or restaurants to stay. The respective person looks for online ratings given, based on cleanliness, hospitality, view, food quality, price etc. If the person is looking for a restaurant or a hotel with a good hospitality and the previous ratings were rewarded based on a good resort view or may be a for cleanliness. The person will face mere dissatisfaction for opting it.

Let us have a look on few more examples such as if a customer wants to buy a mobile phone online and the person is looking for phone with a good display quality. The mobile phone generally consists of other features also such as camera quality (front and back), RAM and internal memory, size, weight, Battery life, Connectivity, Performance etc. There are chances of awarding online ratings based on other features. Hence a person purchases a product for which he wasn't looking for. Similarly, if a person is looking for a best hospital for treatment around the place. The person will do a survey and go fora hospital with good online rating. The rating is rewarded based on only one feature. The person might be looking for a hospital which is affordable. The ratings are awarded based on different criteria. It creates a sense of dissatisfaction to a person and hence can also result in a low rating. The ratings which are awarded are not specific. Hence the problem which arises commonly these days is that the customers who wants to buy product online cannot completely depend on online ratings for the product as the ratings may be given based on other feature performances. The opinions are discrete and hence online ratings cannot be trusted.

4. Proposed Work/Methodology

By analyzing all the existing review and rating processes we have observed that there is no clarity on the products rating and there is no feature specific rating of the product which helps the customers to buy a product based on a specific feature. In our proposed approach we are trying to concentrate on the feature-based rating process of the product by considering the reviews given by the users.

Let us take examples to understand it better. The attribute specific rating makes it easy for people to take decisions. Figure 5 shows the online rating of a hotel where the ratings are feature specific. The ratings are given below for Hospitality (4 stars), Food Quality (4 stars), Taste (4 stars), Cleanliness (5 stars), Price (4 stars), Spatial comfort (3 stars). The overall rating awarded is 4 stars. It shows if the person is looking for a hotel with specific feature like Food Quality, according to 546 votes it is clear that the restaurant serves delicious food or a person is looking for affordability the person can form a decision accordingly. It makes easy for travelers to form an opinion on the restaurant according to their need. If a person visits the restaurant, the person awards ratings for specific features accordingly. For food quality the opinion would be different from hospitality and so the ratings as well. Similarly, for the remaining features.

Fig 5: Online rating on hotel

The biggest source for communication Mobile phone all over the world consists of many exciting features. Figure 6, depicts a phone with feature specific online rating Operating system (4 stars), Connectivity (4 stars), Performance (4 stars), Camera quality (5 stars), Display (4 stars) and Battery life (3 stars) depending on the various votes. It shows the feature-based rating of a product and makes the decision of the customer easy. If a customer wants to buy mobile phone with good battery life the person can take decision accordingly. The overall rating is not specific and hence the customers are left with mere dissatisfaction.

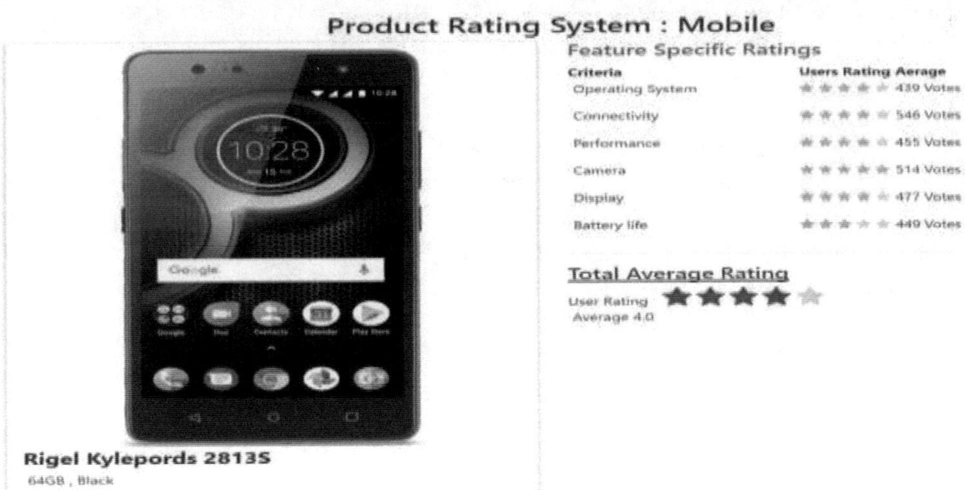

Fig 6: Rating with feature specific

Figure 7 shows the rating system of a hospital. The care quality or the hospitality (4 stars), expertise and competence (5 stars), healing environment (4 stars), cleanliness (4 stars), affordability (3 stars) based on the various votes from the customers who have visited it. The total overall rating awarded is 4 stars. If a person focuses more on hygiene, can form the decision accordingly.

IOP Conf. Series: Materials Science and Engineering 1074 (2021) 012035 doi:10.1088/1757-899X/1074/1/012035

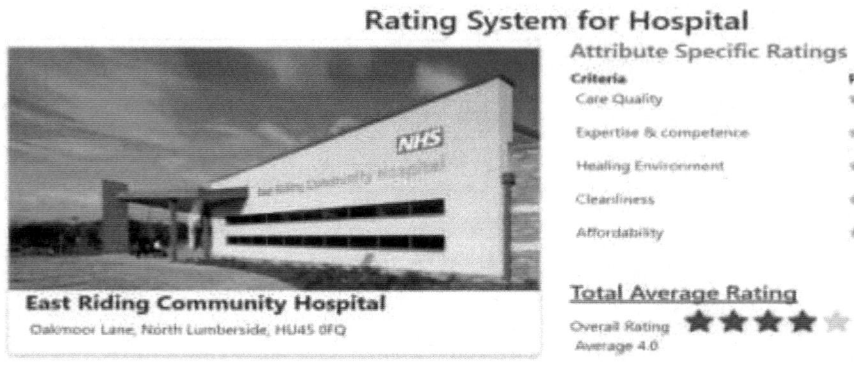

Fig 8: Rating of a hospital with features

The feature specific online rating system for any website such as for hotels, movies, e-commerce websites, hospitals etc. makes it easy for the customers to take decisions or to form opinions. The customers cannot rely upon overall rating system because they cannot conclude that on which basis the ratings have been awarded. Here feature based online rating system shows you a bigger picture.

5. Conclusion

Online ratings and reviews are a crucial part of business. These are the feedback given by the customers and hence are helpful to increase the quality of service. Customer satisfaction is generally the main motive in any business. The overall rating which has been awarded after purchasing and using any product in any e-commerce website, after visiting any restaurant or hotel, or after having treatment in any hotel is considered as a feedback from the people. The overall rating cannot be trusted completely as it can never be concluded that on which basis it has been given. Feature based online rating is reliable, different people have different opinions on different attributes thus the overall ratings is not justified. It contributes in the growth of the business as it can be derived easily from the ratings that the business is lagging in which part. That particular part can be modified to get the better outcomes in a business. It shows that your product excels in which portion and attract more and more customers towards it. Therefore, feature based online rating system in any website is considered to give better outcomes than normal rating system.

6. References

[1] Hao Liao, An Zeng,Rui Xiao, Zhuo-Ming Ren, Duan-Bing Chen, Yi-Cheng Zhang: Ranking Reputation and Quality in Online Rating Systems-method to effectively recognize the unreliable users and reduce their weight in judging the quality of objects, finally leads to a more accurate evaluation of the online products May 12, 2014

[2] WenjingDuan ,Bin Gu , Andrew B.Whinston: Do online reviews matter? — An empirical investigation of panel data- examines the persuasive effect and awareness effect of online user reviews on movies' daily box office performance 2008

[3] D. R. Kumar Raja, and Dr. S. Pushpa, “Feature level review table generation for E-Commerce websites to produce qualitative rating of the products", *Future Computing and Informatics Journal* (2017)118-124.

[4] Narahari P Rao et al., "Real Time Opinion Mining of Twitter Data", (IJCSIT) International Journal of Computer Science and Information Technologies, vol. 6, no. 3, pp. 2923-2927, 2015.

[5] D. R. Kumar Raja, and Dr. S. Pushpa, “An Approach To Construct MDRT To Produce Qualitative Ratings For E-commerce Websites”, International Journal of Computer Science Engineering and Information Technology Research (IJCSEITR) Vol. 7, Issue 1, Feb 2017, 9-14, 2016.

[6] Ravendra Ratan and Singh Jandail, "A proposed Novel Approach for Sentiment Analysis and Opinion Mining", India International Journal of UbiComp (IJU), vol. 5, no. 1/2, April 2014.
[7] D.R. Kumar Raja, S Pushpa, BS Naveen Kumar, "Multidimensional distributed opinion extraction for sentiment analysis - a novel approach", 2nd International Conference on Advances in Electrical, Electronics, Information, Communication and Bio-Informatics (AEEICB), IEEE 2016.
[8] Piyush Gupta et al., "Sentiment Analysis on Hadoop with Hadoop Streaming", International Journal of Computer Applications (0975-8887), vol. 121, no. 11, July 2015
[9] Sunil B. Mane et al., "Real Time Sentiment Analysis of Twitter Data Using Hadoop", IJCSIT) International Journal of Computer Science and Information Technologies, vol. 5, no. 3, pp. 3098-3100, 2014, ISSN 0975-9646.

IOP Conf. Series: Materials Science and Engineering 1074 (2021) 012036 doi:10.1088/1757-899X/1074/1/012036

Design and Development of Adaptive filter for Real Time Signal Processing

Dr Krishna Samalla[1], S.P.V SubbaRao[2], G Mallikarjuna Rao [3], B.N.Jagadeesh[4]
1,2.Sreenidhi Institute of Scinece and Technology,Hyderabad,Telangana (State).
3.RCI DRDO Hyderabad,Telangana (State)
4.Srinivasa Institute of Engineering&Technology,Amalapuram,A.P
krishnas@sreenidhi.edu.in

Abstract:

In Communication the signal is distorted due to noise and the notch filters are used for eliminating unwanted frequencies For filtering the signal in a specific band Notch filter is used. Notch filter can pass the signal below and above notch frequency. In most of the signal processing applications frequency estimation of narrow band signals in noisy environment are required For frequency detection and tracking Adaptive IIR filter or Adaptive FIR filters can be used. Adaptive FIR Filter is more stable than Adaptive IIR filter. In most practical applications Adaptive IIR Notch Filter (AIIRNF)is used The advantage of Adaptive IIR notch filter over FIR Filter is less number of filter coefficients and less complexity in design

In this paper Adaptive IIR Notch filter is designed for real time applications. The performance of designed adaptive IIR Notch filter is evaluated based on Magnitude and Frequency response.

Key words: FFT, IIR filter , Real Time Signal Processing ,Adaptive filter

I. Introduction

In Communications a digital filter is used to filter noise. The input signals sensed by the sensors or from the microphone will have signal with noise. If filter specified frequency is band stop a band stop filter or Notch filter can be used. In Notch filter certain frequency components can be eliminated and other frequency component cannot be effected. Adaptive filters are used to vary the filter coefficients in order to get optimum response. In most practical applications Adaptive IIR Notch filter is used because of less complexity in design. The adaptive mechanism allows automatic detection and removal of un wanted noise signal that is present along with original signal.

Adaptive Notch filters [ANFs]are used in howling canceller [2] many practical applications for removal of interference and noise [3].Mostly used [ANFs] are based on Second order IIR Filter and two methods foe designing the transfer functions of the filters are Constrained Poles and Zeros-based ANF (CPZ-ANF) [4],and All-Pass Filter-based ANF (APF-ANF) [5], [6]. These two methods are widely used.Steady state analysis adaptive algorithm is proposed base on the above .

IOP Conf. Series: Materials Science and Engineering 1074 (2021) 012036 doi:10.1088/1757-899X/1074/1/012036

In this paper Adaptive IIR Notch Filter (AIIRNF) is described Section II describes the functional description of Adaptive IIR Notch filter section III describes the Design of Adaptive IIR Notch filter. Section IVResult Analysis and Section Vconclusion.

II :Functional Description of Adaptive IIR Notch Filter (AIIRNF)

In the functional diagram shown in figure 2.1 the input signal from the sensors are given to analog to digital converter and the output of Analog to Digital converter is given to Fast Fourier Transform to extract the frequency components. The frequency components from the output of FFT are given to Adaptive IIR Notch filter for filtering noise. The output of filter is given to Digital to Analog converter for getting the analog output .

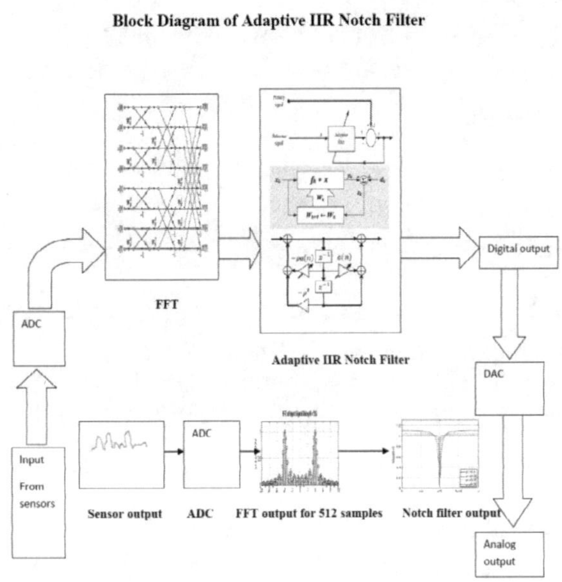

Figure2.1: Functional Description of Adaptive IIR Notch Filter (AIIRNF)

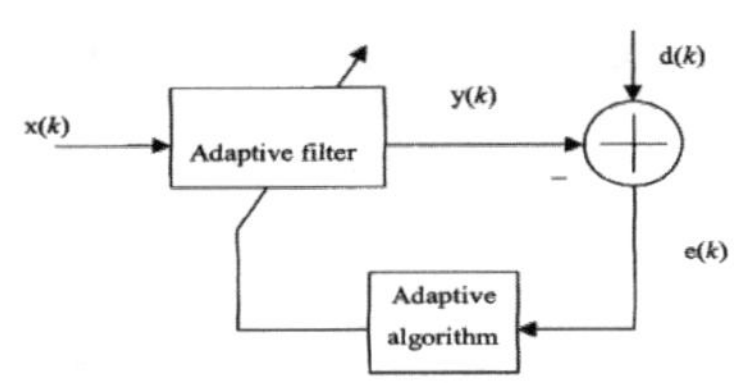

Figure 2.2: Block diagram of Adaptive notch filter

As shown in figure 2.2: x(k) is the input for the filter and y(k) is the filter output and d(K) is the adapted e desired output. And as shown in figure 2,3 Y(K) is the output for the input x(k) and W(k) is the coefficient resulted from the adaptive algorithm and converts f(k) co-efficient for x(k) to the output y(k).

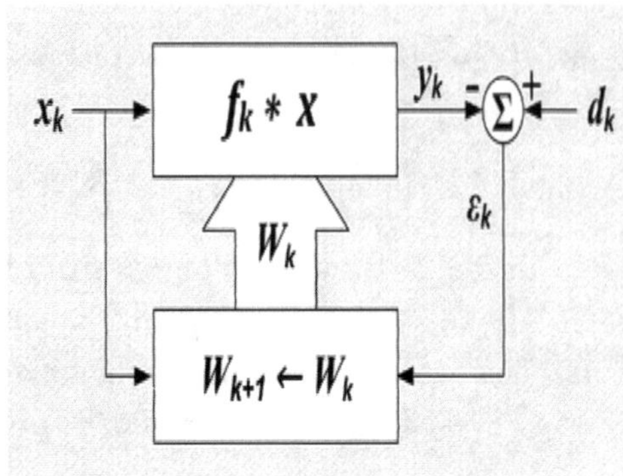

Figure 2.3 : Adaptive Loop.

III. Design of AdaptiveNotch IIR Filter

The following are the three basic steps involved in design of Adaptive Notch IIR Filter

a) Moving average Filter.
b) FFT
c) IIR based ANF.

3.1 Real Time Signal Processing: Moving AverageFilter .

The moving Average is the best and simole method. It averages number of input signal samples and results single output sample value..

3.2 Fast Fourier Transform (FFT)

FFT samples the signals and divide the signal into different frequencycomponents..DiscreteFourier Transform (DFT) of the input signal is computed using FFT and the inverse FFT convertssignal form frequency domain to original domain. Decimation InTime FFT (DIT FFT) and Decimation in Frequency FFT (DIF FFT) are used for computing DFT.Figure 3.1 shows Radix 2 DIT FFT

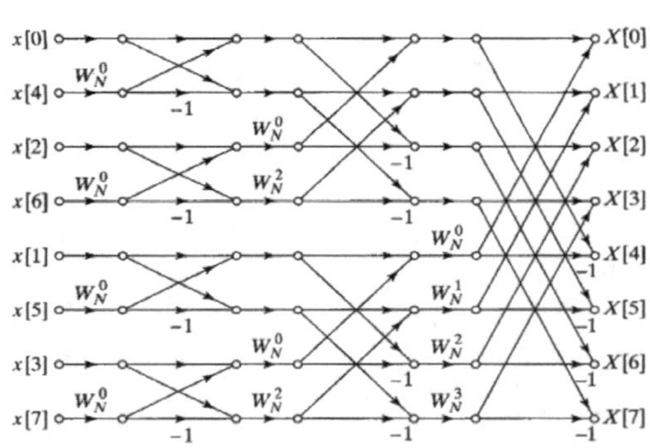

Figure 3.1 : DIT FFT

In DIT FFT the input will be in bit reversal order and the output will be in the normal order. DFT is defined as

The DFT equation $$X[k]=\sum_{n=0}^{N-1}x[n]W_N^{nk}$$

3.3 IIR Filter:IIR filter is implemented using Direct Form –I structure Figure 3.1 shows the IIR DF-1digitl filter

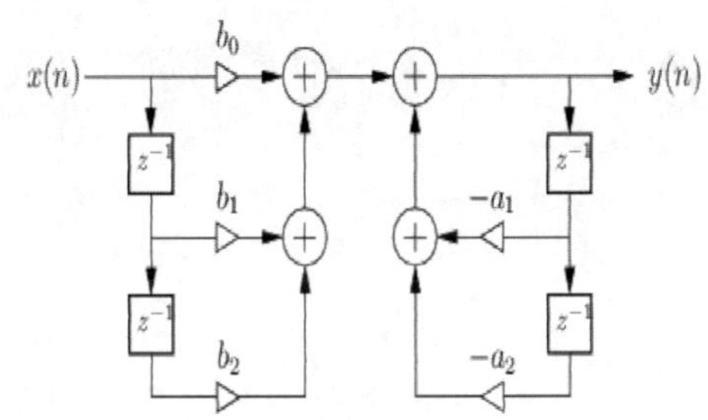

Figure 3.1: Direct-Form-I IIR of a 2nd-order digital filter.

Butterworth or Chebyshev filter IIR filter can be designed ..The advantage of Butterworth filter is flat response but more transition width and the advantage of Chebyshev filter is having less transition band but ripples in stop band and pass band .Design of IIR filter is explained in flow chart

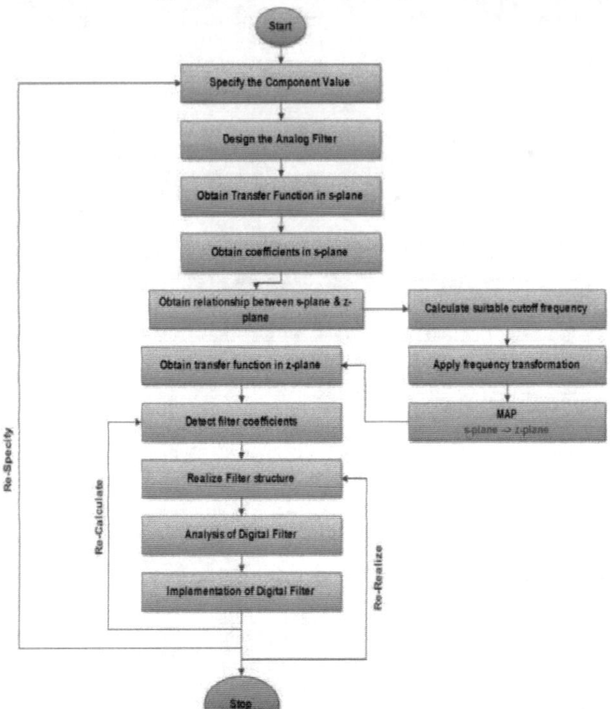

The following are the steps followed for design of IIR filter. Filter transfer function is obrained in analog domain for the given specifications and the filter coefficients are obtained in S Domain A Cut off frequency is calculatedand the transfer function from s plane to z plane is obtained using Bilinear transformation or Mapping technique Again, Filter coefficients are obtained and the digital filter is realized

IV.Result Analysis: Adaptive NotchIIR filter is Simulated and implemented in MAT lab and the following are the result analysis.

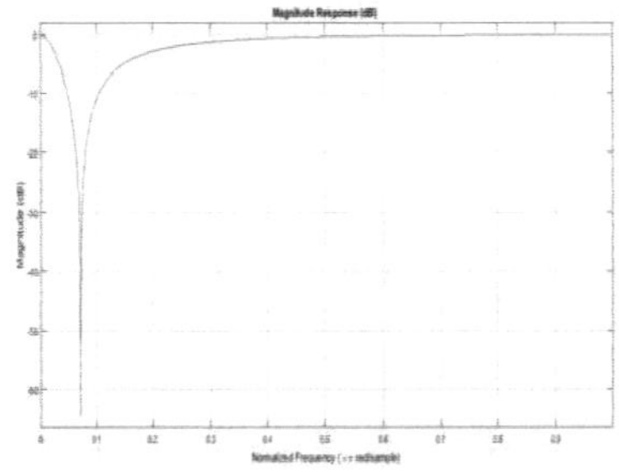

Figure 4.1:Basic response of Adaptive IIR Notch filter

.IIR notch filter is shown in figure 4.1 and the notch is observed at 50Hz for the frequency variationfrom 0-50Hz.

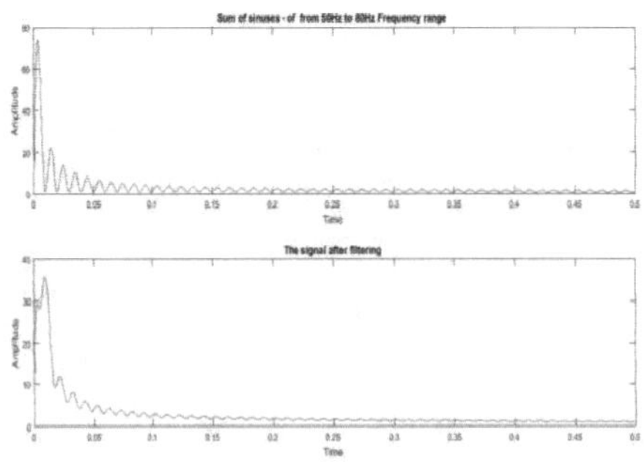

Figure 4.2: Sinusoidal response of Adaptive IIR Notch filter.

For sinusoidal input from 0-90Hz the filter characterics are shown in figure 4.2 .It is observed that except at 50Hz other frequencies are allowed by the filter.

IOP Conf. Series: Materials Science and Engineering 1074 (2021) 012036 doi:10.1088/1757-899X/1074/1/012036

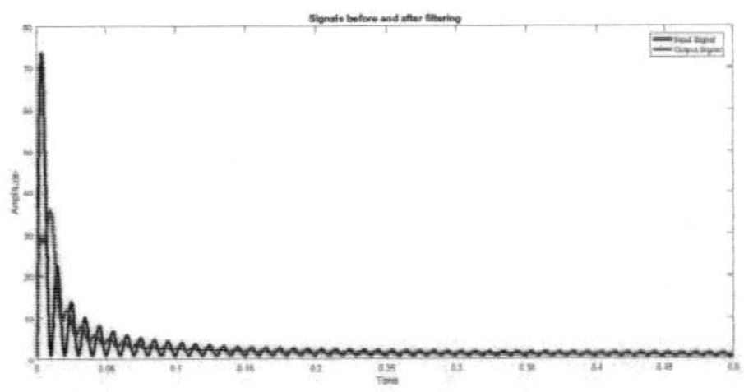

Figure 4.3:Signal before and after filtering

Signal before and after filtering of the Notch filter is shown in figure 4.3

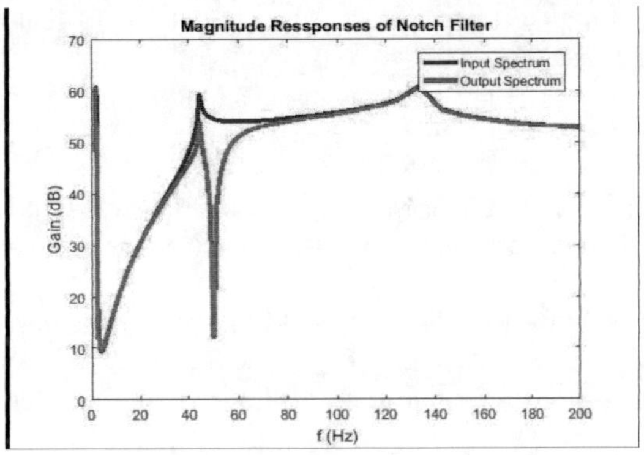

Figure 4.4:Magnitude response of Adaptive IIR Notch filter.

The desired response and actual magnitude response of Adaptive IIR Notch filter is almost same is shown in figure in 4.4.

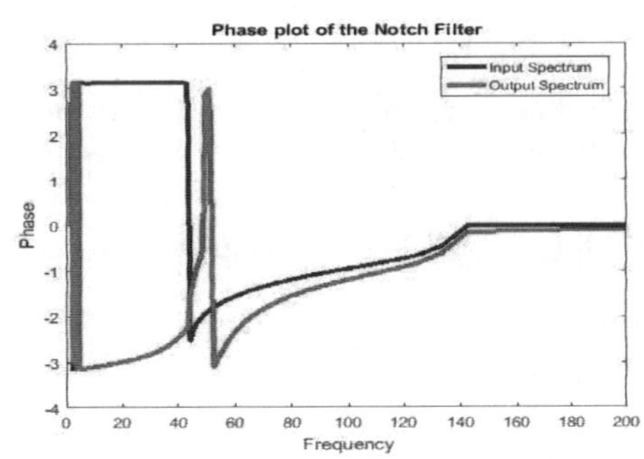

Figure 4.5:Phase response of Adaptive IIR Notch Notch filter

The desired response and actual phase response of Notch filter is almost same is shown in figure in 4.5.and it is observed that there is a deviation in the phase for lower frequency component and there is no deviaton for higher frequency.

V.Conclusion :In this paper, basic Adaptive IIR notch filter is designed and simulated in MAT lab. For basic response of IIR notch filter the notch is observed at 50Hz for the frequency variation from 0-50Hz. The desired response and actual magnitude response of Notch filter is almost same .The desired

IOP Conf. Series: Materials Science and Engineering 1074 (2021) 012036 doi:10.1088/1757-899X/1074/1/012036

response and actual phase response of Notch filter is observed that there is a deviation in the phase for lower frequency component and there is no deviation for higher frequency.

5.REFERENCES

[1] S. Haykin, "Adaptive4th ed ed., Pearson Education: Prentice Hall, 2013.
[2] R. Chassaig, "Digital Signal Processing and Applica-tions with the C6713 and C6416 DSK", 2nd ed. Uni-ted States of America, New Jersey: John Wiley & Sons, 2008.
[3] N. Kehtarnavaz, N. Kim, and I. Panahi, "Digital signal processing system design: using LabVIEW and TM-S320C6000," New York: the 11th IEEE Digital Signal Processing Workshop and The 3rd IEEE Signal Proces-sing Education Workshop, pp. 10-14, 2004.
[4] W. S. Gan, "Teaching and learning the hows and whys of real-time digital signal processing," IEEE Trans. on Education, vol. 45, no. 4, pp. 336-343, 2002.
[5] A. Spanias, V. Berisha, H. Kwon, C. Huang, A. Natara-jan and R. Ferzili, "Using the Java-DSP real-time hard-ware interface in undergraduate classes," San Diego: presented at the 36th ASEE/IEEE Annual Frontiers in Education Conference, pp. 12-17, 2006.
[6] M. Galanis, A. Papazacharias, and E. Zigouris, "A DSP course for real-time systems design and implementa-tion based on the TMS320C6211 DSK," presented at the IEEE International Conf. on Digital Signal Proces-sing, vol. 2, pp. 853-856, 2002.
[7] D. Orofino, "Rapid prototyping of a surveillance video compression system," Matlab Digest, The MathWorks, Inc. Massachusetts, vol. 11, n. 5,
[8] G. Budura, "Nonlinear systems identification using the volterra model," Romania: Scientific Bulletin of the Po-litehnica University of Timinoara, 2005.
[9] L. Gulfo and J. Valencia, "Identificación y modelamien-to de sistemas de audio API mediante filtros adapta-tivos". Medellín, Colombia: Ingeniería de Sonido, Uni-versidad de San Buenaventura, 2012.
[10] D. Millan, "Estudio y comparativa de diferentes algorit-mos adaptativos para la identificación de sistemas,". Barcelona, España: Ingeniería Técnica de Telecomuni-caciones, Universidad Politécnica de Catalunya, 2012.
[11] P. S. Diniz, "Adaptive filtering: algorithms and practical implementation," 3rd ed. New York, NY, USA: pringer, 2008.
[12] H. Quanzhen, G. Zhiyuan, G. Shouwei, S. Yong and Z. Xiaojin, "Comparison of LMS and RLS algorithm for ac-tive vibration control of smart structures," presented at Third International Conference on Measuring Te-chnology and Mechatronics Automation ICMTMA´11, vol.1, no. 1, pp. 745–748, 2011.
[13] S. Agraval and V. Gopal, "Performance analysis of adaptive filtering algorithms for system identification," International Journal of Electronics and Communica-tion Engineering, vol. 5, no. 2, pp. 207-217, 2012.
[14] E. Soria, J. Calpe, J. Chambers, M. Martínez, G. Camps and J. D. Guerrero, "A novel approach to introducing adaptive filters based on the LMS algorithm and its variants," IEEE Trans. On Education, vol. 47, pp. 127-133, 2008.
[15] L. N. Reyes, "Análisis de Filtros adaptativos de la fami-lia SM aplicados para el diseño de un cancelador de eco acústico," Sangolquí, Ecuador: Ingeniería en Elec-trónica, Escuela Politécnica del Ejército, 2013.
[16] E. H. Krishna, M. Raghuram, K. V. Madhav and K. A. Reddy, "Acoustic echo cancellation using a computa-tionally efficient transform domain LMS adaptive fil-ter," presented at 10th International Conf. on Informa-tion sciences signal processing and their applications (ISSPA), pp. 409-412, 2010.
[17] H. Zhao, S. Hu, L. Li and X. Wan, "NLMS adaptive FIR filter design method," presented at IEEE Region 10 Conference TENCON, pp. 1- 5, 2013.
[18] C. Paleologu, J. Benesty, S. L. Grant and C. Osterwise, "Variable step-size NLMS algorithms for echo cancella-tion," presented at Conf. Record of the forty-third Asilo-mar Conference on Signals, Systems and Computers, pp. 633-637, 2009.

AUTHOR INDEX

Alhamdani, Wasim 253
Anandarao, S. 133
Anuradha, S. 192
Apparao, Giduturi. 192
Arun M 36
Babu, K. Srinivasa 164
Bharath, Mullaguri Venkata 260
Bhoir, Smita Vinit 303
Bindu, Ch. Hima 244
Ch, Satyanarayana 60
Chamundeswari, G 216
Dara, Raju 224
Devarakonda, Nagaraju 133
Devi, M Shyamala 260, 379
Devi, M. Shyamala 336
Dhage, Sudhir 89
Dhanusha, C. 155
Divakar, Ch. 82
Gadde, Sridevi 184
Haritha, D. 1
Hazarika, Mreenmoy 336
Indira, Dnvsls 369
Indukuri, Hemalatha 253
Jagadeesh, B. N. 394
Jayalakshmi, G. 169
Jayaprada, S. 169
Kamarajugadda, Raviteja 133
Kanyakumari, L. 169
Kavitha, A. V. 14
Kavitha, K. 234
Kavitha, V. 270
Koul, Nimrita 52
Kumar, A. V. Senthil 155
Kumar, B. Srinivas 178
Kumar, J M S V Ravi 139
Kumar, K. Kranthi 178
Kumar, N. V. Narendra 148
Kumar, Sankarpu Sateesh 164
Kundale, Jyoti 89
Lakshmanarao, A. 184
Lalitha, J. Divya 234
Lalitha, R. V. S. 234
Leelavathi, N 139
Madhu, T. 164
Mahankali, Preethi 379
Mancha, Venkata Ramana 60
Manvi, Sunilkumar S 52
Markapudi, Babu Rao 369
Mohan, T Murali 224
Muppalaneni, Naresh Babu 107
Nareshkumar, B 322
Navya, Boggula 379
Patel, Harish 379
Patil, Sunita 303
Poornima, A 260, 379
Potluri, Avinash 148
Prathima, Ch. 107
Praveenkumar, S. 36
Raja, D. R. Kumar 387
Rajakumar, P. S. 36
Rajasekar, P. 359
Ramareddy, T 234
Rameshwaraiah, K. 164
Ramu, M. 387
Rao, Ch. Seshadri 184
Rao, G Appa 347
Rao, G Mallikarjuna 394
Reddy, P Kavitha 322
Reddy, P Vijayapal 224
Reddy, R Anirudh 322
Reddy, T Raghunadha 224
Reddy, T Vasudeva 322
Roy, Rita 347
Sagar, Ankita 336
Sah, Nitesh Kumar 260
Sai Sravanthi, G. 359
Samalla, Dr Krishna 394
Santhi, S. G. 178
Sarankumar, R. 359
Sarkar, Shatadru 379
Sathishkumar, S 260
Satyanandam, N 71
Satyanarayana, Ch. 14, 82
Satyanarayana, Ch 71
Sekhar, Akula Chandra 107, 387
Sharma, Nakul 116, 289
Sirisha, Katikireddy Phani 82
Sirisha, P. G. K. 1
Sivakkolundu, R. 270
Sridevi, G. 244
Srikrishna, A. 14
Srinivas, Rayudu 234
Srinivasa, Reddy E 60
Subbarao, S. P. V 394
Sujana, Challapalli 234
Sujatha, B 139

Sumalatha, L....369
Sumathi, K....359
Sumithra, M....260, 379
Swathi, P....336
Swathi....260
Thamizhikkavi, P....36
Thapa, Karan....336
Upadhyay, Shubham Santosh....260
Valavala, Mounicasri....253
Venkata Srinivas, Akana Chandra Mouli....82
Venkateswarlu, B....359
Vuyyuru, Veera Ankalu....192
Yalla, Prasanth....116, 289
Yogesh, K....260

Institute of Physics
Dirac House, Temple Back
Bristol BS1 6BE UK

ISSN: 1757-8981
ISBN 978-1-7138-2694-1
9 781713 826941